Graduate Texts in Physics

Graduate Texts in Physics publishes core learning/teaching material for graduate- and advanced-level undergraduate courses on topics of current and emerging fields within physics, both pure and applied. These textbooks serve students at the MS- or PhD-level and their instructors as comprehensive sources of principles, definitions, derivations, experiments and applications (as relevant) for their mastery and teaching, respectively. International in scope and relevance, the textbooks correspond to course syllabi sufficiently to serve as required reading. Their didactic style, comprehensiveness and coverage of fundamental material also make them suitable as introductions or references for scientists entering, or requiring timely knowledge of, a research field.

Pedro Pereyra

Contemporary Theory of Semiconductors and Devices

Semiclassical and Quantum Physical Principles and Applications

Springer

Pedro Pereyra
Área de Física Teórica y Materia
Condensada
CBI Universidad Autónoma
Metropolitana-Azcapotzalco
Mexico City, Mexico

ISSN 1868-4513 ISSN 1868-4521 (electronic)
Graduate Texts in Physics
ISBN 978-3-032-00295-2 ISBN 978-3-032-00296-9 (eBook)
https://doi.org/10.1007/978-3-032-00296-9

This Springer imprint is published by the registered company Springer Nature Switzerland AG
The registered company address is: Gewerbestrasse 11, 6330 Cham, Switzerland

To my wife
Liuddys

And to the memory of my parents
Alberto and Beatriz

Preface

This book began as many books do, with short texts and article excerpts prepared for students. In this case, for students of an elective course on semiconductor physics that I intermittently offer as part of the physical engineering degree program at UAM Azcapotzalco. A course designed for students who are generally familiar with the electromagnetic theory, materials science, and the principles of quantum physics, many of whom also have a background in standard solid state physics. The primary goal of the course has been to introduce the fundamentals of semiconductor physics and the emblematic optoelectronic devices, such as light emitting diodes, the bipolar transistor and solar cells. Over time, these initial objectives evolved, enriched by insights and results from my own research work that enabled me to contribute with new formalisms, results and perspectives to the study of semiconductor physics.

An important starting point in my approach to semiconductor physics was the work I carried out with Pier A. Mello and Narendra Kumar, on disordered quantum conductors, which led to the DMPK equation. This experience laid the groundwork for my later efforts to apply the transfer matrix method, from the point of view of the *'scattering' theory* approach, to finite periodic systems, as an alternative to Bloch's theory, rigorously valid for infinite periodic systems.

Studying the transfer matrix representations for the various universality classes and exploring the Saxon-Hutner conjecture for sequences of potential barriers, emerged the Jones-Abelès transfer matrix for a *finite sequence* o barriers and wells, a matrix that was initially found in the realm of the electromagnetic theory, and was later independently rediscovered at least a dozen of times over the past century in the context of one-dimensional periodic systems and a single propagating mode. This approach soon became a central focus of my research. The aim to extend its applicability to higher dimensional systems led me, after a few years of straggling with non-commutative recurrence relations, to a general solution: a formalism valid for an arbitrary number of propagating modes, unit cells and potential profiles. This formalism, exact, and yet simple and compact, has proven to be a powerful tool for explaining a wide range of quantum and electromagnetic open problems. Some of the new results obtained through this approach are included in this book.

My interest in the physics of semiconductors has also been shaped by a personal life experience, one that I believe is similar to others experience too. A few years after the discovery of the transistor by J. Bardeen, W. H. Brattain and W. Shockley, the 'transistor radio' reached all corners of the world, including the foothills of the eastern mountains chain of the Bolivian Andes. My father owned a farm located near Mandinga, and around 1957, shortly before a vacation, he showed up with a Sony transistor pocket radio, "for when we go to San Jacinto", he said. There, in San Jacinto, where the civilization and electric current had not yet arrived, where the only thing that could be heard were the whisper of the wind and the roar of the storms, we listened in astonishment and wonder to the words and music that came forth from that tiny thing, a hundred times smaller than a bulb radio. Glued to that radio we heard the transmissions of the first manned rocket launches, satellites that we tirelessly searched in the starry sky. The magic of that device, and of the term *transistor* was installed in the minds of the entire world. Alongside terms like nuclear physics and atomic bomb, the transistor became a symbol of a new era.

From the early days following the discovery of transistors and first semiconductor devices up to the discovery of the quantum Hall effect, the experimental methods evolved continuously and the mesoscopic devices gradually transformed into fully quantum structures: As a result, the range of research topics and publications expanded dramatically, at times becoming overwhelming. Fortunately, much of the experimental and theoretical work developed over these decades, which had significantly enhanced our understanding of semiconductors, particularly in determining their fundamental parameters, are extensively discussed and well-referenced in excellent and widely recognized books, among others in books on semiconductors by Seeger [1], Sze [2], Yu and Cardona [3], Grove [4], Shalimova [5], Enderlein and Horing [6], as well as solid-state physics textbooks by Madelung [7], Hook and Hall [8] and Kittel [9], where the reader will find detailed explanations of key topics essential for understanding semiconductor physics. Additionally, the comprehensive and well-curated database of semiconductor material parameters maintained by the Ioffe Institute [10] remains an invaluable resource for researchers.

In the final decades of the 20th century, several areas of physics emerged as hot areas of research, some new, others representing long-standing open problems. Among them, the hight temperature superconductivity, the quantum Hall effect, the tunneling time, the blue laser diode, and spintronics. While some led to actual developments, other remain elusive challenges. Meanwhile, the semiconductor technology of field effect devices, advanced steadily, with continuous progress in miniaturization and increasing memory capacity. Although I address some of the newly emerged topics, I discuss with more detail on two basic subjects where I could contribute to develop further, the optical response in superlattices (blue laser) and the analytical solution of the nonlinear Poisson equation in the inversion layer of MOSs and MOSFETs.

During my sabbatical stay at Ohio University in Athens (1997–1998), I was impressed by the sense of urgency within the experimental physics community and the optoelectronics industry to uncover the 'know-how' behind the blue laser diode pioneered in Japan, among others, by S. Nakamura, using GaN-based superlattices in the active region. It was not the know-how problem that caught my attention, but rather the fact that the active zone of the blue laser was a semiconductor superlattice. At that time, a rigorous theoretical description of the optical response of such superlattices was a pending topic and a subject that led me to additional developments of the theory of finite periodic systems (TFPS), particularly toward the derivation of analytical expressions for the eigenvalues and eigenfunctions in spatially bounded periodic structures. In collaboration with A. Kunold and Fernanda Avila, we initiated the first detailed calculations of the optical response of GaN superlattices. Finally, from the analysis of the symmetries of the eigenfunctions of the superlattices, a *true* set of selection rules for optical transitions in heterostructures and superlattices was obtained, together with a complete formalism to describe the optical response of these devices, a framework that diverges significantly from conventional approaches. The theoretical and practical importance of this topic in optoelectronics is one of the motivations behind this book. To support a through understanding, the first part includes a dedicated chapter on the TFPS.

The metal-oxide-semiconductor (MOS) structure lies at the core of modern electronic technology, forming the basis of the metal-oxide-semiconductor field-effect-transistor (MOSFET), the most widely used device in integrated circuits today. Within the MOS structure, the inversion layer is both an operational phenomenon and virtually a physical system. The population inversion is an essential process in the operation of a MOSFET, enabling the switching behavior of the MOSFET. Since the MOS was introduced in the 1950s, the basic phenomenology of field-effect devices has been well understood. Seminal theoretical contributions, most notably by William Shockley and his contemporaries, provided a sufficient framework to quickly pivot to the field of applications. Nevertheless, certain theoretical problems remained unresolved. Among these was the analytical solution of the nonlinear Poisson equation in the inversion layer. In the absence of an analytical solution, the field long relied on numerical methods, empirical models and approximate formalisms were used with great success. With billions of transistors fabricated annually at the nanometer scale, this theoretical issue was largely overshadowed and remained buried beneath layers of technological success. In recent years, however, I have succeeded in deriving analytical solutions to the nonlinear Poisson equation in the inversion layer. These include explicit expressions for the electric potential distribution, the effective thickness of the inversion layer and the channel currents in the MOSFETs. The physical argument that made possible the solution of the nonlinear equation and the solution is presented in the second part of this book.

Based on this experience, writing a book on semiconductors from the perspective of quantum and electromagnetic theories applied to *finite* systems, where the reference to band structure and other physical properties of periodic systems *do not require* the use of Bloch theory, has been both a challenge and a central motivation.

In addition to the topics directly related to my own research work, the remainder of the material presented in this book covers foundational aspects of the fairly well-established semiconductor physics.

I thank the Universidad Autonoma Metropolitana at Azcapotzalco, for the facilities that the university provides. To the National System of Researchers (SNI) of CONACyt, Mexico, for the financial support. To my collaborators, friends and students: Fernanda Avila-Ortega, Jose Luis Cardoso, Alejandro Kunold, Herbert P. Simanjuntak, Dieter Weiss, Jaime Grabinsky, Victor Ibarra, Jürgen Reiter, Arturo Robledo, Juan Carlos Sandoval, Carlos J. Avila, and Sergio Ulloa. To the unnamed students, to my wife Liuddys for her constant love and support, to my daughters Naira and Jaina, to my son Sebastian, and to my family in Bolivia and Cuba for their love and encouragement.

Mexico City, Mexico
June 2024

Pedro Pereyra

Competing Interests The author has no competing interests to declare that are relevant to the content of this manuscript.

Introduction

Since Bardeen, Brattain and Shockley [11–13] introduced the doped semiconductor three-layer transistor, the most revolutionary invention of the 20th century, experimental and applied physics, along with electronics and optoelectronics, have experienced a rapid and uninterrupted evolution. The same group of physicists who developed the transistor [14–16], also laid the foundations of the semiclassical theory of semiconductors that allowed to understand the phenomenology and operation of transistors and an important part of the current electronic devices. In the decades that followed this technological breakthrough, particularly during the 1950s and 1960s, further innovations emerged, including the silicon solar cell, announced in 1954 by Chapin, Fuller, and Pearson, of Bell Labs [17], and the semiconductor laser [18, 19]. The development of the Schottky junction and the metal-oxide-semiconductor (MOS) structure proved essential for field-effect transistors, such as the MESFET (for metal-semiconductor-field-effect-transistor) and the MOSFET (MOS-field-effect-transistor) [20], as well as for a wide range of contemporary optoelectronic devices, many of which are controlled by external and induced electric fields within the inversion region of the MOS structure.

The development of advanced semiconductor growth techniques—such as metal-organic chemical vapor deposition (MOCVD) and molecular beam epitaxial (MBE) [22]—in the 1970s enabled the fabrication of heterostructures with remarkable precision. As early as 1957 Kroemer [23] suggested that by alternating semiconductor layers with different energy gaps could produce tailored potential profiles, eliminating the need for conventional doping [24–26]. This concept laid the foundation for a new class of quantum devices in which the primary distinguishing parameter between layers is the bandgap. From this innovation emerged quantum wells (QWs), multiple quantum wells (MQWs), and superlattices (SLs) structures that significantly expanded the landscape of electronics and optoelectronics devices design [27–31]. These artificial systems marked the beginning of a new era in semiconductor physics, characterized first by mesoscopic and later by nanoscopic dimensions, offering an endless variety of combinations, new physical properties, and far-reaching technological applications.

In the rich and evolving history of semiconductor physics, numerous problems and challenges have arisen, some resolved through rigorous theoretical and experimental efforts, others addressed by adopting simplified models and approximations. Over the decades, a wide array of novel phenomena and technological applications have been investigated. Key topics of interest have been the electronic transport under forward and reverse bias, tunneling [25, 32], semiconductor lasers [30, 33–42], injection currents thresholds [28, 43, 44], the influence of temperature on crystal growth [45], charge carrier mobilities [46, 47], absorption-recombination and stimulated emissions processes [48–51], exciton size in quasi-two-dimensional devices [52–56], and the impact of impurities on photoluminescence [57], as well as on photoreflectance and related optical phenomena. In more recent years, research in periodic systems has advanced further. Notably, tunneling times of photons and optical pulses were measured directly; [58–60] and the blue laser diode based on GaN superlattices in the active zone was successfully developed [56, 61–64]. The 1990s and early 2000s saw a surge of activity in spintronics [65–74], and in the study of transport properties in magnetic superlattices [75–95]. Although quantum dots, graphene, become fields of great activity [96, 97], they will be addressed only briefly in this book. Topological insulators and more recent advances on two-dimensional materials, while highly significant, lie beyond the scope of this work and will not be discussed.

MQWs and SLs represent systems of intermediate complexity, more intricate than textbook-like simple systems, yet less demanding than fully three-dimensional molecular systems, which often require heavy numerical calculations. Multi quantum wells and superlattice structures gained significant attention due to their relevance in optical and electronic device applications. To describe these structures of quantum dimensions, various theoretical approaches have been developed. Some, such as the envelope function theory [98–101] emerged by adjusting known theories originally formulated for infinite periodic systems. Others, like the transfer matrix method, rooted in the scattering theory and widely used in nuclear physics, electromagnetism, and particle physics, have been adapted and evolved into a rigorous formalism suitable for superlattices. The understanding of the impact of the *additional periodicity* in superlattices has also matured. The early interpretations that focused on the emergence of minigaps within the continuous dispersion relations, has evolved into the calculation of truly discrete subbands, with discrete dispersion relations and well-defined energies, including those of surface states. The Bloch-type functions that were once regarded as the correct wave functions can now be replaced by the actual eigenfunctions derived from the theory of finite periodic systems.

The semiclassical formalism of semiconductors remains the backbone for describing the fundamental properties of semiconductor devices. However the breadth of research and the variety of topics encompassed within semiconductor physics are so vast that providing a comprehensible review of them all, or even a complete enumeration, is virtually impossible.

The primary objective of this book is to present, in a comprehensive and tutorial style, the fundamentals of the standard (semiclassical) theory developed to understand semiconductors phenomenology and the operation of the most representative semiconductor devices. Since numerous excellent textbooks on the fundamentals of semiconductors are available [1–9], we discuss some topics briefly and some in detail. The aim is to extend the theoretical framework to incorporate quantum approaches for analyzing the physical properties of mesoscopic and nanoscopic semiconductor heterostructures.[1] The underlying physical principles, grounded in the seminal formalism of Shockley and Bardeen, are presented in a self-contained, and exhaustive manner, providing the reader with a solid academic foundation for understanding the behavior of *np* junctions, bipolar transistors, solar cells, Schottky junctions, metal-oxide-semiconductor (MOS) and MOSFETs. While the book covers the core concepts typically found in standard semiconductor texts, it also addresses a number of advanced topics that have emerged over the last three decades and are closely related to the authors's research. These include: (1) a brief introduction to the theory of finite periodic systems; (2) a detailed analysis of the inversion layer in a MOS structure, including the analytical solution of the nonlinear Poisson equation, which had remained unsolved for over seventy years; (3) the optical response of semiconductor lasers with superlattices in their active regions, for which no prior solutions existed; (4) a comprehensive treatment of the time distribution of stimulated transition probabilities, encompassing ultrafast photoluminescence and related relaxation mechanisms; (5) a reinterpretation of the integer quantum Hall effect, resulting in a new formula that accurately reproduces resistance plateaus as functions of the magnetic field; (6) the tunneling time and a summary of the space-time evolution of electron and photon wave packets through semiconductors and silica-titanium superlattices, and (7) magnetic superlattices and spin-polarized injection at ferromagnetic/normal (F/N) semiconductor interfaces.

The book is organized into two parts. Part I presents the fundamental theoretical framework, while Part II is dedicated to the analysis of emblematic semiconductor devices. In the initial chapters of Part I, we introduce the semiclassical formalism along with the essential variables, definitions and concepts needed to characterize the *n*-type and *p*-type semiconductors, and to understand the operating principles of optoelectronic devices. The content of the first three chapters serves as a concise *rsum* of basic definitions, intended especially for readers who may not have completed a formal course in solid state physics or materials science. This introductory section provides the background required to engage with the more advanced topics that follow. To prepare for the analysis of superlattice lasers, wave packet transport, spin transport, and graphene, a dedicated chapter is included on the theory of finite periodic systems. The Part I concludes with the quantum formalism of stimulated and spontaneous transition probabilities and the time distribution of these transitions. As a complement to the analysis of currents in terms of transmission coefficients, we present a brief discussion of the Landauer conductances. Part II focuses on the theoretical understanding of the most representative semiconductor devices, including light-emitting diodes, solar cells, bipolar transistors, Schottky junctions, MOS and MOSFETs, Additional topics include the optical response of

[1] This is much in the spirit of earlier works such as [102–104].

superlattices, point contacts and conductance quantization, quantum Hall effect, quantum dots, disordered conductors, graphene, tunneling and spintronics.

The theoretical formalisms and applications are presented in a clear and self-contained manner to ensure accessibility for advanced undergraduate, graduate students in physics and engineering, and researchers in physics, materials science, and related technologies. This book aims to deepen the reader's understanding of systems relevant to contemporary semiconductor physics and to provide the necessary tools for calculating transport and optical properties in both simple and complex optoelectronic devices. It also offers analytical solutions to the Poisson equation in the inversion layer, along with simplified relations between key parameters governing the behavior of MOS structures and MOSFETs.

References

1. K. Seeger, *Semiconductor Physics*, (Springer, Berlin, 1996).
2. S. Sze and K. K. Ng, *Physics of Semiconductor Devices*, (Wiley Interscience and John Wiley & Sons, New Jersey 2007)
3. P. Y. Yu and M. Cardona *Fundamentals of Semiconductors* 1ra Ed. (Springer, Berlin, 1996).
4. A. S. Grove, *Physics and Technology of Semiconductor Devices*, (John Wiley, New York 1967).
5. K. V. Shalímova, *Física de los Semiconductores*, (MIR, Moscú, 1975).
6. R. Enderlein and N. J. M. Horing, *Fundamentals of Semiconductor Physics and Devices*, (World Scientific, Singapure, 1997).
7. O. Madelung, *Introduction to Solid-State Theory* 2nd Ed. (Springer, Berlin, 1981).
8. J. R. Hook and H. E. Hall Solid State Physics (John Wiley, Chichester 1991 2nd Ed.)
9. C. Kittel *Introduction to Solid State Physics* 7ma Ed. (J. Wiley & Sons, N. York, 1995).
10. Ioffe Institute, https://www.ioffe.ru/SVA/NSM/Semicond/.
11. J. Bardeen and W. H. Brattain *Phys. Rev.* **74**, 230 (1948).
12. W. H. Brattain and J. Bardeen *Phys. Rev.* **74**, 230 (1948).
13. W. Shockley and G. L. Pearson *Phys. Rev.* **74**, 230 (1948).
14. W. Shockley, *Bell System Tech. J.* **28**, 435 (1949).
15. J. Bardeen and G. L. Pearson *Phys. Rev.* **75**, 865 (1949).
16. W. Shockley, *Electrons and Holes in Semiconductors*, (D. Van Nostrand, New York, 1950).
17. D.M. Chapin, C.S. Fuller and G.L. Pearson, *J. Appl. Phys.* **25**, 676 (1954).
18. H. Kroemer: 'Semiconductor laser with electric pumping'. US Patent No. 3309553 (1967) (Filed August 16, 1963)
19. Zh.I. Alferov, R.F. Kazarinov: 'Semiconductor laser with electric pumping'. Inventor's Certificate No. 181737 in Russian, Application No. 950840, priority as of March 30, 1963
20. B.L. Altshuler, A.G. Aronov, D.E. Khmelnitskii and A.I. Larkin, in Quantum Theory of Solids (Ed. I.M. Lifshits, MIR, URSS, 1982).
21. Atalla, M. and Kahng, D. (November 1962). "A new "Hot electron" triode structure with semiconductor-metal emitter". *IRE Transactions on Electron Devices* **9**, 507 (1962) ITED....9..507 A. 10.1109/T-ED.1962.15048.
22. A.Y. Cho, *Appl. Phys. Lett.* **19**, 467 (1971).
23. H. Kroemer, *Proc. Rad. Rad. Eng.* **45**, 1535 (1957)
24. L. V. Keldysh *Fiz. Tverd. Tela* **4**, 2265 (1962), English translation, in *Sov. Phys. -Solid State* **4**, 1658 (1963).
25. L. Esaki, R. Tsu, *IBM J. Res. Develop.* **14**, 61 (1970).
26. R. D. Dupuis, P. D. Dapkus, Nick Holonyak, E. A. Rezek, R. Chin, *Appl. Phys. Lett.* **32**, 295 (1978).

27. H. Kroemer, *Proc. IEEE* **51**, 1782 (1963).
28. Zh. I. Alferov, V. M. Andreev, V. I. Korolkov, D. N. Tratyakov, V. M. Tuchkevich, *Fiz. Tekh. Poluprovodn.* **1**, 1579 (1967). English translation, in *Sov. Phys. Semiconduc.* **1**, 1313 (1968). H. Kroemer: Proc. IRE **45**, 1535 (1957); RCA Rev. **28**, 332 (1957).
29. H. Rupprecht, J. M. Woodall, G. D. Pettit, *Appl. Phys. Lett* **11**, 81 (1967).
30. Zh. I. Alferov, V. M. Andreev, D. Z. Garbuzov, Yu. V. Zhilyaev, E. P. Morozov, E. L. Portnoi, V. G. Trofim, *Fiz. Tekh. Poluprovodn.* **4**, 1826 (1970). English translation, in *Sov. Phys. Semiconduc.* **5**, 174 (1971).
31. I. Hayashi, M. B. Panish, P. W. Foy, S. Sumski, *Appl. Phys. Lett.* **17**, 109 (1970).
32. B Movaghart, J Leo, A MacKinnon, *Semicond. Sci. Technol.* **3**, 397 (1988).
33. Z. Y. Xu, V. G. Kreismanis, C. L. Tang, *Appl. Phys. Lett.* **43**, 415 (1983).
34. M. J. Ludowise, W. T. Dietze, and C. R. Lewis, *Appl. Phys. Lett.* **42**, 487 (1983).
35. H. Sakaki, J. Yoshino, Y. Sekiguchi, *Electronics Letts.* **20**, 320 (1984).
36. YaoHwa Wu, Michael Werner, Shyh Wang, *Appl. Phys. Letts.* **45**, 606 (1984).
37. I. Hayashi, *IEEE Trans. elect. Dev.* **31**, 1630 (1984).
38. D. Kirillov, J. L. Merz, P. D. Dapkus, J. J. Coleman, *J. of Appl. Phys.* **55**, 1105 (1984)
39. B. Cai, A. J. Seeds, J. S. Roberts, *IEEE Photon. Tech. Lett.* **6**, 496 (1994).
40. Zh. I. Alferov, A. I. Vasilev, S. V. Ivanov, P. S. Kopev, N. N. Ledentsov, M. E. Lutsenko, B. Ya. Meltser, V. M. Ustinov, *Pisma Zh. Tekh. Fiz.* 1988, 14, 1803 *Sov. Tech. Phys. Lett.* **14**, 782 (1988).
41. G. Ji, D. Huang, U. K. Reddy, T. S. Henderson, R. Houdre, H. Morkoe; *J. Appl. Phys.* **62**, 3366 (1987).
42. Yukio Narukawa, Yoichi Kawakami, Shizuo Fujita, Shigeo Fujita, Shuji Nakamura *Phys. Rev. B* **55**, R1938 (1997).
43. W. T. Tsang, C. Weisbuch, R. C. Miller, R. Dingle *Appl. Phys. Lett.* **35**, 673 (1979).
44. M. Zuniga, J. A. Kong, L. Tsang, *J. Appl. Phys.* **51**, 2315 (1980).
45. R. D. Dupuis, P. D. Dapkus, *Appl. Phys. Lett.* **31**, 466 (1977).
46. N. Holonyak, Jr., R. M. Kolbas, E. A. Rezek, R. Chin, R. D. Dupuis, P. D. Dapkus, *J. Appl. Phys.* **49**, 5392 (1978).
47. R. Dingle, H. L. Störmer, A. C. Gossard, W. Wiegmann, *Appl. Phys. Lett.* **33**, 665 (1978).
48. R, F, Kazarinov, R. A. Suris, *Sov. Phys,- Semicond.* **6**, 120 (1972).
49. B. A. Vojak, N. Holonyak Jr., R. Chin, E. A. Rezek, R. D. Dupuis, P. D. Dapkus, *J. Appl. Phys.* **50**, 5835 (1979).
50. R. M. Kolbas, N. Holonyak Jr., B. A.VojakK. K. Hess, M. Altareli, R. D. Dupuis, P. D. Dapkus, *Solid. Stat. Commun.* **31**, 1033 (1979).
51. J. S. Smith, L. C. Chiu, S. Margalit, A. Yariv, A. Y. Cho, *J. Vacuum Sci. Tech. B* **1**, 376 (1983).
52. N. Peyghambarian, H. M. Gibbs, J. L. Jewell, A. Antonetti, A. Migus, D. Hulin, A. Mysyrowicz, *Phys. Rev. Lett.* **53**, 2433 (1984).
53. D. A. B. Miller, D. S. Chemla, D. J. Eilenberger, P. W. Smith, A. C. Gossard, W. T. Tsang, *Appl. Phys. Lett.* **41**, 679 (1982).
54. F. F. So, S. R. Forrest, *Phys. Rev. Lett.* **66**, 2649 (1991).
55. D. S. Citrin, *Phys. Stat. Sol. (b)* **188**, 43 (1995).
56. Y. Narukawa, Y. Kawakami, S. Fujita S. Nakamura, *Phys. Rev. B* **59**, 10283 (1999).
57. P. M. Petroff, C. Weisbuch, R. Dingle, A. C. Gossard, W. Wiegmann, *J. Vacuum Sci. Tech.* **19**, 571 (1981).
58. A. M. Steinberg, P. G. Kwiat and R. Chiao, *Phys. Rev. Lett.* **71**, 708 (1993).
59. C. Spielmann, R. Szipöcs, A. Stingl, F. Krausz, *Phys. Rev. Lett.* **73**, 2308 (1994).
60. P. Pereyra, *Phys. Rev. Lett.* **84**, 1772 (2000).
61. S. Nakamura, S. Pearton, G. Fasol, *The Blue Laser Diode. The complete history* (Springer-Verlag, Berlin Heidelberg 1997). See page 247 of.
62. S. Nakamura, M. Senoh, S. Nagahama, N. Iwasa, T. Yamada, T. Matsushita, H. Kiyoku and Y. Sugimoto, *Appl. Phys. Lett.* **68**, 3269 (1996).
63. Y. Nawakami, S. Fujita, S. Fujita and S. Nakamura, *Phys. Rev. B* **55**, R1938 (1997).
64. P. Pereyra, *Ann. Phys.* **397**, 159 (2018).

65. S. Datta, B. Das, *Appl. Phys. Lett.* **56**, 665 (1990).
66. M. N. Baibich, J. M. Broto, A. Fert, F. N. Van Dau, F. Petroff, P. Eitenne, G. Creuzet, A. Friederich, J. Chazelas, *Phys. Rev. Lett.* **61**, 2472 (1988).
67. H. Ohno, *Science* **281**, 951 (1998).
68. E. Johnston-Halperin, D. Lofgreen, R. K. Kawakami, D. K. Young, L. Coldren, A. C. Gossard, and D. D. Awschalom, *Phys. Rev. B* **65**, 041306 (2002).
69. M. Kohda, Y. Ohno, K. Takamura, F. Matsukura, H. Ohno, *Jpn. J. Appl. Phys., Part 2* **40**, L1274 (2001).
70. T. Dietl, H. Ohno, F. Matsukura, J. Cibert, D. Ferrand, *Science* **287**, 1019 (2000).
71. A. Oiwa, Y. Mitsumori, R. Moriya, T. Supinski, H. Munekata, *Phys. Rev. Lett.* **88**, 137202 (2002).
72. I. Zutić, J. Fabian, S. Das Sarma, *Rev. Mod. Phys.* **76**, 323 (2004).
73. J. Fabian, A. Matos-Abiague, C. Ertlera, P. Stano, I. Zutić, *Act. Phys. Slov.* **57**, 565 (2007).
74. A. Van Esch, L. Van Bockstal, J. De Boeck, G. Verbanck, A. S. van Steenbergen, P. J. Wellmann, B. Grietens, R. Bogaerts, F. Herlach, and G. Borghs, *Phys. Rev. B* **56**, 13103 (1997).
75. H. Itoh, J. Inoue, S. Maekawa, *Phys. Rev. B.* **47**, 5809 (1993).
76. R. Schad, P. Belien, G. Verbanck, et ál., *Phys. Rev. B* **59**, 1242 (1999).
77. I. K. Schuller, S. Kim, C. Leighton, *J. Magnetism Mag. Mats.* **200**, 571 (1999).
78. Y. Asano, A. Oguri, S. Maekawa, *Phys. Rev. B* **48**, 6192 (1993).
79. B. Hillebrands, A. Boufelfel, C. M. Falco, et ál. *J. Appl. Phys.* **63**, 3880 (1988).
80. V. Leiner, K. Westerholt, A. M. Blixt, et ál., *Phys. Rev. Lett.* **91**, 037202 (2003).
81. Y. Guo, B.L. Gu, Z.Q. Li, et ál., *J. Appl. Phys.* **83**, 4545 (1998).
82. Z.Y. Zeng, L. D. Zhang, X. H. Yan, et ál. *Phys. Rev. B* **60**, 1515 (1999).
83. J.L. Cardoso, P. Pereyra, A. Anzaldo-Meneses, *Phys. Rev. B* **63**, 153301 (2001).
84. J.L. Cardoso, P. Pereyra, *Europhys. Lett.* **83**, 38001 (2008).
85. P. M. Levy, S. F. Zhang, *J. Magnetism Mag. Matls* **89**, 67 (1991).
86. N. I. Polushkin, S. A. Michalski, L. Yue, et ál., *Phys. Rev. Lett.* **97**, 256401 (2006).
87. Q. H. Huo, R. Z. Wang, H. Yan, *Appl. Phys. Lett.* **101**, 152404 (2012).
88. Z.Y. Zeng, F. Claro, *Phys. Rev. B* **65**, 064207 (2002).
89. Z.L. Ji, D.W.L. Sprung, *Phys. Rev. B* **56**, 1045 (1997).
90. J. Barnas, *Phys. Rev. B* **45**, 10427 (1992).
91. M. Mardaani, A. A. Shokri, *Chem. Phys.* **324**, 541 (2006).
92. W.-T. Lu, W. Li, Y.-L. Wang, et ál., *J. Appl. Pkys.* **112**, 083712 (2012).
93. C.-Z. Ye, R. Xue, Y.-H. Nie, J.-Q. Liang, *Phys. Lett. A* **373**, 1290 (2009).
94. R.-H. Zhu, H.-Y. Peng, M.-H. Zhang, et ál., *Physica B-Cond. Matt.* **404**, 2086 (2009).
95. P. Pereyra, V.G. Ibarra-Sierra, J. L. Cardoso, *Microelectronics J.* **40**, 779 (2009)
96. M. Barbier, P. Vasilopoulos, F. M. Peeters, *Phys. Rev. B* **81**, 075438 (2010).
97. M. Killi, S. Wu, A. Paramekanti, *Phys. Rev. Lett.* **107**, 086801 (2011).
98. G. Bastard, *Phys. Rev. B* **24**, 5693 (1981)
99. S. R. White, L. J. Sham, *Phys. Rev. Lett.* **47**, 879 (1981).
100. G. Bastard, *Phys. Rev. B* **25**, 7584 (1982).
101. M. Altarelli, F. Bassani, Handbook on semiconductors vol I Band Theory and Transport Properties vol 1 Editor W. Paul (Amsterdam: North-Holland 1982).
102. C Weisbuch and B. Vinter, *Quantum Semiconductor Structures, Fundamentals and Applications* (Academic Press Inc., Boston 1991).
103. G. Bastard Wave Mechanics Applied to Semiconductor Heterostructures (Halsted Press, N. York 1988).
104. P. Pereyra, *Fundamentals of Quantum Physics*, (Springer, Berlin, 2011).

Contents

Part I

Classical and Quantum Physical Properties of Semiconductors

Basic Physical Properties

1

Abstract

This chapter introduces several fundamental concepts, physical quantities, and approximations essential for describing the behavior of semiconductors. It begins with a brief overview of the zincblende and wurtzite crystal structures, followed by a discussion of the electronic band structure and the effective mass approximation. The processes of electron-hole excitation and the associated density of states are presented for one-, two-, and three-dimensional systems, along with the joint density of states relevant to optical transitions. The chapter also covers the occupation probability of energy levels through Fermi-Dirac statistics and provides a concise introduction to the influence of donor and acceptor impurities on the position of the Fermi level.

1.1 Crystal Structures

Semiconductors are materials whose atoms, in the solid state, are in crystalline configurations, that is, in periodic arrangements. Most semiconductors have the zincblende or wurtzite configurations. Both related to the fcc structure shown in Fig. 1.1.

The zincblende crystal configuration is formed when a face-centered cubic (fcc) lattice of atoms A is superimposed onto another fcc lattice of atoms B, such that is displaced from the other by $\sqrt{3}a/4$, along the main body diagonal. When both atoms A and B in the zincblende structure are carbon atoms, one has a diamond structure. Figure 1.2 illustrates both the zincblende structure of ZnS and the diamond structure. In Fig. 1.2a the tetrahedral bonds that connect each atom to its first four nearest neighbor are underlined. Each atom shares a valence electron with each of its neighbors, forming covalent bonds. Like many crystals, these structures exhibit symmetry. While a detailed analysis of their symmetries is important in crystallography, it is beyond the scope of this book. Notably, as shown in Fig. 1.3b when viewed along the main diagonal, the diamond structure, exhibits a hexagonal symmetry that resembles that of the wurtzite structure, an alternative crystalline form of ZnS. Although similar

P. Pereyra, *Contemporary Theory of Semiconductors and Devices*,
Graduate Texts in Physics, https://doi.org/10.1007/978-3-032-00296-9_1

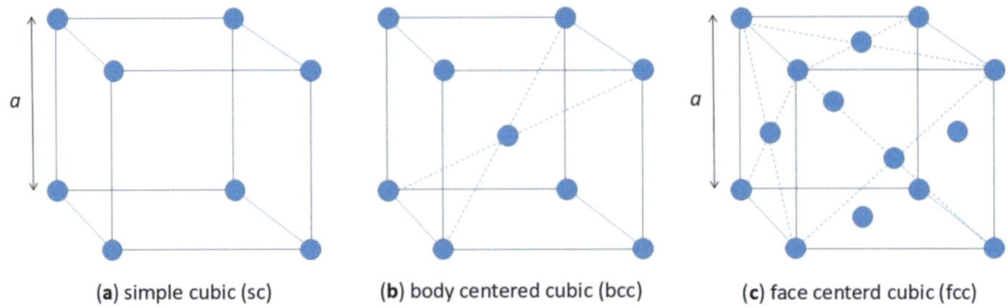

Fig. 1.1 Cubic lattices. In **a** the simple cubic (sc) with cell parameter a, in **b** the body-centered cubic (bcc) and in **c** the face-centered cubic (fcc)

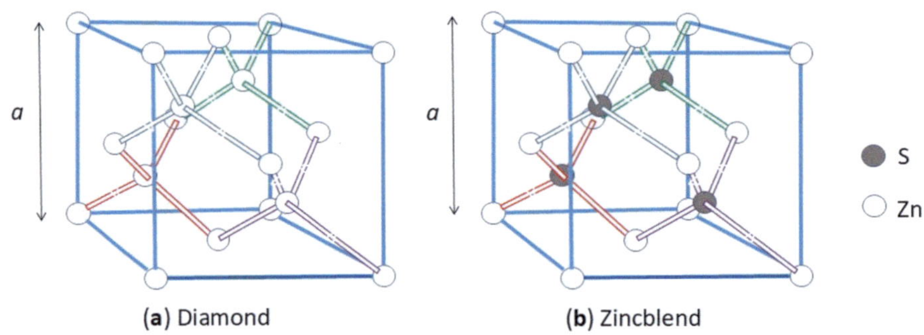

Fig. 1.2 If an fcc lattice of carbon atoms is superimposed on another fcc lattice, also of carbon atoms, displaced by $\sqrt{3}a/4$ on the main diagonal, we have the diamond structure shown in (**a**). If one of the fcc structures is made of A atoms and the other is made of B atoms, we have a zincblende structure. In **b** the zincblende structure of ZnS is shown

to zincblende, the wurtzite structure (Fig. 1.4) is distinct. In wurtzite, the bonds a and b are equal in length, but differ from the length of bond c. The tetrahedra, that form the basic of this structure (highlighted by darker atoms in Fig. 1.4), possess different symmetry axes (shown in magenta), indicating that the wurtzite crystal lacks inversion symmetry. This absence of inversion symmetry prevents complete cancellation of internal dipole moments, giving rise to piezoelectric properties and local Stark effects. Features that can be advantageous for some applications.

Silicon (Si) and germanium (Ge) are natural or elemental semiconductors with the symmetry of diamond. Some semiconductors composed of two or more elements are listed in Table 1.1. In general, compound semiconductors result from associations with other elements of the periodic table in such a way that the total valence band electrons of the compound add up to 8.[1] In III–V semiconductors, such as gallium arsenide GaAs, there are three valence band electrons from Ga and five from As. In II–VI semiconductors, the group II atom contributes 2 valence band electrons and the group VI atom contributes 6; In IV–IV as in SiC, each atom contributes 4 valence band electrons. In ternary semiconductors such as $Al_xGa_{1-x}As$, a fraction x of the

[1] Octant or Lewis rule.

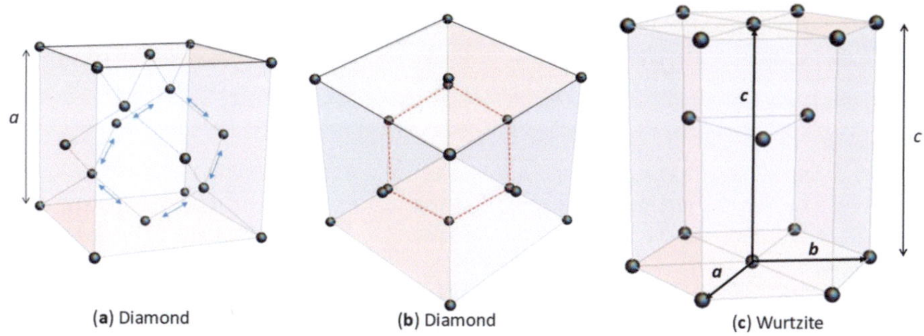

Fig. 1.3 When the diamond structure is rotated as shown in (**b**), the hexagonal symmetries shown in (**c**) are perceived

Fig. 1.4 Wurtzite structure of a compound semiconductor containing atoms A and B. The basic structure, with no inversion symmetry, is shown with darker atoms and tetrahedra, with their own symmetry axes the length; u is generally of the order of $3c/8$

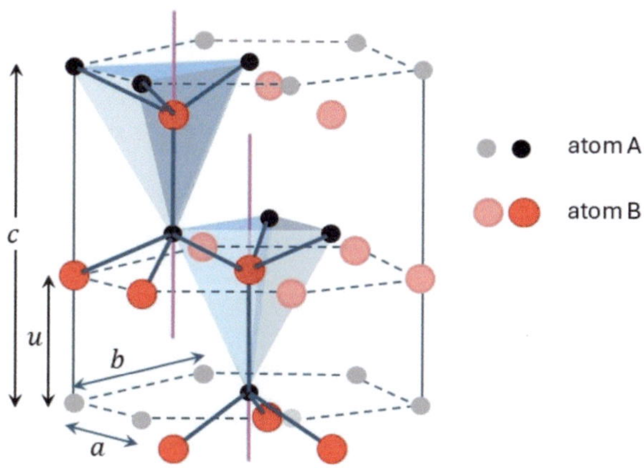

Ga atoms are replaced by Al atoms that are in the same group as Ga in the periodic table, that is, one atom is replaced by another with the same number of electrons in the valence band (Tables 1.2 and 1.3).

1.2 Band Structure, the Fundamental Gap

Due to the crystalline structure of semiconductors, the valence electrons of each atom experience the collective potential of the entire lattice, a periodic potential, which may vary in periodicity along different crystallographic directions. When viewed along a specific crystal direction, this potential appears, to a first approximation, similar to the periodic potential shown in Fig. 1.5. The Schrödinger equation cannot be solved analytically for such complex periodic potentials. However, approximate and numerical methods reveal interesting features of the energy eigenvalues and their corresponding eigenfunctions. Of particular importance in semiconductor physics is the energy eigenvalue spectrum, commonly referred to as the band structure. This band structure is fundamental for analyzing and understanding the electronic

Table 1.1 Gap energy, crystalline structure and lattice parameters

Clasification	Symbol	E_g [eV]† 0 K, 300 K	Structure	a [Å]	c [Å]	Dielectric constant (ϵ_s/ϵ_0)	Density gr cm^{-3}
(1) Elemental							
	Si	1.17, 1.11 i	Di	5.43		11.7	2.32
	Ge	0.74, 0.66 i	Di	5.658		16.2	5.323
	C	5.47 i	Di	3.567		5.70	3.515
(2) Compound							
(a) IV–IV	SiC	2.3 i	Wu	3.08	5.05	9.72	3.166
(b) III–V	AlP	2.45 i	ZB	5.4635		9.8	2.401
	AlAs	3.29, 2.16 i	ZB	5.66		10.06	3.76
	AlN	6.28 d	Wu	3.111	4.981	$\bar{\epsilon} = 9.14$	3.255
	AlSb	1.65 i/2.2 d					
	GaN	3.47, 3.40 d	Wu	3.175	5.158	$\epsilon_{\parallel} = 10.4$	6.15
	GaP	2.32, 2.25 i	ZB	5.4505		11.11	4.138
	GaAs	1.52, 1.42 d	ZB	5.653		13.18	5.318
	GaSb	0.81, 0.68 d	ZB	5.6137		15.69	5.6137
	InP	1.42, 1.27 d	ZB	5.8687		12.56	4.81
	InN	1.89 d	Wu	3.545	8.70	15.3	6.81
	InAs	0.43, 0.36 d	ZB	6.058		15.15	5.667
	InSb	0.23, 0.17 d	ZB	6.479		16.8	5.775
(c) II–VI	CdS	2.42 d	Wu	4.13	6.75	$\bar{\epsilon} = 9.38$	4.82
	CdSe	1.84, 1.74 d	Wu	4.30	7.02	$\epsilon_{\parallel} = 10.16$	5.81
	CdTe	1.61, 1.44 d	ZB	6.482		10.2	5.87
	ZnO	3.44, 3.2 d	Wu	3.25	5.21		
	ZnS	3.91, 3.6 d	ZB	5.41		8.9	4.079
	ZnSe	2.82 d	Wu	3.98	6.53	9.1	5.266
	ZnTe	2.25 d	Wu	4.27	6.99	6.104	5.636
(d) IV–VI	PbS	0.37 d	NaCl	5.936		169	7.597
	PbSe	0.27 d	NaCl	6.117		210	8.26
	PbTe	0.32 d	NaCl	6.462		414	8.219
(3) Alloys							
(a) Binary	Si_xGe_{1-x}						
(b) Ternary	$Al_xGa_{1-x}As$	1.424 + 1.247x					
	$Al_xGa_{1-x}N$	6.28 d					
	$Ga_xAs_{1-x}Sb$	0.7 d					
	$In_xGa_{1-x}N$	3.4 d					

† d = gap direct, i = indirect, Di = diamond, ZB = Zincblende, Wu = Wurtzite

Table 1.2 Frequently used physical parameters

Physical variable	Symbol	Value
Planck's constant	h	6.626×10^{-34} J s
	$\hbar = \dfrac{h}{2\pi}$	1.055×10^{-34} J s
Light velocity	c	2.998×10^{8} m/s
Electron charge	e	1.602×10^{-19} C
Electron volt	eV	1.602×10^{-19} J
Electron mass	m_o	9.109×10^{-31} kg
Vacuum permittivity	$\epsilon_o = \dfrac{10^7}{4\pi c^2}$	8.85×10^{-12} F m^{-1}
Boltzmann constant	k_B	8.617×10^{-5} eV K^{-1}
Thermal energy at 300 K	$k_B T$	0.0258 eV

Table 1.3 Useful relations

Wave length-energy relation: $\lambda\,(\text{nm}) = \dfrac{1240}{E(\text{eV})}$
Line width $\delta E = 1\,\text{meV} \Longrightarrow \delta\nu = 0.243\,\text{THz}$
Line width $\delta E = 1\,\text{meV} \Longrightarrow \begin{cases} \delta\lambda = 19.4\,\text{Å} & \lambda = 1.55\,\mu\text{m} \\ \delta\lambda = 6.2\,\text{Å} & \lambda = 0.88\,\mu\text{m} \end{cases}$

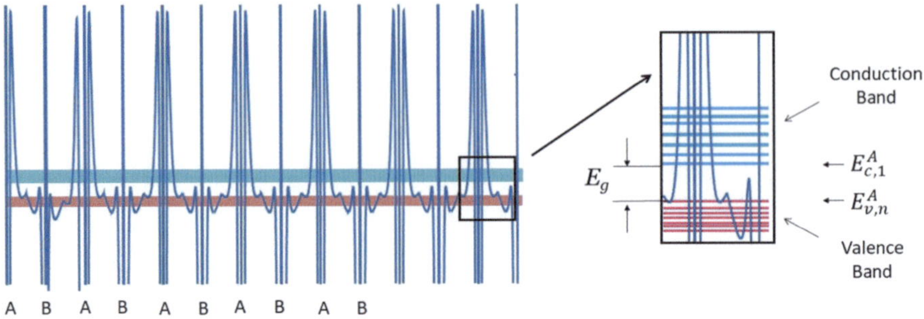

Fig. 1.5 Periodic potential along a crystalline axes of a semiconductor structure of two atomic elements A and B, and the valence and conduction energy bands. The enlarged figure indicates the energy gap E_g, which, generally, is of the order of 1 eV

properties of semiconductors and other periodic materials. The formation of energy bands is a direct consequence of quantum phase coherence. In a double quantum well, for example, phase coherence leads to energy level splitting. Similarly, in a periodic array of atoms, the discrete energy levels of isolated atoms, starting from the valence level and extending to higher energies, split into a number of closely spaced levels equal to the number of atoms in the chain. These discrete levels collectively form energy bands. If the Schrödinger equation for a periodic potential such as that in Fig. 1.5 were solved, the resulting spectrum would show these split levels grouped into bands. This phenomenon will be revisited in Chap. 6, particularly in the context of sequences of potential wells and barriers, where bands appear when the number of unit cells increases, as shown in Fig. 6.2. In some cases, bands may overlap, but typically they remain well-separated. A key property of these bands is that as the size of the periodic system increases. i.e., as the number of atoms grows, the band width remains constant, while the density of energy levels within each band increases.

The most critical bands for semiconductor behavior are the valence band (the highest fully occupied energy band) and the conduction band (the nearest unoccupied band of higher energy). While numerical methods exist to calculate band structures, optical spectroscopy remains the most accurate experimental technique for determining the energy difference between the conduction and valence bands.

In the standard theory of periodic systems based on Bloch's theorem, in which the system is assumed infinite,[2] the bands are continuous. A simplified but illustrative version of the continuous band structure, induced by periodicity and Bloch functions, is the Kronig-Penney model [2]. Real periodic systems are finite. In the theory of finite periodic systems that we will introduce in Chap. 6, the bands are discrete and contain a finite number of levels. The energy levels of the conduction band are the levels in which electrons move or the levels to which electrons are excited in optical phenomena. Unlike what happens in individual atoms, where the wave functions of the valence electrons and their excitations are *localized*, in the valence and conduction

[2] A good book on solid state physics with a comprehensive discussion of this topic is Madelung's book, see [1].

bands of periodic systems, the wave functions are *extended*, that is, they are finite throughout the periodic system.

According to the Pauli exclusion principle, each energy level in a solid can accommodate at most two electrons (one with spin $s_z = +1/2$ and one with spin $s_z = -1/2$). In a one-dimensional periodic system with N unit cells, each energy band contains N discrete energy levels, and can therefore hold up to $2N$ electrons. The energy intervals that separate the these allowed bands are known as energy gaps or forbidden bands. The energy gap between the valence band (the highest fully occupied band) and the conduction band (the nearest unoccupied band) is called the fundamental *gap*, typically denoted as E_g. As we will see, this gap plays a central role in determining the electrical and optical properties of semiconductors.

The electronic behavior of a material, whether it is an insulator, semiconductor, or conductor, depends on how the energy bands are filled and on the size of the fundamental gap. In semiconductors, the number of valence electrons is even, resulting in completely filled valence band and empty conduction band at low temperatures. In metals, the number of valence electrons is typically odd, leaving the highest band only partially filled, which allows electrons to move freely and conduct electricity. Insulators can be either amorphous with highly localized states, or crystalline, with completely filled valence bands and very large gap between valence and conduction bands. In such materials, a significant amount of energy is required to excite electrons from the valence band to the conduction band, if it exists. The energy gaps generally vary with temperature. For example, at $T = 300\,\mathrm{K}$ the gaps of Si, Ge and GaAs are approximately $1.12\,\mathrm{eV}$, $0.66\,\mathrm{eV}$ and $1.42\,\mathrm{eV}$, respectively, and at $T = 4\,\mathrm{K}$ they are $1.17\,\mathrm{eV}$, $0.74\,\mathrm{eV}$ and $1.519\,\mathrm{eV}$, respectively. By lowering the temperature, the average inter-atomic distances are reduced and the energy levels are pushed to larger values with the consequent increase in the width of the allowed and forbidden bands.

1.3 The Effective Mass

The electrons and holes that participate in electronic transport and optical transitions in semiconductors are primarily those located near the top of the valence band and the bottom of the conduction band. These bands, along with their extended wave functions and the quantum phase coherence, result from the conjunction of crystalline periodicity and quantum tunneling. As long as the potential is periodic, the energy levels in the bands spread and remain constant throughout the periodic system and the charge carriers in the valence and conduction band behave as if they were moving in a constant potential. This key property enables a crucial simplification in the theoretical treatment of periodic systems: the effective mass approximation. This approximation significantly simplifies the description of carrier dynamics while preserving the essential physics near the band edges, where most transport and optical phenomena occur. The effective mass approximation is also useful for studying transport and optical properties of heterostructures that are created by combining different semiconductors but with similar lattice parameters.

In this approach, the complex and rapidly varying atomic potential is replaced by a constant potential, which coincides with the upper edge of the valence band or with the lower edge of the conduction band, as required, and, simultaneously, in return for this simplification, the free electron mass is replaced by an effective mass m_e^* defined, in its simplest version, by the following relation

$$\frac{1}{m_e^*} = \frac{1}{\hbar^2}\frac{d^2E(k)}{dk^2}, \tag{1.1}$$

Here $E(k)$ is the energy that would be obtained if the Schrördinger equation were solved, as a function of the wave number k. The effective mass is directly related to the curvature of E with respect to k. Any expression that relates the energy or frequency of a particle or wave to its wave number k is called the dispersion relation. This function is, in general, different in the valence and conduction bands, and also varies depending on the crystallographic direction being considered. The function $E(k)$ may or may not be a symmetric function with respect to the components k_x, k_y and k_z. If the function is asymmetric, particularly in anisotropic materials, a more general definition of the effective masses is[3]

$$\left(\frac{1}{m_e^*}\right)_{ij} = \frac{1}{\hbar^2}\frac{\partial^2E(k)}{\partial k_i \partial k_j}, \tag{1.2}$$

known as the effective mass tensor, which accounts for the direction-dependent behavior. Virtually all known values of effective masses have been experimentally determined. The most common method used to determine an effective mass is by cyclotron resonance. To obtain the effective mass, the sample is placed in an axially symmetric cavity in which a magnetic field **B** parallel to the symmetry axis is applied. Due to the interaction with this field, the electrons in the extended bands move with the cyclotron frequency $\omega_c = eB/m$. The objective is to compare this frequency with the theoretical prediction, and infer the magnitude of the mass. The measurement of ω_c is not a direct process. In addition to the field **B**, a radio-frequency electric field $E_{rf} = E_x e^{i\omega t}$, perpendicular to the direction of **B**, is applied. The frequency ω is varied until the real part of the conductivity, given by the expression

$$\begin{aligned}\sigma &= \frac{e^2N\tau}{m}\frac{1+i\omega\tau}{1+(\omega_c^2-\omega^2)\tau^2+2i\omega\tau},\\ &= \frac{e^2N\tau}{m}\frac{1+(\omega_c^2+\omega^2)\tau^2 - i\left(\omega\tau-(\omega_c^2-\omega^2)\tau^2\right)}{\left(1+(\omega_c^2-\omega^2)\tau^2\right)^2+4\omega^2\tau^2},\\ &= \sigma_R + i\sigma_I,\end{aligned} \tag{1.3}$$

[3] This topic is discussed in more detail in solid-state physics books. See for example the books of Hook and Hall [3] and of Kittel [4].

Table 1.4 Gap energies $E_g(0)$, α, β and effective masses of Si, Ge and GaAs

Semiconductor	$E_g(0)$ [eV]	α_T [eV K^{-1}]	β_T [K]	m_l^* longit	m_t^* transv	m_c stat. dens.	m_{hh}^* heavy h	m_{hl}^* light h
Si	1.17	4.37×10^{-4}	636	$0.98m_e$	$0.19m_e$	$0.33m_e$	$0.49m_e$	$0.16m_e$
Ge	0.74	4.77×10^{-4}	235	$1.64m_e$	$0.082m_e$	$0.22m_e$	$0.29m_e$	$0.044m_e$
GaAs	1.519	5.8×10^{-4}	204			$0.85m_e$	$0.51m_e$	$0.08m_e$
GaN (Wu)	3.47					$1.5m_e$	$1.4m_e$	$0.3m_e$
GaN (ZB)	3.28					$1.4m_e$	$1.3m_e$	$0.19m_e$
Semiconductor	$E_g(0)$	α_X	β_X	m_l^*	m_t^*	Δm_{ex}^*	Δm_{hhx}^*	Δm_{hlx}^*
$Al_xGa_{1-x}As$	1.519	1.155	0.37			$0.08x$ m_e	$0.25x$ m_e	$0.068x$ m_e
$In_xGa_{1-x}N$	3.4	−2.45	1.0			$0.09x$ m_e	$0.8x$ m_e	

resonates. The resonance is detected by measuring the energy absorbed from the cavity's radiation field, which is proportional to σ_R.[4] Because the collision time τ is very small ($\sim 10^{-13}$ s), the experiments are carried out with ultra-pure samples and at temperatures of the order of 4 K.

In general, as mentioned before the value of the effective masses depends on the semiconductor, the crystallographic orientation, and whether the particle is in the valence or conduction band. Likewise, the effective masses in the longitudinal and transverse directions are different, as shown in Table 1.4. The definition (1.1) that allows (in first approximation) to transform a periodic potential problem with mass m_e into one with constant potential but mass m_e^*, makes it possible to deal with the Schödinger equations for semiconductor layers using rather simple methods. That is, if originally the Schrödinger equation of a layer X (X represents layer A or B or ...), is

$$\left(\frac{-\hbar^2 \hat{\mathbf{p}}^2}{2m} + U_X(\mathbf{r})\right)\Phi^X(\mathbf{r}) = E\Phi^X(\mathbf{r}), \tag{1.4}$$

where $U_X(\mathbf{r})$ is the rapidly varying periodic potential, in the effective mass approximation, it is transformed into the equation

$$\left(-\hbar^2 \frac{\Delta^2}{2m^*} + V_X\right)\varphi^X(\mathbf{r}) = E\varphi^X(\mathbf{r}), \tag{1.5}$$

Here the potential V_X of layer X is constant, while the particle is restricted to layer X. This is an important simplification. If one has a single layer and the potential

[4] For more information see for example: [5].

is constant, the solution to the Schrödinger equation is easy to obtain. If we have a multilayer system, the potential V_X varies along the direction of growth, with values sectionally constant and effective masses that changes with the semiconductor material and the type of charge carrier particle. In this class of systems it is necessary to distinguish the direction of growth, normally chosen as the z direction, from the transverse direction. If it is a semiconducting layer that extends from z_L to z_R with effective potential $V_X = V_0$ in $0 \leq x \leq L_x$ and $0 \leq y \leq L_y$ and $V_X = \infty$ outside $\{0 \leq x \leq L_x, 0 \leq y \leq L_y\}$, the Schrödinger equation can be separated into

$$-\frac{\hbar^2}{2m_t^*}\left(\frac{\partial^2}{\partial x^2}+\frac{\partial^2}{\partial y^2}\right)\varphi_T(\mathbf{r}_T) = E_T\varphi_T(\mathbf{r}_T) \quad \text{with} \quad \mathbf{r}_T = x\hat{\mathbf{x}} + y\hat{\mathbf{y}} \tag{1.6}$$

and

$$-\frac{\hbar^2}{2m_l^*}\frac{\partial^2}{\partial z^2}\varphi_l(z) = (E_l - V_0)\varphi_l(z) \quad \text{with} \quad E_l = E - E_T, \tag{1.7}$$

with the solutions

$$\varphi_T(\mathbf{r}_T) = A_t e^{i\mathbf{k}_T\cdot\mathbf{r}_T} \quad \text{and} \quad \varphi_l(z) = A_t e^{ik_z z} \quad \text{with} \quad k_x = \frac{n_x\pi}{L_x}, \quad k_y = \frac{n_y\pi}{L_y}. \tag{1.8}$$

Here $k_z = \sqrt{2m_l^*(E_l - V_0)/\hbar^2}$. The longitudinal solution must also satisfy continuity conditions and also boundary conditions. Although the total stationary solution of the equation is of the form

$$\varphi(\mathbf{r}) = Ae^{i\mathbf{k}_T\cdot\mathbf{r}_T}e^{ik_z z} = \varphi_T(\mathbf{r}_T)\varphi_l(z), \tag{1.9}$$

it is normal to refer to the z dependent function only. In almost all semiconductor devices the transverse dimensions are much larger than the longitudinal ones. For this reason, it is expected that in these systems the physics is practically invariant under translations in the transverse plane, that is, that the physics is independent of $\mathbf{r}_T$, and that the transverse momentum is conserved. This invariance argument is behind another important and widely used approximation: *the one-dimensional approximation*. In the 1D approach the focus is on the dynamics along the z growth direction. In this case, it is common to eliminate the subscript l in the wave function, effective masses, etc.

In Chap. 5 we will study quantum structures with more than one semiconductor layer. Solving these systems in the effective mass approximation, we will have Schrödinger equations as in (1.5), generally with different effective mass in each layer. As for any quantum problem, we will require solutions fulfilling continuity conditions and the flux conservation principle. These requirements on particles whose masses change from one layer to another, will demand small changes in the continuity requirement of the first derivative of the wave functions.

1.4 Dispersion Relations, Parabolic Approximation

In general the energies are functions of the components k_x, k_y and k_z. The minima of these functions are specially important because they define the energy gaps of the semiconductor. Minima generally occur in crystalline directions related to some symmetry. Since the periodicity is not the same in any direction inside the crystal, the minima are not the same, and as said before they also differ from semiconductor to semiconductor. For the purposes of this course, it is sufficient to keep this information in mind. The values of the gaps and other properties of the semiconductors are known and tabulated. One of the most complete pages on semiconductor parameters is that of the Ioffe Institute [6]. In the vicinity of these minima, the energy $E(k)$ can be approximated by a parabolic function (see Fig. 1.6), whose second derivative with respect to k defines the inverse of the effective mass. If the zero of energy is chosen at the top of the valence band, the energy at the low levels of the conduction band will be given by

$$E_c(k) = E_g + \frac{\hbar^2 k^2}{2m_{ec}^*}, \tag{1.10}$$

and for the higher levels of the valence band (see Fig. 1.6), by

$$E_v(k) = -\frac{\hbar^2 k^2}{2|m_{ev}^*|}. \tag{1.11}$$

In these expressions m_{ec}^* and m_{ev}^* are the corresponding effective masses of the electron. Note that if the definition of effective mass (1.1) is used, we have

$$\frac{1}{m_{ev}} = \frac{1}{\hbar^2}\frac{d^2E_v(k)}{dk^2} = -\frac{1}{|m_{ev}^*|}. \tag{1.12}$$

The effective mass of electrons in the valence band is negative. Normally, semiconductors are not pure. This has consequences on the band structure. Some lines below we will see that the artificial doping of semiconductors is an important procedure that gives them new and important physical properties. There is a big difference between the effects of randomly located impurities or defects and those that result

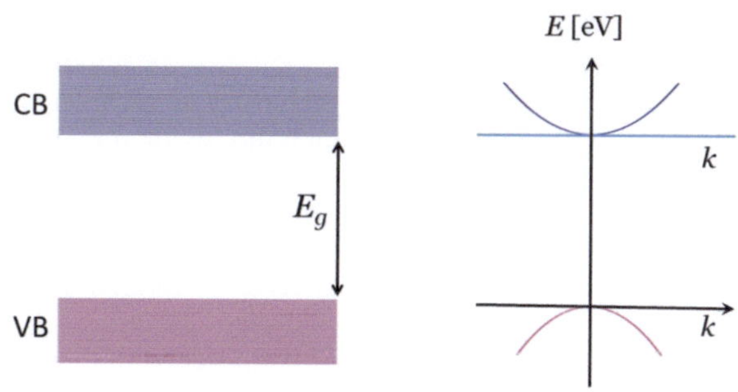

Fig. 1.6 The band structure. The conduction and valence bands, and the energy gap E_g. In the right panel, parabolic dispersion relations, with minimum and maximum at the edges of the bands

from impurities introduced homogeneously into the semiconductors, and in a controlled manner. In fact, the semiconductors most widely used in electronics and optoelectronics are intentionally doped with *donor* atoms or *acceptors*. We will refer to these semiconductors in detail in the following chapters.

1.5 Other Important Properties

To understand the physical meaning of some properties that will be considered later, we will comment in the following on the exciton (or electron-hole excitation *e-h*), the level density $g(E)$, and the occupation probability $f(E)$ of a quantum state.

1.5.1 Electron-Hole Excitations, and Physical Properties of Holes

At absolute zero temperature, all electrons in a semiconductor occupy states in the valence band, but when the temperature increases, thermal energy excites electrons, some of which may reach the conduction band, leaving vacancies in the valence band. Electrons in the conduction band behave distinct from when they resided in the valence band. Similarly, the electronic properties of the valence band change in the presence of one or more empty states. The contribution of the valence band to the physical properties of semiconductors can be understood more intuitively in terms of the behavior of the vacancies rather than in terms of the electrons that remain. The vacancies display particle-like properties, such as mobility, charge, mass, momentum and energy. These quasi-particles are known as holes. A hole h is created when an electron is excited of the valence band, and it is annihilated when the electron returns to fill the vacancy. Holes play a central role in charge transport. The pair *e-h*, that is, the electron-hole excitation, is called an exciton. Because of the importance that holes and excitons have in semiconductor physics, we will now define some properties of holes and excitons. We shall first refer to holes and, in the next section, we will briefly refer to the binding energy of an excitons in the semiconductor layered structures.

The properties of holes are easily recognized if we define them as those of the valence band with a vacancy. In fact, the hole is defined precisely as the quasi-particle introduced to represent the set of electrons that remain in the valence band when an electron is excited. Consequently, the hole properties will be obtained by referring to the valence band with a vacancy. To determine, for example, the momentum of a hole, we argue as follows: (i) before the excitation of the electron, the band was full and the total momentum added up to zero. (ii) if an electron with momentum $\hbar\mathbf{k}_e$ is excited to the conduction band, the residual (uncompensated) momentum of the valence band will be $-\hbar\mathbf{k}_e$. This momentum left in the valence band is the one assigned to the hole, that means that:

$$\hbar\mathbf{k}_h = -\hbar\mathbf{k}_e, \quad \text{then} \quad \mathbf{k}_h = -\mathbf{k}_e. \tag{1.13}$$

To obtain the energy of the hole we proceed in the same way. The energy left in the valence band as a consequence of the electron excitation will be the hole energy.

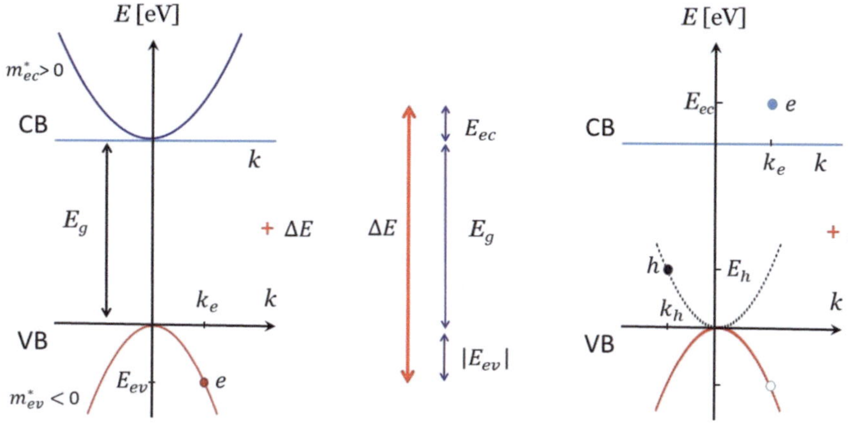

Fig. 1.7 The creation of an excitonic state. When an electron in the valence band (left panel) with energy E_{ev} and momentum $\hbar\mathbf{k}_e$ receives an energy $\Delta E = |E_{ev}| + E_g + E_{ec}$, an exciton, consisting of the electron e in the conduction band and the hole h in the valence band, is formed, as illustrated in the right panel

Suppose that, as illustrated in Fig. 1.7, an electron in the valence band with energy $E_{ev} = -\hbar^2 k_e^2/2|m^*_{ev}|$, receives an amount of energy ΔE that excites it to the energy level $E_{ec} = E_g + \hbar^2 k_e^2/2m^*_{ec}$, in the conduction band. For this process to occur, the energy supplied to the valence band needs to be

$$\Delta E = E_{ec}(k) - E_{ev}(k) = E_g + \frac{\hbar^2 k_e^2}{2m^*_{ec}} + \frac{\hbar^2 k_e^2}{2|m^*_{ev}|}. \tag{1.14}$$

Since the first two terms represent the energy possessed by the excited electron, the last term, due to conservation of energy, is the energy remaining in the valence band. This is then the energy attributed to the hole, that is:

$$E_h(k) = \frac{\hbar^2 k_e^2}{2|m^*_{ev}|} = -E_{ev}(k), \tag{1.15}$$

which can be rewritten in the form

$$E_h(k) = \frac{\hbar^2 k_h^2}{2m^*_h}, \tag{1.16}$$

where m^*_h is the effective mass of the hole, which is equal to $|m^*_{ev}|$. If we remember the relationship (1.12), the hole mass is equal to $-m^*_{ev}$, the effective mass of the hole is positive. It is important to note that the hole energy described by the function $E_h(k)$ in (1.16) is positive, equal to the negative of the electron energy in the valence band (see right panel of Fig. 1.7). In what follows, only the effective masses of holes in the valence band and of electrons in the conduction band will be used. For simplicity we will denote the effective mass of an electron in the conduction band as m^*_e. Both masses, m_h^* and m_e^*, are positive. Taking into account that $m_h^*\mathbf{v}_h = \hbar\mathbf{k}_h, m_e^*\mathbf{v}_e = \hbar\mathbf{k}_e$, and that also $\mathbf{k}_h = -\mathbf{k}_e$, we conclude also that

$$m_h^*\mathbf{v}_h = -m_e^*\mathbf{v}_e. \tag{1.17}$$

The speed of a hole is antiparallel to the speed that electron had before it was excited. However their magnitudes are not necessarily equal, since the effective masses are generally different. Finally, what is the charge of a hole? By promoting an electron to the conduction band, the valence band is less negative, as if it had received a positive charge. Therefore the charge of a hole is positive, that is:

$$q_h = -q_e = e. \tag{1.18}$$

This relationship is also reached if we consider the equations of motion in an electric field.

$$m\frac{d\mathbf{v}}{dt} = q\mathbf{E}, \tag{1.19}$$

which for hole and electron are

$$m_h^*\frac{d\mathbf{v_h}}{dt} = q_h\mathbf{E} \quad \text{and} \quad m_e^*\frac{d\mathbf{v_e}}{dt} = q_e\mathbf{E}, \tag{1.20}$$

respectively. If $\mathbf{v_e}$ is replaced by $-m_h^*\mathbf{v_h}/m_e^*$ in the last of these equations, we have

$$m_h^*\frac{d\mathbf{v_h}}{dt} = -q_e\mathbf{E}. \tag{1.21}$$

Comparing this equation with the first of (1.20), we have that indeed $q_h = -q_e$.

It is known that in general the current density $\mathbf{j}$ of a concentration n of charge carriers q, moving with speed $\mathbf{v}$, is given by the product $nq\mathbf{v}$. The current density of electrons is then $\mathbf{j}_e = nq_e\mathbf{v_e}$ while the hole current density is $\mathbf{j}_h = nq_h\mathbf{v_h}$. To establish a relationship between these currents, we consider the current density of holes and replace q_h with $-q_e$ and $\mathbf{v_h}$ with $-m_e^*\mathbf{v_e}/m_h^*$ it follows that

$$\mathbf{j}_h = nq_e m_e^*\mathbf{v_e}/m_h^* = \mathbf{j_e} m_e^*/m_h^*. \tag{1.22}$$

Since the masses are positive, it is clear that $\mathbf{j}_h$ and $\mathbf{j}_e$ are parallel. Both types of carriers contribute with currents in the same direction! Generally $m_e^* < m_h^*$, therefore the current $\mathbf{j}_e$ of the electrons is generally greater than $\mathbf{j}_h$.

1.5.2 Exciton Binding Energy

When we refer to an excited electron, we typically assume that both the electron and the hole occupy energy levels of a system defined by the system's external potentials. However, in low-dimensional quantum structures, the Coulomb interaction between the electron and hole can led to significant binding energies, which must be taken into account for an accurate physical description. In these structures, commonly referred to as quasi-two-dimensional (2D) quantum wells (QWs), quasi-one-dimensional (1D) quantum wires (QWrs), and quasi-zero-dimensional (0D) quantum dots (QDs), the charge carriers are spatially limited in one, two or all three dimensions, respectively. This spatial confinement gives rise to quantization effects, often referred to as size quantization. The calculation of the binding energy for excitons has been an important issue. In quantum wells, double quantum wells SLs, and heterostructured devices, the binding energies play an important role to identify and to fit optical-resonances positions, relaxation mechanisms, transport properties, efficiency of optoelectronic devices, etc. The *e-h* Coulomb interaction is an essential quantity in determining the excitonic spectra in quantum dots.

The concept of the excitonic spectrum as analogous to the hydrogen atom spectrum is, in principle, straightforward to visualize. However, calculating it rigorously, especially in low-dimensional systems, can be quite challenging. One of the main difficulties arises from the actual asymmetry of the confining potential, not only in quantum wells and multiple quantum wells, this problem appears also in quantum dots and quantum wires. The lack of spherical symmetry led to search different procedures to solve the exciton Schrödinger equation. A substantial body of research has addressed this problem using various techniques, including fractional-dimension approaches as well as variational and perturbative methods. These models allow for approximate solutions by treating the exciton as a pseudo spherical object confined in a space with an effective dimensionality D where $2 < D < 3$. In this framework, the exciton binding energy in quantum wells is often expressed as a function of the well width, providing useful insights into how spatial confinement affects excitonic properties.

We will present here one of the simplest variational procedures to calculate the *e-h* binding energy when only the Coulomb interaction is present, which is known as the free exciton binding energy. We will refer to exciton binding energies later, specifically in relation to the optical response of superlattices and with respect to quantum dot spectra.

In the center of mass and relative coordinates system, and assuming cylindrical symmetry, the relative exciton coordinate can be written as

$$\mathbf{r} = \boldsymbol{\rho}_e + \mathbf{z}_e - \boldsymbol{\rho}_h - \mathbf{z}_h. \tag{1.23}$$

In this representation, the radial part of the quasi-two-dimensional Schrödinger equation, with a Coulomb potential and azimuthal quantum number m=0, takes the form

$$\left[-\frac{\hbar^2}{2\mu_\parallel} \frac{1}{\rho} \frac{d}{d\rho} \rho \frac{d}{d\rho} - \frac{e^2}{\epsilon\sqrt{\rho^2 + z^2}} \right] \phi_n(\rho, z) = E_n^{eh} \phi_n(\rho, z) \tag{1.24}$$

where ρ and z are the in-plane $\rho_e - \rho_h$ and perpendicular $z_e - z_h$ electron-hole distances, and $\mu_\parallel$ the in-plane relative e-h effective mass. In the Appendix A a variational model to calculate the binding energy is summarized. Based on exact 2D and trial wave functions, the expectation values of the dimensionless energies of the ground and the first excited states, defined as

$$\langle w_{\lambda_0}(\zeta)\rangle = -\frac{\langle E_0^{eh}\rangle}{R_{eh}} \quad \text{and} \quad \langle w_{\lambda_1}(\zeta)\rangle = -\frac{\langle E_1^{eh}\rangle}{R_{eh}} \tag{1.25}$$

where

$$R_{eh} = \frac{\mu_\parallel e^4}{2\epsilon^2\hbar^2}, \quad \zeta = \frac{z_e - z_h}{a_0}, \quad \text{and} \quad \lambda_k = \frac{2}{2k+1+m+2\sqrt{\zeta}} = \frac{1}{\eta_k}, \tag{1.26}$$

have been obtained, and are explicitly given as

$$\langle w_{\lambda_0}(\zeta)\rangle = -\lambda_0^2 + \frac{4\lambda_0 + 4\zeta^2\lambda_0^4}{1+2\sqrt{\zeta}} e^{2\zeta\lambda_0}\Gamma(0, 2\zeta\lambda_0), \tag{1.27}$$

and

$$\begin{aligned}\langle w_{\lambda_1}(\zeta)\rangle &= -4e^{2\zeta\lambda_1}\zeta^2\lambda_1^4(3+2\zeta\lambda_1)\Gamma(0, 2\zeta\lambda_1) \\ &+ \frac{4\lambda_1 - \lambda_1^2\Big(3 + 2\zeta\lambda_1(5 + 2\zeta\lambda_1(5 + 2\zeta\lambda_1))\Big)}{3+2\zeta\lambda_1}.\end{aligned} \tag{1.28}$$

respectively. R_{eh} is the effective Rydberg energy, formally similar to Rydberg energy R_y, but much smaller because the exciton reduced mass $\mu_\parallel$ is much smaller than the reduced mass of the hydrogen atom; a_0 is Bohr's radio and λ_k a variational parameter, whose inverse η_k is the principal quantum number of the exciton. For the ground state $\lambda_0 = 2/\left(1+2\sqrt{\zeta}\right)$ and for the first excited state $\lambda_1 = 2/\left(3+2\sqrt{\zeta}\right)$. In Fig. A.1 the dimensionless binding energies $\langle w_{\lambda_0}(\zeta)\rangle$ and $\langle w_{\lambda_1}(\zeta)\rangle$ are plotted as functions of ζ. In Fig. 1.8 we plot the binding energies. In Fig. 1.8a in units of the effective Rydberg energy and in Fig. 1.8b the ground state binding energies in eVs, for excitons e-hh and e-lh in GaAs. In the limit $\zeta = 0$, the exciton is $2D$, and $3D$ when $\zeta = 1$. In the 2D limit, the exciton binding energy is given by

$$E_k^{eh} = -\frac{\alpha_{rk}^2\hbar^2}{8\mu_\parallel} \quad \text{with} \quad \alpha_{rk} = \frac{\mu_\parallel e^2}{2\pi\epsilon\hbar^2\eta_k}, \tag{1.29}$$

where $\epsilon = \epsilon_s\epsilon_0$ is the semiconductor dielectric constant. The quantum number η_k values for the ground and excited 2D states are $\eta_0 = 1/2 = 1/\lambda_0$ and $\eta_1 = 3/2 = 1/\lambda_1$, respectively.

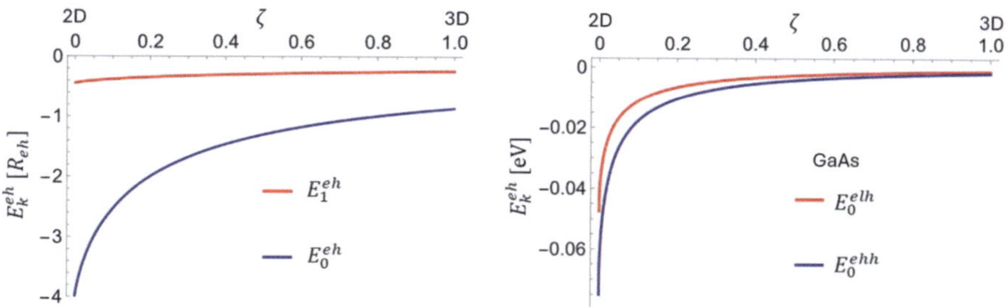

Fig. 1.8 Dimensionless exciton-binding energy for the ground and the first excited states. The magnitude of the binding energy reduces as the exciton dimension goes from $D = 2$ to $D = 3$

1.5.3 Density of Levels

If one is able to evaluate the number of levels as a function of the energy, represented by $N(E)$, one can also determine the density of levels $g(E)$, and the density of states. The density of levels or number of levels per unit of energy is given by

$$g(E) = \frac{dN}{dE}, \tag{1.30}$$

The density of levels is a fundamental physical quantity that frequently appears in the definition of other key variables, particularly when analyzing systems composed of regions or components with differing properties. As we will see below, the dimensionality of the system plays a crucial role in determining the density of levels: the analytical expressions vary significantly between one-dimensional, two-dimensional and three-dimensional systems.

One has also to distinguish between the densities of levels in the conduction band from those of the valence band, and these densities from the joint density of levels, which involves the densities of both bands. The joint density is essential in the study of optical transitions. The number of levels and the density of levels can be easily calculated when the dispersion relation is parabolic, that is, when the energies depend on the wavenumber in the form

$$E = E_o + \frac{\hbar^2 k^2}{2m^*}, \tag{1.31}$$

where E_o is the distance from the band edge to the zero energy level, and $k = |\mathbf{k}|$. We will determine the density of levels for systems with different dimensionality, but first it is important to comment that the density of levels and the density of states are related. In non-degenerate systems, they are the same, and differ when the energy levels are degenerate. For example, if the density of energy levels of an electron system is $g(E)$ and in each energy level we have two spin states, the degeneracy is two and the density of states will be $2g(E)$.

Suppose now a 3D system with dimensions $L_x = L_y = L_z = L$ and volume $V = L^3$. The quantum particles inside this cube are described by stationary eigenfunctions with wave number

$$\mathbf{k} = k_x\hat{x} + k_y\hat{y} + k_z\hat{z} = \frac{n_x\pi}{L}\hat{x} + \frac{n_y\pi}{L}\hat{y} + \frac{n_z\pi}{L}\hat{z}. \tag{1.32}$$

Here n_x, n_y and n_z are integers. Alternative derivations are also obtained by imposing periodicity conditions on the propagating modes, with spacings of $2\pi/L$ and positive and negative wave numbers. The final results are similar. To obtain $g(E)$ it is convenient to first deduce the density of levels $g(k)$ in the space k. Since the spacing between the eigenvalues of k_i is π/L, the number of energy levels in the space k, from $k = 0$ to k, with positive components, can be evaluated as follows. Since k depends on the squares of k_x, k_y and k_z, the energy is the same at all 8 points $(\pm k_x, \pm k_y, \pm k_z)$, therefore, the number of energy levels will correspond to the number of points in an octant of a sphere of radius k. This number is equal to the volume of the octant divided by the volume $(\pi/L)^3$ associated with each point, that is

$$N_k^{3d} = \frac{1}{8}\frac{4\pi k^3}{3}\frac{1}{(\pi/L)^3} = \frac{L^3}{6\pi^2}k^3. \tag{1.33}$$

If we express k as a function of energy, we can directly obtain the density of energy levels $g(E)$. Let us proceed determining beforehand the density of levels $g(k)$ per unit of wave number k. In this case, the 3D density $g_{3d}(k)$ is given by the relation

$$g_{3d}(k) = \frac{dN_k^{3d}}{dk} = \frac{L^3}{2\pi^2}k^2. \tag{1.34}$$

For two-dimensional and one-dimensional systems, the numbers of levels are

$$N_k^{2d} = \frac{1}{4}\pi k^2\frac{1}{(\pi/L)^2} = \frac{L^2}{4\pi}k^2 \quad \text{and} \quad N_k^{1d} = k\frac{1}{\pi/L} = \frac{L}{\pi}k, \tag{1.35}$$

and the corresponding level densities per unit of k are

$$g_{2d}(k) = \frac{L^2}{2\pi}k \quad \text{and} \quad g_{1d}(k) = \frac{L}{\pi}. \tag{1.36}$$

Given the level density $g(k)$, the level density $g(E)$ can easily be obtained when the dispersion relation is known. In fact, it is clear from the following identity

$$g(E) = \frac{dN_k}{dk}\frac{dk}{dE} = g(k)\frac{dk}{dE}, \tag{1.37}$$

that, among them, bear a simple relation, with the factor dk/dE defined by the dispersion relation $E(k)$. If we rewrite (1.37) in the form

$$\frac{dN_k}{dk}dk = \frac{dN_k}{dE}dE \tag{1.38}$$

we can see that the product $g(E)dE$, which corresponds to the number of levels between E and $E + dE$, is equal to $g(k)dk$, the number of levels between k and $k + dk$.

The dispersion relation depends on the system under study. For photons the dispersion relation is $E = \hbar\omega = \hbar ck$. For electrons and holes, in the parabolic approximation, the dispersion relations are as indicated in (1.31). For conduction band electrons, it is

$$E_{ec} = E_c + \frac{\hbar^2 k_e^2}{2m_{ec}^*}, \tag{1.39}$$

in which case

$$k_e = \left(\frac{2m_{ec}^*}{\hbar^2}\right)^{1/2} \sqrt{E_{ec} - E_c} \quad \text{if} \quad E_{ec} > E_c. \tag{1.40}$$

Here E_c is the energy of the conduction band edge. When the energies are measured from the top edge of the valence band, E_c is equal to E_g. Using the dispersion relation (1.39), we have

$$\frac{dE_{ec}}{dk_e} = \frac{\hbar^2 k_e}{m_{ec}^*}, \quad \text{and} \quad \frac{dk_e}{dE_{ec}} = \frac{m_{ec}^*}{\hbar^2 k_e}. \tag{1.41}$$

Replacing this derivative, the wave number k_e, and the densities $g_{nd}(k)$ into the identity (1.37), we have the corresponding dimensionality-dependent densities of energy levels

$$g_{1d}^c(E_{ec}) = \frac{L}{2\pi}\left(\frac{2m_{ec}^*}{\hbar^2}\right)^{1/2} \frac{1}{\sqrt{E_{ec} - E_c}}; \tag{1.42}$$

$$g_{2d}^c(E_{ec}) = \frac{L^2}{4\pi}\left(\frac{2m_{ec}^*}{\hbar^2}\right)^{2/2}; \tag{1.43}$$

$$g_{3d}^c(E_{ec}) = \frac{L^3}{4\pi^2}\left(\frac{2m_{ec}^*}{\hbar^2}\right)^{3/2} \sqrt{E_{ec} - E_c}. \tag{1.44}$$

If the spin of the electron is taken into account, the density of states will be twice the density of levels.

In the valence band, the electrons energy, measured with respect to edge E_v of the valence band, is given by the dispersion relation

$$E_{ev} = E_v - \frac{\hbar^2 k_e^2}{2m_h^*}. \tag{1.45}$$

Therefore, the density of levels in the valence band of a three-dimensional system is

$$g_{3d}^v(E_{ev}) = \frac{L^3}{4\pi^2}\left(\frac{2m_h^*}{\hbar^2}\right)^{3/2} \sqrt{E_v - E_{ev}}, \quad E_{ev} < E_v. \tag{1.46}$$

In Fig. 1.9b, the densities $g_{3d}(E_{ec})$ and $g_{3d}(E_{ev})$ are plotted.

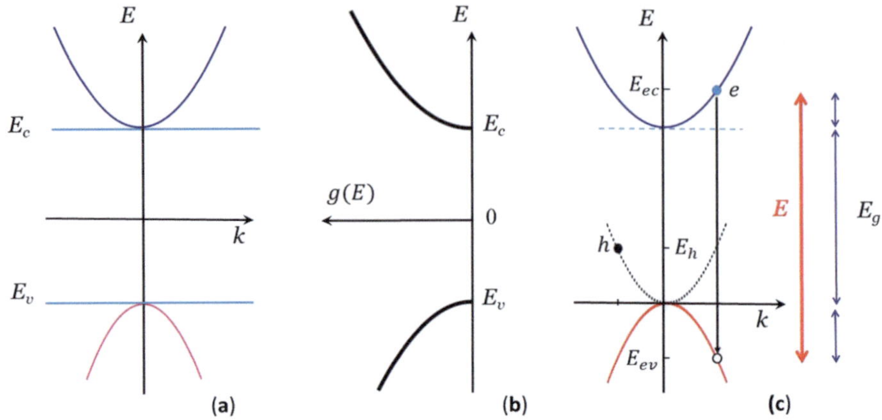

Fig. 1.9 Densities of energy levels and transitions from one level to another. In **b** the level densities in the conduction and valence band of (1.44) and (1.46) are plotted. In **c** the energy $E = \hbar\omega_t$ is associated with the transition of an electron from the conduction band to the valence band

1.5.4 Joint Density of Levels

The joint density is defined with the purpose of representing with a single expression the densities of levels of the two bands. In the excitation of an electron (see right panel of Fig. 1.9) or in the inverse process, the recombination, the absorbed (or emitted energy) is

$$\begin{aligned} E &= \hbar\omega_t = E_{ec} - E_{ev}, \\ &= E_c + \frac{\hbar^2 k_e^2}{2m_{ec}^*} - E_v + \frac{\hbar^2 k_h^2}{2m_h^*}. \end{aligned} \tag{1.47}$$

Notice that $E_g = E_c - E_v$. These processes involve bands in which the densities of energy are different. For this reason a joint (optical) density of energy, associated with the energy $E = \hbar\omega_t$, is a convenient quantity. In a transition process, without transfer of linear momentum, i.e., when $\hbar k_e = \hbar k_h$, the previous equation can be written in the form

$$\hbar\omega_t = E_c - E_v + \frac{\hbar^2 k_e^2}{2m_r}, \tag{1.48}$$

with m_r the reduced mass $m_r = m_e^* m_h^*/(m_e^* + m_h^*)$. If we solve for k_e in the last equation and substitute k_e into (1.39), we have

$$E_{ec} = E_c + \frac{\hbar^2}{2m_{ec}^*}\frac{2m_r}{\hbar^2}(\hbar\omega_t - E_c + E_v). \tag{1.49}$$

This equation establishes a one-to-one relationship between E_{ec} and $\hbar\omega_t$ and we can use it to relate the density of levels in the conduction band with the *joint* density of levels. For a two-dimensional system, the joint density of states $g_{cv}^{2d}(\hbar\omega_t)$ for the process $v \rightarrow c$ is given by

$$g_{cv}^{2d}(\hbar\omega_t) = g_{2d}^{c}(E_{ec})\frac{dE_{ec}}{d(\hbar\omega_t)}, \tag{1.50}$$

If we combine this equation with the density in (1.43) where $L^2 = L_x L_y$, we have

$$g_{cv}^{2d}(\hbar\omega_t) = \frac{L_x L_y}{2\pi}\frac{m_r}{\hbar^2}. \tag{1.51}$$

In 3D systems, the relationship is

$$g_{cv}^{3d}(\hbar\omega_t) = g_{3d}^{c}(E_{ec})\frac{dE_{ec}}{d(\hbar\omega_t)}, \tag{1.52}$$

and the joint density of levels becomes

$$g_{cv}^{3d}(\hbar\omega_t) = \frac{L^3}{4\pi^2}\left(\frac{2m_r}{\hbar^2}\right)^{3/2}\sqrt{\hbar\omega_t - E_g}. \tag{1.53}$$

If the spin is taken into account, the joint density of states has an additional factor of 2.

1.6 Density of Levels in a Quantum Well

Semiconductor heterostructures were developed a few dozen of years after the initial use of semiconductors in diodes and transistors began. Currently, semiconductors layers can be grown with atomic-layer precision, achieving thicknesses on the order of nanometers, just a few atomic layers thick. One of the most important structures enabled by this level of control is the quantum well, which is formed by stacking alternating layers of semiconductors with different energy gaps. This gap difference causes the edges of the conduction and valence bands to have discontinuities. Thus barriers and wells, and other kind of potential profiles can be formed. For instance, if on top of a layer of material A, such as $Al_xGa_{1-x}As$ with $x = 0.3$ and gap energy $Eg_1 = 1.75\,\text{eV}$, another layer B is grown, say of $GaAs$ with gap energy $Eg_2 = 1.422\,\text{eV}$, and above this, again a layer A, a quantum well is formed in the conduction band, see Fig. 1.10. Within this well electron energy levels E_μ become quantized. On the valence band side, the band alignment creates a potential barrier, which acts as a quantum well for holes, with quantized energy levels $E_{\mu'}$ (see the right panel of Fig. 1.10). The quantum confinement occurs in the z direction. In the transverse plane, with $L_x, L_y \gg L_z$, the energies form almost a continuum spectrum, due to the much larger spatial extent. To obtain the density of levels for the conduction band in a quantum well, it is essential to recognize two key aspects: 1. The system is inherently three-dimensional, and 2. Quantization occurs in the z direction, due

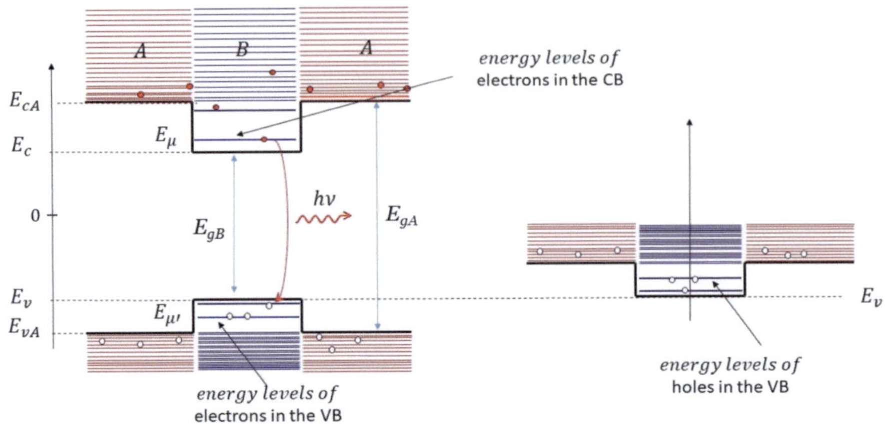

Fig. 1.10 Energy levels in a quantum well. When layers of semiconductors A, B, ..., with different energy gaps, alternate, a heterostructure is formed. With a sequence $A/B/A$ where $E_{gB} < E_{gA}$, we have a well. When the layers' widths in the growth direction is comparable to the de Broglie electron wavelength $\lambda_B \simeq 10\,\text{nm}$, the structure is a quantum well. One part of the difference $\Delta E_g = E_{gA} - E_{gB}$ ($\sim(2/3)\Delta E_g$) forms the well for electrons in the conduction band and the other part ($\sim(1/3)\Delta E_g$) forms a well for the holes in the valence band

to quantum confinement. In this direction the energy $E_{z\mu}$ and the momentum $k_{z\mu}$ are discrete. If the total energy depended only on the quantized z-component and not on the in-plane momentum, the allowed energy levels would appear as isolated levels, as shown in Fig. 1.11a. However, this is not the case: the transverse momentum components k_x and k_y vary continuously, resulting in a continuous distribution of transverse energy. We assume that the total energy varies parabolically as a function of the momentum components k_x and k_y, as illustrated in Fig. 1.11b. As the total energy increases, it eventually reaches the onset (or threshold) of the next energy level in the z-direction. At this point the energy jumps from $E_{z\mu}$ to $E_{z\mu+1}$ and this new subband begins to contribute to the density of levels as transverse energy continues to accumulate. Only the components, k_x and k_y, have almost continuous variation and their spacings, π/L_x and π/L_y are much smaller than those of $k_{z\mu}$, if $L_z \ll L_x, L_y$. In this system the vector $\mathbf{k}_e$ is

$$\mathbf{k}_e = k_{ex}\hat{\mathbf{x}} + k_{ey}\hat{\mathbf{y}} + k_{ez\mu}\hat{\mathbf{z}} = \mathbf{k}_{eT} + \sqrt{\frac{2m_e^* E_\mu}{\hbar^2}}\hat{\mathbf{z}}, \quad k_e = \sqrt{k_{Te}^2 + k_{ez\mu}^2} \tag{1.54}$$

where $\mathbf{k}_{eT}$ is the transverse component of the wave vector (in the μ-th Brillouin zone). This structure leads to a step-like density of states, characteristic of two-dimensional systems.

As discussed in the previous section, calculating the density of levels, requires first the number of available quantum levels from $k_e = 0$ to a given wave vector k_e. We now follow this procedure using a two-dimensional approximation, appropriate for quantum wells. In the solved example at the end of this chapter, the full three-dimensional density, corresponding to the envelope curve in Fig. 1.11c, is derived.

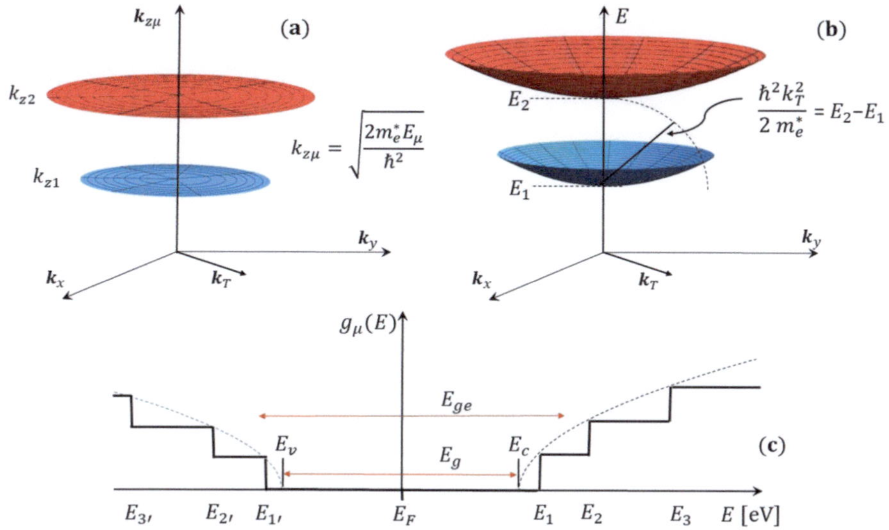

Fig. 1.11 The wave vector in a quantum well and the density of levels. In **a** the components of $\mathbf{k_e}$. In **b** the energy levels and the transverse energy $E_{T\mu} = \hbar^2 k_{T\mu}^2 / 2m_e^*$. It is seen that when the transverse energy $E_{T\mu}$ is equal to the difference $E_{\mu+1} - E_\mu$, it reaches the onset (or threshold) of the next energy level in the z-direction. At this point the energy jumps from $E_{z\mu}$ to $E_{z\mu+1}$ and the energy density takes a jump as shown in panel (**c**)

Based on that result, we will now extract the density $g_{QW}^{2d}(k)$ expressed per unit of wave vector k_e.

Assume that the quantized z-component of the wave vector is fixed, for instance, $k_{z\mu} = k_{z0} = 0$. In this case, the number of levels characterized by the triplets $(k_{xn_x}, k_{yn_y}, 0)$ is the same as the number of levels in a pure 2D system, represented by the pairs (k_{xn_x}, k_{yn_y}), and is equal to N_k^{2d}. As the total energy increases and reaches the threshold for the next quantized energy level E_{z1}, a new component k_{z1} becomes allowed, giving rise to an additional set of states $(k_{xn_x}, k_{yn_y}, k_{z1})$. This effectively doubles the number of available states. With each new subband (corresponding to a higher quantized $k_{z\mu}$), the number of levels increases proportionally. Hence, the number of levels grows with the energy, and is equal to that of a 2D system multiplied by the quantum number $\mu + 1$ of the occupied subbands that is

$$N_{k\mu}^{2d} = (\mu + 1)\frac{L_x L_y}{4\pi} k_T^2 = (\mu + 1)\frac{L_x L_y}{4\pi}(k_e^2 - k_{ez\mu}^2), \quad \mu = 1, 2, 3, \ldots \tag{1.55}$$

With this number we can evaluate the density of levels $g_{QW}^{2d}(k)$ per unit of k_e. Taking into account that $k_T = (k_e^2 - k_{ez\mu}^2)^{1/2}$, we obtain

$$g_{QW}^{2d}(k_e) = (\mu + 1)\frac{\partial N_k^{2d}}{\partial k_e} = (\mu + 1)\frac{L_x L_y}{2\pi} k_e. \tag{1.56}$$

Thus, the density of levels per unit of energy becomes

$$g_{QW}^{2d}(E_e) = g_{QW}^{2d}(k_e)\frac{dk_e}{dE_e} = (\mu+1)\frac{L_x L_y}{2\pi}k_e\frac{m_e^*}{\hbar^2 k_e}, \tag{1.57}$$

which reduces to

$$g_{QW}^{2d}(E_e) = (\mu+1)\frac{m_e^* L_x L_y}{2\pi\hbar^2}, \quad \mu = 1, 2, 3, ... \tag{1.58}$$

This density looks like the density $g_{2d}^c(E_{ec})$ in (1.43), and likewise that density, it is constant. This density changes when the quantum number changes, as seen in Fig. 1.11c, whose envelope is the density $g_{QW}^{3d}(E_e)$. The same density is obtained in a solved problem at the end of this chapter.

Figure 1.11c illustrates schematically the behavior of the density $g_{QW}^{2d}(E_e)$ compared to the density in the three-dimensional system (dotted line).

An important property that must be taken into account in the quantum well, especially in applications, is that the effective gap E_{ge}, defined as the energy separation between the first levels in the bands (see Fig. 1.11c), is larger than the original gap E_g. This difference can be controlled at will by adjusting the thickness L_z of the quantum well. It will be seen later that the position of the energy levels depends strongly on L_z.

Since this system is widely used in quantum well lasers, it is convenient to determine the joint density of levels of a quantum well, also called optical joint density because it involves the initial and final states of the optical transitions.

1.6.1 Joint Density of Levels in a Quantum Well

Given the density of levels in the wells, it is easy to obtain the joint density of levels associated with the transition from the conduction band to the valence band. The electron energy at level μ, in the conduction band, is

$$E_e = E_c + E_\mu + \frac{\hbar^2 k_{eT\mu}^2}{2m_e^*}, \tag{1.59}$$

and the energy at level μ' in the valence band is

$$E_h = E_v - E_{\mu'} - \frac{\hbar^2 k_{hT\mu'}^2}{2m_h^*}. \tag{1.60}$$

In the transition from the conduction to valence band, the emitted photon energy $\hbar\omega_t$ is given by

$$\hbar\omega_t = E_e - E_h = E_c + E_\mu + \frac{\hbar^2 k_{eT\mu}^2}{2m_e^*} - E_v + E_{\mu'} + \frac{\hbar^2 k_{hT\mu'}^2}{2m_h^*}. \tag{1.61}$$

Again, if the transverse momentum is conserved, that is, if $k_{eT\mu} = k_{hT\mu'}$, the emitted energy can be written as

$$\hbar\omega_t = E_g + E_\mu + E_{\mu'} + \frac{\hbar^2 k_{eT\mu}^2}{2}\left(\frac{1}{m_e^*} + \frac{1}{m_h^*}\right) = E_g + E_\mu + E_{\mu'} + \frac{\hbar^2 k_{eT\mu}^2}{2m_r}, \tag{1.62}$$

where m_r is the reduced effective mass. Solving for $k_{eT\mu}^2$ and substituting into (1.59), we have

$$\begin{aligned} E_e &= E_c + E_\mu + \frac{\hbar^2}{2m_e^*}\frac{2m_r}{\hbar^2}(\hbar\omega_t - E_g - E_\mu - E_{\mu'}), \\ E_e &= E_c + E_\mu + \frac{m_r}{m_e^*}(\hbar\omega_t - E_g - E_\mu - E_{\mu'}). \end{aligned} \tag{1.63}$$

This relation connects the energy of the electron in the initial state E_e and the energy of the emitted photon $\hbar\omega_t$. Since

$$g_{\mu\mu'}(\hbar\omega_t)d(\hbar\omega_t) = g_\mu(E_e)dE_e, \tag{1.64}$$

the joint density of levels, per unit of energy, is

$$g_{\mu\mu'}(\hbar\omega_t) = (\mu + 1)\frac{L_x L_y}{\pi}\frac{m_r}{\hbar^2}. \tag{1.65}$$

When the spin is taken into account, we have the joint density of states

$$\mathcal{D}_{\mu\mu'}(\hbar\omega_t) = (\mu + 1)\frac{L_x L_y}{\pi}\frac{2m_r}{\hbar^2}. \tag{1.66}$$

1.6.2 Occupation Probability of Levels and Thermal Excitation

At temperature $T = 0$ and in the absence of impurities, all energy levels in the valence band are fully occupied, so, the occupation probability of these levels is 1, and the occupation probability in the conduction band is zero. An important physical quantity in conductors and semiconductors is the Fermi energy E_F, which at $T = 0$ represents the highest occupied energy level. In conductors, the Fermi level lies within the partially filled valence band. In semiconductors, however, it is generally located within the fundamental energy gap, between the valence and conduction bands. At zero temperature and in the absence of impurities, the occupation probability is described by the step function (see Fig. 1.12)

$$f(E, T = 0) = \theta(E - E_F) = \begin{cases} 1 & \text{for} \quad E < E_F, \\ 0 & \text{for} \quad E > E_F. \end{cases} \tag{1.67}$$

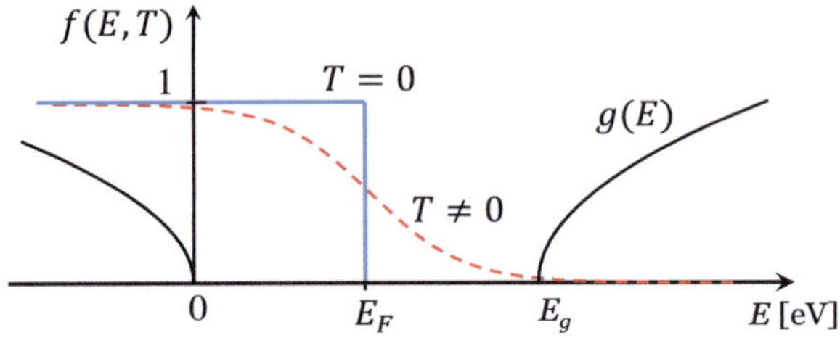

Fig. 1.12 Occupation probability. At temperature $T = 0$ the function $f(E, T)$ is a step function: it equals 1 if $E < E_F$ and 0 if $E > E_F$, where E_F is the Fermi energy. For $T \neq 0$, $f(E, T)$ becomes the Fermi-Dirac distribution $f_F(E, T)$, shown as the dashed curve. In this case, the Fermi energy E_F is replaced by the Fermi level μ, defined such that $f_F(\mu, T) = 1/2$

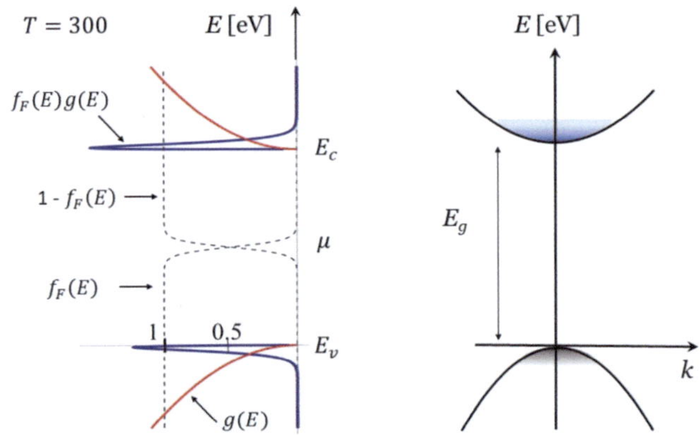

Fig. 1.13 Level densities and occupation probabilities. The Fermi-Dirac distribution $f_F(E)$ gives the probability that the energy level E is occupied by an electron, and $1 - f_F(E)$ the probability that that level is not occupied by an electron. The blue curves correspond, for $E > \mu$, to the product $f_F(E)g^c_{3d}(E)$ which gives the number of excited electrons as a function of E and, for $E < \mu$, to the product $\left(1 - f_F(E)\right)g^v_{3d}(E)$ which gives the number of holes as a function of E

We will show later that in pure semiconductors, the Fermi energy E_F is exactly in the center of the fundamental energy gap. When the temperature is above zero, the probability that an energy level E is occupied by an electron is given by the Fermi-Dirac distribution

$$f_F(E, T) = \frac{1}{e^{(E-\mu)/k_B T} + 1} \tag{1.68}$$

This function is illustrated in Fig. 1.13. At nonzero temperature, the Fermi energy E_F is replaced by the chemical potential or Fermi level μ, which is defined such that the occupation probability of the energy level $E = \mu$ is exactly $1/2$, that is, $f_F(\mu, T) = 1/2$, consequently, the probability that this level is unoccupied, or equivalently occupied by a hole, is also $1/2$.

By multiplying the Fermi-Dirac distribution with the appropriate density of levels, we obtain the number of thermally excited electrons in the conduction band and the number of holes in the valence band. Due to charge conservation, these two numbers must be equal, that is:

$$N_v = \int_{-\infty}^{E_v} (1 - f_F(E))g_{3d}^v(E)dE = \int_{E_c}^{\infty} f_F(E)g_{3d}^c(E)dE = N_c \tag{1.69}$$

In Fig. 1.13 the number of holes per unit of energy, given by the product $(1 - f_F(E))g_{3d}^v(E)$, and the number of electrons per unit of energy in the conduction band, given by the product $f_F(E)g_{3d}^c(E)$, are plotted as functions of the energy when $T = 300$ K. At this temperature, the therm)al energy k_BT is $\simeq$0.026 eV, and as shown in this figure, the electrons and holes are very close to the band edges.

When semiconductors contain impurities, the Fermi level shifts upwards or downwards depending on whether the impurities are donor or acceptor-like atoms. We will discuss this issue in the next section, and with more detail in the next chapter.

If the Fermi level is at a distance of the order of $4k_BT$ ($\simeq$0.1 eV when T = 300 K) or greater from the edge of the conduction and valence bands, the system is said to be *non-degenerate*. In this case it is justified to approximate the Fermi-Dirac distribution by the Maxwell-Boltzmann form:

$$f_F(E, T) \xrightarrow{E_b - \mu >> 4k_BT} f(E, T) = \frac{1}{e^{(E-\mu)/k_BT}} = e^{-(E-\mu)/k_BT}. \tag{1.70}$$

1.7 Semiconductors with Donor and Acceptor Impurities

When an atom in the crystal structure of a semiconductor is replaced by an atom of a different element, one has a substitutional impurity. If the impurity atom has a different number of valence electrons than the atom it replaces, there will be a greater or fewer number of electrons than required for equilibrium. When the impurity has more valence electrons than the host atom, it is called a *donor*. The excess electron is less tightly bound and can move into a larger orbit. Since the impurity is initially neutral, the loss of this loosely bound electron leaves behind a positively charged ion, referred to as the ionized impurity. To roughly estimate the effect of this charge configuration on the band structure of the semiconductor, we will assume that the host semiconductor belongs to group IV (with four valence electrons per atom, as in Si or Ge), and assume that the impurity belongs to group V (with five valence electrons, such as arsenic As or phosphorus P). In this case, the extra electron introduced by the impurity does not participate in bonding with neighboring atoms, but remains weakly bound to the impurity site. This electron and the rest of the impurity, that is, the ionized impurity, can be modeled as if it were a hydrogen atom. The energies of the weakly bound electron, in the hydrogen-like atom approximation, are given by

$$E_n^i = -\frac{m_e^* e^4}{2\epsilon^2\hbar^2 n^2 (4\pi\epsilon_0)^2}, \tag{1.71}$$

and the radii of their "orbits" are

$$r_n^i = \frac{\epsilon n^2 \hbar^2}{m_e^* e^2} 4\pi\epsilon_0. \tag{1.72}$$

Here m_e^* is the effective mass in the semiconductor and ϵ the dielectric constant (defined as ϵ_s/ϵ_0, with ϵ_s the semiconductor electric permittivity). At the lowest level (for $n = 1$)

$$E_1^i = -\frac{m_e^*}{m_e \epsilon^2} \frac{m_e e^4}{2\hbar^2 (4\pi\epsilon_0)^2} = -\left(\frac{m_e^*}{m_e \epsilon^2}\right) 13.6\,\text{eV}, \tag{1.73}$$

and

$$r_1^i = \left(\frac{\epsilon m_e}{m_e^*}\right) \frac{\hbar^2 4\pi\epsilon_0}{m_e e^2} = \left(\frac{\epsilon m_e}{m_e^*}\right) 0.53\,\text{Å}. \tag{1.74}$$

In cases where the electron motion is neither purely longitudinal or purely transverse, the conductivity effective mass is used, defined as $m_c = (9m_l^* m_t^{*2})^{1/3}$. This expression combines the longitudinal m_l^* and the transverse m_t^* effective masses. For germanium Ge, the ratio of the conductivity effective mass to the free electron mass is approximately $m_c/m_e \simeq 0.22$ and the dielectric constant is $\epsilon \simeq 16.2$, so

$$E_1^i \approx -0.01\,\text{eV} \quad \text{and} \quad r_1^i \approx 40\,\text{Å}. \tag{1.75}$$

This means that the effective radius of a donor impurity within the semiconductor is significantly larger than that of an isolated atom. In the hydrogen-like model, the zero of energy is taken at the edge of the conduction band. If the impurity electron had positive energy, it would behave like a free particle. However, its energy is negative, slightly below the edge of the conduction band. It also means that the excess electrons are very weakly bound and their wave functions are relatively extended, and that a minimum energy of E_1^i is required to excite them into the conduction band. Consequently, donor impurities introduce energy levels within the band gap, close to the conduction band edge. These levels can be partially occupied by the weakly bound, quasi-free electrons, shifting the Fermi level toward the donor level. The donor energy level is given by $E_D = -|E_1^i|$ with respect to the conduction band edge, or $E_D = E_g - |E_1^i|$ if measured from the top of the valence band, as illustrated in Fig. 1.14. The hydrogen-like atom model remains valid as long as the donor impurity concentration N_D is low enough that the impurity wave functions do not overlap. This happens when the impurity concentration N_D multiplied by the volume $4\pi(r_1^i)^3/3$ of the impurity atom does not exceed the number of atoms n_{at} in the volume V of the semiconductor, that is, does not exceed the number $n_{at} \simeq N_o 4\pi r_o^3/3$. As $N_o \sim 10^{22}$ atoms/cm^3 and $(r_o/r_1^i) \approx 10^{-1}$, the model will be valid when the concentration of donors is

$$N_D \leq N_o \left(\frac{r_o}{r_1^i}\right)^3 \simeq 10^{19} \left[\frac{\text{imp}}{\text{cm}^3}\right]. \tag{1.76}$$

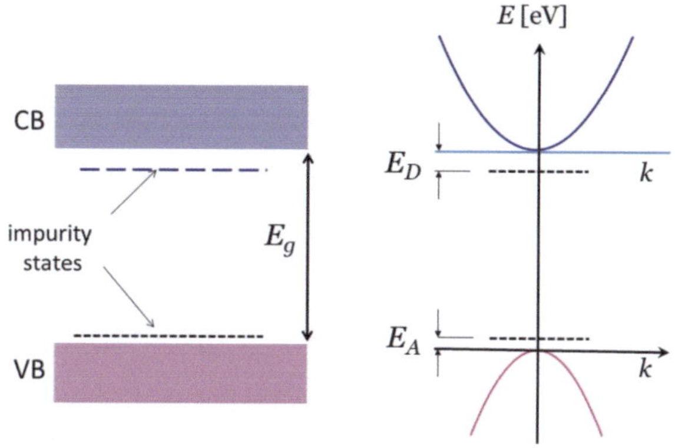

Fig. 1.14 Donor impurities provide additional electrons that remain loosely bound in quasi-localized states, and energies slightly below the conduction band. Acceptor impurities participate with fewer electrons and as a consequence of this deficit, the impurity and its immediate neighbors tend to attract the missing electron but cannot accommodate it in the valence band, so the acceptor states are located slightly above the valence band

If the impurity is an acceptor the effect on the semiconductor properties is different. The impurity offers fewer electrons than those required by its neighboring atoms. If there were any extra electrons, provided by a donor impurity, or by thermal excitations when $T \neq 0$, the electron will be attracted to the impurity, but the potential of the acceptor impurity in insufficient to confine this electron to the semiconductor valence band, partly because the neighborhood it reaches is neutral and partly because the attracted electron still feels the positive ion it leaves behind. The acceptor impurity cannot create a bound level in the valence band but it can create an attractor level near the valence band edge. This attractor level behaves in a very similar way to the vacancies left by the excited electrons of the valence band. To evaluate the order of magnitude of the energy of this configuration, a hydrogenic atom model can also be used but with the sign of the charges reversed. The acceptor levels will be close to the valence band edge. The energy of the attractors is denoted as E_A. Figure 1.14 shows the donor and acceptors, immersed in the gap and close to the conduction and valence bands, respectively.

1.8 Solved Problem

1. An alternative to determine the levels density of a quantum well is to recognize that the QW is a 3D system. Show that starting from the 3D density $g^{3d}_{QW}(k)$, and taking into account that, in fact, $k_{z\mu} = \mu\pi/L_z$, the 3D density becomes the QW density $g^{2d}_{QW}(k)$ in (1.58).

   Solution In the 3D case the number of levels is given by

$$N^{3d}_{k\mu} = \frac{L_x L_y L_z}{6\pi^2} k_e^3 = \frac{L_x L_y L_z}{6\pi^2} (k^2_{Te} + k^2_{ez\mu})^{3/2}, \tag{1.77}$$

Given this number of levels the density of levels per unit of k_e is

$$g_{QW}^{3d}(k_e) = \frac{\partial N_{k\mu}^{3d}}{\partial k_e} = \frac{L_x L_y L_z}{2\pi^2} k_e^2, \tag{1.78}$$

and the density of levels per unit of energy is obtained as follows

$$g_{QW}^{3d}(E_e) = g_{QW}^{3d}(k_e)\,\frac{dk_e}{dE_e} = \frac{L_x L_y L_z}{2\pi^2} k_e^2 \frac{m_e^*}{\hbar^2 k_e}, \tag{1.79}$$

which reduces to

$$g_{QW}^{3d}(E_e) = \frac{m_e^* L_x L_y L_z}{2\pi^2 \hbar^2} k_e. \tag{1.80}$$

If the energies are measured from the valence band and E_c is the edge energy of the conduction band, that is, if

$$k_e = \sqrt{\frac{2m_e^*}{\hbar^2}\,(E - E_c)} \tag{1.81}$$

the density in (1.80) will look like the density $g_{3d}^c(E_{ec})$ in (1.44). In what follows of this solution and only to simplify the expressions, without physical consequences, we will assume that $E_c = 0$.
We are interested in using the density $g_{QW}^{3d}(E_e)$ to derive the density $g_{QW}^{2d}(E_e)$ in (1.58). The density $g_{QW}^{3d}(E_e)$ can be rewritten as

$$g_{QW}^{3d}(E_e) = \frac{m_e^* L_x L_y L_z}{2\pi^2 \hbar^2} \sqrt{k_{eT\mu}^2 + k_{ez\mu}^2}. \tag{1.82}$$

As discussed before, the energy E_μ moves to $E_{\mu+1}$ when the transverse momentum at μ level reaches a threshold to change the energy $E_{z\mu}$ to $E_{z(\mu\pm1)}$. At this point

$$E_{\mu\pm1} = \frac{\hbar^2 k_{ez(\mu\pm1)}^2}{2m_e^*} = \frac{\hbar^2 (k_{ez\mu}^2 + k_{eT\mu}^2)}{2m_e^*} \tag{1.83}$$

and

$$k_{eT\mu}^2 = k_{ez(\mu+1)}^2 - k_{ez\mu}^2 \quad \text{then} \quad k_e = \sqrt{k_{ez(\mu+1)}^2 - k_{ez\mu}^2 + k_{ez\mu}^2} = k_{ez(\mu+1)} \tag{1.84}$$

If we substitute this into $g_{QW}^{3d}(E_e)$, we have

$$g_{QW}^{3d}(E_e) = \frac{m_e^* L_x L_y L_z}{2\pi^2 \hbar^2} k_{ez(\mu+1)} \tag{1.85}$$

This density reduces to $g^{2d}_{QW}(E_e)$ if

$$k_{ez\mu} = \frac{\mu\pi}{L_z} \quad \mu = 1, 2, 3, \ldots \tag{1.86}$$

This shows that both densities are compatible. Figure 1.11c shows schematically the behavior of the density $g^{2d}_{QW}(E_e)$ compared to the density in the three-dimensional system (dotted line).

1.9 Problems

1. The volume in k-space depends on whether the physical system is one-dimensional (1D), two-dimensional (2D), or three-dimensional (3D). Obtain the levels densities $g_{1D}(k)$ and $g_{2D}(k)$.
2. Obtain the levels densities $g_{1D}(E)$ and $g_{2D}(E)$ and show that the levels density in a system of dimension d (=1, 2 or 3), except for a factor of 2π in $g_{1D}(E)$, can be written in general as

$$g_{dD} = \frac{\pi}{4}\left(\frac{L}{\pi\hbar}\right)^d (2m^*)^{d/2}\, E^{(d-2)/2} \tag{1.87}$$

3. Using the results of the previous problem, obtain the densities of levels $g_{nc}(k)$ and $g_{nv}(k)$ in the conduction and valence bands for $n = 1, 2$.
4. Plot the levels densities $g_{nc}(k)$ and $g_{nv}(k)$ for 1D, 2D and 3D systems.
5. Plot the occupation probability as a function of energy for different temperatures (T = 0, 100, 200 and 300 K) assuming $\mu = E_g/2$.
6. Calculate the wave number and momentum of an electron excited to the conduction band of GaAs with an energy of 1.6 eV, measured from the upper edge of the valence band, and $T = 300$ K.
7. Calculate the number of electronic levels in the conduction band of GaAs between 0.1 and 0.2 eV. The sample size is 1 μm^3 and the effective electron mass is $m^*_{ec} = 0.067m_e$.
8. The effective mass of an electron in the conduction band of a semiconductor is $0.2m_e$. Calculate the energy of this electron if its wave number k is (a) $k = 0.4\,\text{Å}^{-1}$, (b) $\mathbf{k} = \frac{2\pi}{a}(1.0, 0.5, 0.1)$, here $a = 5.43$ Å.

References

1. O. Madelung, *Introduction to Solid-State Theory*, 2nd edn. (Springer, Berlin, 1981)
2. R.L. Kronig, W.G. Penney, Proc. R. Soc. A **499** (1931)
3. J.R. Hook, H.E. Hall, *Solid State Physics*, 2nd edn. (Wiley, Chichester, 1991)

4. C. Kittel, *Introduction to Solid State Physics*, 7th edn. (Wiley, New York, 1995)
5. G. Dresselhaus et al., Phys. Rev. **98**, 368 (1955)
6. Ioffe Institute, https://www.ioffe.ru/SVA/NSM/Semicond/

2 Intrinsic and Extrinsic Properties of Semiconductors

Abstract

In this chapter we introduce explicit derivations of charge carrier concentration formulas in semiconductor materials. It covers both intrinsic carrier concentrations, arising from thermal excitation, and extrinsic concentrations, resulting from intentional doping with donor or acceptor impurities. The law of mass action is introduced, along with a discussion of impurity energy levels within the band gap and the Fermi level as a function of impurity concentration and temperature. Criteria for classifying semiconductors as n-type or p-type are explained, and the concepts of majority and minority carriers are defined for each case. The chapter concludes with a set of solved problems to reinforce the key concepts. Two important quantities in the analysis of the physical characteristics of semiconductors are the electron concentration n and the hole concentration p. These quantities can vary significantly in the presence of impurities. To distinguish different regimes, we define intrinsic and extrinsic carrier concentrations.

2.1 Intrinsic Concentrations

In pure semiconductors, isolated and at temperature T, the population of electrons (in the conduction band) and holes (in the valence band) is the product mainly from thermal excitation. The concentrations of electrons and holes are called in this case intrinsic concentrations. With the densities and occupation probabilities that were introduced in the previous chapter, one can get these numbers easily. When $T \to 0$, n and p must reduce to zero.

If $f_F(E)g_c(E)$ gives the number of electrons per unit of energy, when the energy is E, the number of electrons per unit volume in the entire conduction band is given by

$$n = \frac{1}{V}\int_{E_g}^{\infty} f_F(E)g_c(E)dE, \tag{2.1}$$

P. Pereyra, *Contemporary Theory of Semiconductors and Devices*,
Graduate Texts in Physics, https://doi.org/10.1007/978-3-032-00296-9_2

where V is the volume of the semiconductor, E_g is the energy gap and also the energy at the lower edge of the conduction band, when energies are measured from the top of the valence band. If we substitute here the level density and the occupation probability from (1.44) and (1.70), we have

$$n = \frac{1}{2\pi^2}\left(\frac{2m_e^*}{\hbar^3}\right)^{3/2}\int_{E_g}^{\infty}(E - E_g)^{1/2}e^{(\mu-E)/k_BT}dE. \tag{2.2}$$

With the change of variable $\xi = E - E_g$, the integration is immediate and results in

$$n = N_c e^{(\mu-E_g)/k_BT}. \tag{2.3}$$

Here the concentration N_c is defined as

$$N_c = 2\left(\frac{m_e^* k_B T}{2\pi\hbar^2}\right)^{3/2}, \tag{2.4}$$

known also as the *effective number of levels per unit of volume in the conduction band.* On the other hand, the number of holes per unit volume in the valence band, with the zero of the energy at the upper edge of this band, is

$$p = \frac{1}{V}\int_{-\infty}^{0}[1 - f_F(E)]\,g_v(E)dE. \tag{2.5}$$

Here $g_v(E)$ is the density of states given in (1.46), with $E < 0$. On the other hand, since $(|E| + \mu) \gg k_BT$ and $1 \gg e^{-(|E|+\mu)/k_BT} = e^{(E-\mu)/k_BT}$, the hole occupation probability can be approximated as follows

$$1 - f_F(E) = \frac{e^{(E-\mu)/k_BT}}{e^{(E-\mu)/k_BT} + 1} \simeq e^{(E-\mu)/k_BT}. \tag{2.6}$$

If we change the variable $\xi = -E$, we have

$$\begin{aligned} p &= \frac{(2m_h^*)^{3/2}}{2\pi^2\hbar^3}\int_0^{\infty}\xi^{1/2}e^{-(\xi+\mu)/k_BT}d\xi, \\ &\simeq N_v e^{-\mu/k_BT}, \end{aligned} \tag{2.7}$$

where

$$N_v = 2\left(\frac{m_h^* k_B T}{2\pi\hbar^2}\right)^{3/2} \tag{2.8}$$

is the *effective number of levels per unit of volume in the valence band.*

A useful relation in the semiconductor theory is the product

$$np = N_c N_v e^{-E_g/k_B T} = 4\left(\frac{k_B T}{2\pi\hbar^2}\right)^3 \left(m_e^* m_h^*\right)^{3/2} e^{-E_g/k_B T}, \tag{2.9}$$

known as the *law of mass action*. This relation is valid regardless of the presence or absence of impurities. Although the presence of donors and acceptors does have an effect on the position of the Fermi level (or chemical potential), it has no effect in the mass action law, which remains *independent of the Fermi level*. In a pure semiconductor, the concentrations of electrons excited at the conduction band and that of holes in the valence band are equal. If the product of these quantities satisfies the law of mass action, the intrinsic concentrations can also be expressed in the form

$$n_i = p_i = (N_c N_v)^{1/2} e^{-E_g/2k_B T}. \tag{2.10}$$

Furthermore, the concentration n_i is also given by (see (2.3))

$$n_i = N_c e^{(\mu - Eg)/k_B T}. \tag{2.11}$$

From these equations, we obtain the relation

$$e^{(2\mu - E_g)/k_B T} = \frac{N_v}{N_c}, \tag{2.12}$$

which allows us to determine the *intrinsic Fermi level* μ shown in Fig. 2.1, which is given by

$$\mu = \frac{1}{2}E_g + \frac{1}{2}k_B T \ln\left(\frac{N_v}{N_c}\right) \equiv e\phi_F, \tag{2.13}$$

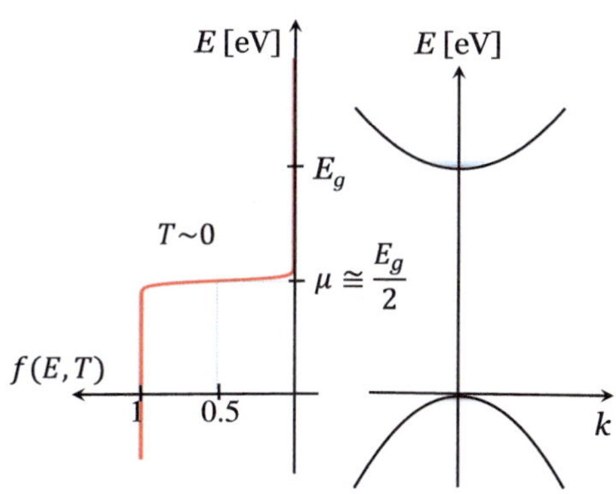

Fig. 2.1 Fermi level in pure semiconductors at $T \simeq 0$. In these semiconductors $\mu \simeq E_g/2$

or, equivalently,

$$\mu = \frac{1}{2}E_g + \frac{3}{4}k_B T \ln\left(\frac{m_h^*}{m_e^*}\right). \tag{2.14}$$

According to this result, the intrinsic Fermi level, denoted also as E_i, lies at the center of the band gap when $T = 0$, and remains close to the center at low temperatures, provided that $k_B T \ll E_g$. For reference typical values of the band gap E_g are of the order of 1 eV, while $k_B T$, for $T = 300$ K, is of the order of 0.026 eV.

2.2 Extrinsic Concentrations

An important characteristic of semiconductors is that the concentrations of charge carriers can be manipulated and modified, almost at will. In the previous chapter we commented that impurities create energy levels in the gap, near the edges of the bands, in which electrons or holes are housed, depending on the nature of the impurity. The impurities introduced by substitution or implantation are, as seen before, other chemical elements with a greater or lesser number of electrons in the valence level, donors and acceptors respectively. If one has a semiconductor with donor impurities and the concentration of donors is much lower than the intrinsic one, that is, if $N_D \ll n_i$, the behavior will be fundamentally intrinsic. To estimate whether or not a doped semiconductor will have intrinsic behavior, based on the previous order relation, it is necessary to evaluate its intrinsic concentration n_i. This is given by

$$n_i = (N_c N_v)^{1/2} e^{-E_g/2k_B T}. \tag{2.15}$$

If the effective masses are, for instance, $\sim 0.1 m_e \sim 10^{-31}$ kg, we have

$$n_i = 2\left(\frac{k_B T}{2\pi\hbar^2}\right)^{3/2} \left(m_e^* m_h^*\right)^{3/4} e^{-E_g/2k_B T} \simeq 2.8 \times 10^{24} m^{-3} e^{-E_g/2k_B T}. \tag{2.16}$$

This means that for temperatures of the order of 300 K and $E_g \sim 1$ eV the behavior is intrinsic when

$$N_D \ll 1 \times 10^{16}\,\mathrm{m}^{-3}. \tag{2.17}$$

Since the atomic concentration is of the order of $5 \times 10^{28}\,\mathrm{m}^{-3}$, the previous relationship indicates that to preserve the intrinsic behavior, the number of impurities must be less than 1 in 10^{12}. This is a pretty strong requirement. With concentrations of donors greater than $10^{16}\,\mathrm{m}^{-3}$ the behavior is extrinsic.

There are various methods for doping a semiconductor. Before doping, the material must first be thoroughly purified. One such purification technique is known as zone refining. This method exploits the fact that impurities are generally more soluble in the liquid phase than in the solid. In this process, a localized heat source moves along the length of a semiconductor rod, temporarily melting it in a narrow region. As the molten zone travels, it “sweeps” impurities along with it, effectively concentrating them at one end, much like a broom of impurities. After purification, the semiconductor is doped in a controlled manner with the desired type and concentration of impurity atoms.

At low temperatures and in the presence of impurities, the Fermi level shifts depend on both the nature and the concentration of the impurities. If acceptor impurities dominate, and the temperature is above absolute zero, the acceptors become ionized: they trap and localize electrons, leaving behind mobile holes. The Fermi level, defined as the energy level with a 50% probability of occupation, shifts toward the acceptor energy level, that is, towards E_A. Conversely, if the dominant concentration is of donors and the temperature is above zero, they become ionized, releasing their weakly bound electrons into the conduction band. In this case, the Fermi level will move towards $E_g - E_D$. Throughout this discussion, we will use for E_D the absolute value convention: $E_D = |E_D|$.

Unlike extended states in the conduction or valence bands, impurity states are localized, and this affects their occupation probabilities. These states also influence the nature and mobility of charge carriers. Impurity states can be labeled as $|n_I, \sigma\rangle$, where $n_I = 0$ or 1 represents the number of electrons occupying the impurity level E_I $(=E_g - E_D$, or $E_A)$ and σ $(=\pm 1/2)$ its spin. The possible impurity states are $|0, 0\rangle$, $|1, 1/2\rangle$, and $|1, -1/2\rangle$ and the probability that an impurity state is occupied by an electron with spin σ is[1]

$$f_\sigma(E_I) = \frac{e^{-(E_I-\mu)/k_B T}}{1 + 2e^{-(E_I-\mu)/k_B T}}. \tag{2.18}$$

The probability that the impurity state is occupied by an electron, regardless of its spin, is

$$f^*(E_I) = f_{+1/2}(E_I) + f_{-1/2}(E_I) = 2\frac{e^{-(E_I-\mu)/k_B T}}{1 + 2e^{-(E_I-\mu)/k_B T}} = \frac{1}{\frac{1}{2}e^{(E_I-\mu)/k_B T} + 1}. \tag{2.19}$$

In the presence of both kinds of impurities, donor and acceptor, the chemical potential is determined by imposing the charge neutrality condition. This requirement, similar to (1.69), is expressed in the form

$$n + N_A^- = p + N_D^+, \tag{2.20}$$

with N_A^- and N_D^+ the concentrations of ionized acceptors and donors. In terms of the Fermi distribution function, these ionized concentrations are

$$N_A^- = N_A f^*(E_A), \tag{2.21}$$

[1] For a more detailed treatment of this topic, see [1]

and

$$N_D^+ = N_D\left(1 - f^*(E_g - E_D)\right). \tag{2.22}$$

Here $f^*(E_A)$ represents the probability that an electron occupies the energy level E_A while the factor $\left(1 - f^*(E_g - E_D)\right)$ gives the probability that the energy level $E_g - E_D$ is empty, or occupied by a hole. Substituting these concentrations in (2.20) yields the general relationship

$$N_c e^{(\mu - E_g)/k_B T} + N_A f^*(E_A) = N_v e^{-\mu/k_B T} + N_D\left(1 - f^*(E_g - E_D)\right), \tag{2.23}$$

whose analytical solution can only be obtained in certain limiting cases. Many of the expressions obtained in this field of physics are approximations sufficiently reliable, simple and useful for analysis. The reader interested in the analysis of concentrations involving the distribution functions $f^*(E_I)$ can consult [1]. We will continue the analysis by assuming that the Fermi levels are μ_n and μ_p and that $\mu_n = E_g - E_D$ and $\mu_p = E_A$. We will also obtain, for these Fermi levels, approximate expressions dependent on the concentrations of impurities N_D and N_A.

2.3 *n* Type and *p* Type Semiconductors

We will see below the consequences of the neutrality condition in the following cases:

1. When $N_D \neq 0$ and $N_A = 0$: corresponds to an n type semiconductor.
2. When $N_D = 0$ and $N_A \neq 0$: corresponds to a p type semiconductor.

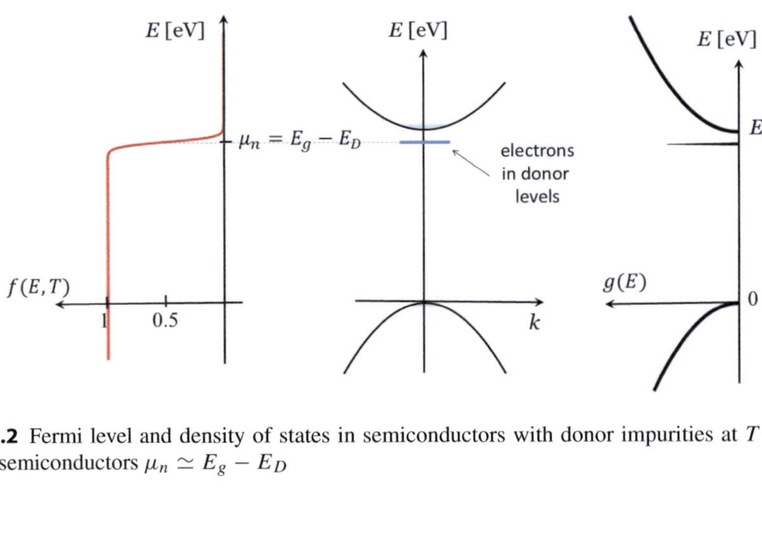

Fig. 2.2 Fermi level and density of states in semiconductors with donor impurities at $T \simeq 0$. In these semiconductors $\mu_n \simeq E_g - E_D$

2.3.1 Semiconductor Type n

Suppose that $N_D \neq 0$, $N_A = 0$ and the temperature is different from zero. In this case, illustrated in Fig. 2.2, the electrons from donor impurities can easily be excited to the conduction band, thereby increasing the carrier concentration beyond the intrinsic value n_i. Under these conditions, almost all or a large fraction of the impurities become ionized. The neutrality equation (2.20) reduces to

$$n - p = N_D^+. \tag{2.24}$$

If the Fermi level moved to μ_n, the extrinsic electron and hole concentrations, n and p, can be calculated in exactly the same way as the intrinsic concentrations n_i and p_i. The results are

$$n = N_c e^{(\mu_n - E_g)/k_B T} \quad \text{y} \quad p = N_v e^{-\mu_n/k_B T}. \tag{2.25}$$

If we assume that $\mu_n = E_g - E_D$, the electron and hole concentrations become

$$n \approx N_c e^{-E_D/k_B T} \quad \text{and} \quad p \approx N_v e^{-(E_g - E_D)/k_B T}. \tag{2.26}$$

Since $k_B T \simeq E_D \ll (E_g - E_D)$, it is easy to conclude that

$$n > n_i, \; p_i > p. \tag{2.27}$$

The semiconductor is said to be *type n* and the concentration n is the *majority concentration*. In this case, it is customary to write n_n instead of n and p_n instead of p, that is:

$$n_n \approx N_c e^{-E_D/k_B T}, \quad \text{and} \quad p_n \approx N_v e^{-(E_g - E_D)/k_B T}. \tag{2.28}$$

There are several representations of these concentrations. We will indicate some of them here. If we add and subtract the intrinsic Fermi energy μ, in the exponents of (2.25), and recall the definitions of n_i and p_i, we have

$$n_n = n_i e^{(\mu_n - \mu)/k_B T} \quad \text{and} \quad p_n = p_i e^{-(\mu_n - \mu)/k_B T}. \tag{2.29}$$

It is also common to denote as E_{Fs} the extrinsic Fermi level, instead of μ_n or instead of μ_p, and the intrinsic Fermi level as E_i, instead of μ. In this notation the concentrations of the previous equation are written as

$$n_n = n_i e^{(E_{Fs} - E_i)/k_B T} \quad \text{and} \quad p_n = p_i e^{-(E_{Fs} - E_i)/k_B T}. \tag{2.30}$$

Other physical quantities that are also defined with the energies E_{Fs} and E_i are the Fermi and thermal potentials [2]

$$\phi_F = \frac{E_{Fs} - E_i}{-e} \quad \text{and} \quad \phi_T = \frac{k_B T}{e}, \tag{2.31}$$

respectively. The Fermi potential measures the shift of the Fermi level due to doping. In terms of these potentials, the carrier concentrations n_n and p_n can be expressed as

$$n_n = n_i e^{-\phi_F/\phi_T} \quad \text{and} \quad p_n = p_i e^{\phi_F/\phi_T}. \tag{2.32}$$

In this case $E_{Fs} > E_i$, thus $\phi_F < 0$. It is clear that $n_n \gg p_n$.

The majority and minority concentrations, n_n and p_n, also satisfy the law of mass action. Indeed

$$n_n p_n = N_c N_v e^{-E_g/k_B T}. \tag{2.33}$$

Before concluding, it should be mentioned that when the impurity concentration is large, it is common to make the approximation

$$n = n_n \simeq N_D \quad \text{then} \quad p = p_n \simeq \frac{n_i^2}{N_D}. \tag{2.34}$$

Thus, in an n type semiconductor, the Fermi potential is

$$\phi_F = \phi_T \ln \frac{n_i}{N_D}. \tag{2.35}$$

2.3.1.1 The Fermi Level μ_n

When $N_D \neq 0$ and $N_A = 0$, the Fermi level μ_n can be determined as a function of the donor concentration N_D. In principle, to find μ_n one must solve the charge neutrality equation (2.24) written in terms of the occupation probabilities $f^*(E_I)$. As shown in solved problem 2, this leads to the following equation

$$n^2(n + n_1) - n_i^2(n + n_1) - N_D n n_1 = 0. \tag{2.36}$$

where $n_1 = N_c \frac{1}{2} e^{-E_D/k_B T}$. This is a cubic equation in n, and solving it exactly can be algebraically involved. However, by applying approximations based on orders of magnitude, the equation can be simplified without significant loss of accuracy. These approximations yield results that are entirely consistent with those obtained when the concentration p is neglected in the original neutrality equation (2.24). Therefore, if the impurities concentration is large and the temperature is high, we can consider the simplified neutrality equation

$$n \simeq N_D, \tag{2.37}$$

which we write as

$$N_c e^{(\mu_n - E_g)/k_B T} \simeq N_D. \tag{2.38}$$

Solving for μ_n we have

$$\mu_n \simeq E_g - k_B T \ln \frac{N_c}{N_D}. \tag{2.39}$$

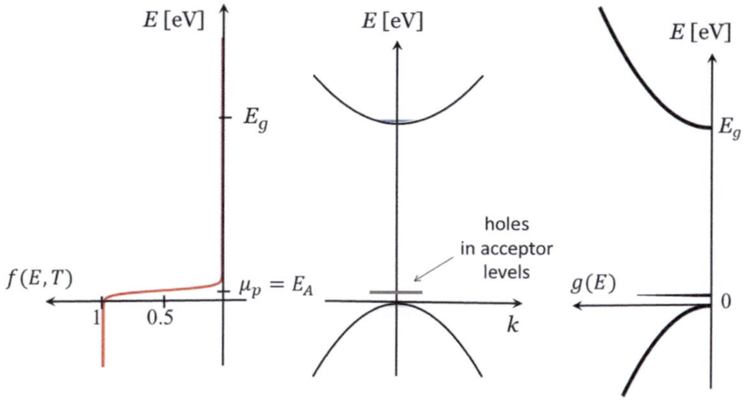

Fig. 2.3 Fermi level and density of states in semiconductors with acceptor impurities at $T \simeq 0$. In these semiconductors $\mu_p \simeq E_A$

This means that

$$E_D \simeq k_B T \ln \frac{N_c}{N_D}. \tag{2.40}$$

2.3.2 Semiconductor Type p

If $N_A \neq 0$ and $N_D = 0$, the neutrality equation (2.20) reduces to

$$n + N_A^- = p, \tag{2.41}$$

If the Fermi level was moved to μ_p, as illustrated in Fig. 2.3, the extrinsic concentrations, n and p, can also be calculated in the same way as the intrinsic ones, and as the electrons and holes concentrations in type n semiconductors. The results are

$$n = N_c e^{(\mu_p - E_g)/k_B T} \quad \text{and} \quad p = N_v e^{-\mu_p/k_B T}. \tag{2.42}$$

For $\mu_p = E_A$ the concentrations simplify to

$$n \approx N_c e^{-(E_g - E_A)/k_B T} \quad \text{and} \quad p \approx N_v e^{-E_A/k_B T}. \tag{2.43}$$

Since $k_B T \simeq E_A \ll (E_g - E_A)$, it is easy to conclude that

$$p > p_i, n_i > n. \tag{2.44}$$

Holes concentration is the majority concentration and the semiconductor is type p. We denote the majority concentration as p_p instead of p and the minority concentration as n_p instead of n, i.e.

$$n_p \approx N_c e^{-(E_g - E_A)/k_B T} \quad \text{and} \quad p_p \approx N_v e^{-E_A/k_B T}. \tag{2.45}$$

If we now use E_{Fs} for μ_p and E_i for the intrinsic Fermi level, the concentrations in the above equation are written as

$$n_p = n_i e^{-(E_i - E_{Fs})/k_B T} \quad \text{and} \quad p_p = p_i e^{(E_i - E_{Fs})/k_B T}. \tag{2.46}$$

In terms of the Fermi potential $\phi_F = (E_{Fs} - E_i)/(-e)$ and the thermal potential ϕ_T, the concentrations n_p and p_p are

$$n_p = n_i e^{-\phi_F/\phi_T} \quad \text{and} \quad p_p = p_i e^{\phi_F/\phi_T}. \tag{2.47}$$

In these semiconductors $E_{Fs} < E_i$, thus, the Fermi potential is $\phi_F > 0$. Therefore, the majority and minority concentrations satisfy the relation

$$p_p \gg p_i = n_i \gg n_p, \tag{2.48}$$

and the law of mass action becomes

$$p_p n_p = N_v N_c e^{-E_g/k_B T}. \tag{2.49}$$

When the impurities concentration is large, it is common to make the approximation

$$p = p_p \simeq N_A \quad \text{then} \quad n = n_p \simeq \frac{n_i^2}{N_A}. \tag{2.50}$$

Thus, in a p type semiconductor the Fermi potential is

$$\phi_F = \phi_T \ln \frac{N_A}{n_i}. \tag{2.51}$$

2.3.2.1 The Fermi Level μ_p

When $N_A \neq 0$ and $N_D = 0$, the Fermi level μ_A can be obtained as a function of the concentration N_A, from the approximate relationship

$$p \simeq N_A, \tag{2.52}$$

which can be written as

$$N_v e^{-\mu_p/k_B T} \simeq N_A. \tag{2.53}$$

Solving for μ_p, we obtain

$$\mu_p \simeq k_B T \ln \frac{N_v}{N_A}. \tag{2.54}$$

Thus,

$$E_A \simeq k_B T \ln \frac{N_v}{N_A}. \tag{2.55}$$

2.3.3 The Fermi Level When $N_A \neq 0$, $N_D \neq 0$ and $T \neq 0$

When the temperature is non-zero and high enough, all the impurities at the donor level and all at acceptor level are ionized, that is: $N_A^- \to N_A$, $N_D^+ \to N_D$. The Fermi levels μ_n and μ_p of a type *n* semiconductor (when $N_D > N_A$), and of type *p* semiconductor (when $N_A > N_D$), can be written as functions of the concentrations N_D and N_A.

In type *n* materials the neutrality condition is

$$n_n + N_A = N_D. \tag{2.56}$$

Substituting the expression for n_n given in (2.3), this condition takes the form

$$e^{(\mu_n - E_g)/k_B T} = \frac{N_D - N_A}{N_c}, \tag{2.57}$$

and solving for μ_n we obtain

$$\mu_n = E_g - k_B T \ \ln \frac{N_c}{N_D - N_A}. \tag{2.58}$$

This Fermi level behaves with temperature as shown in Fig. 2.4. When temperatures are high enough, the concentrations of carriers generated by thermal excitation become the dominant ones and $n_n \approx p_n$.

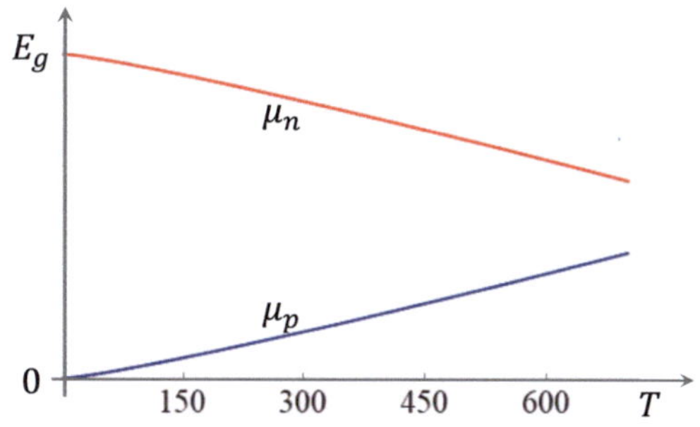

Fig. 2.4 Fermi level in semiconductors type *n* and type *p* when $T \simeq 0$. If $N_D \simeq 10^{22}$ and $\gg N_A$, the Fermi level μ_n decreases monotonically as a function of temperature. If, on the other hand, $N_A \simeq 10^{22}$ and $\gg N_D$, the Fermi level μ_p grows monotonically with temperature

In semiconductors type p and high temperature the neutrality condition can be written in the form

$$N_A = N_D + p_p. \tag{2.59}$$

Substituting the expression for p_p in the form (2.7) we have

$$N_v e^{-\mu_p/k_B T} = N_A - N_D, \tag{2.60}$$

and solving for the Fermi level we obtain

$$\mu_p = k_B T \ \ln \frac{N_v}{N_A - N_D}. \tag{2.61}$$

Figure 2.4 also shows the behavior of μ_p as a function of the temperature. As in the n type case, the concentrations of carriers generated by thermal excitation are the dominant and $n_n \approx p_n$. The semiconductor tends towards intrinsic behavior.

The expressions derived in this chapter provide fundamental relationships for the equilibrium carrier populations in semiconductors. Yet, the physical properties of semiconductor devices under operation, i.e. in non-equilibrium, are equally important.

Semiconductors are unique in that certain physical quantities, particularly the concentrations of charge carriers, can be easily altered and driven out of equilibrium. This characteristic is essential, as it is precisely the ability to sustain out-of-equilibrium processes that enables the functioning of semiconductor devices in transport and optoelectronic applications.

In the next chapter, we will examine transport and optoelectronic properties using both semiclassical and quantum approaches. We will discuss the most relevant non-equilibrium processes, which may be triggered by external influences, such as electric fields or carrier concentration gradients (as in semiconductor junctions), or by internal mechanisms, such as the generation and recombination of electron-hole e-h pairs, scattering events, and collisions.

Additionally, we will introduce key concepts from the quantum formalism, including electric currents, transmission coefficients, and conductance.

2.3.3.1 Solved Problems

1. Suppose a Ge semiconductor has impurities and the temperature is $T = 300\,\text{K}$. Determine the Fermi level μ_n as a function of concentration N_D when the impurities are donors. Determine also the Fermi level μ_p as a function of concentration N_A when the impurities are acceptors. Finally, compare with the level calculated in the hydrogenic atom model.

   Solution

   For the Fermi levels, the functions obtained in Sect. 2.3 are

$$\mu_n = E_g - k_B T \ \ln \frac{N_c}{N_D} \quad \text{and} \quad \mu_p = k_B T \ \ln \frac{N_v}{N_A}. \tag{2.62}$$

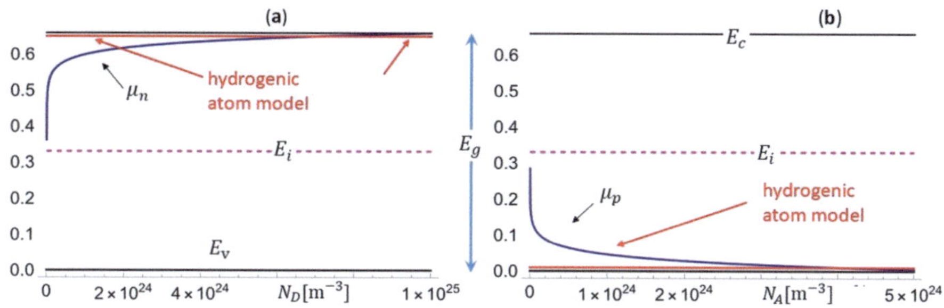

Fig. 2.5 Fermi levels in n-type and p-type germanium semiconductors. Panel **a** shows the Fermi level μ_n as a function of N_D, while panel **b** shows the Fermi level μ_p as a function of N_A. The magenta line is the intrinsic Fermi level. The black lines are the edges of the valence and conduction bands. The red lines are the predictions of the Fermi levels using the hydrogenic impurity model. The temperature is $T = 300\,\text{K}$

The effective concentrations and energy gap of germanium are $N_c = 1.04 \times 10^{25}/\text{m}^3$, $N_v = 6.0 \times 10^{24}/\text{m}^3$ and $E_g = 0.66\,\text{eV}$, respectively. We show in Fig. 2.5a the behavior of μ_n as a function of N_D (blue curve) and in Fig. 2.5b the behavior of μ_p as a function of N_A (blue curve). Both graphs show the intrinsic level (dotted magenta), the levels of donors and acceptors calculated in the hydrogenic impurity model (red lines) and the edges of the conduction and valence bands (black lines). When the impurity concentrations N_D and N_A approach the effective concentrations, the levels μ_n and μ_p approach the edges of the conduction and valence bands, respectively.

2. In Sect. 2.3 the occupation probability in a localized impurity state was introduced. This probability, evaluated at the energy level $E = E_I$, is

$$f^*(E_I) = \frac{1}{\frac{1}{2}e^{(E_I-\mu)/k_BT} + 1}, \tag{2.63}$$

with $E_I = Eg - |E_1^i|$ when the impurity is a donor, and $E_I = |E_1^i|$ when the impurity is an acceptor. The energy E_1^i is the binding energy calculated in chapter 1, using the hydrogenic atom model. We will write the binding energies as E_B. Notice that when $E_I = \mu$ the probability to be occupied by an electron, independent of its spin, is

$$f^*(E_I = \mu) = \frac{2}{3}, \tag{2.64}$$

which is different from $1/2$. This is because the spin states $s_z = \pm 1/2$ are degenerate at the impurity level.

Taking into account the neutrality condition (2.20), the localized occupation probabilities $f^*(E)$, the binding energies E_B, and assuming strongly doped systems, obtain:

(a) the Fermi level μ_n and the concentration n when the impurities are donors, plot μ_n as a function of the concentration N_D for $T = 300$ K;

(b) the Fermi level μ_p and the concentration p when the impurities are acceptors, plot μ_p as a function of the concentration N_D for $T = 300$ K.

Solution

(a) If there are only donor impurities, the continuity equation is (2.24), which, by taking into account the occupation probability $f^*(E_I)$, can be written as

$$n - p = N_D f^*(E_I). \tag{2.65}$$

If $E_I = E_g - E_B$ and $\mu = \mu_n$, we have

$$n - p = N_D \left(1 - \frac{1}{\frac{1}{2}e^{(E_g - E_B - \mu_n)/k_B T} + 1}\right). \tag{2.66}$$

We can write this expression as

$$n - p = N_D \left(\frac{\frac{1}{2}e^{-E_B/k_B T}}{\frac{1}{2}e^{-E_B/k_B T} + e^{(\mu_n - E_g)/k_B T}}\right). \tag{2.67}$$

When the quotient on the right is multiplied and divided by N_c, and we recall that $n = N_c e^{(\mu_n - E_g)/k_B T}$, and define the concentration $n_1 = N_c e^{-E_B/k_B T}/2$, we obtain

$$n - p = N_D \frac{n_1}{n_1 + n}. \tag{2.68}$$

If we use the law of mass action, and replace the concentration p by $p = n_i^2/n$, this equation takes the form

$$n^2 = n_i^2 + \frac{N_D n n_1}{n_1 + n}. \tag{2.69}$$

It is evident from this relation that $n > n_i > p$. Although the cubic equation in n can be solved, we can take advantage of this order relation to simplify a bit. If the concentration N_D is large, a good approximation to this equation is

$$n \simeq N_D \frac{n_1}{n_1 + n}, \tag{2.70}$$

and one of the solutions is

$$n \simeq -\frac{n_1}{2} + \left(\frac{n_1}{2}^2 + N_D\right)^{1/2}. \tag{2.71}$$

Since $n_1 << N_D$ it is easy to verify that an approximate solution for μ_n is

$$\mu_n = E_g - \frac{E_B}{2} - \frac{1}{2} k_B T \ln \frac{N_c}{N_D}. \tag{2.72}$$

This means that the distance to the conduction band is

$$E_D^* = \frac{E_B}{2} + \frac{1}{2}k_B T \ln \frac{N_c}{N_D}. \tag{2.73}$$

Figure 2.6a shows the Fermi level μ_n as a function of the donor concentration.

(b) If the impurities are only acceptors, the neutrality equation (2.41), with the probability that the impurity level $E_I = E_B$ is unoccupied, is

$$p = n + N_A\big(1 - f^*(E_B)\big). \tag{2.74}$$

Replacing the explicit functions in the right hand side, we have

$$p = n + N_A \left(\frac{\frac{1}{2}e^{E_B/k_B T}}{\frac{1}{2}e^{E_B/k_B T} + e^{\mu_p/k_B T}} \right). \tag{2.75}$$

Multiplying and dividing by N_v, taking into account the definition of p, defining $p_1 = N_v e^{-E_B/k_B T}$, and using the law of mass action $n = p_i^2/p$, we obtain

$$p^2 = p_i^2 + N_A \frac{p p_1}{p_1 + p}. \tag{2.76}$$

In this case the concentration p is much larger than p_i and we can approximate this equation by

$$p^2 = N_A \frac{p p_1}{p_1 + p}. \tag{2.77}$$

If we proceed as in the previous section, we obtain

$$\mu_p = \frac{E_B}{2} + \frac{1}{2}k_B T \ln \frac{N_v}{N_A}. \tag{2.78}$$

In this case, the distance to the valence band will be

$$E_A^* = \frac{E_B}{2} + \frac{1}{2}k_B T \ln \frac{N_c}{N_D}. \tag{2.79}$$

These Fermi levels are similar to those obtained in Sect. 2.3, with an additional term $E_B/2$.

In Fig. 2.6b the Fermi level μ_p is shown as a function of acceptor concentration. The behavior of μ_n and μ_p are similar to the behavior of these quantities in the previous problem, with a small displacement towards the center of the gap.

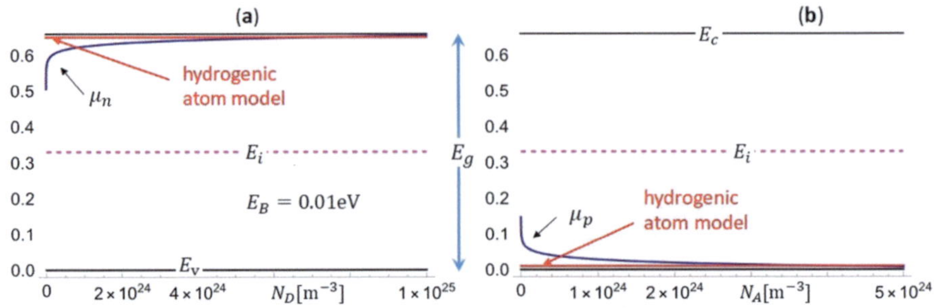

Fig. 2.6 Fermi levels in n-type and p-type germanium semiconductors, taking into account the occupation probabilities of localized impurity states. In **a** the Fermi level μ_n as a function of N_D and in **b** the Fermi level μ_p as a function of N_A. The magenta line is the intrinsic Fermi level. The black lines are the edges of the valence and conduction bands. The red lines are the predictions of the Fermi levels with the hydrogenic atomic model. For these graphs we consider $E_B = 0.01$ eV and $T = 300$ K

3. If $N_D \gg N_A$, with $N_D^+ \simeq N_D \approx 10^{22}/\mathrm{m}^3 \simeq n_n$, and $\sigma = 10^3\ \Omega_{-1}\,\mathrm{m}^{-1}$ and $T \simeq 50$ K, determine the mean free path and the collision diameter when the kinetic energy is $mv_e^2/2 \simeq 3k_bT/2$.
   When the minority concentration is much smaller than the majority concentration, $p_n \ll n_n$, the current is mostly electron current and

$$\sigma \simeq \sigma_e = \frac{n_n e^2 \tau_e}{m}, \tag{2.80}$$

so

$$\tau_e = \frac{\sigma_e}{e^2 n_n} \simeq 3.55\ 10^{-12}\ \mathrm{s}. \tag{2.81}$$

This result with $v_e = (3k_BT/m)^{1/2}$, allows us to obtain the mean free path from

$$l_e = v_e\tau_e = \left(\frac{3k_BT}{m_e}\right)^{1/2}\tau_e. \tag{2.82}$$

Substituting, we have

$$l_e \simeq 0.04781\ 10^6\tau_e \simeq 1.5\ 10^{-7}\ \mathrm{m}, \tag{2.83}$$

so that the collision diameter will be: $D \sim 3 \times 10^{-7}$ m $= 300$ nm.

2.4 Problems

1. Plot the intrinsic concentrations n_i of Si, Ge and GaAs as functions of temperature, for temperatures between 300 K and 400 K. Assume that the effective

masses do not change with temperature and that, at $T = 300\,\text{K}$, the product np is, respectively, $2.25 \times 10^{20}\,\text{cm}^{-6}$, $5.76 \times 10^{26}\,\text{cm}^{-6}$ and $3.24 \times 10^{12}\,\text{cm}^{-6}$ for Si, Ge and AsGa. Also assume that the energy gap is given by

$$E_g(T) = E_g(0) - \frac{\alpha_T T^2}{T + \beta_T}. \tag{2.84}$$

$E_g(0)$, α_T, β_T and the effective masses are given in Tables 1.4 and 3.1.

2. Calculate the effective number of levels N_c of the conduction band, and N_v of the valence band for Si and GaAs at $T = 300\,\text{K}$. Consider for GaAs $m_e^* = 0.067 m_e$ and $m_h^* = 0.44 m_e$, for Si assume that $m_e^* = 0.32 m_e$ and $m_h^* = 0.5 m_e$.
3. Calculate the intrinsic carrier concentration of $In_xGa_{1-x}As$ as a function of x, for x between 0 and 0.5, assuming that the energy gap and the effective masses of the electron are given by (see Tables 1.4 and 3.1)

$$\begin{aligned} E_g(x) &= E_g(0) + \alpha_x x + \beta_x x^2, \\ m_j^*(x) &= m_j^* + \Delta m_{jx}. \end{aligned} \tag{2.85}$$

4. Calculate the intrinsic carrier concentration of InAs at $T = 300\,\text{K}$, assuming that the energy gap is 0.35 eV, and the effective masses of the electron and hole are $0.0027 m_e$ and $0.4 m_e$, respectively. What will be the concentration if the temperature is 600 K?
5. Calculate the donor energies E_D of GaAs and Si using the equation

$$E_1^i = -\frac{m_e^*}{m_e \varepsilon^2} \frac{m_e e^4}{2\hbar^2 (4\pi\epsilon_0)^2} = -\left(\frac{m_e^*}{m_e \varepsilon^2}\right) 13.6\,\text{eV}, \tag{2.86}$$

and compare with Table 2.1. For silicon use the electron effective mass

$$m_\sigma^* = \frac{3 m_l^* m_t^*}{2 m_l^* + m_t^*}. \tag{2.87}$$

6. Suppose that Si semiconductor is heavily doped, with phosphor (P), in one case, and with boron (B) in another. Obtain the majority and minority concentrations in each case when $T = 300\,\text{K}$, and when $T = 320\,\text{K}$. How do the concentrations change?
7. Obtain the majority and minority concentrations of $GaAs$ heavily doped with Si atoms acting as acceptors (i.e., replacing the As atoms), when $T = 300\,\text{K}$, and when $T = 320\,\text{K}$. How do the concentrations change?
8. A Si semiconductor has a doping of acceptor atoms with concentration $N_A = 10^{21}\,\text{cm}^{-3}$. Determine the energy level E_A when the temperature is 300 K, and determine the concentrations p_p and n_p.
9. A Si semiconductor has a doping of donor atoms with concentration $N_D = 10^{21}\,\text{cm}^{-3}$. Determine the energy level E_D when the temperature is 300 K, and determine the concentrations n_n and p_n.

Table 2.1 Energies E_D and E_A of Si, Ge and GaAs with different impurities

Semiconductor	Donor impurity	Energy [meV] E_D	Acceptor impurity	Energy E_A [meV]
Si	Li	33	B	45
	As	54	Ga	72
	Sb	39	Al	67
	P	45	In	160
Ge	Li	9.3	B	10
	As	13.0	Ga	11
	Sb	9.6	Al	10
	P	12.0	In	11
GaAs	Si	5.8	C	26
	Ge	6.0	Be	28
	Sn	6.0	Si	35
	S	6.0	Mg	28

10. What are E_A and n_p (or E_D and p_n) when instead of Si you have GaAs in the previous problems?

References

1. R. Enderlein, N.J.M. Horing, *Fundamentals of Semiconductor Physics and Devices* (World Scientific, Singapure, 1997)
2. Y. Tsividis, *Operation and Modelling of MOS Transistor* (The Oxford Series in Electrical and Computer Engineering, Oxford, 2003)

Transport and Optical Properties. Semiclassical Description

3

Abstract

In this chapter we explore the drift and diffusion transport equations governing the motion of electrons and holes in the presence of electric and magnetic fields. We begin with a review of the Drude model and the classical Hall effect. A dedicated section provides a concise overview of the key mechanisms influencing charge carrier transport in semiconductors, including scattering by acoustic phonons, ionized impurities, and Coulomb interactions. The impact of phonon modes on mobility and the Einstein relations are also discussed. In the second part of the chapter, we introduce the Boltzmann transport equations for non-equilibrium electron and hole distributions, along with the continuity equations and the processes of carrier generation and recombination. Finally, we examine the fundamental principles of stimulated absorption and emission, providing a brief introduction to optical gain, population inversion, and the concepts of coherent and incoherent emission in diodes and lasers.

3.1 Introduction

In the previous chapters, we introduced the fundamental properties that characterize both the medium and the particles that move through it or undergo transitions. Among the medium properties, the most important macroscopic quantum feature distinguishing semiconductors from other solids, is the band structure. It is within these bands that charge carriers move, or where the particles are excited or make optical transitions. Using the band structure and occupation probabilities described by the Fermi-Dirac distribution, we have derived important results such as the concentration of electrons and holes in both intrinsic and doped semiconductors. These properties and concepts are essential for understanding the behavior of semiconductor devices, which typically consists of layers with varying doping levels and energy gaps.

P. Pereyra, *Contemporary Theory of Semiconductors and Devices*,
Graduate Texts in Physics, https://doi.org/10.1007/978-3-032-00296-9_3

In this chapter we shall present the basic principles governing transport influenced by internal and external fields and the properties of the medium.

3.2 Drift and Magnetoelectric Currents

In the presence of electric and magnetic fields, the motion of a charge carrier in a conducting media can be described as a drift process. In the classical framework, this motion is governed by the Lorentz force, the resistive forces exerted by the medium, and the effective forces arising from concentration gradients. We will start studying to drift and magnetoelectric currents, which are basically described by the equations of motion

$$m_e^* \frac{d\mathbf{v}_q}{dt} = q\mathscr{E} + q\mathbf{v}_q \times \mathscr{B} - \frac{\Delta \mathbf{p}_q}{\Delta t_q}, \tag{3.1}$$

Here the term $-\Delta \mathbf{p}_q/\Delta t_q$ represents the resistive forces, being Δt_q the momentum relaxation time. This quantity results from the interaction of the moving particle with the medium. Charge carriers moving in a semiconductor crystal undergo both elastic and inelastic collisions with impurities, lattice particles and lattice vibrations. These vibrations can be of low frequency (acoustic) or high frequency (optical), the latter correspond to interactions with phonon modes. These interactions have been extensively studied in other texts.[1] In the next section, we shall briefly discuss the effect of some interactions on the charge-carrier mobilities.

Returning to the dynamics of charged particles in a semiconductor, under the influence of electric and magnetic fields, we will start analyzing their motion in the presence of an electric field alone, and we will then consider the motion when a magnetic field is also applied.

3.2.1 Drift Currents. The Drude Model

Let us assume that only an electric field is applied, and the motion of a particle in a resistive medium reaches a steady state. In this case, the time derivative in (3.1) vanish, and the equations of motion for electrons and holes, reduce to

$$\Delta \mathbf{p}_q = q \Delta t_q \mathscr{E} \quad \text{with} \quad q = \mp e. \tag{3.2}$$

Long before the interest in studying semiconductors became the focus of research, the motion of electrons in metals was a major issue, and several models were developed. Notably the Drude's model [2] and Kubo's linear response theory [3].

In Drude's model, it is assumed that electrons move freely between collisions inside the solid, experiencing scattering events at random intervals. In this model, the momentum relaxation time Δt_q, is replaced by the average collisions time τ_q, and

[1] A comprehensive analysis of this topic is presented in the book by Seeger [1].

the momentum $\Delta \mathbf{p}_q$ by $m_q^* \langle \mathbf{v}_q \rangle$, where the average velocity $\langle \mathbf{v}_q \rangle$ is given by ℓ_q / τ_q, with ℓ_q the mean free path. In terms of these quantities, the equation of motion (3.2) becomes

$$\langle \mathbf{v}_q \rangle = \frac{q\tau_q}{m_q^*}\mathscr{E}. \tag{3.3}$$

This equation, written for electrons, is

$$\langle \mathbf{v}_e \rangle = -\frac{e\tau_e}{m_e^*}\mathscr{E} \equiv -\mu_e \mathscr{E}, \tag{3.4}$$

where the *electron mobility* μ_e has been defined. For holes, we have

$$\langle \mathbf{v}_h \rangle = \frac{e\tau_h}{m_h^*}\mathscr{E} \equiv \mu_h \mathscr{E}, \tag{3.5}$$

with μ_h the *hole mobility*. Both, the electron and hole mobilities, are important quantities in the analysis of carrier transport in semiconductors. We will refer to these quantities later.

If we have a concentration n_q of particles and electric charge q moving through a semiconductor with an average drift velocity $\langle \mathbf{v}_q \rangle$, the resulting *electric current density* is

$$\mathbf{j}_q = q n_q \langle \mathbf{v}_q \rangle = \frac{n_q q^2 \tau_q}{m_q^*}\mathscr{E}. \tag{3.6}$$

Thus, the Drude conductivity is

$$\sigma_q = \frac{n_q q^2 \tau_q}{m_q^*}. \tag{3.7}$$

In semiconductors, both electrons and holes contribute to transport. Hence, the total drift or drag current is given by the sum

$$\mathbf{j}_d = \mathbf{j}_{ed} + \mathbf{j}_{hd} = -e n_e \langle \mathbf{v}_e \rangle + e n_h \langle \mathbf{v}_h \rangle. \tag{3.8}$$

Applying the Drude's model and considering the velocities given in (3.4) and (3.5), this current can also be written in terms of mobilities and conductivities. Substituting the velocities, the drift current becomes

$$\mathbf{j}_d = \left(n_e\, e\, \mu_e + n_h\, e\, \mu_h\right)\mathscr{E}, \tag{3.9}$$

As observed before, the current densities of electrons and holes, $\mathbf{j}_{ed}$ and $\mathbf{j}_{hd}$, are vectors pointing in the same direction, aligned with the applied electric field $\mathscr{E}$, even though electrons and holes move in opposite directions. Below we will adopt the common semiconductor notation n and p for electron and hole concentrations. In this notation, the drift current is

$$\mathbf{j}_d = \big(n\, e\, \mu_e + p\, e\, \mu_h\big)\boldsymbol{\mathscr{E}}. \tag{3.10}$$

In the more compact form, consistent with the well known Ohm's law, the current is also written as

$$\mathbf{j}_d \equiv \sigma\boldsymbol{\mathscr{E}}. \tag{3.11}$$

where the semiconductor *conductivity* is defined and given in Drude's model by

$$\sigma = n\, e\mu_e + p\, e\mu_h = \sigma_e + \sigma_h = \frac{n_e e^2 \tau_e}{m_e^*} + \frac{n_h e^2 \tau_h}{m_h^*}. \tag{3.12}$$

This quantity plays a central role in the electronic transport in semiconductors. The aim, generally, is to determine the conductivity σ. As defined in the previous equation, we need to know the concentrations and the mobilities. The precise calculation of these quantities, for each type of semiconductor and the various interaction mechanisms that interfere with the charge carriers motion, has been a crucial objective from both theoretical and experimental points of view. We will briefly refer to mobilities in the next section.[2] It is important to keep in mind that if electrons are the majority carriers, holes must be minority carriers, and vice versa, and the concentrations generally satisfy the law of mass action $np = n_i^2 = N_c N_v e^{-E_g/k_B T}$.

Kubo's linear response theory (LRT) is another well-known approach in the electronic transport theory.[3] This approach treats the electronic transport as a non-equilibrium phenomenon, and the response to the external field *as a first order perturbation*. Starting from a system in equilibrium, adds the field that drives the system out of equilibrium. We will briefly refer to this approach in Chap. 5.

In Drude's model, as well as in models where other types of interaction and relaxation mechanisms are considered, the collision times are essential quantities to determine the charge carriers mobilities. The collision time is defined, in general, as the inverse of the collision probability per unit time. Thus, if the probability of collision and transfer of linear momentum per unit time is $W_{kk'}$, the collision time is given by

$$\tau = \frac{1}{W_{kk'}}. \tag{3.13}$$

Important collisions in the semiconductor medium are those with the crystalline oscillations, the phononic oscillations. The average time $\tau_f(T)$ between successive collisions with the phononic oscillations depends in general on the temperature T

[2] For a review of electron and hole mobilities see for instance: Yu and Cardona [4], Chap. 5, and [5,6].
[3] See [3] and Chap. 4 of [7].

and tends to infinity when the temperature tends to zero, that is, when the amplitudes of the oscillations in the crystal also tend to zero. In addition to collisions with phononic modes, the charge carriers, moving in a metal or semiconductor, encounter impurities whose configuration is fixed, hence the average time τ_0 between successive collisions with impurities is basically independent of the temperature. In the presence of different interaction mechanisms, acting independently, the total collision or transition probability is the sum of the collision probabilities. In the presence of phonon oscillations and impurities, the total transition probability, in metals, is generally written as

$$\frac{1}{\tau} = \frac{1}{\tau_f(T)} + \frac{1}{\tau_0}. \tag{3.14}$$

Replacing in Drude's formula, the inverse of the conductivity, *the resistivity*, is given by

$$\varrho = \frac{m_e}{n|e|^2\tau} = \frac{m_e}{n|e|^2u}\left(\frac{1}{\tau_f(T)} + \frac{1}{\tau_0}\right) = \varrho_f(T) + \varrho_0. \tag{3.15}$$

The resistivity ϱ_0 is known as the residual resistivity, i.e. the resistance that remains at $T = 0$.[4]

Unlike in metals, the conductivity of semiconductors exhibits a much more complex dependence on temperature. In addition to the important role that the temperature plays in the electrons and holes concentrations, temperature also affects the relaxation times, due to interactions with phonons and impurities. We will refer to these interactions in the next section. Before we comment on these relaxation mechanisms, let us consider the effect of the magnetic fields on the electronic currents. There are several effects related to this interaction. All important as experimental methods to characterize semiconductors. Here we shall refer only to the Hall effect, an important tool to determine the sign and concentration of charge carriers in semiconductors.

3.2.2 Magnetoelectric Currents. The Hall Effect

In the presence of an electric and a magnetic field, the equations of motion of electrons and holes, assuming in general collision times τ_e and τ_h, are

$$m_e^*\frac{d\mathbf{v}_e}{dt} = -e\mathscr{E} - e\mathbf{v}_e \times \mathscr{B} - m_e^*\frac{\mathbf{v}_e}{\tau_e}, \tag{3.16}$$

and

$$m_h^*\frac{d\mathbf{v}_h}{dt} = e\mathscr{E} + e\mathbf{v}_h \times \mathscr{B} - m_h^*\frac{\mathbf{v}_h}{\tau_h}. \tag{3.17}$$

[4] For additional information on this topic, see for example [8].

To simplify the analysis and the notation, we shall consider a generic particle with charge q moving with velocity $\mathbf{v}_q$ in the plane x-y, subject to an electric field, lying in the same plane, and to a magnetic field $\boldsymbol{\mathscr{B}} = \mathscr{B}\hat{\mathbf{z}}$. In this case, the equation of motion is

$$m_q^* \frac{d\mathbf{v}}{dt} = q\boldsymbol{\mathscr{E}} + q\mathbf{v}_q \times \boldsymbol{\mathscr{B}} - m_q^* \frac{\mathbf{v}_q}{\tau}, \tag{3.18}$$

Again, we shall assume that the motion reaches the stationary regime. Thus, the last equation, in Cartesian components, reads

$$v_{qx} = \frac{q\tau_q}{m_q^*}\mathscr{E}_x + \frac{q\tau_q}{m_q^*} v_{qy}\mathscr{B}, \tag{3.19}$$

$$v_{qy} = \frac{q\tau_q}{m_q^*}\mathscr{E}_y - \frac{q\tau_q}{m_q^*} v_{qx}\mathscr{B}. \tag{3.20}$$

Defining the cyclotron frequency

$$\omega_{qc} = \frac{e\mathscr{B}}{m_q^*}, \tag{3.21}$$

it is not difficult to show that the velocity components can be written as

$$v_{qx} = \frac{q\tau_q}{m_q^*}\left(\frac{1}{1+\tau_q^2\omega_{qc}^2}\right)\mathscr{E}_x \pm \frac{q\tau_q}{m_q^*}\left(\frac{\tau_q\omega_{qc}}{1+\tau_q^2\omega_{qc}^2}\right)\mathscr{E}_y, \tag{3.22}$$

$$v_{qy} = \mp\frac{q\tau_q}{m_q^*}\left(\frac{\tau_q\omega_{qc}}{1+\tau_q^2\omega_{qc}^2}\right)\mathscr{E}_x + \frac{q\tau_q}{m_q^*}\left(\frac{1}{1+\tau_q^2\omega_{qc}^2}\right)\mathscr{E}_y, \tag{3.23}$$

with the upper sign for holes and lower sign for electrons. If we multiply these equations by $n_q q$, where n_q is the carrier concentration, we have in the left sides the current density components. In matrix representation, equations (3.22) and (3.23) can be written as

$$\begin{pmatrix} j_x \\ j_y \end{pmatrix} = \frac{n_q q^2 \tau_q}{m_q^*}\begin{pmatrix} \dfrac{1}{1+\tau_q^2\omega_{qc}^2} & \pm\dfrac{\tau_q\omega_{qc}}{1+\tau_q^2\omega_{qc}^2} \\ \mp\dfrac{\tau_q\omega_{qc}}{1+\tau_q^2\omega_{qc}^2} & \dfrac{1}{1+\tau_q^2\omega_{qc}^2} \end{pmatrix}\begin{pmatrix} \mathscr{E}_x \\ \mathscr{E}_y \end{pmatrix} \equiv \sigma \begin{pmatrix} \mathscr{E}_x \\ \mathscr{E}_y \end{pmatrix}. \tag{3.24}$$

The matrix σ is known as the conductivity tensor and is, generally, written as

$$\sigma = \begin{pmatrix} \sigma_{xx} \; \sigma_{xy} \\ \sigma_{yx} \; \sigma_{yy} \end{pmatrix} = \begin{pmatrix} \sigma_{xx} & \sigma_{xy} \\ -\sigma_{xy} & \sigma_{xx} \end{pmatrix}, \tag{3.25}$$

whose inverse is the resistivity tensor

$$\rho = \frac{1}{\sigma_{xx}^2 + \sigma_{xy}^2}\begin{pmatrix} \sigma_{xx} & -\sigma_{xy} \\ \sigma_{xy} & \sigma_{xx} \end{pmatrix} = \begin{pmatrix} \rho_{xx} & \rho_{xy} \\ -\rho_{xy} & \rho_{xx} \end{pmatrix}. \tag{3.26}$$

Notice that when $\mathscr{B} = 0$, both matrices are diagonal.

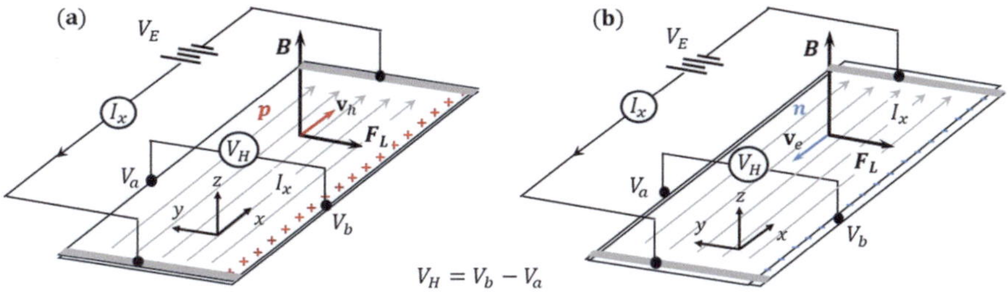

Fig. 3.1 Hall effect in type p and type n semiconductors. In panel **a** a 2D type p semiconductor, and in panel **b** a 2D type n semiconductor in a four probe configuration to measure the longitudinal current I_x and the Hall voltage V_H. In both systems an electric field $\boldsymbol{\mathscr{E}} = \mathscr{E}\hat{\mathbf{x}}$, parallel to axis x, and a magnetic field $\boldsymbol{\mathscr{B}} = \mathscr{B}\hat{\mathbf{z}}$, perpendicular to the x-y plane, are applied

3.2.2.1 Classical Hall Effect

We now turn to the classical Hall effect discovered by Edwin Hall in 1879. This effect has become an extremely useful tool to characterize semiconductors, in which the majority charge carriers sign may be positive or negative. This effect makes it possible to determine not only the sign of the majority charge carriers, but also their concentration.

In Fig. 3.1 we consider two possible experimental configurations. In panel (a) a 2D type p semiconductor, and in panel (b) a 2D type n semiconductor. In both systems, an electric field $\boldsymbol{\mathscr{E}} = \mathscr{E}\hat{\mathbf{x}}$ parallel to axis x, and a magnetic field $\boldsymbol{\mathscr{B}} = \mathscr{B}\hat{\mathbf{z}}$ perpendicular to the x-y plane, are applied. With the four-contact configuration, shown in the figure, the longitudinal current I_x and the Hall voltages V_H are measured. This system is a particular realization of the more general case discussed lines above, now with $\mathscr{E}_y = 0$. Taking into account this difference, the particles velocities given in (3.22) and (3.22) become

$$v_{qx} = \frac{q\tau_q}{m_q^*}\left(\frac{1}{1+\tau_q^2\omega_{qc}^2}\right)\mathscr{E}_x, \quad \text{and} \quad v_{qy} = \mp\frac{q\tau_q}{m_q^*}\left(\frac{\tau_q\omega_{qc}}{1+\tau_q^2\omega_{qc}^2}\right)\mathscr{E}_x. \tag{3.27}$$

In this system, due to the finite width of the semiconductor slabs, there is a piling up of charges at the edges of the slab. Eventually, the accumulation of charges reach a maximum and the field that these charges produce, equilibrates the Lorentz force. On the other side, the Hall potential V_H, measured by probes in opposite sides of the slab, is the potential produced by the accumulated charges at the edges. We shall now write the last equations in terms of the measured quantities.

Let us start with the physical relations for holes in panel (a). Since $j_{hx} = pev_x$ and $I_x = j_{hx}L_y$, we have

$$v_{hx} = \frac{I_x}{peL_y} = \frac{e\tau_h}{m_h^*}\left(\frac{1}{1+\tau_h^2\omega_{hc}^2}\right)\mathscr{E}_x. \tag{3.28}$$

Thus, the velocity in response to Lorentz force can be written as

$$v_{hy} = -\frac{I_x}{peL_y}\tau_h\omega_{hc}. \tag{3.29}$$

On the other hand, the velocity related with the Hall field $\mathscr{E}_H = V_H/L_y$, can be written as

$$v^h_{Hy} = \frac{e\tau}{m^*_h}\mathscr{E}_H = \frac{e\tau}{m^*_h}\frac{V_H}{L_y}. \tag{3.30}$$

Since v^h_{Hy} is the negative of v_{hy}, we end up with the following relation

$$V_H = \frac{1}{pe}\mathscr{B}I_x \equiv R_H\mathscr{B}I_x. \tag{3.31}$$

Here the Hall coefficient $R_H = 1/pe$ for type p semiconductors was defined. A similar analysis for the electron velocities leads to obtain the following relations

$$V_H = -\frac{1}{ne}\mathscr{B}I_x \equiv R_H\mathscr{B}I_x. \tag{3.32}$$

Here the Hall coefficient is $R_H = -1/ne$. This means that we can in general write the Hall voltage and the Hall coefficients as

$$V_H = \frac{1}{qn_q}\mathscr{B}I_x \quad \text{and} \quad R_H = \frac{1}{qn_q}, \tag{3.33}$$

where $q = \pm e$ and n_q the corresponding concentration.

3.2.2.2 Hall Coefficient in Compensated Semiconductors

In the previous analysis of the Hall effect we assumed that we had only one type of charge carrier, either holes or electrons. This is of course an ideal case. A more realistic analysis should consider the presence of both types of charge carriers. Semiconductors where acceptors and donors are present are called *compensated semiconductors*.

We will now derive the expression for the Hall coefficient in compensated semiconductors. At the end, we will briefly mention a thermomagnetic effect that appears simultaneously with the Hall effect.

To simplify the results let us assume that the magnetic field is small. If we neglect quadratic order terms in $\mathscr{B}$, the velocities in (3.27) become

$$v_{qx} = \frac{q\tau_q}{m^*_q}\mathscr{E}_x = \mu_q\mathscr{E}_x, \quad \text{and} \quad v_{qy} = \mp\frac{q\tau_q}{m^*_q}\tau_q\omega_{qc}\mathscr{E}_x = \mu_q^2\mathscr{B}\mathscr{E}_x. \tag{3.34}$$

We must consider the Hall velocities

$$v^h_{Hy} = \mu_h\mathscr{E}_H \quad \text{and} \quad v^e_{Hy} = -\mu_e\mathscr{E}_H. \tag{3.35}$$

Since the total currents along the x and y axes are

$$j_x = j_{ex} + j_{hx} = -env_{ex} + epv_{hx} \tag{3.36}$$

and

$$j_y = j_{Ly} - j_{Hy} = -nev_{ey} + pev_{hy} - \left[-nev^e_{Hy} + pev^h_{Hy}\right] = 0, \tag{3.37}$$

replacing the velocities we have

$$j_x = e\mathscr{E}_x \left(n\mu_e + p\mu_h\right), \tag{3.38}$$

and

$$e\mathscr{E}_H \left(n\mu_e + p\mu_h\right) = -e\mathscr{B}\mathscr{E}_x \left(p\mu_h^2 - n\mu_e^2\right).$$

Combining these equations we obtain

$$\mathscr{E}_H = \frac{\mathscr{B}\left(p\mu_n^2 - n\mu_h^2\right) j_x}{e\left(n\mu_e + p\mu_h\right)^2}. \tag{3.39}$$

Therefore, the Hall coefficient for compensated semiconductors, to first order of the magnetic field is

$$R_H = \frac{p\mu_n^2 - n\mu_h^2}{e\left(n\mu_e + p\mu_h\right)^2}. \tag{3.40}$$

This expressions shows that the Hall coefficient reflects the competition between electron and hole contributions. Soon after the discovery of the Hall effect, A. Ettinghausen and W. Nernst observed that at the same time that the charge accumulates in the Hall experiment, a temperature gradient along the y axis appears [9]. This thermomagnetic effect occurs as a consequence of the increased number of collisions in the medium. The experimental results have shown that the temperature gradient is given by the following relation

$$\frac{dT}{dy} = P\mathscr{B} j_x. \tag{3.41}$$

To conclude this section on the Hall effect, it is worth mentioning that frequently the semiconductor samples cannot be shaped into the ideal rectangular Hall bar geometry, as shown in Fig. 3.1. A well known contacts configuration for arbitrary shapes is the van der Pauw method [10].

Almost hundred years after the discovery of the classical Hall effect, the quantum Hall effect was reported by Klaus von Klitzing in 1980 [11]. An important characteristic of this finding is that the Hall conductance is quantized in integer multiples of e^2/h, where h is the Planck constant. The high precision with which this quantization occurs, led to introduce h/e^2 as a resistance unit, named the Klitzing constant. In

Chap. 12 we will refer to this effect, we will review the well-known Landau theory for a charged particle in a homogeneous magnetic field, and we will provide a new interpretation of the integer quantum Hall effect and the resistance plateaus, based on the quantization of edge states and the Büttiker-Landauer approach.

3.2.3 Diffusion Currents

When charge carriers concentrations vary as functions of position, a diffusion phenomenon of charges occurs, a diffusion that tends to establish a homogeneous distribution of charges. The diffusion current densities of electrons and holes are proportional to the gradients of the respective concentrations, and by Fick's law [12] are given by

$$\mathbf{j}_{ed} = eD_e \nabla n \quad \text{and} \quad \mathbf{j}_{hd} = -eD_h \nabla p, \tag{3.42}$$

where D_e and D_h are the corresponding diffusion coefficients. These coefficients depend on the diffusion velocities v_q, which are defined by the ratio of the average displacement or diffusion lengths l_q and the diffusion time τ_q, in which the particles experience the displacement l_q. The diffusion coefficients are defined as

$$D_e = \frac{l_e^2}{\tau_e}, \quad \text{and} \quad D_h = \frac{l_h^2}{\tau_h}. \tag{3.43}$$

Therefore, the diffusion currents can be written as

$$\mathbf{j}_{ed} = e\frac{l_e^2}{\tau_e}\nabla n \quad \text{and} \quad \mathbf{j}_{hd} = -e\frac{l_h^2}{\tau_h}\nabla p. \tag{3.44}$$

To understand the functional dependence of the diffusion coefficients, we are going to deduce the first of (3.42). Suppose we have a semiconductor medium in which the electron concentration is $n(\mathbf{r})$. For simplicity, we will assume that n varies only along the x-axis, and that electrons moving to the left and right have the same average speed. If however the number of electrons moving to the left differs from the number moving to the right, there will be a net flow of electrons. The net number N_e of electrons crossing an imaginary surface A, perpendicular to the x-axis and located at some point x_0, during a time τ_e, will be

$$N_e = A\int_{x_0-l_e}^{x_0} n(x)dx - A\int_{x_0}^{x_0+l_e} n(x)dx. \tag{3.45}$$

If we express $n(x)$ by its series expansion around the point x_0, as

$$n(x) = n(x_0) + \left.\frac{dn}{dx}\right|_{x=x_0}(x - x_0) + \cdots \tag{3.46}$$

and replace $n(x)$ by the first two terms of this series, and integrate, we have

$$N_e = -Al_e^2 \left.\frac{dn}{dx}\right|_{x=x_0}. \tag{3.47}$$

Since x_0 is any point inside the semiconductor, we can write

$$N_e = -Al_e^2 \frac{dn}{dx}. \tag{3.48}$$

The total charge passing through surface A will then be $-eN_e$. Therefore the electron diffusion current is

$$I_n = -e\frac{N_e}{\tau_e} = e\frac{Al_e^2}{\tau_e}\frac{dn}{dx}, \tag{3.49}$$

and the diffusion current density

$$j_{nd} = e\frac{N_e}{A\tau_e} = e\frac{l_e^2}{\tau_e}\frac{dn}{dx} = e\langle v_e\rangle l_e\frac{dn}{dx} = eD_e\frac{dn}{dx}. \tag{3.50}$$

This confirms that the electron and hole diffusion coefficients can be written as $D_e = l_e^2/\tau_e$ and $D_h = l_h^2/\tau_h$, respectively.

3.3 Conductivity and Relaxation Times

In previous sections we referred to charges undergoing collisions with the phononic modes (especially with the acoustic modes) and with the impurities. In Sect. 3.2, the conductivity was introduced as a function of mobility

$$\sigma = nq\mu_q = nq\frac{q\langle\tau_h\rangle}{m_q^*}. \tag{3.51}$$

We shall now, briefly, refer to some models that have been introduced to determine collision times and mobilities in semiconductors. As mentioned before, the collision times are just the inverse of the collision probabilities, which can be estimated from golden rule expressions of the form $Nv_q\sigma_c$, where N is the density of scattering centers, v_q the particle velocity, and σ_c the collision cross section.

3.3.1 Collision with Acoustic Modes. Shockley and Bardeen Model

Shockley and Bardeen, using a perturbative approach, modeled the interaction of charge carriers with acoustic oscillations in a semiconductor as a momentum transfer mechanism. In this formalism, the probability per unit of time that a carrier interacts with transfer of momentum, written as the inverse of the relaxation time is defined as

$$\frac{1}{\tau} = \frac{V}{\pi\hbar}\int k'^2 dk' \left|H_{kk'}\right|^2 \delta(E_k - E_{k'}). \tag{3.52}$$

Here

$$k'^2 dk' = \frac{m_q^2 v_q}{\hbar^3} dE \quad \text{and} \quad H_{kk'} = \frac{|e|\varepsilon_{ac} q_l A_l}{V}\int e^{i(\mathbf{k}-\mathbf{k}'\pm\mathbf{q}_l)\cdot\mathbf{r}} dr^3 \simeq \frac{\varepsilon_{ac}(k_B T)^{1/2}}{(2Vc_l)^{1/2}} \tag{3.53}$$

represent the density of states and the matrix elements of the interaction, respectively. The subscript q refers to electron or hole. For collision with the acoustic phonon modes one obtains

$$\begin{aligned}\frac{1}{\tau_{ac,q}} &= \frac{V}{\pi\hbar}\int k'^2 dk' \left|H_{kk'}\right|^2 \delta(E_k - E_{k'}),\\ &= \frac{m_q^2 v_q k_B T}{\pi\hbar^4}\frac{\varepsilon_{ac}^2}{c_l}.\end{aligned} \tag{3.54}$$

Here ε_{ac} is the deformation constant and c_l the longitudinal elastic constant of the crystal. If

$$v_q = \sqrt{\frac{2k_B T}{m_q}}, \tag{3.55}$$

the mean free path is

$$\ell_{ac,q} = \frac{\pi\hbar^4}{m_q^2 k_B T}\frac{c_l}{\varepsilon_{ac}^2}, \tag{3.56}$$

the mobility, when carriers interact only with the acoustic phonons, is

$$\mu_{ac,q} = \frac{e}{m}\tau_{ac,q}. \tag{3.57}$$

3.3.2 Interaction with Ionized Impurities

In semiconductors, an important interaction is with ionized impurities whose charge is $Z|e|$. Since the interaction is electrostatic, the strength depends on the average distance $\langle r_I\rangle$ between ionized impurities. This distance can be estimated from

$$\frac{3}{2}k_B T = \frac{1}{4\pi\epsilon_r\epsilon_0}\frac{Ze^2}{\langle r_I\rangle}, \tag{3.58}$$

where $\epsilon_s = \epsilon_r\epsilon_0$ is the dielectric permittivity of the semiconductor. The transition probability per unit time can be approximated as

$$W_{I,q} = S\langle v_{I,q}\rangle N_I, \tag{3.59}$$

where N_I is the impurities concentration, S is the cross section $\sim\pi\langle r_I\rangle^2$, and $\langle v_{I,q}\rangle$ the average velocity obtained from

$$\frac{1}{2}m_q^*\langle v_{I,q}\rangle^2 = \frac{3}{2}k_B T. \tag{3.60}$$

Replacing these quantities, we have

$$\frac{1}{\tau_{I,q}} = \frac{4\pi N_I}{3^{3/2}\sqrt{m_q}(k_B T)^{3/2}}\left(\frac{Ze^2}{4\pi\epsilon_s}\right)^2. \tag{3.61}$$

Therefore, the mobility $\mu_{I,q}$ associated with electrons interacting, only with impurities, is

$$\mu_{I,q} = \frac{e}{m}\tau_{I,q}. \tag{3.62}$$

3.3.3 Relaxation Time due to Coulomb Interaction

When the interaction is purely Coulombic, the relaxation time derived from perturbation theory, is

$$\frac{1}{\tau_{Ip,q}} = \frac{\pi^{3/2}N_I}{2^{7/2}\sqrt{m_q}\left(k_B T\right)^{3/2}}\left(\frac{Ze^2}{4\pi\epsilon_s}\right)^2 f(\beta_{BH}). \tag{3.63}$$

Here

$$f(\beta_{BH}) = \ln(1+\beta_{BH}^2) - \frac{\beta_{BH}^2}{1+\beta_{BH}^2} \tag{3.64}$$

is a nearly constant function of the Brooks-Herring function [13]

$$\beta_{BH} = 2\frac{m_q}{\hbar}\left(\frac{2}{m_q}3k_B T\right)^{1/2} L_{Dq}, \tag{3.65}$$

with L_{Dq} the Debye length. For electrons, this length is $L_{Dn} = \sqrt{D_n\tau_e} = \sqrt{k_B T/e^2}$ $\sqrt{\epsilon_r\epsilon_0/n}$, with a similar expression for holes. The mean free path in this case is

$$\langle\ell_{I,q}\rangle = \frac{16}{\pi^{3/2}}\left(\frac{k_B T 4\pi\epsilon_s\epsilon_0}{Ze^2}\right)^2 \frac{f(\beta_{BH})}{N_I}. \tag{3.66}$$

Note that the transition probabilities $\tau_{I,q}$ and $\tau_{Ip,q}$ are practically the same, differing only by a constant.

In Fig. 3.2 the electron mobilities and conductivities for Ge are plotted as functions of temperature, considering interactions only with acoustic modes (red curve) or only with impurities (blue curve). For these plots it was assumed that the concentration of impurities is $N_I = 10^{19}\,\text{cm}^{-3}$. Although the conductivities tend, in both cases, to grow, the mobilities go in opposite directions. As T increases, the system tends towards an intrinsic behavior.

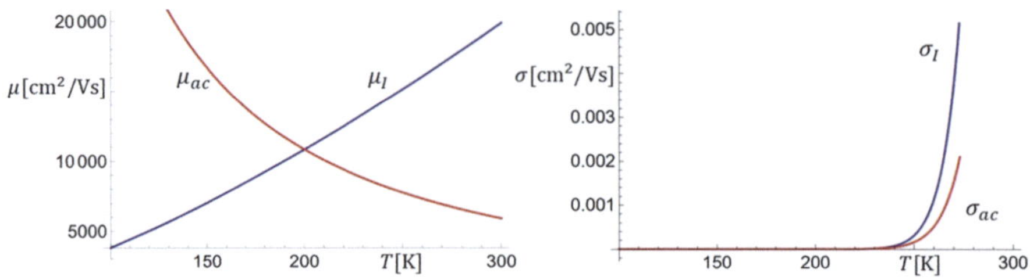

Fig. 3.2 Mobility and conductivity of electrons in Ge. On the left the mobilities $\mu_{ac,e}$ and $\mu_{I,e}$, for interaction with acoustic modes and impurities, as functions of temperature, and on the right, the conductivities $\sigma_{ac,e} = en_i\mu_{ac,e}$ and $\sigma_{I,e} = en_i\mu_{I,e}$. For this plot we considered $N_I = 10^{19}\,\text{cm}^{-3}$, and the effective masses in Table 3.1. The conductivities increase exponentially with temperature, but the mobilities do not. Moreover, as functions of temperature, $\mu_{ac,e}$ and $\mu_{I,e}$ show opposite trends

3.3.4 Mobility

In real systems, in addition to the interactions discussed in this section, other interactions also influence the mobility of charge carriers. Concerning mobility, the most relevant information is its order of magnitude. The reader interested in this topic should see specialized literature and books (for instance the book by Seeger [1]).

If the semiconductor is intrinsic, the most important momentum-relaxation mechanism is the collision with phonon modes, the intrinsic mobility in these cases is simply

$$\mu_{i,q} = \frac{|e|}{m_q}\tau_{ac,q}(T), \tag{3.67}$$

but if the semiconductor is impurified, the total transition probability is the sum of the transition probabilities induced by interaction with the phonon modes and by the collisions with the ionized impurities, i.e.:

$$\frac{1}{\tau_q} = \frac{1}{\tau_{ac,q}(T)} + \frac{1}{\tau_{I,q}}. \tag{3.68}$$

In this case, the mobility will be given by

$$\mu_q = \frac{|e|}{m_q}\tau_q = \frac{|e|}{m_q}\frac{\tau_{ac,q}(T)\tau_{I,q}}{\tau_{ac,q}(T) + \tau_{I,q}}. \tag{3.69}$$

In Fig. 3.3 the intrinsic and the extrinsic mobility are shown. With these mobilities replaced into (3.12), together with the parameters that correspond to the specific semiconductor, the conductivity can be evaluated.

3.3.5 Einstein Relations

When both fields and concentration gradients are present we have the drift and diffusion currents that may or may not be in equilibrium. To establish a condition of

Table 3.1 Effective number of levels and intrinsic concentrations

Semi-conductor	Effective density conduction band (N_c)	Effective density valence band (N_v)	Intrinsic concentration ($n_i = p_i$)	$E_g(300)$ [eV]	m_e^*/m_e $6^{2/3}m_c$	m_h^*/m_e $4^{2/3}m_c$
Si (300 K)	$2.78 \times 10^{19}\,\text{cm}^{-3}$	$9.84 \times 10^{18}\,\text{cm}^{-3}$	$1.5 \times 10^{10}\,\text{cm}^{-3}$	1.12	1.07	0.535
Ge (300 K)	$1.04 \times 10^{19}\,\text{cm}^{-3}$	$6.0 \times 10^{18}\,\text{cm}^{-3}$	$2.33 \times 10^{13}\,\text{cm}^{-3}$	0.661	0.56	0.385
GaAs (300 K)	$4.45 \times 10^{17}\,\text{cm}^{-3}$	$7.72 \times 10^{18}\,\text{cm}^{-3}$	$1.84 \times 10^{6}\,\text{cm}^{-3}$	1.42	0.067	0.52
GaN (Wu)	$4.3 \times 10^{14}\,\text{cm}^{-3}\,T^{3/2}$	$8.9 \times 10^{15}\,\text{cm}^{-3}\,T^{3/2}$		3.2	0.2	1.5
GaN (ZB)	$2.3 \times 10^{14}\,\text{cm}^{-3}\,T^{3/2}$	$8.0 \times 10^{15}\,\text{cm}^{-3}\,T^{3/2}$		3.39	0.13	1.4

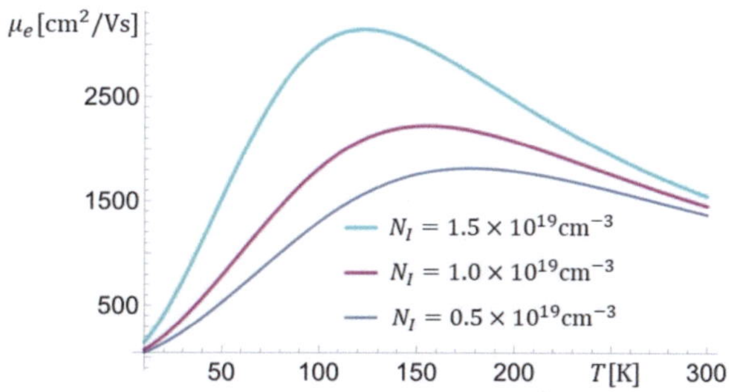

Fig. 3.3 Electron mobility in Ge when the interactions are with phonon modes and with impurities, for different N_I concentrations. In all three cases, mobility tends to behave like intrinsic mobility when temperature increases

equilibrium in the currents, a relationship must be established between the parameters that characterize the drift and diffusion currents. For example, between the mobility and the diffusion coefficient. These relations are known as the Einstein relations.

If a semiconductor, say type n, is under the action of an electric field $\mathscr{E}$, the electric force on each electron is

$$\mathscr{F}(\mathbf{r}) = -e\mathscr{E}(\mathbf{r}). \tag{3.70}$$

It is known that the electric force and the electric field can be expressed as gradients of a potential. The electric force is the gradient of the potential energy $U(\mathbf{r})$:

$$\mathscr{F}(\mathbf{r}) = -\nabla U(\mathbf{r}), \tag{3.71}$$

while the electric field is the gradient of the electric potential $\phi(\mathbf{r})$:

$$\mathscr{E}(\mathbf{r}) = -\nabla\phi(\mathbf{r}). \tag{3.72}$$

Using the relationship (3.71) one can rewrite (3.70) as

$$-e\mathscr{E}(\mathbf{r}) = -\nabla U(\mathbf{r}). \tag{3.73}$$

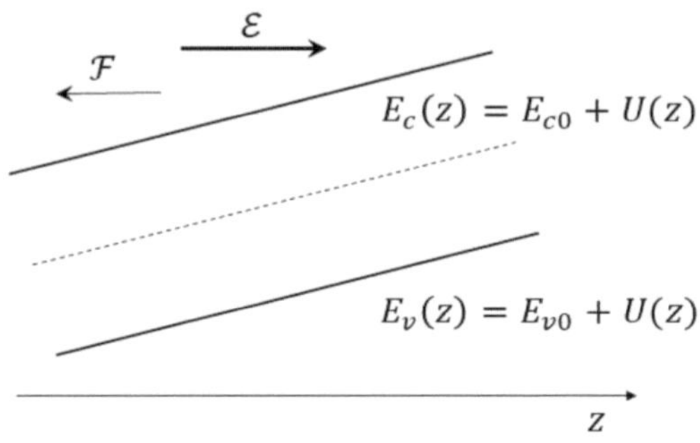

Fig. 3.4 Electron in uniform electric field. If an electron is in a field $\mathscr{E}$, as indicated if the figure, it feels a force $\mathscr{F} = -e\mathscr{E}$ and the potential energy $U(z)$ grows linearly with z. In this case, electrons tend to move to the left

For the purpose of this section, we will assume that the field is uniform and the potential energy only depends on z, i.e.

$$-e\mathscr{E} = -\frac{U(z)}{dz}. \tag{3.74}$$

This means that if U(0) = 0, the potential energy will be

$$U(z) = e\mathscr{E}z \quad \text{while} \quad \phi(z) = -\mathscr{E}z. \tag{3.75}$$

Thus, the electric potential energy $U(z)$ grows linearly with z. This energy adds to that of the electrons. This causes the band edges to tilt as shown in Fig. 3.4. If we denote the band-edge energies before applying the electric field as E_{c0} and E_{v0}, the band-edge energies in the presence of a field will be

$$E_c(z) = E_{c0} + e\mathscr{E}z \quad \text{and} \quad E_v(z) = E_{v0} + e\mathscr{E}z. \tag{3.76}$$

Now suppose that the distribution of electrons is given by the Boltzmann distribution

$$n(z) = n_0 e^{-(K+U(z))/k_BT}. \tag{3.77}$$

If the currents are in equilibrium, we have

$$\mathbf{j}_{ea} = -\mathbf{j}_{nd}. \tag{3.78}$$

This means that

$$ne\mu_e\mathscr{E} = -eD_e\frac{dn}{dz} = ne\frac{D_e}{k_BT}\frac{dU(z)}{dz}. \tag{3.79}$$

Simplifying and substituting, we have

$$\mu_e = \frac{D_e}{k_BT}e \quad \text{or} \quad \frac{D_e}{\mu_e} = \frac{k_BT}{e}. \tag{3.80}$$

These are known as the *equilibrium Einstein relations*.

3.4 Non-equilibrium Carrier Densities

In thermodynamic equilibrium, the Fermi-Dirac distribution and the carrier concentrations are straightforward functions of the macroscopic parameters, temperature and chemical potential. Out of equilibrium, however, these parameters lose their direct physical meaning. It is no longer possible to express the distribution function or the concentrations using simple expressions dependent on such parameters.

Instead of closed-form solutions, one must work with differential equations that describe the evolution of carrier distributions and concentrations. These equations are particular cases of the Boltzmann equation, whose full solution lies beyond the scope of this book. However, we will explore the phenomenological equations that govern current densities and the continuity (or conservation) of charge.

During the operation of electronic devices, the charge carrier concentrations deviate from equilibrium. In what follows, we will derive the equations that describe how these perturbed concentrations evolve over time. We will denote by n_o and p_o the concentrations at $t = 0$ and for an arbitrary time t by

$$\begin{aligned} n(\mathbf{r}, t) &= n_o + \delta n(\mathbf{r}, t), \\ p(\mathbf{r}, t) &= p_o + \delta p(\mathbf{r}, t). \end{aligned} \tag{3.81}$$

It will be assumed that

$$\delta n(\mathbf{r}, t), \delta p(\mathbf{r}, t) \ll \begin{cases} n_o & \text{in semiconductors type n,} \\ p_o & \text{in semiconductors type p.} \end{cases} \tag{3.82}$$

There exist a number of different processes that may modify the concentrations, and the form of the differential equations depends only on which are the relevant processes. Among the most common processes, with effects on the distributions, are those of diffusion, generation of excitonic pairs, recombination and the action of external fields. In general, the equations obtained here will be particular cases of the Boltzmann equation that we will briefly refer to next.

3.4.1 The Boltzmann Equation

A generalization of the distribution function $f(E)$, in thermodynamic equilibrium, is the non-equilibrium distribution $f(E(\mathbf{k}), \mathbf{r}, t)$. This function depends, in general, on $\mathbf{r}$ and time t, and on external and internal factors. These factors determine the evolution of the distribution function. Taking into account the origin of the factors, we distinguish them as external and internal factors and we denote the corresponding temporal variations of the distribution function as $(df(E(\mathbf{k}), \mathbf{r}, t)/dt)_{ext}$ and $(df(E(\mathbf{k}), \mathbf{r}, t)/dt)_{int}$, respectively. According to Liouville's theorem, the total change in the distribution function at a point $(\mathbf{r}, \hbar\mathbf{k})$ of the phase space that moves with the particle, must cancell, that is:

$$\left(\frac{df(E(\mathbf{k}), \mathbf{r}, t)}{dt}\right)_{ext} + \left(\frac{df(E(\mathbf{k}), \mathbf{r}, t)}{dt}\right)_{int} = 0. \tag{3.83}$$

If the time derivative of the first term is developed, we have the Boltzmann equation

$$\left(\frac{\partial}{dt} + \frac{d\mathbf{r}}{dt}\cdot\nabla_r + \frac{d\mathbf{k}}{dt}\cdot\nabla_k\right) f(E(\mathbf{k}),\mathbf{r},t)|_{ext} = -\left(\frac{df(E(\mathbf{k}),\mathbf{r},t)}{dt}\right)_{int} . \quad (3.84)$$

We will rewrite this equation below for the distributions $n(\mathbf{r},t)$ and $p(\mathbf{r},t)$, expressed in terms of current densities. To identify the components of the currents, we will refer to diffusion and drift currents. The second member involves internal processes such as exciton generation and exciton recombination (or annihilation) processes that modify the distributions.

3.4.2 The Boltzmann Equations for $n(\mathbf{r},t)$ and $p(\mathbf{r},t)$

Suppose now that the charges in the semiconductors are in the presence of fields and the concentrations of charge carriers can vary from one point to another of the semiconductor, that is,

$$n(\mathbf{r},t) = n(\mathbf{r}_0,t) + \nabla_r n|_{r=r_0}\cdot(\mathbf{r}-\mathbf{r}_0) + \cdots \quad (3.85)$$

and

$$p(\mathbf{r},t) = p(\mathbf{r}_0,t) + \nabla_r p|_{r=r_0}\cdot(\mathbf{r}-\mathbf{r}_0) + \ldots . \quad (3.86)$$

In the absence of other mechanisms such as the generation and recombination of charges, the Boltzmann equation (3.84) for the concentration $n(\mathbf{r},t)$ (similar equations can be written for $p(\mathbf{r},t)$) takes the form

$$\left(\frac{\partial}{dt} + \frac{d\mathbf{r}}{dt}\cdot\nabla_r + \frac{d\mathbf{k}}{dt}\cdot\nabla_k\right) n(\mathbf{r},t) = 0. \quad (3.87)$$

Let us assume that the term $\nabla_k n(\mathbf{r},t) = 0$. So, the above equation can be written as

$$\frac{\partial n(\mathbf{r},t)}{dt} + \langle\mathbf{v}_e\rangle\cdot\nabla_r n(\mathbf{r},t) = 0 \quad (3.88)$$

where $d\mathbf{r}/dt$ was replaced by an average velocity $\langle\mathbf{v}_e\rangle$. If the average speed of the carriers does not change much from one point to another, we have

$$\nabla_r\cdot\langle\mathbf{v}_e\rangle n(\mathbf{r},t) = \langle\mathbf{v}_e\rangle\cdot\nabla_r n(\mathbf{r},t). \quad (3.89)$$

Taking the second member to the first and vice versa, and writing the concentration $n(\mathbf{r},t)$ as it appears in (3.85), we have

$$\begin{aligned}\langle\mathbf{v}_e\rangle\cdot\nabla_r n(\mathbf{r},t) &= \nabla_r\cdot\left[\langle\mathbf{v}_e\rangle n(\mathbf{r}_0,t) + \langle\mathbf{v}_e\rangle\nabla_r n|_{r=r_0}\cdot(\mathbf{r}-\mathbf{r}_0)\right],\\ &= \nabla_r\cdot\langle\mathbf{v}_e\rangle\nabla_r n|_{r=r_0}\cdot(\mathbf{r}-\mathbf{r}_0). \quad (3.90)\end{aligned}$$

The first term in the square parenthesis of this equation corresponds to the drift current density and the second to the diffusion current. Notice that both forms of the current are considered. Assuming that $(\mathbf{r} - \mathbf{r}_0) \simeq -\mathbf{l}_e$, i.e., that the diffusion displacement is oriented in the opposite direction to the growth of n, and taking into account that $\mathbf{r}_0$ is any point inside the semiconductor, we write

$$\langle \mathbf{v}_e \rangle \cdot \nabla_r n(\mathbf{r}, t) = \frac{1}{q} \nabla_r \cdot q \langle v_e \rangle l_e \nabla_r n(\mathbf{r}, t). \tag{3.91}$$

The function on which the operator ∇_r acts, in the second member, is the current, therefore

$$\langle \mathbf{v}_e \rangle \cdot \nabla_r n(\mathbf{r}, t) = \frac{1}{q} \nabla_r \cdot \mathbf{j}_n. \tag{3.92}$$

With this relationship, the Boltzmann equation (3.88) becomes

$$\frac{\partial n(\mathbf{r}, t)}{dt} - \frac{1}{e} \nabla_r \cdot \mathbf{j}_n = 0. \tag{3.93}$$

Similarly we have

$$\frac{\partial p(\mathbf{r}, t)}{\partial t} + \frac{1}{e} \nabla \cdot \mathbf{j}_p(\mathbf{r}, t) = 0, \tag{3.94}$$

for holes. It must be emphasized that (3.93) and (3.94) are nothing else than the continuity equations for electrons and holes, respectively.

In extrinsic semiconductors of type n or type p the currents can also be expressed as j_e or j_h, respectively. It is common to analyze some transport properties in terms of the mobilities μ_e and μ_h. The commonly used unit of mobility is $\mathrm{cm^2\,V^{-1}\,s^{-1}}$. If the mobility is $1\ \mathrm{cm^2\,V^{-1}\,s^{-1}}$ in a field of $1\,\mathrm{V\,cm^{-1}}$, the drift velocity will be

$$|\mathbf{v}_e| = |\mu_e \mathscr{E}| = 1\,\frac{\mathrm{cm^2}}{\mathrm{Vs}} \times 1\,\frac{\mathrm{V}}{\mathrm{cm}} = 1\,\frac{\mathrm{cm}}{\mathrm{s}}. \tag{3.95}$$

At room temperature, the mobilities μ_e of electrons in Si, Ge and $GaAs$ are 1350, 3900 and 8800 $\mathrm{cm^2\,V^{-1}\,s^{-1}}$, respectively. Those of the holes are 500, 1900 and 400 $\mathrm{cm^2\,V^{-1}\,s^{-1}}$, respectively. It is clear that the mobilities depend on the concentrations of impurities. They decrease when these increase.

3.5 Generation, Recombination and Decay Time

As indicated before, the currents $\mathbf{j}_n$ and $\mathbf{j}_p$ consist of drift and diffusion contributions. For electrons

$$\mathbf{j}_e(\mathbf{r}, t) = \sigma_e \mathscr{E}(\mathbf{r}, t) + e D_e \nabla n(\mathbf{r}, t), \tag{3.96}$$

and of holes

$$\mathbf{j}_h(\mathbf{r},t) = \sigma_h \mathscr{E}(\mathbf{r},t) - eD_h \nabla p(\mathbf{r},t). \tag{3.97}$$

In what follows we will use $\mathbf{j}_e$ and $\mathbf{j}_n$ interchangeably for the electron currents and $\mathbf{j}_h$ and $\mathbf{j}_p$ for the hole currents.

The concentrations of electrons and holes also depend on the excitation of electrons and its inverse process, the recombination. If we denote the generation rate of excited electrons as $\partial n/\partial t|_g$ and the recombination as $\partial n/\partial t|_r$, and similarly for p, we write the continuity equations as

$$\frac{\partial n(\mathbf{r},t)}{\partial t} - \frac{1}{e}\nabla \cdot \mathbf{j}_e(\mathbf{r},t) = \left.\frac{\partial n}{\partial t}\right|_g - \left.\frac{\partial n}{\partial t}\right|_r, \tag{3.98}$$

for electrons, and

$$\frac{\partial p(\mathbf{r},t)}{\partial t} + \frac{1}{e}\nabla \cdot \mathbf{j}_h(\mathbf{r},t) = \left.\frac{\partial p}{\partial t}\right|_g - \left.\frac{\partial p}{\partial t}\right|_r. \tag{3.99}$$

for holes. Multiple mechanisms contribute to carrier generation and recombination in semiconductors. Some processes involve the generation of only electrons or only holes, typically through transitions to or from deep levels associated with defects, or via charge injection mechanisms. Other processes involve both carriers simultaneously. For instance, when a photon with energy $\hbar\omega > E_g$ is absorbed by an electron in the valence band, the electron is excited to a level in the conduction band, leaving behind a hole in the valence band. This results, as mentioned in previous chapters, in the creation of an electron-hole pair, commonly referred to as an exciton. Since excitons have finite lifetimes, the electron eventually recombines with the hole, returning to the valence band and releasing the absorbed energy $\Delta E = \hbar\omega$. This energy may be emitted as a photon of the same energy, with frequency $\omega = \Delta E/\hbar$. Processes of this kind, where the electron and hole concentrations n and p change, are described by the generation and recombination terms on the right-hand sides of (3.98) and (3.99).

To understand the meaning of some relationships that arise in the processes of generation and recombination, without being very specific yet, let us assume that we have a system in which the carriers make transitions between two energy levels while they are also interacting with a radiation field confined in a resonant cavity, in which photons have a frequency ω. Suppose that this frequency corresponds to the energy difference between levels 1 and 2 of Fig. 3.5, i.e. that $E_2 - E_1 = \hbar\omega$ and the electron concentrations are n_1 and n_2, respectively. The rate at which n_2 changes, due to excitation of electrons from the valence band, depends on n_1, the absorption coefficient $\alpha(\omega)$ and the intensity $\mathtt{I}(\omega)$ of the light in the cavity. $\mathtt{I}(\omega)$ is defined as $\hbar\omega c\phi$, where ϕ is the number of photons per unit volume. The units of $\mathtt{I}(\omega)$ are energy per unit of time and area. Dividing $\mathtt{I}(\omega)$ by $\hbar\omega$, one has the number of photons per cm^2 and unit of time, capable of exciting an electrons to level 2. This means that if the surface area exposed to light is S, we will have

$$\left.\frac{\partial n_2}{\partial t}\right|_g = S\frac{\mathtt{I}(\omega)\alpha(\omega)}{\hbar\omega}n_1. \tag{3.100}$$

This expression gives the number of electrons that reach the level 2 per unit of time, through the absorption mechanism. In the inverse process, the recombination, there are two contributions: the emission mediated by the photonic field (by the interaction with light), whose intensity is $\mathtt{I}(\omega)$, and the spontaneous emission. If $\gamma(\omega)$ is the emission coefficient of a photon of energy $\hbar\omega$ and τ_{21} is the lifetime of an excited electron, the rate of change of the concentration n_2, due to recombination, will be

$$\left.\frac{\partial n_2}{\partial t}\right|_r = S\frac{\mathtt{I}(\omega)\gamma(\omega)}{\hbar\omega}n_2 + \frac{n_2}{\tau_{21}}. \tag{3.101}$$

The absorption and emission coefficients are essentially the same, deffering only when the degeneracies g_1 and g_2 of levels 1 and 2 are different. In general the relation between $\alpha(\omega)$ and $\gamma(\omega)$ is

$$\alpha(\omega) = \frac{g_2}{g_1}\gamma(\omega). \tag{3.102}$$

The total rate of change, by absorption and emission, of n_2 can then be written as

$$\frac{dn_2}{dt} = \left.\frac{\partial n_2}{\partial t}\right|_g - \left.\frac{\partial n_2}{\partial t}\right|_r = \left(n_1\frac{g_2}{g_1} - n_2\right)S\frac{\mathtt{I}(\omega)\gamma(\omega)}{\hbar\omega} - \frac{n_2}{\tau_{21}}. \tag{3.103}$$

The factor $\left(n_1 g_2/g_1 - n_2\right)$ is the *gain factor*, a central quantity in device operation since it determines whether stimulated processes dominate or not. It is important to determine the gain when a device is operating and the carrier concentrations are being modified, either because the system is subject to an external field or because charges are injected into the system. We will discuss this topic with more detail in the next chapter. If there is also a pumping of electrons from level n_1, we have

$$\frac{dn_2}{dt} = \left(n_1\frac{g_2}{g_1} - n_2\right)S\frac{\mathtt{I}(\omega)\gamma(\omega)}{\hbar\omega} - \frac{n_2}{\tau_{21}} + W_p n_1. \tag{3.104}$$

The first term describes the gain and loss by stimulated transitions, the second the loss by spontaneous emission and the last term accounts for pumping. If the number of particles n_i in the energy level E_i is proportional to the probability of occupation of that level, which to a good approximation is given by the Boltzmann distribution $e^{-E_i/k_BT}/\sum e^{-E_j/k_BT}$, and to the degeneracy g_i, the ratio n_2/n_1 will be given by

$$\frac{n_2}{n_1} = \frac{g_2}{g_1}e^{-(E_2-E_1)/k_BT} = \frac{g_2}{g_1}e^{-\hbar\omega/k_BT}. \tag{3.105}$$

If we substitute this relation in (3.103), we have

$$\frac{dn_2}{dt} = n_2 \left(e^{\hbar\omega/k_B T} - 1\right) S \frac{\mathrm{I}\gamma(\omega)}{\hbar\omega} - \frac{n_2}{\tau_{21}} + W_p n_1. \tag{3.106}$$

Alternative expressions in terms of the Einstein coefficients, will be introduced in the following section, and with more detail in Chap. 7. Before concluding this section, let us make a brief digression related to Planck's distribution.

In thermodynamic equilibrium, and zero pumping, the generation rate is equal to the recombination rate and we have,

$$\left.\frac{\partial n_2}{\partial t}\right|_g - \left.\frac{\partial n_2}{\partial t}\right|_r = n_2\left(e^{\hbar\omega/k_B T} - 1\right)S\frac{\mathrm{I}\gamma(\omega)}{\hbar\omega} - \frac{n_2}{\tau_{21}} = 0, \tag{3.107}$$

which reduces to

$$\mathrm{I}\gamma(\omega) = \frac{1}{S\tau_{21}} \frac{\hbar\omega}{e^{\hbar\omega/k_B T} - 1}. \tag{3.108}$$

This is, essentially, the relation that Einstein obtained to re-derive the Planck distribution. In this relation, as in Planck's distribution, the intensity is proportional to the photon occupation number

$$n_\phi(\omega) = \frac{1}{e^{\hbar\omega/k_B T} - 1}. \tag{3.109}$$

From (3.108) one can obtain the decay time in terms of the emission coefficient $\gamma(\omega)$. If we define this coefficient as $\gamma(\omega) = \gamma_o \rho(\omega)$, we have

$$\frac{1}{\tau_{21}} = \mathrm{I} S \gamma_o \frac{\omega^2}{\pi^2 c^3}. \tag{3.110}$$

In equilibrium and assuming that a pumping contribution is present, we have

$$n_2\left(e^{\hbar\omega/k_B T} - 1\right)S\frac{\mathrm{I}\gamma(\omega)}{\hbar\omega} - \frac{n_2}{\tau_{21}} + W_p n_1 = 0. \tag{3.111}$$

From this equation, it is clear that the minimum injection, or minimum pumping to keep the radiation process in balance, is

$$W_p = \frac{n_2}{n_1}\left(\frac{1}{\tau_{21}} - \left(e^{\hbar\omega/k_B T} - 1\right)S\frac{\mathrm{I}\gamma(\omega)}{\hbar\omega}\right). \tag{3.112}$$

3.6 Einstein Coefficients for Stimulated Emission and Generation

The generation and recombination rates induced by the interaction with radiation (photon field) have been defined in terms of absorption coefficients, emission coefficients and decay times, but they can also be written in terms of the *Einstein coefficients*.

We will now write the recombination generation rates of (3.98) in terms of the Einstein coefficients and of the occupation probabilities of the energy levels. The *electron generation rate* in the conduction band can be expressed as[5]

$$\left.\frac{\partial n_2}{\partial t}\right|_g = B_{21} f_{21} \rho(\omega) n_1 + W_b n_1. \tag{3.113}$$

where B_{12} is the Einstein absorption coefficient proportional to the excitation probability from level 1 to 2, $\rho(\omega)$ is the spectral distribution of the radiation, $f_{12}(\omega)$ is the probability that energy level E_1 is occupied and energy level E_2 is free, i.e.

$$f_{12}(\omega) = f(E_1, T)\big(1 - f(E_2, T)\big), \tag{3.114}$$

with $f(E_j, T)$ the Fermi-Dirac occupation probability of the state with energy E_j. Both energies, E_1 and E_2, are measured from the same reference level, ω is the frequency of the absorbed or emitted photon whose relation to the energies E_1 and E_2 is $\hbar\omega = |E_1 - E_2| \geq E_g$. For intra-band transitions, the analysis is formally similar, with the corresponding changes. If the emitting system is a heterostructure, the dimensions are not equal in the three directions; they are generally larger in the transverse direction and smaller in the growth direction z. The last term of (3.113) describes the rate at which electrons are injected (or pumped) from the valence band, whose concentration is n_1, to the conduction band with concentration n_2.

Similarly, the *recombination rate* can be expressed as

$$\left.\frac{\partial n_2}{\partial t}\right|_r = B_{21} f_{21} \rho(\omega) n_2 + A_{21} f_{21} n_2, \tag{3.115}$$

where B_{21} is the Einstein stimulated emission coefficient from level 2 to 1. This coefficient is practically equal to the coefficient B_{12}. f_{21} is the same as f_{12} with the indices swapped and A_{21} is the Einstein spontaneous transition coefficient, proportional to the probability of spontaneous transition per unit time from level 2 to 1. Notice that the coefficients B_{ab} and A_{ab} do not have the same units. The first term of the second member describes stimulated recombination processes and the last term is the spontaneous decay processes, generally with very short decay times τ_{21}.

---

[5] The density of levels accessible in the excitation and emission processes will be taken into account in Chap. 7. In Chap. 7 also, the coefficients A_{ab} and B_{ab} will be determined.

If we substitute the previous expressions in (3.98), for $n = n_2$, and assume that no currents are present, we have

$$\frac{\partial n_2(\mathbf{r}, t)}{\partial t} = n_2 B_{21} \rho(\omega)(f_{21} - f_{12}) - A_{21} f_{21} n_2 + W_b n_1, \tag{3.116}$$

Notice that

$$\begin{aligned} f_{12} - f_{21} &= f(E_1, T)\big(1 - f(E_2, T)\big) - f(E_2, T)\big(1 - f(E_1, T)\big) \\ &= f(E_1, T) - f(E_2, T). \end{aligned} \tag{3.117}$$

It can be shown that since $E_1 < \mu$ and $E_e - \mu \gg k_B T$, the last difference reduces to

$$f_{12} - f_{21} \simeq 1 - e^{-\hbar\omega/k_B T} \quad \text{and} \quad f_{21} \simeq e^{-\hbar\omega/k_B T}. \tag{3.118}$$

Substituting these approximate factors, (3.116) becomes

$$\frac{\partial n_2(\mathbf{r}, t)}{\partial t} = n_2 B_{12} e^{-\hbar\omega/k_B T} \rho(\omega)\left(e^{\hbar\omega/k_B T} - 1\right) - A_{21} e^{-\hbar\omega/k_B T} n_2 + W_b n_1. \tag{3.119}$$

If the system is in thermodynamic equilibrium and there is no pumping

$$n_2 B_{12} e^{-\hbar\omega/k_B T} \rho(\omega)\left(e^{\hbar\omega/k_B T} - 1\right) - A_{21} e^{-\hbar\omega/k_B T} n_2 = 0, \tag{3.120}$$

we obtain Planck's spectral distribution

$$\rho(\omega) = \frac{A_{21}}{B_{12}} \frac{1}{e^{\hbar\omega/k_B T} - 1}, \tag{3.121}$$

when

$$A_{21} = \frac{\hbar\omega^3}{\pi^2 c^3} B_{12}, \tag{3.122}$$

It is worth making an additional observation here that will help us understand the physical meaning of some variables introduced in later chapters. The factor that multiplies n_2 in (3.120) or in (3.119) is the transition probability per unit of time and per particle. The previous analysis is restricted to transitions between two levels, E_1 and E_2, so it only refers to the emission and absorption of photons with the frequency $\omega = (E_2 - E_1)/\hbar$. This is not the most general case. Almost every physical system has more than one energy level in the conduction and valence bands, and the possible transitions are many more and must be added or integrated, if the continuous limit is justified. If energy level E_1 is in band a and level E_2 is in band b, the transition rate (or transition probability per unit time) between energy levels E_1 and E_2 (with $\omega_{12} = E_{12}/\hbar \equiv |E_1 - E_2|/\hbar$) is

$$w_{ab}(\omega_{12}) = B_{12} e^{-\hbar\omega_{12}/k_B T} \rho(\omega_{12})\left(e^{\hbar\omega_{12}/k_B T} - 1\right) - A_{12} e^{-\hbar\omega_{12}/k_B T}. \tag{3.123}$$

The total transition rate that takes into account the totality of possible transitions between the levels of band a and band b, in the limit $E_{12} \to E$, will be

$$\begin{aligned}\mathscr{W}_{ab} &= \int_{E_g}^{\infty} w_{ab}(\omega) g_{ab}(E) dE \\ &= \int_{E_g}^{\infty} dE g_{ab}(E) \left[B_{ab} e^{-\hbar\omega/k_B T} \rho(\omega) \left(e^{\hbar\omega/k_B T} - 1 \right) - A_{ab} e^{-\hbar\omega/k_B T} \right].\end{aligned} \tag{3.124}$$

Here $g_{ab}(E)$ is the joint density of levels in bands a and b. The integrand $\mathscr{R}_{ab} = w_{ab}(\omega) g_{ab}(E)$ is the transition rate per unit time and unit energy.

3.7 Charge Injection, Quasi-Fermi Energy Levels and Gain

In Chap. 2, we have seen that the Fermi level, in a type n semiconductor, approaches the conduction band, and in type p semiconductors, it approaches the valence band. When the system is operating and the electron and hole populations are not in equilibrium, it is not obvious where the Fermi level lies. To overcome this problem, a model that has been useful is the model in which electrons and holes are assumed to be, separately, in equilibrium with distribution functions

$$\begin{aligned}f_n^e(E) &= \frac{1}{e^{(E-\mu_n)/k_B T} + 1} \\ f_p^h(E) &= 1 - f_p^e(E) = 1 - \frac{1}{e^{(E-\mu_p)/k_B T} + 1} = \frac{1}{e^{(\mu_p - E)/k_B T} + 1}.\end{aligned} \tag{3.125}$$

It is clear that in equilibrium, $\mu_n = \mu_p$, but if electrons are injected into the conduction band and holes into the valence band, we have two populations with quasi-Fermi levels, and we can obtain the concentrations n and p of

$$\begin{aligned}n &= \int_{E_c}^{\infty} g_e(E) f_n^e(E) dE = N_c e^{(\mu_n - E_c)/k_B T} \\ p &= \int_{-\infty}^{E_v} g_h(E) f_p^h(E) dE = N_v e^{(E_v - \mu_p)/k_B T},\end{aligned} \tag{3.126}$$

where μ_n and μ_p are the quasi-Fermi levels

$$\begin{aligned}\mu_n &= E_c - k_B T \ln \frac{N_c}{n} \\ \mu_p &= E_v + k_B T \ln \frac{N_v}{p}.\end{aligned} \tag{3.127}$$

If the system is emitting and absorbing, the probability of occupation in the photon emission process with frequency ω is $\hat{f}_{21}(\omega) = f_n^e(E_2)\big(1 - f_p^e(E_1)\big)$ and the occupation probability in the excitation process is $\hat{f}_{12}(\omega) = f_p^e(E_1)\big(1 - f_n^e(E_2)\big)$. When the difference between the number of emitted photons and the number of absorbed ones is positive, the system is in a population inversion regime and the gain $\mathcal{G}(\hbar\omega)$ is defined as

$$\mathcal{G}(\hbar\omega) \propto f_n^e(E_2)\big(1 - f_p^e(E_1)\big) - f_p^e(E_1)\left(1 - f_n^e(E_2)\right), \tag{3.128}$$

which can be written as

$$\mathcal{G}(\hbar\omega) \propto f_n^e(E_2) f_p^h(E_2) - \left(1 - f_p^h(E_1)\right)\left(1 - f_n^e(E_2)\right), \tag{3.129}$$

and reduces to

$$\mathcal{G}(\hbar\omega) \propto f_p^h(E_1) + f_n^e(E_2) - 1. \tag{3.130}$$

Consequently, the condition for positive gain, or condition for *population inversion* requires that

$$f_n^e(E_2) + f_p^h(E_1) > 1 \tag{3.131}$$

3.8 Light-Emitting Diode and Lasers

Due to the characteristics of the excitation and recombination mechanisms and also due to the characteristics of the emitted spectrum, it is common to distinguish light emitting devices or LEDs (acronym for "light emitting devices"), and lasers (acronym for "light amplification by stimulated emission of radiation"). The first occur in structures made up of *pn* junctions or metal-semiconductor junctions. Light-emitting diodes are characterized because the emission is spontaneous, the light is incoherent and has higher frequency dispersion for two reasons: due to the uncertainty relationship $\hbar\Delta\omega \geq \hbar/\tau_{12}$, with τ_{12} small, and due to the spread in the initial and final energies, which at temperature T is of the order of $\Delta\omega \simeq k_B T/\hbar \sim 10^{12}$ Hz. This corresponds to $\Delta\lambda \simeq \lambda^2 \simeq k_B T/hc \sim 5$ nm. Lasers, on the other hand, are systems that emit coherent and almost monochromatic light. It can be a *pn* junction or a semiconductor heterostructure with one or multiple quantum wells, which, as indicated in Fig. 3.5, are in a resonant cavity, in which a monochromatic electromagnetic field interacts with electrons in well-defined energy states and induces transitions by exciting electrons and recombining them by emitting photons whose frequency is equal to the resonant cavity frequency. The light produced is practically monochromatic. We will deal with these devices in the second part of the book.

In interband transitions, from the conduction to the valence band, and direct band semiconductors, the moment $\hbar k$ is conserved. The energies of the electrons in the conduction band and the holes in the valence band are, respectively:

$$E^e = E_c + \frac{\hbar^2 k^2}{2m_e^*} \quad \text{and} \quad E^h = E_v - \frac{\hbar^2 k^2}{2m_e^*}, \tag{3.132}$$

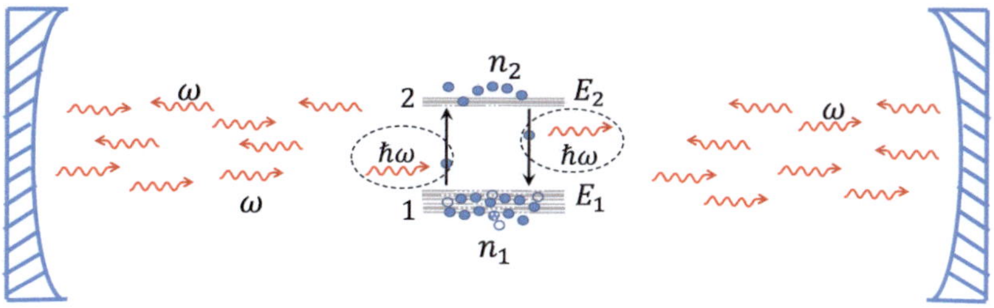

Fig. 3.5 Two-level system interacting with the field of a resonant cavity, which selects photons with a resonant energy $\hbar\omega$

so the energy of the emitted photon can be written as

$$\begin{aligned}\hbar\omega &= E_g + \frac{\hbar^2k^2}{2}\left(\frac{1}{m_e^*} + \frac{1}{m_h^*}\right)\\ &= E_g + \frac{\hbar^2k^2}{2m_r^*},\end{aligned} \tag{3.133}$$

where m_r^* is the reduced mass of the exciton e-h.

Solved Problem Use the approximation for quasi-Fermi levels

$$\begin{aligned}E_n &= E_c + k_BT\left(\ln\frac{n}{N_c} + \frac{n}{N_c\sqrt{8}}\right)\\ E_p &= E_v - k_BT\left(\ln\frac{p}{N_v} + \frac{p}{N_v\sqrt{8}}\right),\end{aligned} \tag{3.134}$$

known as the Joyce-Dixon approximation, to obtain the concentrations n and p. Obtain the concentration n for the population inversion to occur at the GaAs band edges, at 300 K, by assuming that $N_c = 4.45\ 10^{17}\,\text{cm}^{-3}$ and $N_v = 7.7\ 10^{18}\,\text{cm}^{-3}$ and that n and p are equal.

The population inversion condition with the electron and hole energies at the GaAs band edges is

$$f^e(E_c) + f^h(E_v) > 1. \tag{3.135}$$

If we consider the distribution functions of (3.125) and the Joyce-Dixon equations, the previous equation takes the form

$$\frac{n(1+\sqrt{8})}{N_c\sqrt{8} + n(1+\sqrt{8})} + \frac{p(1+\sqrt{8})}{N_v\sqrt{8} + p(1+\sqrt{8})} > 1. \tag{3.136}$$

If also $n = p$, this equation reduces to

$$n > \left(\frac{8}{1+\sqrt{8}}\right)^{1/2}\sqrt{N_cN_v} \sim \sqrt{2\,N_cN_v} \sim 2.6\,10^{18}\,\text{cm}^{-3} \tag{3.137}$$

3.9 Problems

1. Show that the units of mobility are (L^2/Vt), example (cm^2/Vs).
2. If the relaxation time of electrons in Si is $\tau_e = 1.9928\,10^{-13}$ s, and its effective mass is $m_e^* = 0.2598$, what is the electron mobility in Si?
3. If the relaxation time of light (heavy) holes in Si is $\tau_{hl} = 1.2272\,10^{-13}$ s ($\tau_{hh} = 3.7583\,10^{-13}$ s), and their effective mass is $m_{hl}^* = 0.16$ ($m_{hh}^* = 0.49$), what is the mobility?
4. If the relaxation time of the electrons in GaAs is $\tau_e = 3.2356\,10^{-13}$ s and its effective mass is $m_e^* = 0.067$, what is the electron mobility in GaAs?
5. If the gap in Si is $E_g = 1.17$ eV and the electron concentration is $n = 10^{21}\,\mathrm{cm}^{-3}$, determine:

   (a) the concentration p when $T = 300$ K,
   (b) the conductivity of Si.

6. If the gap in GaAs is $E_g = 1.51$ eV and the electron concentration is $n = 10^{21}\,\mathrm{cm}^{-3}$, determine:

   (a) the concentration p when $T = 300$ K,
   (b) the conductivity of GaAs.

7. Show that the units of the diffusion constant are (L^2/t), for example (cm^2/s).
8. In a type n semiconductor the electron concentration varies with z according to the function

$$n(z) = n_o e^{-z/L}$$

   If $n_o = 10^{16}\,\mathrm{cm}^{-3}$ and $L = 1\,\mu$m, determine the diffusion coefficient if the diffusion current is $j_n = 4000\,\mathrm{A\,cm}^{-2}$.
9. Obtain the Einstein relation of holes.
10. Derive the Boltzmann equation (3.94) for the concentration $p(\mathbf{r}, t)$.

References

1. K. Seeger, *Semiconductor Physics* (Springer, Berlin, 1996)
2. P. Drude, Ann. Phys. **306**, 566 (1900)
3. R. Kubo, J. Phys. Soc. Jpn. **12**, 570 (1957)
4. P.Y. Yu, M. Cardona, *Fundamentals of Semiconductors*, 1st edn. (Springer, Berlin, 1996)
5. D.L. Rode, in *Semiconductors and Semimetals*, ed. by R.K. Willardson, A.C. Beer, vol. 10 (Academic Press, New York, 1982), pp. 1–89
6. J.D. Wiley, in *Semiconductors and Semimetals*, ed. by R.K. Willardson, A.C. Beer, vol. 10 (Academic Press, New York, 1982), pp. 91–174
7. P.A. Mello, N. Kumar, *Quantum Transport in Mesoscopic Systems. Complexity and Statistical Fluctuations* (Oxford University Press, Oxford, 2004)

8. J.R. Hook, H.E. Hall, *Solid State Physics*, 2nd edn. (Wiley, Chichester, 1991)
9. A. Ettienghausen, W. Nernst, Ann. Phys. Chem
10. L. van der Pauw, Philips Res. Rep. **13**, 1 (1958)
11. K. von Klitzing, G. Dorda, M. Pepper, Phys. Rev. Lett. **45**, 494 (1980)
12. A. Fick, Ann. Phys. **94**, 59 (1855)
13. H. Brooks, Phys. Rev. **83**, 879 (1951)

Physical Properties of the *pn* Junction 4

Abstract

This chapter introduces the fundamental physical properties of the simplest semiconductor heterostructure: the *pn* junction. Both non-polarized and polarized *pn* junctions are examined. The discussion begins with the initial charge redistribution at the interface, which leads to equilibration, the formation of the contact potential, and the determination of carrier concentrations within and outside the transition (depletion) region. The effects of forward and reverse bias on minority carrier concentrations are then analyzed. The chapter continues analyzing the *pn* junction as a rectifying diode, we also review briefly the Esaki barrier and the Zener effect, calculate the injected carrier currents in the generation-recombination processes. Finally, we introduce the concept of injection efficiency and its role in the operation of light-emitting diodes (LEDs).

4.1 Properties of Non-polarized *pn* Junction

The simplest semiconductor device is possibly the *p-n* diode. To understand the basic physical properties of semiconductor structures, we will analyze the behavior of carriers at the interface *p-n* and in its vicinity.

When a *p-n* junction is produced by growing, for instance, a layer of semiconductor type p, on top a layer of semiconductor type n, or vice versa, a transient flow of electrons and holes occurs that ends when the electric forces due to the displaced charges and concentration gradients are balanced. Simultaneously, the chemical potentials μ_n and μ_p align to each other as shown schematically in Fig. 4.2. This alignment of the chemical potentials forces the bending of the valence and conduction bands. A little later we will see how the band edges are really modified. When the equilibrium is reached, the regions near the interface become charged as illustrated in the lower panel of Fig. 4.1. For a more quantitative description of this process, let us assume that on the p side the acceptors concentration is N_A per unit volume, and on the n side, the concentration of donor atoms is N_D. In this system,

P. Pereyra, *Contemporary Theory of Semiconductors and Devices*,
Graduate Texts in Physics, https://doi.org/10.1007/978-3-032-00296-9_4

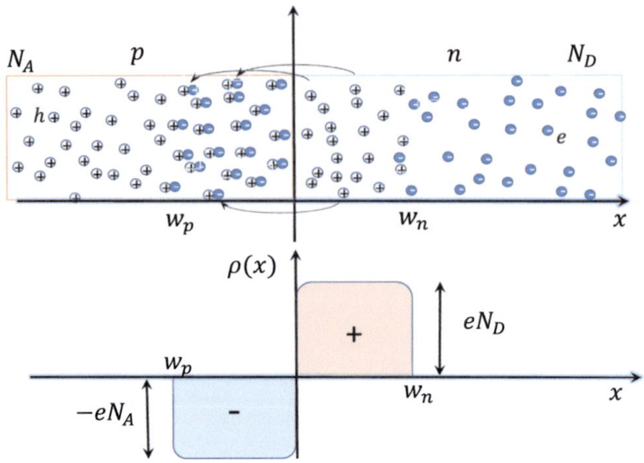

Fig. 4.1 If a type n semiconductor is joined with a type p, some electrons from the donor atoms in the n region move towards the p region. This transient process stops when the field strengths of the displaced charges and the force resulting from the concentration gradient, balance. At equilibrium, close to the interface, the charge distributions shown in the lower panel remain

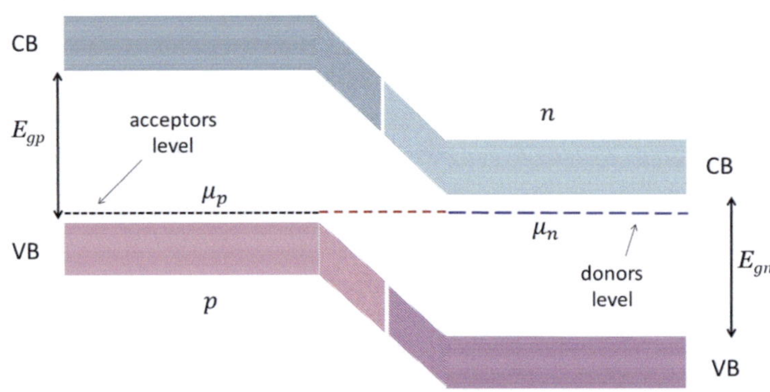

Fig. 4.2 The transient process of charge displacement is accompanied by the leveling of the Fermi levels μ_p and μ_n. For this it is necessary that the bands of region p move with respect to the bands of region n, as shown schematically here. This band bending phenomenon is very important and is discussed and explained better below

electrons from donors side are attracted to acceptors side. If all acceptors capture one electron, the resulting space charge density on the p side will be $-eN_A$. On the n side, donor atoms lose electrons and become positively ionized, with the same density of the donor atoms. Therefore, the charge density on the donor side is eN_D. Since the system as a whole is neutral, the total charge on the left plus the total charge on the right must sum to zero, that is:

$$-eN_A A w_n + eN_D A w_p = 0 \quad \rightarrow \quad -eN_A w_n + eN_D w_p = 0 \tag{4.1}$$

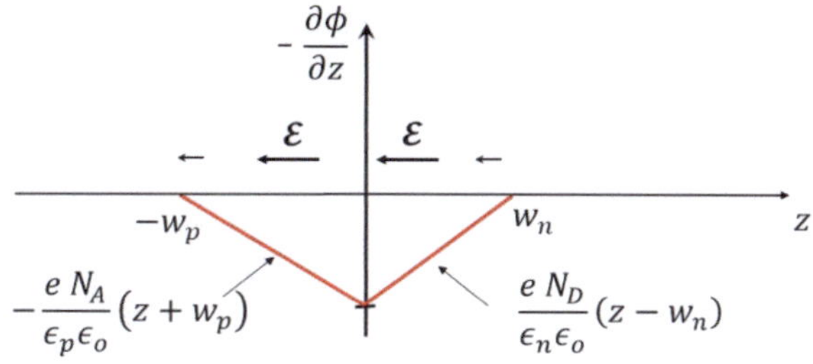

Fig. 4.3 The electric field due to the charge densities $-eN_A$ and eN_D. Its magnitude changes linearly, it takes its maximum value at the interface. Its negative value means that it is oriented from the positive charge towards the negative charge

Here A is the cross-sectional surface, while w_p and w_n are the widths of the ionized layers due to the displacement of the charges, (see Fig. 4.3). The charge densities on the left and right side produce electric fields $\mathscr{E}_n$ and $\mathscr{E}_p$ and modify the potential energy. These potentials and fields are obtained by solving the Poisson equation

$$\frac{d^2\phi(z)}{dz^2} = -\frac{\rho(z)}{\epsilon\epsilon_0} \tag{4.2}$$

with

$$\rho(z) = \begin{cases} -eN_A & -w_p < z < 0 \\ eN_D & 0 < z < w_p \end{cases} \tag{4.3}$$

By integrating the Poisson equation with the condition

$$\frac{d\phi(-w_p)}{dz} = \frac{d\phi(w_n)}{dz} = 0, \tag{4.4}$$

we obtain the linear behavior of the field

$$\mathcal{E}(z) = -\frac{d\phi(z)}{dz} = \begin{cases} -\dfrac{eN_A}{\epsilon\epsilon_0}(z + w_p) & -w_p < z < 0 \\ \dfrac{eN_D}{\epsilon\epsilon_0}(z - w_n) & 0 < z < w_p \end{cases} \tag{4.5}$$

as is illustrated in Fig. 4.3. We have here assumed that $\epsilon_n = \epsilon_p = \epsilon$. The continuity of the field at $z = 0$ is fulfilled if

$$N_A w_p = N_D w_n. \tag{4.6}$$

This relation is compatible with the charge neutrality condition in the transition zone. If we integrate once again, now with the conditions

$$\phi(-w_p) = 0 \qquad \text{and} \qquad \phi(w_n) = \Delta\phi_0, \tag{4.7}$$

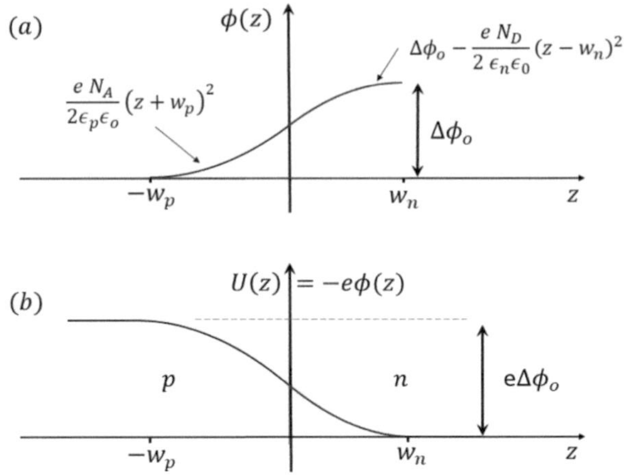

Fig. 4.4 The electric potential $\phi(z)$ (**a**) and the electric potential energy $U(z)$ (**b**). The behavior of the potential energy reflects the bending of the energy bands

we finally obtain

$$\phi(z) = \begin{cases} \dfrac{eN_A}{2\epsilon\epsilon_0}(z+w_p)^2 & -w_p < z < 0 \\ \\ \Delta\phi_0 - \dfrac{eN_D}{2\epsilon\epsilon_0}(z-w_n)^2 & 0 < z < w_n. \end{cases} \tag{4.8}$$

In Fig. 4.4 we show each of the results obtained for the transition zone. The electric potential $\phi(z)$ varies parabolically as a function of z, as seen in the top panel, in the lower panel we have the electron's potential energy $U(z) = -e\phi(z)$. The results obtained so far are important in the analysis of processes that occur in the vicinity of the *pn* junction interface. In Fig. 4.5 we illustrate the curved valence and conduction bands at the *pn* junction. In what follows, instead of depicting the full band structure shown in panel (a), we will represent only the band edges, as shown in panel (b). Electrons on the n type side, see a kind of step barrier. The same thing happens to holes on the p side. We are now in a position to evaluate the thickness of the transition region and the carrier concentrations $n(z)$ and $p(z)$ within it. We will begin by determining the thickness of this transition zone, also known as the space-charge region. The potentials' continuity at $z = 0$ implies the relation

$$\Delta\phi_0 = \frac{e}{2\epsilon\epsilon_0}\left(N_A w_p^2 + N_D w_n^2\right). \tag{4.9}$$

This potential drop is also known as the *contact potential*. From this equation and (4.6) we obtain, on the one hand

$$w_p = \left(\frac{2\epsilon\epsilon_0 N_D \Delta\phi_0}{eN_A(N_A+N_D)}\right)^{1/2}, \tag{4.10}$$

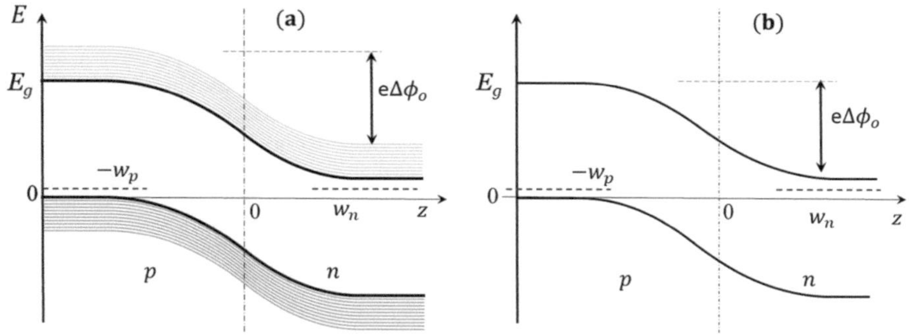

Fig. 4.5 Bands bending in the *pn* junction. In the following, instead of the band structure in (**a**) only the band edges will be drawn as shown in (**b**)

and on the other

$$w_n = \left(\frac{2\epsilon\epsilon_0 N_A \Delta\phi_0}{eN_D(N_A + N_D)}\right)^{1/2}. \tag{4.11}$$

It is evident from these equations that the higher the concentration of impurities, the narrower is the transition region. To have a quantitative idea of the widths of the transition zones, we will assume that $N_A \simeq N_D \simeq 10^{21}\,\text{m}^{-3}$, and $\epsilon = 11$. Therefore $\Delta\phi_0 = 0.41$, and $w_p + w_n \simeq 1\,\mu\text{m}$. On the other hand, if $N_A \simeq N_D \simeq 10^{23}\,\text{m}^{-3}$, the thickness is $w_p + w_n \simeq 0.1\,\mu\text{m}$. This shows that, certainly, one has a narrower transition zone with higher impurity concentrations. Keeping this information in mind, is important. The transition region that goes from $-w_p$ to w_n, is also called the *depletion layer*. The height of the step is generally denoted as $e\Delta\phi_o$. Let us now obtain $e\Delta\phi_o$ from the chemical potentials derived in the previous chapter.

In *n* and *p* type the semiconductors, the Fermi levels, measured from the edge of their respective valence bands, are μ_n and μ_p. The band bending , and the ensuing potential energy drop $e\Delta\phi_o$, are, as explained before, a consequence of the alignment of the chemical potentials μ_n and μ_p. Therefore, it is not difficult to see that, in fact, $e\Delta\phi_o$ is equal to the difference between μ_n and μ_p, before the alignment, i.e.:

$$e\Delta\phi_o = \mu_n - \mu_p. \tag{4.12}$$

Let us assume that we have both types of impurities on each side. When $N_D \gg N_A$, in the *n* type semiconductor, and $N_A \gg N_D$, in the *p* type semiconductor, the chemical potentials given in (2.58) and (2.61), written again here as

$$\mu_n \simeq E_g - k_B T \ln\frac{N_c}{N_D}, \quad \text{and} \quad \mu_p \simeq k_B T \ln\frac{N_v}{N_A}, \tag{4.13}$$

allow us to express the height of the step as

$$e\Delta\phi_o = E_g + k_B T \ln\frac{N_D N_A}{N_c N_v}. \tag{4.14}$$

If we use the law of mass action, written in the form

$$E_g = -k_B T \ \ln \frac{n_i^2}{N_c N_v}, \tag{4.15}$$

the contact potential becomes

$$\Delta\phi_o = \frac{k_B T}{e} \ln \frac{N_D N_A}{n_i^2}. \tag{4.16}$$

It is evident here that the height of the step increases with the temperature and also with the concentrations of donors and acceptors. In the transition region the height of the step is a function of z.

4.2 Carrier Concentration in the Transition Zone

We have just seen that due to the displacement of charges, the concentrations in the transition zone change. We expect that these concentration change as functions of z. To obtain the explicit functions we can use the same procedures used to obtain the intrinsic concentrations (2.3) and (2.7). The Fermi levels are given and the integration limits that correspond to the edges of the conduction and valence bands are now functions of z in the transition region. To visualize this, let us consider, as a reference, the band structure in the p type semiconductor, to the right of point $z = -w_p$, with the zero of the energy at the upper edge of the valence band. For the calculation of $n(z)$ the lower integration limit in the transition region is modified as

$$E_g \rightarrow E_g - e\phi(z), \tag{4.17}$$

while for $p(z)$ the upper limit is modified as

$$0 \rightarrow -e\phi(z). \tag{4.18}$$

With these changes, since the integrands are the same as before, we easily obtain

$$n(z) = N_c e^{(\mu_p - E_g)/k_B T} e^{e\phi(z)/k_B T} = n_p \, e^{e\phi(z)/k_B T} \tag{4.19}$$

and

$$p(z) = N_v e^{-\mu_p/k_B T} e^{-e\phi(z)/k_B T} = p_p \, e^{-e\phi(z)/k_B} T. \tag{4.20}$$

Here n_p and p_p are, respectively, the minority and majority concentrations in the p type semiconductor. These values are obtained when $n(z)$ and $p(z)$ are evaluated at $z = -w_p$, where $\phi = 0$. On the other hand, if we evaluate at $z = w_n$ where $\phi = \Delta\phi_0$, and recalling that $\mu_n = \mu_p + e\Delta\phi_0$, we have

$$n(w_n) = n_p \, e^{e\Delta\phi_0/k_B T} = N_c e^{(\mu_n - E_g)/k_B T} = n_n \tag{4.21}$$

and

$$p(w_n) = p_p\, e^{-e\Delta\phi_0/k_BT} = N_v e^{-\mu_n/k_BT} = p_n. \tag{4.22}$$

It is clear that these functions represent the majority n_n and minority p_n concentrations in the n type semiconductor. We can see, using (4.21) and (4.22), that the relation between the minority and majority electron concentrations is

$$n_p = n_n\, e^{-e\Delta\phi_0/k_BT}. \tag{4.23}$$

Similarly, one can show that the relation between the minority and majority hole concentrations is

$$p_n = p_p\, e^{-e\Delta\phi_0/k_BT} \tag{4.24}$$

4.3 Relation of Concentrations at Different Points

In the previous section, the concentrations $n(z)$ and $p(z)$ were obtained in the region in which the electric potential ϕ and the potential energy $-e\phi$ vary as functions of z. The relationships (4.19) and (4.20) can be used to obtain a relation between concentrations at different points, within the transition region of the system. With

$$n_1 = n(z_1) = n_p e^{e\phi(z_1)/k_BT} \quad \text{and} \quad n_2 = n(z_2) = n_p e^{e\phi(z_2)/k_BT} \tag{4.25}$$

it is clear that

$$n(z_2)e^{-e\phi(z_2)/k_BT} = n(z_1)e^{-e\phi(z_1)/k_BT}. \tag{4.26}$$

If $U_1 = -e\phi(z_1)$ and $U_2 = -e\phi(z_2)$, and $\Delta U_{21} = U_2 - U_1 = -e\Delta\phi_{12}$, the last relation can be written as

$$n(z_2)e^{U_2/k_BT} = n(z_1)e^{U_1/k_BT}, \tag{4.27}$$

and also in the form

$$n_2 = n_1 e^{\Delta\phi_{21}/\phi_T}. \tag{4.28}$$

Here ϕ_T is the thermal potential.

4.4 Minority Concentrations Outside the Transition Zone

An important result of the previous section is the existence of minority concentrations n_p and p_n at the edges of the transition region and, naturally, beyond the edges. We will see now that these concentrations vary as one moves away from the transition zone. To determine how they change with z, we have to consider the drift and diffusion currents, responsible of any change in these regions.

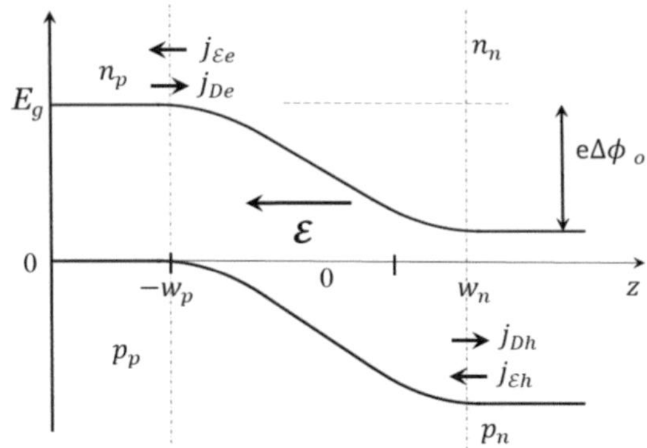

Fig. 4.6 Diffusion and drift currents at the left and right boundaries of the transition zone

4.4.1 Diffusion and Drift Currents in Thermodynamic Equilibrium

Under conditions of thermodynamic equilibrium, the diffusive and drift currents of electrons and holes cross the junction in both directions and balance each other in such a way that the net current is zero. We will analyze here the electron currents and obtain results for electron concentrations. For holes the results are analogous. To determine the currents that move in one direction or another, let us stand on the side p of the junction (to the left of the transition zone in Fig. 4.6), at $z = -w_p$. At this point the minority concentration is n_p. Part of the electrons that reach this point come from the intrinsic concentration and another part from the electrons that crossed the interface of the pn junction. The electrons in the transition zone are attracted by the electric field with a drift current $j_{\mathcal{E}e}$ from right to left, as indicated in Fig. 4.6. In addition, there is a diffusion current that responds to the concentration gradient. If the electrons move with average speed $\langle v \rangle \sim L_e/\tau_e$, where L_e is the diffusion length, $\tau_e \equiv L_e^2/D_e$ is the diffusion time, and A is the cross section, the intensity of the drift current of the minority electrons is

$$I_{n0}^{\mathcal{E}} = en_p \frac{L_e}{\tau_e} A = en_p \frac{D_e}{L_e} A. \tag{4.29}$$

This current is numerically equal to the diffusion current $\mathbf{j}_{Dq} = qD_e(-\nabla n)$ that tends to move the charges towards the region of lower concentration. With the charge $q = -e$, we have the diffusion current of electrons

$$I_{n0}^{D} = eD_e \frac{\partial n_p(z)}{\partial z} A. \tag{4.30}$$

which, as indicated in Fig. 4.6, moves electrons from right to left. In thermodynamic equilibrium the net electron current is zero at every point z. We can use this condition to obtain the concentration $n_p(z)$ of minority carriers outside the transition zone. At any point z to the left of $-w_p$, the following equality should be satisfied

$$eD_e \frac{\partial n_p(z)}{\partial z} A = en_p(z) \frac{L_e}{\tau_e} A. \tag{4.31}$$

On the left side we have the diffusion current and on the right the drift current. Simplifying, the previous equation can be rewritten in the form

$$\frac{\partial n_p(z)}{\partial z} = n_p(z)\frac{1}{L_e}. \tag{4.32}$$

We integrate this differential equation from $z < -w_p$ to $z = -w_p$ (where $n(-w_p) = n_p$), and we obtain the minority electron concentration

$$n_p(z) = n_p e^{(z+w_p)/L_e} \quad \text{for} \quad z < -w_p. \tag{4.33}$$

This concentration diminishes as we move away from the transition zone.

Similarly, the hole drift and diffusion currents, through a surface A (perpendicular to z) located at the point $z = \omega_n$, defined by the expressions

$$I^{\mathcal{E}}_{p0} = ep_n\frac{L_h}{\tau_h}A = ep_n\frac{D_h}{L_h}A, \tag{4.34}$$

and

$$I^{D}_{p0} = -eD_h\frac{\partial p_n(z)}{\partial z}A, \tag{4.35}$$

sum to zero under equilibrium conditions, and satisfy the differential equation

$$\frac{\partial p_n(z)}{\partial z} = p_n(z)\frac{1}{L_h}. \tag{4.36}$$

Integrating with $p_n(w_n) = p_n$ at the end of the transition zone gives the minority concentration of holes as

$$p_n(z) = p_n e^{-(z-w_n)/L_h} \quad \text{for} \quad z > w_n. \tag{4.37}$$

This concentration also diminishes as one moves away from the transition zone.

The currents $I^{\mathcal{E}}_{n0}$ and $I^{\mathcal{E}}_{p0}$ contribute to the total drift current

$$I^{\mathcal{E}}_0 = I^{\mathcal{E}}_{n0} + I^{\mathcal{E}}_{p0} = en_p\frac{D_e}{L_e}A + ep_n\frac{D_h}{L_h}A \tag{4.38}$$

which is compensated by the total diffusion current I^D_0, of equal magnitude but opposite direction, so that in equilibrium the total current will be

$$I_0 = I^{\mathcal{E}}_0 + I^D_0 = 0 \tag{4.39}$$

4.5 Properties of Polarized *pn* Junctions

If a potential difference is applied between a point z_L on the p side and another point z_R on the n side of the pn junction, changes occur in the profiles of the band edges. There are two possible configurations of the external potential source. One is called *direct polarization* or forward bias, with the positive pole connected to the p type semiconductor, and the other, the *inverse polarization* or reverse bias, with the positive pole connected to the n type semiconductor. Let us assume that the applied potential is $V = V(z_L) - V(z_R)$, the potential energy of the electrons that are in $-w_p$ undergoes a change equal to

$$\Delta U_e = -eV, \tag{4.40}$$

and consequently, the original potential-energy drop $e\Delta\phi_o$ experiences the transformation:

$$e\Delta\phi_o \quad \rightarrow \quad e\Delta\phi_o^{'} = e\Delta\phi_o - eV. \tag{4.41}$$

If the applied potential V is positive, i.e. a direct polarization, the height of the potential barrier that the electrons "see" from the n side decreases, as seen in Fig. 4.7a; but if $V < 0$, i.e. the inverse polarization, the height of the barrier increases, as seen in Fig. 4.7b. The physical variables expressed as a function of $\Delta\phi_0$, change according to the transformation (4.41). The widths of the transition zone are also transformed as follows:

$$w_n \quad \rightarrow \quad w_{nV} = \left[\frac{2\epsilon\epsilon_0}{eN_D}(\Delta\phi_o - V)\frac{N_A}{N_D + N_A}\right]^{\frac{1}{2}} \tag{4.42}$$

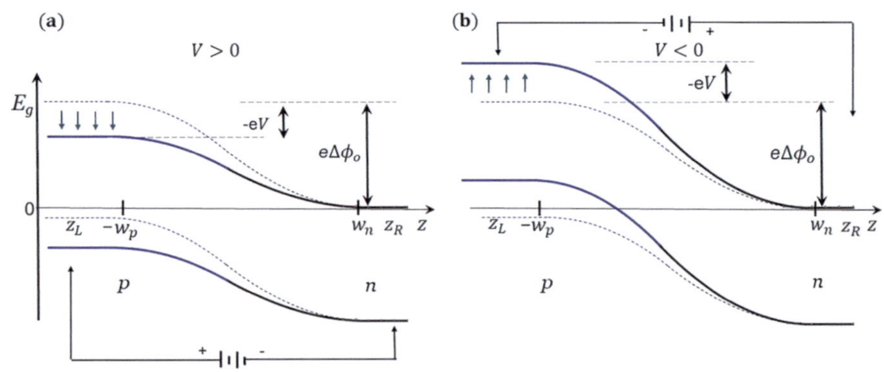

Fig. 4.7 The effect of connecting the pn junction to a battery: in **a** the connection is referred to as the forward or direct bias ($V > 0$ on the p side) and in **b** as the reverse bias ($V < 0$ on the p side). In forward bias the barrier flattens, and a greater number of minority carriers n_p arrive at the p side, and p_n carriers at the n side. In reverse bias the height of the barrier increases and the minority concentrations decrease

and

$$w_p \quad \rightarrow \quad w_{pV} = \left[\frac{2\epsilon\epsilon_0}{eN_A}(\Delta\phi_o - V)\frac{N_D}{N_A + N_D}\right]^{\frac{1}{2}}. \tag{4.43}$$

In forward bias ($V > 0$), the widths decrease, while in reverse bias ($V < 0$), they increase. The expressions for the minority concentrations are transformed as follows:

$$n_p \quad \rightarrow \quad n_{pV} = n_n\ e^{(-e\Delta\phi_o + eV)/k_BT} = n_p\ e^{eV/k_BT} \tag{4.44}$$

and

$$p_n \quad \rightarrow \quad p_{nV} = p_p\ e^{(-e\Delta\phi_o + eV)/k_BT} = p_n\ e^{eV/k_BT}. \tag{4.45}$$

If $V > 0$ (forward polarization), the minority concentrations increase exponentially; in reverse bias decrease exponentially. This are also key results that should be kept in mind. The majority concentrations remain practically unmodified by the external potential. The extreme sensitivity of minority populations to external potential is an important and transcendent property, which has allowed semiconductor structures to be used in diverse applications, without the need for large applied potentials. Variations in the transition zone widths and minority concentrations are fundamental in the operation of the *pn* system as a *light-emitting diode*. In the following sections we will discuss this issue. If the minority concentrations n_p and p_n at $z = -w_p$ and $z = w_n$, respectively, change when a potential is applied, there will also be changes in the concentrations outside the transition zone and, of course, also in the transition zone itself. In the external region, we have

$$n_p(z) \quad \rightarrow \quad n_{pV}(z) = n_p e^{(z+w_p)/L_e} e^{eV/k_BT} \quad \text{for} \quad z < -w_p \tag{4.46}$$

and

$$p_n(z) \quad \rightarrow \quad p_{nV}(z) = p_n e^{-(z-w_n)/L_h} e^{eV/k_BT} \quad \text{for} \quad z > w_n. \tag{4.47}$$

For the transition zone, the concentrations $n(z)$ and $p(z)$ in the (4.19) and (4.20) become

$$n(z) \quad \rightarrow \quad n_V(z) = n_p\, e^{e\left(\phi(z)+\phi_V(z)\right)/k_BT} \quad \text{for} \quad -w_p < z < w_n, \tag{4.48}$$

and

$$p(z) \quad \rightarrow \quad p_V(z) = p_p\, e^{-e\left(\phi(z)+\phi_V(z)\right)/k_BT} \quad \text{for} \quad -w_p < z < w_n. \tag{4.49}$$

The electric potential $\phi(z)$ in the transition region and the external potential $\phi_V(z)$ outside this region, are respectively

$$\phi(z) = \begin{cases} \dfrac{eN_A}{2\epsilon\epsilon_o}(z + w_p)^2 & -w_p < z < 0 \\ \\ \Delta\phi_o - \dfrac{eN_D}{2\epsilon\epsilon_o}(z - w_n)^2 & 0 < z < w_n \end{cases}, \tag{4.50}$$

and

$$\phi_V(z) = V\frac{w_R - z}{w_L + w_R}. \qquad (4.51)$$

Here $-w_L$ and w_R are the coordinates of the contact points on the left and right of the transition zone.

The purpose of applying a potential difference is not only to modify the minority populations, it is also to maintain a current through the junction and the process taking place, for instance the generation-recombination (or generation-annihilation) of *e-h* pairs, and consequently the light emission, a characteristic phenomenon of *p-n* junction.

Before analyzing the generation and recombination processes in the presence of an external potential and minority populations outside the transition region, we will consider the currents through the junction *pn*, in the presence of an external field.

4.6 The *pn* Junction as a Current Rectifier: Rectifier Diode

We saw in Sect. 4.4.1 that in thermodynamic equilibrium and in the absence of an external potential, electron currents by diffusion $I_{n0}^D \equiv \mathbf{j}_{nD} \cdot \mathbf{A}$ and by drift $I_{n0}^{\mathcal{E}} \equiv \mathbf{j}_{n\mathcal{E}} \cdot \mathbf{A}$, compensate and, consequently, the total current I_{n0} is zero. Similarly, diffusion hole currents $I_{h0}^D \equiv \mathbf{j}_{hD} \cdot \mathbf{A}$ and hole drift currents $I_{h0}^{\mathcal{E}} \equiv \mathbf{j}_{h\mathcal{E}} \cdot \mathbf{A}$, compensate, and so I_{p0} is also zero. However, in the presence of an external potential, we will have a net current $I(V)$ related to additional concentrations, δn and δp, of electrons and holes injected to each side of the junction, due to the increase of minority charges when a potential V is applied. Before obtaining $I(V)$ we will determine these concentrations, outside the transition zone. The net currents are fundamentally related to the variation of minority concentrations in the external regions. The currents will now depend not only on z but also on V. Minority concentrations outside the transition region were obtained earlier. The expressions for the excess minority concentrations δn_p and δp_n are similar to those obtained in Sect. 4.4.1 for n_p and p_n. If we use (4.32) for $\delta n_p(z)$ instead of $n_p(z)$, we have

$$\frac{\partial \delta n_p(z)}{\partial z} = \frac{\delta n_p(z)}{L_e}, \qquad (4.52)$$

whose solution is

$$\delta n_p(z) = \delta n_p(-w_p)e^{(z+w_p)/L_e} \quad \text{for} \quad z \leq -w_p. \qquad (4.53)$$

For the factor $\delta n_p(-w_p)$ we use the relationship

$$\delta n_p(-w_p) = n_{pV}(-w_p) - n_p(-w_p) = n_p\big(e^{eV/k_BT} - 1\big). \qquad (4.54)$$

Substituting, we have

$$\delta n_p(z) = n_p\big(e^{eV/k_BT} - 1\big)e^{(z+w_p)/L_e} \quad \text{for} \quad z < -w_p. \qquad (4.55)$$

Similarly, we have

$$\delta p_n(z) = p_n\left(e^{eV/k_BT} - 1\right)e^{-(z-w_n)/L_h} \qquad \text{for} \qquad z > w_n. \tag{4.56}$$

These are important results that allow us to easily obtain the current densities $\mathbf{j}_n = eD_e\nabla\delta n_p$ and $\mathbf{j}_p = -eD_h\nabla\delta p_n$ and with them the currents

$$I_n(z) = \mathbf{j}_n(z)\cdot\mathbf{A} = \frac{eD_e n_p A}{L_e}\left(e^{eV/k_BT} - 1\right)e^{(z+w_p)/L_e}, \tag{4.57}$$

and

$$I_p(z) = \mathbf{j}_p(z)\cdot\mathbf{A} = \frac{eD_h p_n A}{L_h}\left(e^{eV/k_BT} - 1\right)e^{-(z-w_n)/L_h}. \tag{4.58}$$

These currents are defined in different regions. The first is for $z < w_p$ and the second is for $z > w_n$. If we evaluate them at the edges of the transition zones, we have the currents of electrons and holes that cross the interface of the n-p junction, which by the principle of conservation of flux must be equal to the total current at any other point. These currents meet this condition only if the widths of the transition zone are small compared to the diffusion lengths. The currents $I_n(-w_p)$ and $I_p(w_n)$ are

$$I_n = \frac{eD_e n_p A}{L_e}\left(e^{eV/k_BT} - 1\right) \tag{4.59}$$

and

$$I_p = \frac{eD_h p_n A}{L_h}\left(e^{eV/k_BT} - 1\right). \tag{4.60}$$

Consequently, the total current through the junction, due to the minority carrier currents, is

$$I(V) = I_0\left(e^{eV/k_BT} - 1\right), \tag{4.61}$$

with

$$I_0 = \left(\frac{eD_e n_p A}{L_e} + \frac{eD_h p_n A}{L_h}\right), \tag{4.62}$$

If we remember the relations $n_p N_A = n_i^2$ and $p_n N_D = n_i^2$, we can write this current in the form

$$I_0 = en_i^2 A\left(\frac{D_e}{L_e N_A} + \frac{D_h}{L_h N_D}\right), \tag{4.63}$$

which can also be expressed as follows

$$I_0 = eN_c N_v A e^{-\frac{E_g}{k_BT}}\left(\frac{D_e}{L_e N_A} + \frac{D_h}{L_h N_D}\right). \tag{4.64}$$

The current in (4.61) has, as a function of the potential V, the behavior shown in the Fig. 4.8. In direct polarization ($V > 0$) the current grows exponentially, and if the polarization is reversed, the current tends to $-I_0$. This asymmetric behavior is what allows the use of this device to rectify alternating currents and to obtain a direct current. It is common to call diode to any device that allows the current to flow in only one direction. The diode is represented in circuit diagrams by the symbol shown in the figure. We will see later other devices with similar properties. One of those is the Zener Diode, another is the Schottky Diode. The actual behavior is somewhat more complex than that described by (4.61) and even differs from some of the trends predicted by this expression. Some of the hypotheses that exist to explain these differences are[1]:

1. The calculation that leads to $I = I_0\left(e^{\frac{eV}{k_B T}} - 1\right)$ ignores processes of recombination and generation of carriers in the transition zone, generally mediated by the so-called deep levels, which are states, due mainly to crystal imperfections, located in the gap and capable of capturing both electrons and holes. Depending on the times associated with the capture and emission processes, there are processes whose final result is a recombination of the electron from the CB with a hole from the VB, or processes in which the deep level emits an electron back to the CB before capturing a hole, etc.
2. To determine the expression of the current, the contribution of the drift currents of the majority concentrations, the ohmic currents and the currents in the regions far from the transition zone were neglected, under the assumption that the potentials are moderate, and the potential gradients are negligible. However, when the potential grows, those contributions should not be ignored. The potential drop in semiconductors can become significant with important effects on the current that could grow to high values. In other words, the effect of the resistance of the semiconductors on the value of the current must be considered. Furthermore, as the negative potential increases, the current tends to a saturation value.
3. Another effect that it is not considered in (4.61) is the sudden growth of the electron population due to the Zener effect [1] that manifests itself in reverse bias when the magnitude of the potential is such that (see Fig. 4.9)

$$e\Delta\phi_0 - eV > E_g \tag{4.65}$$

   In this case, the energy of the upper edge of the valence band of the p type semiconductor is greater than that of the lower edge of the n type conduction band, see Fig. 4.9. The electrons of the valence band of the p-type region are energetically capable of tunneling to the conduction band of the n-type region. When this occurs there is a sudden growth of the negative current, which is not contemplated by the previous equations. This phenomenon is known as the reverse breakdown or Zener breakdown.

[1] It is worth noticing that the literature and the number of specialized books on these topic is abundant, and a detailed discussion is beyond the purpose of this book.

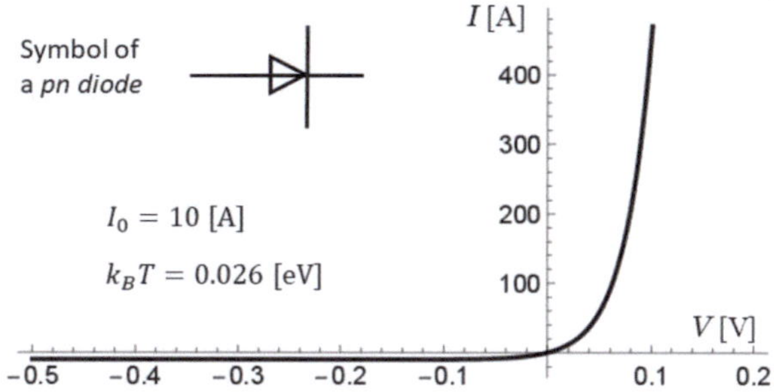

Fig. 4.8 The current-potential $I - V$ characteristic across the pn junction. The exponential growth with $V > 0$ is very fast even when the temperature is $T \simeq 300$ K. For this graph $I_0 = 10$ A, which is the limiting value of the current when $V < 0$. This system, as part of a circuit, is a diode represented by the symbol shown here

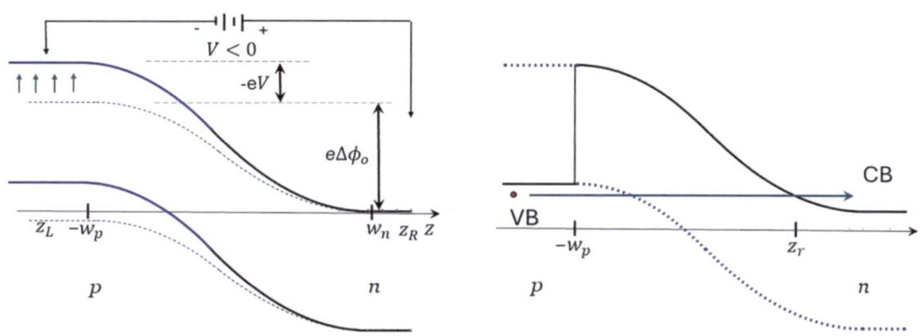

Fig. 4.9 In the reverse polarization of the Fig. 4.7b, the upper edge of the valence band (VB) is above the lower edge of the conduction band (CB), the electrons on the p side tunnel to n through the barrier highlighted in the right panel. This barrier is known as the Esaki barrier, and the quantum tunneling effect at the pn interface as the Zener effect [1]

4.6.1 Current in Short Diodes, and of Arbitrary but Finite Length

So far it has been assumed that the lengths of the semiconductors are much greater than the diffusion lengths L_e and L_h, and for this reason, some terms in the solutions of the second-order differential equations were canceled. However, most of the systems are really short in length and there is no risk of divergency in the exponential functions of the solutions. Therefore, to be more accurate, all solutions must be considered. To avoid confusions, let us discuss separately the case of short diodes and that of diodes with arbitrary length.

4.6.1.1 Short Diodes

In some devices, in particular bipolar transistors, the length of the device is smaller than the diffusion lengths. In these diodes, the ohmic boundary condition must be considered. According to this condition the minority concentrations tend linearly to zero at the contacts, that means

$$\delta n_p(z_L) = 0, \qquad \delta p_n(z_R) = 0. \tag{4.66}$$

If the diffusion current of minority electrons, on the p side, is given by

$$I_n(z) = eD_eA\frac{\partial\left(\delta n_p(z)\right)}{\partial z}, \tag{4.67}$$

we can also write this equation as

$$I_n(z) = eD_eA\frac{\delta n_p(z+\Delta z) - \delta n_p(z)}{\Delta z}, \tag{4.68}$$

and evaluating at $z = -w_p$ and $z + \Delta z = z_L$, we have

$$I_n(-w_p) = eD_eA\frac{\delta n_p(z_L) - \delta n_p(-w_p)}{z_L + w_p}. \tag{4.69}$$

If we now take into account the boundary condition (4.66), the previous equation reduces to

$$I_n(-w_p) = -\frac{eD_eA}{z_L + w_p}\delta n_p(-w_p). \tag{4.70}$$

By substituting the expression for $\delta n_p(-w_p)$ in (4.54), this current becomes

$$I_n(-w_p) = -\frac{eD_eA}{z_L + w_p}n_p\left(e^{eV/k_BT} - 1\right). \tag{4.71}$$

Similarly, we have

$$I_p(w_n) = \frac{eD_hA}{z_R - w_n}p_n\left(e^{eV/k_BT} - 1\right). \tag{4.72}$$

So, the total current in the short diode is

$$I_c(V) = \left(-\frac{eD_eA}{z_L + w_p}n_p + \frac{eD_hA}{z_R - w_n}p_n\right)\left(e^{eV/k_BT} - 1\right). \tag{4.73}$$

Comparing with (4.61) we see that in a short diode the coefficient I_0 transforms into

$$I_{0c} = -\frac{eD_eA}{z_L + w_p}n_p + \frac{eD_hA}{z_R - w_n}p_n. \tag{4.74}$$

4.6.1.2 Arbitrary Length Diodes

If the length of the semiconductors in the diode is finite, and there is no risk that the solution that grows exponentially diverges, some expressions of the concentrations outside the zone of transition and also the currents evaluated at the edges of this zone, will change. Let us assume that, in general, an external potential is applied. To obtain the concentration injected by the presence of the external potential, we start from the diffusion equation

$$\frac{\partial^2 \delta p_n(z)}{\partial z^2} - \frac{\delta p_n(z)}{D_h \tau_h} = 0 \qquad w_n < z < z_R, \tag{4.75}$$

whose general solution is

$$\delta p_n(z) = c_1 e^{-z/L_h} + c_2 e^{z/L_h}. \tag{4.76}$$

The coefficients c_1 and c_2 are easily determined if we use the boundary conditions

$$\delta p_n(w_n) = p_n(e^{eV/k_B T} - 1) \quad \text{and} \quad \delta p_n(z_R) = 0, \tag{4.77}$$

the concentration in (4.76) takes the form

$$\delta p_n(z) = \frac{\sinh\left[(z_R - z)/L_h\right]}{\sinh\left[(z_R - w_n)/L_h\right]} \delta p_n(w_n). \tag{4.78}$$

Given this carrier concentration and defining the current

$$I_p(w_n) = -\, e A D_h \left.\frac{\partial\big(\delta p(z)\big)}{\partial z}\right|_{z=w_n}, \tag{4.79}$$

the hole current becomes

$$I_p(w_n) = eA\frac{D_h}{L_h} p_n \coth\Big(\frac{z_R - w_n}{L_h}\Big)\big(e^{eV/k_B} - 1\big). \tag{4.80}$$

Therefore, the total current

$$I = I_p(w_n) + I_n(-w_p) \tag{4.81}$$

is expressed, for systems of arbitrary but finite length, in the form

$$I = eA\left(\frac{D_h p_n}{L_h} \coth\Big(\frac{z_R - w_n}{L_h}\Big) + \frac{D_e n_p}{L_e} \coth\Big(\frac{z_L - w_p}{L_e}\Big)\right)\big(e^{eV/k_B T} - 1\big). \tag{4.82}$$

Note that when the diode is made narrow, such that $z_R - w_n \ll L_h$ and $z_L + w_p \ll L_e$, this reduces to (4.74).

4.6.2 Esaki Barrier and Zener Effect, the Zener Diode

Very shortly after the invention of the transistor, Esaki and others [2] from the Sony company showed, in 1957–1958, that a *pn* diode in reverse bias reaches a point where the edge of the valence band on the *p* side appears at higher energy than the conduction band edge and the electrons can cross the (Esaki) barrier, thanks to the quantum tunneling effect. C. M. Zener predicted the emergence of currents when the fields are intense [1], hence, the effect bears his name. The effect is particularly intense when the semiconductors are heavily doped because the width of the Esaki barrier is smaller and the probability of tunneling, given approximately by

$$T \simeq e^{-2\int_{-w_p}^{z_r} q(z)dz} = e^{-2\int_{-w_p}^{z_r} \sqrt{(2m/\hbar^2)\left(U(z)-E\right)}dz}, \tag{4.83}$$

increases. The Zener diode became the first device whose operation was explained by an effect of quantum nature. The impact of this system on device development and the understanding of the physics of semiconductors and solid-state electronics was profound. A few years later, the metal-oxide-semiconductor structure (MOS), that was introduced in the 1950s, also became an alternative device. This device allows the formation, at the semiconductor-oxide interface (at the discontinuity of the conduction band), of a quasi-two-dimensional gas of electrons capable of moving parallel to the interface. Simultaneously, thanks to the bending of the bands, the formation of a confining potential in a single heterojunction. This system is the basis of the most widely used device in electronic circuits, the MOSFET (for MOS field-effect transistor). The recognition that band discontinuity could be exploited to create confinement potentials is possibly the most significant discovery in the history of semiconductor devices. We will return to MOSs and MOSFETs in later chapters.

4.7 Generation-Recombination and the Continuity Equation

We mentioned before that in thermodynamic equilibrium the rate of pair generation *e-h* is equal to that of its annihilation. However, if a potential difference is maintained that constantly supplies electrons to the band of conduction at one end and extracts them (from the valence band) at the other, a stationary situation is established, on the one hand, in the minority populations, and on the other, an increase in the rate of recombination generation that favors the application of this system as a light-emitting diode.

When minority concentrations increase, as (4.46) and (4.47) indicate, a state of equilibrium is also reached, both on the left and on the right sides of the *pn* interface. For application purposes it is convenient to increase the minority concentrations and transform them into the illumination intensity.

The generation depends fundamentally on the population of electrons in the valence band and the temperature. Recombination, on the other hand, depends on the populations of electrons excited and supplied to the conduction band, and on the temperature. As mentioned before, the electron population in the valence band is insensitive to external potentials. If we denote as $g_0(T)$ the rate of pair generation, and at the p side the recombination rate is $r(T) = a_{pV}(T)n_p$, we can then express the change in electron concentration by unit of time as

$$\left.\frac{\partial n}{\partial t}\right|_g - \left.\frac{\partial n}{\partial t}\right|_r = g_0(T) - a_{pV}(T)n_p. \tag{4.84}$$

To obtain $a_{pV}(T)$, we use the following argument. In equilibrium, when $V = 0$, the recombination rate must reduce to $a_{p0}(T) = g_0(T)/n_p(z)$. This suggests that when $V \neq 0$ the recombination rate a_{pV} can be expressed as

$$a_{pV} = \frac{g_0(T)}{n_p} + \frac{e^{eV/k_BT} - 1}{\tau_n}, \tag{4.85}$$

so that if we substitute in (4.84), the change of the electron concentration associated with the generation-recombination process takes the form

$$\left.\frac{\partial n}{\partial t}\right|_g - \left.\frac{\partial n}{\partial t}\right|_r = -\frac{n_p\left(e^{eV/k_BT} - 1\right)}{\tau_n} = -\frac{n_{pV}(z) - n_p(z)}{\tau_n} = -\frac{\delta n_p(z, V)}{\tau_n}. \tag{4.86}$$

Similarly, for holes we have

$$\left.\frac{\partial p}{\partial t}\right|_g - \left.\frac{\partial p}{\partial t}\right|_r = -\frac{p_{nV}(z) - p_n(z)}{\tau_p} = -\frac{\delta p_n(z, V)}{\tau_p}, \tag{4.87}$$

These equations show that in the presence of a external potential the balance between the recombination rate and the generation rate is broken and is proportional to the variation in the concentration of minority carriers, and inversely proportional to the recombination time. When the polarization is direct, $n_{pV}(z) > n_p(z)$ and also $p_{nV}(z) > p_n(z)$. This means that recombinations dominate in forward polarization. For these processes to remain stationary in the presence of external potentials, it is necessary that the time derivatives of the concentrations, in the continuity equations (3.98) and (3.99), cancel out, and therefore that the currents and the generation-recombination rates satisfy the following equations

$$-\frac{1}{e}\nabla \cdot \mathbf{j}_n = -\frac{\delta n_p(z, V)}{\tau_n} \tag{4.88}$$

and

$$\frac{1}{e}\nabla \cdot \mathbf{j}_p = -\frac{\delta p_n(z, V)}{\tau_p}. \tag{4.89}$$

These steady-state operation requirements depend especially on the generation-recombination processes in the transition zone. Given these conditions, we will derive the generation-recombination currents.

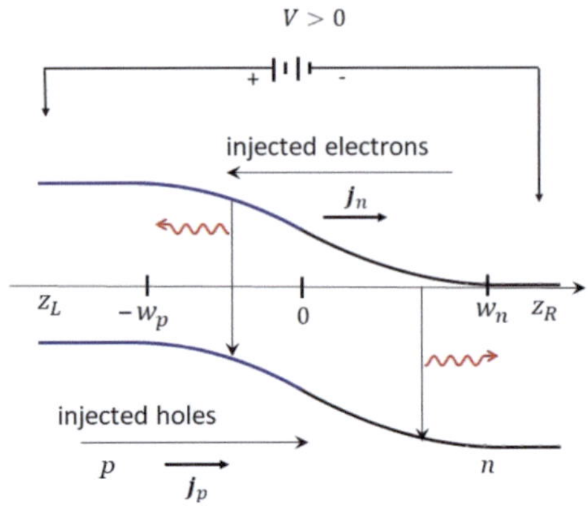

Fig. 4.10 The pn junction in forward bias. The injected electrons and holes contribute to the currents $\mathbf{j}_n$ and $\mathbf{j}_p$. Part of these currents passes through the system and part is recombined

4.7.1 Injection and Generation-Recombination Currents

Relations (4.88) and (4.89) define the currents that must be injected to maintain a stationary generation-recombination process in a pn diode that can not only be used as a current rectifier but also as a light emitter.

To simplify a little the expressions let us define the generation-recombination rates

$$r^n_{gr}(z, V) = \left.\frac{\partial n}{\partial t}\right|_g - \left.\frac{\partial n}{\partial t}\right|_r = -\frac{\delta n_p(z, V)}{\tau_n} \tag{4.90}$$

and

$$r^p_{gr}(z, V) = \left.\frac{\partial p}{\partial t}\right|_g - \left.\frac{\partial p}{\partial t}\right|_r = -\frac{\delta p_n(z, V)}{\tau_p}. \tag{4.91}$$

and the stationary operating equations are written now as

$$-\frac{1}{e}\nabla \cdot \mathbf{j}_n = r^n_{gr}(z, V) \quad \text{and} \quad \frac{1}{e}\nabla \cdot \mathbf{j}_p = r^p_{gr}(z, V). \tag{4.92}$$

If we integrate over the entire volume in which the generation-recombination process occurs and use the divergency theorem we have

$$\frac{1}{e}\oint \mathbf{j}_n \cdot d\mathbf{a} = -A\int dz\, r^n_{gr}(z, V) \quad \text{and} \quad \frac{1}{e}\oint \mathbf{j}_p \cdot d\mathbf{a} = A\int dz\, r^p_{gr}(z, V). \tag{4.93}$$

Here A is the cross-sectional area of the diode. The integration domain in the z direction of the first equation is the domain where electrons are minority, i.e. $-w_p < z < 0$. In the second equation the integration domain is $0 < z < w_n$. The surfaces $d\mathbf{a}$ are oriented outside the corresponding volumes. The currents are as shown in Fig. 4.10. If the generation-recombination rates are approximately constant in the integration interval of z, which we will assume is in the transition zone (between

$-w_p$ and w_n), and the currents that flow through the lateral surfaces are zero, we have

$$\frac{A}{e}\big(j_n(0) - j_n(-w_p)\big) = -Aw_p\, r^n_{gr}(z_0, V) \tag{4.94}$$

and

$$-\frac{A}{e}\big(j_p(w_n) - j_p(0)\big) = -Aw_n\, r^p_{gr}(z_0, V). \tag{4.95}$$

In each of these equations, the current differences, in the left hand sides, are equal to the currents in right hand sides defined by the generation-recombination rates $r^n_{gr}(z, V)$ and $r^p_{gr}(z, V)$. If we add these equations we have

$$\big(j_n(0) + j_p(0)\big)A = \big(j_n(-w_p) + j_p(w_n)\big)A - eA\big(w_p\, r^n_{gr}(z_0, V) + w_n\, r^p_{gr}(z_0, V)\big) \tag{4.96}$$

On the left hand side, the sum $A\big(j_n(0) + j_p(0)\big)$ is the total current $I_T = I_{Tn} + I_{Tp}$, that passes through the surface at z_0. By conservation of flux it is equal to the current passing through a similar surface at any other point. On the right hand side we have, in the first term, the total injected current $I_I = A\big(j_n(-w_p) + j_p(w_n)\big) = I_{In} + I_{Ip}$, and, in the second term, the generation-recombination current $I_{gr} = I_{grp} + I_{grn}$. In a compact representation, we have

$$I_T = I_I - I_{gr} \tag{4.97}$$

with

$$I_{gr} = eA\big(w_p\, r^n_{gr}(z_0, V) + w_n\, r^p_{gr}(z_0, V)\big) \tag{4.98}$$

the generation-recombination current.

4.7.2 Injection Efficiency

In previous sections (see (4.57) and (4.58)) the currents $I_n(z) = j_n(z)A$ and $I_p(z) = j_p(z)A$ were obtained of minority populations. If they are evaluated at $z = -w_p$ and $z = w_n$ we have the injection currents

$$j_n = \frac{eD_e n_p}{L_e}\big(e^{eV/k_BT} - 1\big) \tag{4.99}$$

and

$$j_p = \frac{eD_h p_n}{L_h}\big(e^{eV/k_BT} - 1\big). \tag{4.100}$$

Furthermore, we saw above that in the transition zone, whose width is $w_T = w_p + w_n$, there are radiative generation-recombination processes that correspond to the generation-recombination current of (4.98) and to the current of (A.8), in the

Appendix A, obtained with the Shockley, Read and Hall model. In this case, the generation-recombination current takes the form

$$j_{gr} = \frac{e n_i w_T}{2\tau_{nr}} \left(e^{eV/2k_B T} - 1 \right), \tag{4.101}$$

where τ_{nr} is the recombination time and n_i the intrinsic concentration.

In the construction of a light-emitting diode (LED), in which the current densities satisfy the relation $j_n > j_p$, the emission is predominantly from the p region. Therefore, in a device covered with some protective layer, the layer p is left as close as possible to the surface while the side n is isolated and at a greater depth, as shown in Fig. 8.1. The quotient of the current j_n and the total current $j_T = j_n + j_p + j_{gr}$ defines the injection efficiency

$$\gamma_{inj} = \frac{j_n}{j_n + j_p + j_{gr}}. \tag{4.102}$$

To maximize the value of j_n the concentration n must be increased[2] to n^+. On a pn^+ device, $n_p \gg p_n$. A system in which j_{gr} is small is also desirable. In previous sections we saw that the higher the concentration, the smaller the width of the transition zone. For the same reason, the width w_T in a pn^+ diode is smaller than that in a pn diode.

4.8 Solved Problems

Some characteristic parameters useful for solving problems are given in the following tables. In all cases the parameters are for $T = 300\,\text{K}$.

Table 4.1 gives the recombination times defined as $\tau_i \propto \dfrac{n_i}{r}$, for high purity semiconductors, intrinsic concentration n_i and the capture section for some semiconductors (Table 4.2).

Here τ_n^*, τ_p^* are the relaxation times of the linear momentum. Some values of these times are given in Table 4.3.

Table 4.1 Recombination time, concentration and capture section

Semiconductor	Gap	τ_i (ms) high purity	n_i (cm^{-3})	σ_n equiv. (cm^2)
Si	Indirect	$1 - 10$	1.5×10^{10}	10^{-15} to 10^{-12}
Ge	Indirect	$0.1 - 1$	2.4×10^{13}	10^{-14} to 10^{-12}
GaP	Direct	10^{-5}	2.7	10^{-13} to 10^{-10}
GaAs	Direct	10^{-5}-5×10^{-5}	1.8×10^{6}	10^{-14} to 10^{-10}
InP	Direct	10^{-6}– 10^{-5}	8.2×10^{6}	10^{-13} to 10^{-10}
InSb	Direct	10^{-6}– 10^{-5}	1.6×10^{16}	10^{-14} to 10^{-11}

[2] Generally, to denote high concentrations, the sign + is added, for example n^+.

Table 4.2 Mobilities of electrons and holes

	Si	Ge	GaAs
$\mu_n = e\frac{\tau_n^*}{m_n^*}$ (cm^2 V^{-1} s^{-1})	1350	3900	8800
$\mu_p = e\frac{\tau_p^*}{m_p^*}$ (cm^2 V^{-1} s^{-1})	500	1900	400

Table 4.3 Diffusion coefficients and diffusion lengths

	D_n (cm^2 s^{-1})	D_p (cm^2 s^{-1})	$\tau_{n,p}$ (s)	L_n (μm)	L_p (μm)
Si	34.9	12.9	10^{-8}	5.9	3.6
Ge	100.8	49.1	10^{-8}	10.0	7.0
GaAs	227.5	10.3	10^{-8}	15	3.2

1. A silicon diode is first made with a type n substrate (with donor concentration $N_D = 10^{16}/\text{cm}^3$) and then a Si layer in which indium impurities are diffused to have a heavily doped p^+ layer (with an acceptor concentration $N_A = 10^{18}/\text{cm}^3$) and a p^+n junction is formed with a very narrow transition zone.

   (a) Calculate the Fermi levels μ_n and μ_p in regions p and n, respectively.
   (b) Determine the contact potential across the diode.
   (c) Calculate the widths of the transition zone.

   Solution Important references are the intrinsic concentrations

$$n = N_c e^{(\mu_i - E_g)/k_B T}, \quad p = N_v e^{-\mu_i/k_B T}, \quad n_i = n = p. \tag{4.103}$$

   To find the Fermi levels we can use any of the following equivalent relations. Assuming complete ionization of the impurities, we will consider

$$n_n = n_i e^{-\frac{(\mu_i - \mu_n)}{k_B T}} = N_c e^{-\frac{(E_g - \mu_n)}{k_B T}} \simeq N_D \tag{4.104}$$

$$p_p = n_i e^{-\frac{(\mu_i - \mu_p)}{k_B T}} = N_v e^{-\frac{\mu_p}{k_B T}} \simeq N_A \tag{4.105}$$

   Here we denote by $\mu_i \simeq E_g/2$ the intrinsic Fermi level. In some texts the notation $\mu_i = e\phi_F$ is also used.

   (a) With these concentrations, we have

$$\begin{aligned}\mu_n &= E_g + k_B T \ln \frac{n_n}{N_c} = E_g - 0.026\,(7.93) \\ &= 1.1 - 0.206\,\text{eV} = 0.894\,\text{eV},\end{aligned} \tag{4.106}$$

$$\mu_p = -k_B T \ln \frac{p_p}{N_v} = 0.026\,(2.286) \\ = 0.0594\,\text{eV}. \tag{4.107}$$

(b) The contact potential will then be

$$e\Delta\phi_o = \mu_n - \mu_p = 0.894 - 0.0594$$

$$e\Delta\phi_o = 0.8346\,\text{eV}. \tag{4.108}$$

(c) The widths of the transition zones are

$$w_p = \left[\frac{2\epsilon\epsilon_0\Delta\phi_o}{e}\frac{N_D}{N_A(N_A+N_D)}\right]^{\frac{1}{2}} \tag{4.109}$$

and

$$w_n = \left[\frac{2\epsilon\epsilon_0\Delta\phi_o}{e}\frac{N_A}{N_D(N_A+N_D)}\right]^{\frac{1}{2}}. \tag{4.110}$$

Substituting, we have

$$w_p = \left[\frac{2(11.9 \times 8.854 \times 10^{-12})(0.8346)}{1.6 \times 10^{-19}} \times \frac{10^{22}}{10^{24}(1.01 \times 10^{24})}\right]^{1/2}, \tag{4.111}$$

$$w_p = 3.29 \times 10^{-9}\,\text{m} = 32.9\,\text{Å}. \tag{4.112}$$

while

$$w_n = \left[\frac{2(11.9 \times 8.854 \times 10^{-12})(0.8346)}{1.6 \times 10^{-19}} \times \frac{10^{24}}{10^{22}(1.01 \times 10^{24})}\right]^{1/2}, \tag{4.113}$$

$$w_n = 3.29 \times 10^{-7}\,\text{m} = 0.329\,\mu\text{m}. \tag{4.114}$$

2. Consider a *pn* diode with a circular cross section diameter of 20 μm, made of Si. The density of donors is $10^{16}/\text{cm}^3$ and of acceptors $10^{18}/\text{cm}^3$. The dielectric constant of Si is $\epsilon = 11.9$. If the diode is at 300 K, calculate:

   (a) the electric potential that would have to be applied to flatten the bands, so that the band edges of the layers to the left and right of the junction are aligned;
   (b) the widths of the transition zone and the profile of electric potential with inverse polarization of 0, 2, 5 and 10 V, and forward bias of 0.2, 0.5 and 0.6 V;

(c) the widths w_p and w_n before and after applying the potentials mentioned in (b).

Solution We first need to obtain the contact potential $\Delta\phi_o$

$$\Delta\phi_o = \frac{k_B T}{e} \ln\left(\frac{p_p}{p_n}\right) = \frac{k_B T}{e} \ln\left(\frac{n_n}{n_p}\right). \tag{4.115}$$

To obtain this potential, the majority and minority concentrations of electrons or holes are needed. Given the density $p_p = 10^{18}/\text{cm}^3$, the minority carriers concentration is

$$p_n = \frac{n_i^2}{n_n} = \frac{N_c N_v e^{E_g/k_B T}}{10^{16}/\text{cm}^3} = \frac{(1.5 \times 10^{10}\,\text{cm}^{-3})^2}{10^{16}\,\text{cm}^{-3}}, \tag{4.116}$$

hence

$$p_n = 2.25 \times 10^4\,\text{cm}^{-3}, \tag{4.117}$$

Therefore,

$$\Delta\phi_o = 0.026\,\text{V} \times 31.5 = 0.819\,\text{V}. \tag{4.118}$$

(a) The potential V that must be applied so that the bands of the semiconductor n are aligned with those of the semiconductor p, is known as the flat-band potential and will later be denoted as V_{G0}. It is evident that

$$V_{G0} = \Delta\phi_o = 0.819\,\text{V} \tag{4.119}$$

(b) In (4.42) and (4.43) the widths were obtained as

$$w_{nV} = \left[\frac{2\epsilon\epsilon_0}{eN_D}(\Delta\phi_o - V)\frac{N_A}{N_D + N_A}\right]^{\frac{1}{2}}$$

and

$$w_{pV} = \left[\frac{2\epsilon\epsilon_0}{eN_A}(\Delta\phi_o - V)\frac{N_D}{N_A + N_D}\right]^{\frac{1}{2}}, \tag{4.120}$$

when a potential V is applied. If we substitute the corresponding constants and potentials mentioned in (b), we have:

$$w_{pV}(V = 0) = 32\,\text{Å}$$

$$w_{nV}(V = 0) = 0.32\,\text{Å}$$

$$w_{pV}(V = -2\text{V}) = 58.9\,\text{Å}$$

$$w_{nV}(V = -2\text{V}) = 0.589\,\text{Å} \tag{4.121}$$

Complete the problem...

3. Consider an ideal silicon diode with $N_D = 10^{16}\,\text{cm}^{-3}$, $N_A = 10^{18}\,\text{cm}^{-3}$, diode area of $10^{-3}\,\text{cm}^2$, and transport properties:
   on the n side:

$$\begin{aligned} D_e &= 35\,\text{cm}^2\,\text{s}^{-1} \quad \mu_e = 1300\,\text{cm}^2\,\text{V}^{-1}\,\text{s}^{-1}, \\ D_h &= 7.8\,\text{cm}^2\,\text{s}^{-1} \quad \mu_h = 300\,\text{cm}^2\,\text{V}^{-1}\,\text{s}^{-1}, \end{aligned} \tag{4.122}$$

   and on the p side:

$$\begin{aligned} D_e &= 7.3\,\text{cm}^2\,\text{s}^{-1} \quad \mu_e = 280\,\text{cm}^2\,\text{V}^{-1}\,\text{s}^{-1} \\ D_h &= 2.6\,\text{cm}^2\,\text{s}^{-1} \quad \mu_h = 100\,\text{cm}^2\,\text{V}^{-1}\,\text{s}^{-1}. \end{aligned} \tag{4.123}$$

   Assuming $\tau_e = \tau_h = 10^{-6}\,\text{s}$, and substituting into

$$L_h = \sqrt{D_h \tau_h}, \tag{4.124}$$

   and

$$I_0 = eA\left[\frac{D_h}{L_h} p_n + \frac{D_e}{L_e} n_p\right], \tag{4.125}$$

   where

$$p_n = \frac{n_i^2}{n_n} \quad \text{and} \quad n_p = \frac{n_i^2}{p_p}, \tag{4.126}$$

   we obtain

$$I_0 = 0.94 \times 10^{-14}\,\text{A}. \tag{4.127}$$

   Determine p_n, n_p, and $e\Delta\phi_o$.
4. A configuration widely used in applications of *pn* diodes like a LED, is the forward polarization. In a LED, it is desirable to inject a single type of charge. In *pn* diodes, the injected electrons into the p region, which is close to the surface, can generate light by recombination with holes. The injected holes to the n layer, which is generally deeper, generate light that cannot escape the device and is therefore absorbed.
   Suppose we have a GaAs *pn* diode with $D_n = 30\,\text{cm}^2\,\text{s}^{-1}$ and $D_p = 15\,\text{cm}^2\,\text{s}^{-1}$ and the semiconductor of the p side, doped with a concentration of impurities $N_A = 5 \times 10^{16}\,\text{cm}^{-3}$, which means $n_p = \frac{n_i^2}{N_A} = \frac{(2 \times 10^6)^2}{N_A} = 8 \times 10^{-5}\,\text{cm}^3$, and on the n side with a concentration of impurities $N_D = 5 \times 10^{17}\,\text{cm}^{-3}$, such that $p_n = 8 \times 10^{-6}\,\text{cm}^{-3}$. Suppose also that $\tau_n = 10^{-8}\,\text{s}$ and $\tau_h = 10^{-7}\,\text{s}$. Calculate:
   (a) the diffusion lengths L_n and L_p;
   (b) the injection efficiency of the diode, defined as

$$\gamma_{iny} = \frac{J_n}{J_{tot}}. \tag{4.128}$$

Solution (a) By direct substitution we have:

$$L_n = \sqrt{D_n \tau_n} = 5.74\,\mu\text{m}, \tag{4.129}$$

$$L_p = \sqrt{D_p \tau_h} = 12.25\,\mu\text{m}. \tag{4.130}$$

(b) Similarly, if we substitute the numerical values into

$$\gamma_{iny} = \frac{\dfrac{eD_n n_p}{L_n}}{\dfrac{eD_n n_p}{L_n} + \dfrac{eD_p p_n}{L_p}}, \tag{4.131}$$

we have an efficiency $\gamma_{iny} \simeq 0.98$.

5. A pn^+ GaAs diode, forward biased, is used as a LED at $T = 300\,\text{K}$. Suppose that all injected electrons into the p region (upper shell) recombine with holes to generate photons. Calculate the generation rate of photons and the optical power of the LED if the device has the parameters of the previous problem, and if $V = 1\,\text{V}$. Assume also that the diode area is $0.1\ \text{mm}^2$.

Solution

The electron current is given by

$$I_n = \frac{eAD_n n_p}{L_n}\left(e^{\frac{eU}{k_B T}} - 1\right). \tag{4.132}$$

Substituting the parameters, we have

$$I_n = \frac{1.6 \times 10^{-19}(0.1 \times 10^{-2}) \times 30 \times (8 \times 10^{-5})}{5.47 \times 10^{-4}}\left(e^{1/0.026} - 1\right),$$

$$I_n = 0.035\,\text{mA} \tag{4.133}$$

For this current, equivalent to 0.000035 C/s, and the electron charge $1.6021 \times 10^{-19}\,\text{C}$, the current, in electrons per second, is

$$\frac{I_n}{e} = \frac{0.035 \times 10^{-3}\,\text{C/s}}{1.6021 \times 10^{-19}\,\text{C}} = 2.18 \times 10^{14}\left[\text{electrons/s}\right] \tag{4.134}$$

If each electron emits one photon, the number of emitted photons per second will also be

$$N_f = \frac{I_n}{e} = 2.18 \times 10^{14}\left[\text{photons/s}\right]. \tag{4.135}$$

Each photon has an energy of 1.43 eV equivalent to

$$E_f = e\,1.43\,[\mathrm{J}] = 2.291 \times 10^{-19}\,[\mathrm{J}] \tag{4.136}$$

the emitted power is then

$$\begin{aligned} P &= \frac{I_n}{e}E_f = \frac{I_n}{e}e\,1.43\,[\mathrm{AV}] \\ &= I_n 1.43\,[\mathrm{J/s}] \end{aligned} \tag{4.137}$$

corresponding to

$$P = 5.005 \times 10^{-5}\,\mathrm{W}. \tag{4.138}$$

6. An important application of a *pn* diode is as an optical radiation detector. The optical signal creates *e*-*h* pairs that contribute to the current, the photo-current, with the largest contribution in the transition zone and up to a diffusion length away from the ends of this zone. Assume that the junction *pn* is based on Si with the following parameter values:
   Area $= 10^4\,\mu\mathrm{m}^2$, $\epsilon = 11.7$, $N_A = 2 \times 10^{16}\,\mathrm{cm}^{-3}$, $N_D = 10^{16}\,\mathrm{cm}^{-3}$, $V = -15$ V, $D_n = 20\,\mathrm{cm}^2/\mathrm{s}$, $D_p = 12\,\mathrm{cm}^2/\mathrm{s}$ and $\tau_n = \tau_p = 10^{-8}$s.
   Calculate the photo-current of the device when the pairs *e*-*h* are generated from an optical signal with a generation rate $r_g = 10^{22}\,\mathrm{cm}^{-3}\,\mathrm{s}^{-1}$ and the photocurrent is $I_f = eA\,r_g\,(w + L_n + L_p)$

   Solution

   We need to calculate L_n, L_p and the width of the transition zone $w = w_n + w_p$. We know that

$$L_n = \sqrt{D_n \tau n} \quad L_p = \sqrt{D_h \tau h} \tag{4.139}$$

$$\Delta\phi_o = \frac{k_B T}{e} \ln \frac{N_A N_D}{n_i^2} \tag{4.140}$$

and

$$w_{nV} = \left[\frac{2\epsilon\epsilon_0}{eN_D}(\Delta\phi_o - V)\frac{N_A}{N_D + N_A}\right]^{\frac{1}{2}}$$

and

$$w_{pV} = \left[\frac{2\epsilon\epsilon_0}{eN_A}(\Delta\phi_o - V)\frac{N_D}{N_A + N_D}\right]^{\frac{1}{2}}. \tag{4.141}$$

Substituting, we have

$$L_n = \sqrt{D_n \tau_n} = 4.5\,\mu\mathrm{m} \tag{4.142}$$

$$L_p = \sqrt{D_h \tau_h} = 3.45\,\mu\text{m} \tag{4.143}$$

$$\Delta\phi_o = \frac{k_B T}{e} \ln \frac{N_A N_D}{n_i^2} = 0.026\,\text{V} \ln \frac{2 \times 10^{32}}{(1.5 \times 10^{10})^2} \tag{4.144}$$

$$\Delta\phi_o = 0.715\,\text{V} \tag{4.145}$$

$$w = \left[\frac{2\epsilon\epsilon_0}{e}\frac{1}{N_A + N_D}(\Delta\phi_o - V)\right]^{\frac{1}{2}} \left(\sqrt{N_A/N_D} + \sqrt{N_D/N_A}\right) \tag{4.146}$$

Since $N_A = 2N_D$, the width w can be written as

$$w = \left[\frac{2\epsilon\epsilon_0}{e}\frac{1}{3N_D}(\Delta\phi_o - V)\right]^{\frac{1}{2}} \left(\sqrt{2} + \sqrt{1/2}\right) = 1.4058\,\mu\text{m}. \tag{4.147}$$

Thus, the photocurrent is

$$I_L = eA\,r_g\,(w + L_n + L_p) = 0.149\,\text{mA} \tag{4.148}$$

The photocurrent increases as the area of transitions increases.

4.9 Problems

1. Assume that the impurity concentrations in semiconductors type n and p of a junction pn are $N_A = N_D = 10^{-21}\,\text{cm}^{-3}$, and that $w_n = w_p = 1\,\mu\text{m}$. What is the potential difference $\Delta\phi_o$ if the relative permittivity is $\epsilon = 11$? Of what order of magnitude is the charge density $\sigma\ (= eN_D w_n)$?
2. Show that by making the integration limits change suggested in (4.17) and (4.18), the concentrations indicated in (4.19) and (4.20) are obtained.
3. If we have a pn junction of Si at temperature $T = 300\,\text{K}$ and $\mu_p \simeq E_G/10$, what is the relation between n_n and p_n
4. If we have a pn junction of Si in thermodynamic equilibrium. Plot the minority concentrations to the left of $-w_p$ and to the right of w_n, and compare them with the concentrations when a potential $V = 0.2\,\text{V}$ is applied in: (i) forward bias, and (ii) reverse bias.
5. What is the recombination rate in a pn junction if $p = n = 10^{21}\,\text{cm}^{-3}$, $T = 300\,\text{K}$, the cross sections are $\sigma_p = \sigma_n \simeq 10^{14}\,\text{cm}^2$ and $N_d = 10^{16}\,\text{cm}^{-3}$?
6. If the concentration of deep impurities N_d of Si is of the order of $= 10^{15}\,\text{cm}^{-3}$ and the cross section is of the order of $10^{14}\,\text{cm}^2$, calculate the capture time of the electrons in the conduction band when (a) $T = 300\,\text{K}$ and (b) $T = 77\,\text{K}$.
7. If $\tau_{n0} = \tau_{n0} = 10^{-9}\,\text{s}$ in Si. Determine the recombination rate at the pn junction when the applied potential is 1.5 V. What is the current associated with recombination through the junction if the area is $A = 10^8\,\text{cm}^2$, and the length of the junction is $w_p + w_n \simeq 1\,\mu\text{m}$?

8. What is the injected current in a *pn* junction with $D_e = D_h = 10^{-2}\,\mathrm{m^2/s}$ and $L_e = L_h = 2\,\mu\mathrm{m}$ if the impurities p and n are $10^{18}\,\mathrm{cm^{-3}}$ and the system is connected to a battery with $V = 1.5\,\mathrm{V}$?

References

1. C. Zener, Proc. R. Soc. (Lond.) **14**(5), 523 (1934); C. Zener, Phys. Rev. **82**, 403 (1951); C. Zener, Phys. Rev. **81**, 440 (1951)
2. L. Esaki, Phys. Rev. **109**, 603 (1958); T. Yajima, L. Esaki, J. Phys. Soc. Jpn. **13**, 1281 (1958)

Electronic Transport in Quantum Heterostructures

5

Abstract

This chapter explores transport properties in semiconductor quantum structures fabricated using epitaxial growth and vapor deposition techniques. We begin by introducing key quantum mechanical concepts relevant to transport, including current density, and transmission and reflection coefficients within the framework of quantum theory. A brief overview of linear response theory is also provided. We then present the transfer matrix method as an effective tool for analyzing layered quantum systems and establish its connection to scattering amplitudes. The Schrödinger equation is solved for both unpolarized and polarized step potentials as well as rectangular barriers, with corresponding transmission coefficients and currents calculated and visualized. Additionally, we review the Landauer formalism and derive the Landauer conductance formulas, which provide a foundational understanding of quantum conductance. The chapter concludes with a discussion of position-dependent effective mass and demonstrates a method for solving the Schrödinger equation in systems with step potentials and rectangular barriers under this condition.

5.1 Introduction

During the first 30 years of semiconductor technology development, device dimensions were on the order of one micron, thus the analysis of transport and optical properties based on semiclassical theories, grounded on classical physics concepts and variables, such as the Boltzmann equation and the Drude model, was well justified. Since the 1970s, the experimental techniques and the semiconductor growth apparatuses, such as the epitaxial molecular beam and chemical vapor deposition methods, have allowed the growth of semiconductor structures with sizes on the order of a few tens of nanometers. Currently the thicknesses of the layered structures are on the order of a nanometer, with a precision of one atomic layer. The sizes of semiconductor systems are in the quantum domain. For this reason, to describe an

P. Pereyra, *Contemporary Theory of Semiconductors and Devices*,
Graduate Texts in Physics, https://doi.org/10.1007/978-3-032-00296-9_5

important part of the transport and optical properties of modern nanoscopic devices, it is necessary to use quantum methods.

One of the most important properties of semiconductors is their crystalline structure. As mentioned in Chap. 1, this is a significant property that characterizes metals and semiconductors. Valence electrons in crystal structures, unlike in many other materials, feel periodic potentials. Two important quantum phenomena that combine in periodic systems are quantum coherence and resonant tunneling. These properties are responsible for the phenomenon of clustering of energy levels into bands. As mentioned above, and will be seen in the next chapter, in the standard theory, where periodic systems are assumed to be infinite, the bands are continuous and are usually reported as functions of the wave number along the crystalline symmetry axes. In finite periodic systems the bands are discrete. In any case, for the purpose of the discussion that follows, the relevant quantity is the energy gap between the valence and conduction bands. Precise experimental methods for measuring the energy gap include optical absorption and photoluminescence.

Also in Chap. 1, related to the fact that the electron energy levels are constant throughout the periodic system, the effective mass approximation was introduced. This approximation allows us to assume that the electrons, in the valence and conduction bands of a semiconductor layer, move in a constant potential. In the specific cases that we will discuss in this chapter, which are simple layered structures, the conduction band edge along the growing direction becomes, due to the differences in the gap energies, a sequence of wells and barriers. For example, in Fig. 5.1 the semiconductors A and B, with different energy gaps and similar lattice parameters, are grown in alternating layers and thus create structures A/B/A, B/A/B, etc., with different potential profiles. In Fig. 5.1, a barrier is created on the left and a potential well on the right.[1] To study the motion of electrons in these potentials, it is sufficient, in general, to consider the one-dimensional (1D) stationary Schrödinger equations in the effective mass approximation

$$\frac{\hbar^2}{2m^*}\frac{\partial^2}{\partial z^2}\psi(z) + U\psi(z) = E\psi(z). \tag{5.1}$$

Here U is a constant and m^* is the effective mass. In this approximation, it is possible to apply the quantum theory to study semiconductor heterostructures in which layers of two or more types of semiconductors are combined. It is important to note that even though the difference in gaps could be enough to create quantum wells and barriers, the reality is a little more complex. In the growth of one layer over another, there are other important factors, one of them is the lattice constant. If the lattice constant of the new layer is very different from that of the layer grown before, tensions and many defects may occur at the interfaces. For this reason, another layer of the same semiconductor, but doped, is generally grown on a layer of semiconductor A. For

[1] An important book where the quantum theory was applied to this type of systems, and numerous references to experimental work is given, is the book of Weisbuch and Vinter [1].

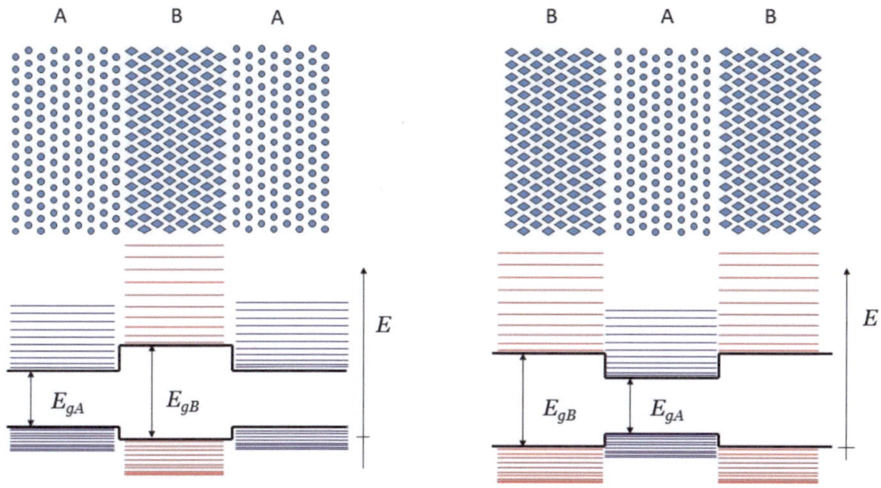

Fig. 5.1 Semiconductor structure. By combining semiconductors with different energy gaps, barrier or well profiles can be created for the electrons and holes of the conduction and valence bands

example, on a layer of $GaAs$, a layer of $Al_xGa_{1-x}As$, in which a fraction of the Ga has been replaced by Al, is grown. In the doped layer the gap $E_g(x)$ is in general different from the gap E_g in the pure semiconductor. There are empirical formulas in the literature to obtain $E_g(x)$ as a function of the concentration x of impurities. We are not going to address this vast topic of semiconductor growth here. We are going to assume that once all the growth problems, which are many, are solved, we have a semiconductor heterostructure. The most important characteristic in a heterostructure is the difference in the energy gaps, a difference responsible for the potential profiles, which are in principle chosen at will. At the end of the chapter, we will briefly comment on one problem that may arise when growing a semiconductor layer on top of another: the diffusion of impurities at the interface. Because of this problem, the discontinuities in the potential profiles are not necessarily abrupt.

Given a heterostructure, the main theoretical problem is the solving of the Schrödinger equation. By solving the Schrödinger equation, we obtain the wave functions $\psi(z)$ and, in systems with confining domains, the eigenvalues of the energy. In the quantum description of the heterostructure, the eigenvalues and eigenfunctions are the central ingredients to describe the transport and optical properties. In this chapter we will introduce essential quantities to describe electronic transport and a couple of examples, the step potential and the rectangular barrier, in the absence and presence of a bias potential. In Chap. 12 we will consider systems with a little more of complexity. Before studying the transport in the quantum structures considered below, we need to introduce the definition of currents in the quantum theory, as well as the definitions of reflection and transmission coefficients. We will then introduce, also at an introductory level, the Landauer conductance and, at the end, a brief analysis of the effects of diffusion of impurities at the interfaces.

5.2 Current Density in Quantum Mechanics

To obtain the current density, in the formalism of the quantum theory, the wave functions $\psi(z)$ or eigenfunctions of the system are essential. In quantum theory, $|\psi(z)|^2$ gives the probability density of finding a particle in the spatial domain in which the wave function is defined. To define the current density in the quantum theory we start from the current defined in classical physics. We have met before a classical equation involving the current and the particle density and current density, the continuity equation

$$\frac{\partial \rho}{\partial t} + \nabla \cdot \mathbf{j} = 0. \tag{5.2}$$

If the classical particle density ρ in this equation is replaced by the quantum probability density $|\psi(\mathbf{r}, t)|^2$, i.e. if

$$\rho = \psi^*(\mathbf{r}, t)\psi(\mathbf{r}, t), \tag{5.3}$$

it is easy to obtain the quantum version of the current density $\mathbf{j}$. In fact, if we differentiate ρ with respect to time

$$\frac{\partial \rho}{\partial t} = \psi \frac{\partial \psi^*}{\partial t} + \psi^* \frac{\partial \psi}{\partial t}, \tag{5.4}$$

and we use the complete Schrödinger equation

$$-\frac{\hbar^2}{2m}\nabla^2\psi + U(\mathbf{r})\psi = i\hbar\frac{\partial \psi}{\partial t}, \tag{5.5}$$

we obtain the equation

$$\frac{\partial \rho}{\partial t} = -\frac{i\hbar}{2m}\left(\psi\nabla^2\psi^* - \psi^*\nabla^2\psi\right), \tag{5.6}$$

which can be re-written in the form

$$\frac{\partial \rho}{\partial t} + \nabla \cdot \left[\frac{i\hbar}{2m}\left(\psi\nabla\psi^* - \psi^*\nabla\psi\right)\right] = 0. \tag{5.7}$$

This equation is similar to the continuity equation, if we identify the quantum current density as

$$\mathbf{j} = \frac{i\hbar}{2m}\left(\psi\nabla\psi^* - \psi^*\nabla\psi\right). \tag{5.8}$$

When the wave function is factored as

$$\psi(\mathbf{r}, t) = \varphi(\mathbf{r})e^{-i\omega t}, \tag{5.9}$$

the particle current density will simply be

$$\mathbf{j} = \frac{i\hbar}{2m}\left(\varphi\nabla\varphi^* - \varphi^*\nabla\varphi\right). \quad (5.10)$$

This is the expression that we were looking for. It is an important expression in the analysis of transport through quantum structures. It is evident that, if the function $\varphi(\mathbf{r})$ is real, the current density $\mathbf{j}$ is zero. A system with real wave functions is the infinite well. Consequently, the current density at each point and in each of the possible quantum states of the infinite well is zero. This does not mean that the particles are static, but that the current to the right is the same as the current to the left.

5.2.1 Free Particle Currents

The simplest quantum system is the free particle. In the effective mass approximation, electrons and holes in a homogeneous semiconductor layer, move in the conduction and valence bands, as if they were in a constant potential with an effective mass m^*. In semiconductor layers, thick enough, they can be modeled as quasi-free particles, confined in a wide quantum well. We will see now some properties of the probability density and current density when particles are described as if they were free particles.

In quantum theory, free particles are described by wave functions, which are solutions of the stationary Schrödinger equation, written as the superposition

$$\varphi(z) = Ae^{ikz} + Be^{-ikz} \quad (5.11)$$

where A and B are normalization constants. In this case and practically for all cases that we will consider in the following, we assume that the total wave functions contain the time dependent factor $e^{-i\omega t}$ and, as a second factor, the functions $\varphi(\mathbf{r})$, solutions of the stationary Schrödinger equation. Therefore, although the complete and correct expression of the free particle wave function is

$$\psi(z,t) = e^{-i\omega t}\varphi(z) = Ae^{i(kz-\omega t} + Be^{-i(kz+\omega t)}, \quad (5.12)$$

we will write only the factor $\varphi(z)$, as in (5.11), and we will refer to Ae^{ikz}, as the function (or probability amplitude) describing particles that move in the positive sense, and to Be^{-ikz} as the function (or probability amplitude) describing particles that move in the negative sense.

The wave function $\varphi(z)$ in (5.11) is complex, but the probability density $|\varphi(z)|^2$ is real. The constants A and B define the magnitudes of the probability amplitudes. When $A = B$, i.e. when the probability amplitudes of particles moving to the right and left are equal, the current density will be zero. The amplitudes do not necessarily have to be equal; their relation depends generally on how the system is prepared. Let us consider, for the particles described by the wave function in (5.11), three possible physical cases: the first in which all particles move to the right and are described by the wave function $\varphi^+(z) = Ae^{ikz}$; the second in which all move to the left described by the wave function $\varphi^-(z) = Be^{-ikz}$; and finally, when particles move to the right and also to the left, described by the wave function $\varphi(z) = Ae^{ikz} + Be^{-ikz}$. The probability densities and particle current densities for each of these cases are:

1. When the particles move towards the right, the probability density is

$$\rho^+ = |Ae^{ikz}|^2 = |A|^2. \tag{5.13}$$

This means that the probability to find the particle is uniformly distributed from $-\infty$ to $+\infty$. When the wave function describes an ensemble of particles, the probability density represents also the particles density. In this case, the particle's current density is

$$\begin{aligned} \mathbf{j}^+(z) &= \frac{i\hbar}{2m}\left(\varphi^+(z)\frac{d}{dz}\varphi^{+*}(z) - \varphi^{+*}(z)\frac{d}{dz}\varphi^+(z)\right)\widehat{\mathbf{z}}, \\ &= \frac{i\hbar}{2m}\left(\varphi^+(z)(-ik)\,A^*e^{-ikz} - \varphi^{+*}(z)(ik)\,Ae^{ikz}\right)\widehat{\mathbf{z}}, \\ \mathbf{j}^+(z) &= \frac{\hbar\mathbf{k}}{m}\rho^+(z). \end{aligned} \tag{5.14}$$

A basic relation in quantum physics is $\hbar\mathbf{k} = \mathbf{p} = m\mathbf{v}$. If we use this relation, the current density of particles moving to the right becomes

$$\mathbf{j}^+(z) = \mathbf{v}\rho^+ = \mathbf{v}|A|^2. \tag{5.15}$$

The constant $|A|^2$ is a norm factor that depends on the normalization criterion that has been chosen. In the Born and Dirac normalizations we have, respectively

$$\mathbf{j}^+(z) = \frac{|A|^2\hbar\mathbf{k}}{m} = \begin{cases} \hbar\mathbf{k}/Lm & A = 1/\sqrt{L}, \\ \mathbf{k}/2\pi m & A = 1/\sqrt{2\pi\hbar}. \end{cases} \tag{5.16}$$

The normalization constant can also be chosen such that the current density is normalized to unity. In this case

$$\mathbf{j}^+(z) = \frac{|A|^2\hbar\mathbf{k}}{m} = 1 \quad \text{and the constant is} \quad A = \sqrt{m/\hbar k}. \tag{5.17}$$

The physical meaning of the normalization constant is essentially of a scaling factor. In some problems it is important, while in others it is not. Current densities are useful in the evaluation of transmission and reflection coefficients.

2. When the particles move to the left with an amplitude B, the particles density is

$$\rho^{-} = |Be^{-ikz}|^2 = |B|^2. \tag{5.18}$$

The probability to find the partcle is also uniformly distributed, and the particle current density is

$$\begin{aligned}
\mathbf{j}^{-}(z) &= \frac{i\hbar}{2m}\left(\varphi^{-}(z)\frac{d}{dz}\varphi^{-*}(z) - \varphi^{-*}(z)\frac{d}{dz}\varphi^{-}(z)\right)\widehat{\mathbf{z}}, \\
&= \frac{i\hbar}{2m}\left(\varphi^{-}(z)(ik)\,A^{*}e^{+ikz} - \varphi^{-*}(z)(-ik)\,Ae^{-ikz}\right)\widehat{\mathbf{z}}, \\
\mathbf{j}^{-}(z) &= -\frac{\hbar\mathbf{k}}{m}\rho^{-}(z).
\end{aligned} \tag{5.19}$$

In this case also $\hbar\mathbf{k} = \mathbf{p} = m\mathbf{v}$, and the current density of particles moving to the left is given by the expression

$$\mathbf{j}^{-}(z) = -\mathbf{v}\rho^{+} = -\mathbf{v}|B|^2. \tag{5.20}$$

3. When particles move to right and left, they are described by the superposition

$$\psi = Ae^{ikz} + Be^{-ikz}. \tag{5.21}$$

In this case, the particles density is

$$\rho = |\psi|^2 = |A|^2 + |B|^2 + 2\Re e\left[A^{*}Be^{-i2kz}\right]. \tag{5.22}$$

This simple example, reveals the origin of one of the most important properties in periodic system, the phase coherence, responsible, as will be seen in the next chapter, of the band structure. Even though the wave function considered here describes free particles, the particles density in this equation shows that the coherent superposition that leads to a non-uniform particles distribution is not a simple sum, and the particles density is not either the sum of ρ^{+} and ρ^{-}. In this case, the quantum phases interference term is responsible for the oscillations shown in Fig. 5.2. On the other hand, the particle current density is given by

$$\mathbf{j}(z) = \frac{\hbar\mathbf{k}}{m}\left[\rho^{+}(z) - \rho^{-}(z)\right] = \frac{\hbar\mathbf{k}}{m}\left[|A|^2 - |B|^2\right]. \tag{5.23}$$

This current does not have an interference term. It is clear from this expression that if $|A|^2$, the probability of finding particles moving to the right, is equal to $|B|^2$, the particle current density is zero. In Fig. 5.2 the probabilty densities $\rho^{+} = |\varphi^{+}|^2 = |A|^2$, $\rho^{-} = |\varphi^{-}|^2 = |B|^2$ and $\rho(z) = |\psi|^2$ are plotted when the particles move with a speed of 10^4 m/s, with A and B real, such that A = 1.2B and the current density $\mathbf{j}(z)$ in (5.23) normalized to 1. It shows that when the wave function describes a superposition of particles moving towards the left and right, the wave functions interfere, and the particles density oscillates around $\rho^{+} + \rho^{-}$.

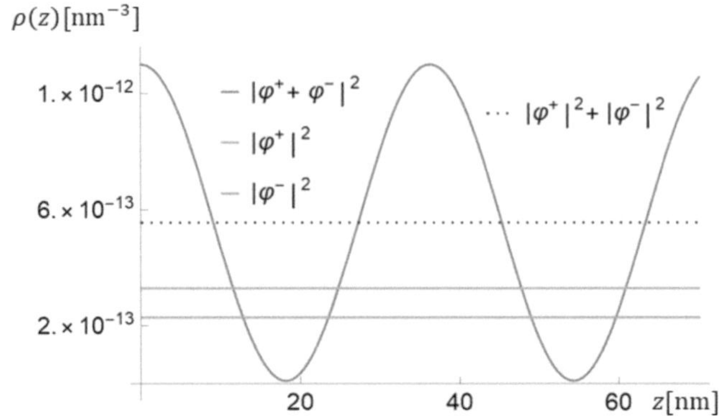

Fig. 5.2 Probability densities. If the wave functions describe particles moving in only one direction, the particles density, or the probability of finding the particles at a point z, is constant, $\rho^+ = |\varphi^+|^2 = |A|^2$ if they move to the right, or $\rho^- = |\varphi^-|^2 = |B|^2$ if they move to the left. But if the wave function describes a superposition of particles moving towards the left and right, the wave functions interfere and the probability density oscillates around $\rho^+ + \rho^-$. For this figure it was assumed that the particles move with a speed of 10^4 m/s, that, in addition A and B are real, that A = 1.2B and the current $\mathbf{j}(z)$ in (5.23) is 1

5.2.2 Linear Response Current and Quantum Corrections

Before we continue with explicit calculation of particle and electronic density currents, based on scattering properties, we will introduce here a brief digression on two issues that, independently, have been important in the calculation of current densities and on the analysis of the conductivity of metals and semiconductors with small mean free path ℓ. We shall briefly refer to Kubo's linear response theory, [2] and to the quantum corrections of the conductivity of electrons, described as quantum wave packets by Altshuler et al. [3]. One of the reasons why we will not discuss these approaches in detail is that they have been amply discussed in other texts. On the other hand, as conducting systems have evolved into the mesoscopic and eventually nanoscopic regime, the focus in the analysis of electronic transport has shifted toward the calculation of global quantities, such as the electrical conductance of microscopic structures. These quantities, and the scattering amplitudes, are analytically tractable within the transfer matrix method and the scattering approach used in this book. In the scattering approach, the entire sample is treated as a single, large-scale scatterer, and the electronic transport problem is recast as one of quantum mechanical scattering or statistical averaging problem. In the Landauer approach, [4] which we will refer to below, the conductance is expressed as simple functions of the transmission and reflection coefficients of the entire conductor. In the DMPK theory [5] that will be briefly introduced in the second part, the statistical distribution of the transfer matrices of the whole conductor allows to calculate averages of the reflection and transmission coefficients and to describe the macroscopic quantum phenomena in terms of a relatively small set of physically meaningful parameters, while the underlying microscopic details fade into the background, serving merely as conceptual scaffolding.

5.2.2.1 Current in the Linear Response Theory

As mentioned before, a widely used approach to calculate the current is Kubo's linear response theory, where the electronic transport is treated as a non-equilibrium problem. Indeed, if a system in equilibrium is in the presence of an external field, the time-dependent Hamiltonian, in the scalar potential gauge,[2] is written as

$$H(t) = H_0 - e \sum_i \phi_{ext}(\mathbf{r}_i, t) \tag{5.24}$$

where H_0 is the equilibrium Hamiltonian, which contains, in general, the electron-medium interactions. In the self-consistent approximation, this Hamiltonian is written as

$$H(t) = H_{sc} - e \sum_i \delta\phi_{ext}(\mathbf{r}_i, t) \tag{5.25}$$

where H_{sc} is the self-consistent field and

$$\delta\phi_{ext}(\mathbf{r}_i, t) = \phi_{ext}(\mathbf{r}_i, t) - \phi_{ind}(\mathbf{r}_i, t) \tag{5.26}$$

the self-consistent effective scalar potential. Here $\phi_{ind}(\mathbf{r}_i, t)$ is a local field induced by the external field. To calculate the current in the LRT, the common procedure is to apply the perturbation theory in the interaction representation, see Appendix D. When the operator $\widehat{Q}$ and the perturbation operator $\widehat{V}'(t_1)$ in

$$\langle \widehat{Q} \rangle_t = \langle \widehat{Q} \rangle_0 - \frac{i}{\hbar} \Big\langle \int_{t_0}^{t} \left[\widehat{Q}, \widehat{V}'(t_1) \right] dt_1 \Big\rangle_0 \tag{5.27}$$

are replaced by the current operator $\widehat{j}(\mathbf{r}, t)$ and the self-consistent effective scalar potential $\delta\phi_{ext}(\mathbf{r}_i, t)$, the current can be written as

$$J(\mathbf{r}, t) = \lim_{\eta \to 0^+} \int_{-\infty}^{\infty} dt' \int d^3\mathbf{r}' \Gamma(\mathbf{r}, \mathbf{r}'; t - t') \delta\phi(\mathbf{r}', t) \tag{5.28}$$

where $\Gamma(\mathbf{r}, \mathbf{r}'; t - t')$ is a response function defined by the correlations of charge and current operator fluctuations in equilibrium. This function is related to the conductivity tensor $\sigma(\mathbf{r}, \mathbf{r}')$ of the system.

A detailed calculation takes into account the particle-medium interactions, as well as the correlations of charge and current operator fluctuations in equilibrium. This task is beyond the purpose of this book. A tutorial derivation of the perturbation current, to first order of $\delta\phi(\mathbf{r}, t)$, is given in [6].

---

[2] For a more detailed discussion of this topic see, for example, Chap. 4 of [6].

5.2.2.2 Quantum Corrections to Conductivity

An important parameter in the classical Drude's conductivity

$$\sigma_q = \frac{n_q q^2 \tau_q}{m_q^*}, \tag{5.29}$$

is the collision time τ_q. This parameter represents the effects of the particle-medium interactions on the electronic transport properties. We have seen that, to include, for example, the electron-impurity interaction effect, the collision time in (3.14) is modified to

$$\frac{1}{\tau} = \frac{1}{\tau_f(T)} + \frac{1}{\tau_0}. \tag{5.30}$$

In the same way, if another relaxation mechanism with a relaxation time τ_m is present, the modified collision time τ' will be

$$\frac{1}{\tau'} = \frac{1}{\tau} + \frac{1}{\tau_m} = \frac{\tau + \tau_m}{\tau \tau_m}. \tag{5.31}$$

this will imply a correction in the conductivity

$$\frac{\delta\sigma}{\sigma} = \frac{\tau}{\tau + \tau_m}. \tag{5.32}$$

Altshuler et al. [3] analyzed the effect that the interference of wave packets, moving along different 'trajectories', has on the conductivity of conductors and semiconductors with small mean free path ℓ. Based on the diffusion equations, they found that the probability density that a trajectory in the space of dimension d beginning at point $\mathbf{r}$, t passes through a point $\mathbf{r}'$, t' is

$$W(|\mathbf{r}' - \mathbf{r}|, t' - t) = \frac{1}{\left(2\pi D(t' - t)\right)^{d/2}} e^{-(\mathbf{r}'-\mathbf{r})^2/2D(t'-t)} \tag{5.33}$$

Therefore, the magnitude of the effect $\delta\sigma/\sigma$, determined by the probability that the trajectory returns to the region of dimensions λ about r, (with $\lambda h/mv$ the transverse dimension of the wave packet), in a time that does not exceed the phase breaking time τ_{phi}, is

$$\frac{\delta\sigma}{\sigma} = \int_{\tau}^{\tau_{phi}} \frac{v dt}{(Dt)^{d/2}} \lambda^{d-1}. \tag{5.34}$$

5.3 The 1D Step Potential

The simplest potential profile, after the constant potential, is the one-dimensional step potential like the one shown in Fig. 5.3, which is essentially a constant potential with a single classical return point at the discontinuity. If this point is at the origin, the potential is described by the function

$$U(z) = \begin{cases} 0, & \text{for} \quad z \leq 0, \\ U_o, & \text{for} \quad z > 0. \end{cases} \tag{5.35}$$

A potential qualitatively similar to this is found by electrons and holes at the junction of two semiconductor layers, for example in the GaAs/AlAs junction. To determine the physical behavior expected for electrons moving through a system like this, we will assume that a beam of particles, all with energy E, comes from the left. The physics that results will depend on whether the energy is such that $E > U_0$ or $E < U_0$. We will review the physics of these two cases separately.

Before we solve the Schrödinger equation in the effective mass approximation it is worth recalling an essential issue. In a layered system, the quantum solutions should satisfy not only the continuity conditions, finiteness and integrability, they should also satisfy the flux conservation principle. Since particles can not appear or disappear anywhere, the flux conservation principle must be fulfilled. The flowing of particles is determined by the current density. The preservation of the density of current defined in (5.10), requires the fulfillment of the equality

$$\begin{aligned} \mathbf{j}_{\rm I}(z_0) &= \frac{i\hbar}{2m_1^*}\left(\varphi_{\rm I}\nabla\varphi_{\rm I}^* - \varphi_{\rm I}^*\nabla\varphi_{\rm I}\right)\Big|_{z=z_0} \\ &= \frac{i\hbar}{2m_2^*}\left(\varphi_{\rm II}\nabla\varphi_{\rm II}^* - \varphi_{\rm II}^*\nabla\varphi_{\rm II}\right)\Big|_{z=z_0} \\ &= \mathbf{j}_{\rm II}(z_0). \end{aligned} \tag{5.36}$$

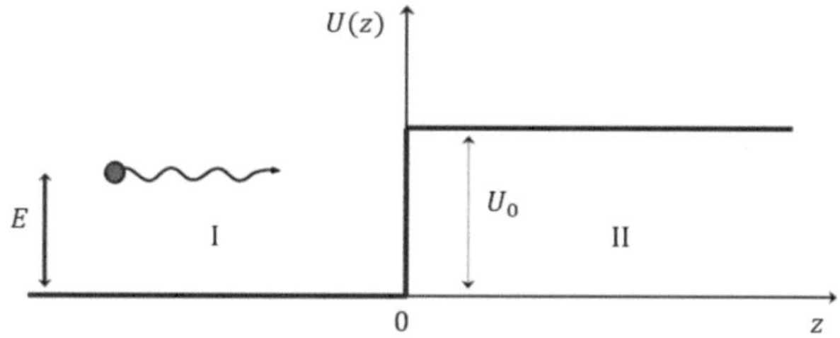

Fig. 5.3 In this figure we have a step potential $U(z)$, the potential parameters and the regions in which the Schrödinger equation is solved

In the layered structures it is justified and common to use the one-dimensional approximation. In this approximation, the last equation takes the form

$$
\begin{aligned}
j_{\mathrm{I}}(z_0) &= \frac{i\hbar}{2m_1^*}\left(\varphi_{\mathrm{I}}\frac{d\varphi_{\mathrm{I}}^*}{dz} - \varphi_{\mathrm{I}}^*\frac{d\varphi_{\mathrm{I}}}{dz}\right)\Big|_{z=z_0} \\
&= \frac{i\hbar}{2m_2^*}\left(\varphi_{\mathrm{II}}\frac{d\varphi_{\mathrm{II}}^*}{dz} - \varphi_{\mathrm{II}}^*\frac{d\varphi_{\mathrm{II}}}{dz}\right)\Big|_{z=z_0} \\
&= j_{\mathrm{II}}(z_0).
\end{aligned}
\tag{5.37}
$$

It is evident that if $m_1^* = m_2^*$, the continuity conditions $\varphi_{\mathrm{I}}(z_0) = \varphi_{\mathrm{II}}(z_0)$ and $d\varphi_{\mathrm{I}}(z_0)/dz = d\varphi_{\mathrm{II}}(z_0)/dz$ imply the conservation of current, but if $m_1^* \neq m_2^*$, the conservation equality (5.37) requires a different condition for the first derivative. The continuity of the magnitude of the wave function at the discontinuity point is an essential property, which implies the continuity of the probability to find the particle. For this reason, we will assume that the equality $\varphi_{\mathrm{I}}(z_0) = \varphi_{\mathrm{II}}(z_0)$ is preserved, and of course of $\varphi_{\mathrm{I}}^*(z_0) = \varphi_{\mathrm{II}}^*(z_0)$, while the derivative of the wave function has to be slightly modified. It is easy to see that, to satisfy the current conservation, it is enough that

$$
\frac{1}{m_1^*}\frac{d\varphi_{\mathrm{I}}}{dz}\Big|_{z=z_0} = \frac{1}{m_2^*}\frac{d\varphi_{\mathrm{II}}}{dz}\Big|_{z=z_0}. \tag{5.38}
$$

This condition is known in the semiconductors literature as the Ben Daniel-Duke condition.

1. *When the energy of the particle is less than U_o*, the Schrödinger equations in regions I and II, in the effective mass approximation, are

$$
-\frac{\hbar^2}{2m_1^*}\frac{d^2\varphi_{\mathrm{I}}(z)}{dz^2} = E\,\varphi_{\mathrm{I}}(z), \quad \text{for} \quad z \leq 0, \tag{5.39}
$$

and

$$
\frac{\hbar^2}{2m_2^*}\frac{d^2\varphi_{\mathrm{II}}(z)}{dz^2} = (U_o - E)\,\varphi_{\mathrm{II}}(z), \quad \text{for} \quad z > 0. \tag{5.40}
$$

In these equations m_1^* and m_2^* are the effective masses. We will start with (5.39). In this, as in the other cases that follow, we will write the solution as a scalar function and as a vector. The solution for (5.39) is

$$
\varphi_{\mathrm{I}}(z) = a_1 e^{ikz} + b_1 e^{-ikz}, \quad \phi_{\mathrm{I}}(z) = \begin{pmatrix} a_1 e^{ikz} \\ b_1 e^{-ikz} \end{pmatrix}, \quad \text{for} \quad z \leq 0, \tag{5.41}
$$

with $k = \sqrt{2m_1^* E/\hbar^2}$. On the other hand, for (5.40) the solutions are the exponential functions, thus

$$\varphi_{\rm II}(z) = a_2 e^{qz} + b_2 e^{-qz}, \quad \phi_{\rm II}(z) = \begin{pmatrix} a_2 e^{qz} \\ b_2 e^{-qz} \end{pmatrix} \quad \text{for} \quad z > 0, \tag{5.42}$$

with $q = \sqrt{2m_2^* (U_o - E)/\hbar^2}$. The solutions $\varphi_{\rm I}(z)$ and $\varphi_{\rm II}(z)$ are continuous in their respective domains, but to be solution of the Schrödinger equation of the step potential, extending from $-\infty$ to ∞, we need the functions $\varphi_{\rm I}(z)$ and $\varphi_{\rm II}(z)$, and their first order derivatives, to be continuous at the discontinuity point.
Before we impose the continuity conditions, let us introduce a quantity extremely useful to simplify the algebraic procedures when the number of layers, and thus the number of coefficients, grows. In (5.41) and (5.42) we have the wave functions $\varphi_j(z)$ (with j = I, II) and the state vectors $\phi_j(z)$ (with j = I, II). Both quantities contain exactly the same information. Given the state vectors at any point the wave functions can also be evaluated at any point of the system. Suppose that z_1 and z_2 are any two points in the system and $\phi(z_1)$ and $\phi(z_2)$ the state vectors evaluated at those points. The quantity that we want to introduce is the matrix $M(z_2, z_1)$ that connects the state vectors $\phi(z_1)$ and $\phi(z_2)$, i.e., the matrix that satisfies the relation

$$\phi(z_2) = M(z_2, z_1)\phi(z_1). \tag{5.43}$$

This matrix is known as the *transfer matrix*. The explicit form of the transfer matrix depends on where the points z_1 and z_2 are. For example, if both points are in the same region of constant potential, for instance in the region I of the step potential, (5.43) takes the explicit form

$$\begin{pmatrix} a_1 e^{ikz_2} \\ b_1 e^{-ikz_2} \end{pmatrix} = \begin{pmatrix} e^{ik(z_2 - z_1)} & 0 \\ 0 & e^{-ik(z_2 - z_1)} \end{pmatrix} \begin{pmatrix} a_1 e^{ikz_1} \\ b_1 e^{-ikz_1} \end{pmatrix}. \tag{5.44}$$

The transfer matrix, in this case, is diagonal and encodes the phase evolution acquired between z_1 and z_2. Similarly, when z_1 and z_2 are both in region II of the step potential, we have

$$\begin{pmatrix} a_2 e^{qz_2} \\ b_2 e^{-qz_2} \end{pmatrix} = \begin{pmatrix} e^{q(z_2 - z_1)} & 0 \\ 0 & e^{-q(z_2 - z_1)} \end{pmatrix} \begin{pmatrix} a_2 e^{qz_1} \\ b_2 e^{-qz_1} \end{pmatrix}, \tag{5.45}$$

which is also a diagonal matrix. However, if z_1 and z_2 are in different regions of the step potential, the transfer matrix is no longer diagonal, since it must incorporate the matching conditions at the discontinuity. In this way, the transfer matrices provide all the information related with the quantum phase evolution, the wave function amplitudes and the coherent interference effects across layered structures. We will see this, some lines below. Let us continue with the continuity conditions for the wave functions and their derivatives at the discontinuity point.

In this example, at $z = 0$. The continuous conditions of the wave functions and their derivatives means

$$\varphi_{\rm I}(0^-) = \varphi_{\rm II}(0^+) \tag{5.46}$$

and

$$\frac{1}{m_1^*}\frac{d\varphi_{\rm I}(0^-)}{dz} = \frac{1}{m_2^*}\frac{d\varphi_{\rm II}(0^+)}{dz}, \tag{5.47}$$

In these equations $0^\pm$ means $\lim_{\epsilon\to 0} 0 \pm \epsilon$. In general, unless it is necessary we will avoid this notation. If we replace the explicit functions $\varphi_{\rm I}(z)$ and $\varphi_{\rm II}(z)$, in these continuity conditions, we obtain the following relations

$$a_1 + b_1 = a_2 + b_2; \tag{5.48}$$

$$i\frac{k}{m_1^*}(a_1 - b_1) = \frac{q}{m_2^*}(a_2 - b_2). \tag{5.49}$$

Solving the Schrödinger equation also means determining all the coefficients that multiply the independent solutions. In this case, four. It is known that with two equations we can reduce the number of unknown coefficients by two. This allows us to express two of the coefficients in terms of the other two, which behave as independent variables. In general, other physical conditions must be invoked to determine the remaining coefficients. The number of coefficients that one has to determine depends on whether we are dealing with an open or a confined potential. One important conditions that one has some times to impose is the finiteness condition. Generally, the quantization of the energy arises from the fulfillment of additional boundary conditions. In the problem at hand, the first term of $\varphi_{\rm I}(z)$, describes the incident particles. For this reason its coefficient a_1 is naturally an independent variable that *can be set at will*. On the other hand, because on the right side we have exponential functions, it is important to take care of the finiteness of the solutions. Since the exponential function with positive exponent in (5.42) diverges when $z \to \infty$, we have to choose $a_2 = 0$. Before we set this coefficient to zero, let us write the continuity conditions in a matrix representation. To this purpose, it is convenient to express the coefficients a_2 and b_2 of (5.49), in terms of the coefficients a_1 and b_1 of region I, as follows:

$$a_2 = \frac{m_2^*}{2q}\left(\frac{q}{m_2^*} + i\frac{k}{m_1^*}\right)a_1 + \frac{m_2^*}{2q}\left(\frac{q}{m_2^*} - i\frac{k}{m_1^*}\right)b_1, \tag{5.50}$$

and

$$b_2 = \frac{m_2^*}{2q}\left(\frac{q}{m_2^*} - i\frac{k}{m_1^*}\right)a_1 + \frac{m_2^*}{2q}\left(\frac{q}{m_2^*} + i\frac{k}{m_1^*}\right)b_1. \tag{5.51}$$

This pair of equations can be written in a matrix form as follows:

$$\begin{pmatrix} a_2 \\ b_2 \end{pmatrix} = \frac{m_2^*}{2q} \begin{pmatrix} \frac{q}{m_2^*} + i\frac{k}{m_1^*} & \frac{q}{m_2^*} - i\frac{k}{m_1^*} \\ \frac{q}{m_2^*} - i\frac{k}{m_1^*} & \frac{q}{m_2^*} + i\frac{k}{m_1^*} \end{pmatrix} \begin{pmatrix} a_1 \\ b_1 \end{pmatrix} \tag{5.52}$$

In this representation, we have explicitly the matrix that connects the state vector $\phi_{\rm I}(z)|_{z=0^-}$ on the left of the discontinuity, and the state vector $\phi_{\rm II}(z)|_{z=0^+}$ on the right of the discontinuity. This is the transfer matrix $M(0^+, 0^-)$ that we mentioned before. In other words, the matrix representation in (5.52) is equivalent to

$$\phi_{\rm II}(0^+) = M(0^+, 0^-)\phi_{\rm I}(0^-)$$

with

$$M(0^+, 0^-) = \frac{m_2^*}{2q} \begin{pmatrix} \frac{q}{m_2^*} + i\frac{k}{m_1^*} & \frac{q}{m_2^*} - i\frac{k}{m_1^*} \\ \frac{q}{m_2^*} - i\frac{k}{m_1^*} & \frac{q}{m_2^*} + i\frac{k}{m_1^*} \end{pmatrix}. \tag{5.53}$$

In this case, the transfer matrix is a *transition matrix* that basically comprises the information of the continuity condition at the potential discontinuity. In this system, with a single point of discontinuity, the advantage of using the matrix representation is not yet evident, but in systems with more discontinuity points and greater complexity, the advantage will be clear.
Let us return to the coefficients problem. From (5.52), with $a_2 = 0$, we have

$$0 = \left(\frac{q}{m_2^*} + i\frac{k}{m_1^*}\right) a_1 + \left(\frac{q}{m_2^*} - i\frac{k}{m_1^*}\right) b_1, \tag{5.54}$$

Of the three coefficients left, two can be determined as functions of a_1, which is an independent variable. From the first of these equations we solve for b_1, obtaining

$$b_1 = \frac{km_2^* - iqm_1^*}{km_2^* + iqm_1^*} a_1. \tag{5.55}$$

The factor that multiplies a_1 is a complex number of magnitude 1. Consequently, when the energy E is less than U_o, the coefficient of the solution that describes reflected particles (moving to the left) is of the same magnitude as the coefficent of the function that describes the incident particles (moving to the right, towards the step). If all the incident flux is reflected, what is the meaning of coefficient b_2 in region II? It is easy to verify that

$$b_2 = \frac{m_2^*}{2q}\left(\frac{q}{m_2^*} - i\frac{k}{m_1^*}\right) a_1 + \frac{m_2^*}{2q}\left(\frac{q}{m_2^*} + i\frac{k}{m_1^*}\right) \frac{km_2^* - iqm_1^*}{km_2^* + iqm_1^*} a_1 \tag{5.56}$$

from which we easily obtain

$$b_2 = \frac{2km_2^*}{iqm_1^* + km_2^*} a_1. \tag{5.57}$$

What implications do the results (5.55) and (5.57) have on the solutions φ_{I} and φ_{II}? If we use (5.55) in (5.41), we have

$$\varphi_{\mathrm{I}} = a_1 e^{ikz} + \frac{km_2^* - iqm_1^*}{km_2^* + iqm_1^*} a_1 e^{-ikz}. \tag{5.58}$$

Since the quotient that appears in the second term is a complex number of magnitude 1, we can write it as

$$\frac{km_2^* - iqm_1^*}{km_2^* + iqm_1^*} = e^{-i2\theta}, \quad \text{with} \quad \theta = \tan^{-1} \frac{qm_1^*}{km_2^*}. \tag{5.59}$$

The function

$$\varphi_{\mathrm{I}} = a_1 e^{ikz} + a_1 e^{-i2\theta} e^{-ikz}, \tag{5.60}$$

can be written in the form

$$\varphi_{\mathrm{I}} = a_1 e^{-i\theta} \left(e^{i\theta} e^{ikz} + e^{-i\theta} e^{-ikz} \right). \tag{5.61}$$

If we define the constant $a = 2ia_1 e^{-i\theta}$, the solution of region I will be

$$\varphi_{\mathrm{I}} = a \cos(kz + \theta), \tag{5.62}$$

which is a stationary function, as shown in the Fig. 5.4, whose current density is zero. This means that for $E < U_0$ the current density to the right is equal to the current density to the left. In this system all incident flux is reflected. What about the function for $z > 0$? In this region $a_2 = 0$, and b_2 can be written as

$$b_2 = \frac{2m_2^* k}{\sqrt{(qm_1^*)^2 + (km_2^*)^2}} e^{-i\theta} a_1. \tag{5.63}$$

If we use the constant defined before, we have

$$\varphi_{\mathrm{II}}(z) = a \frac{km_2^*}{\sqrt{(qm_1^*)^2 + (km_2^*)^2}} e^{-qz}. \tag{5.64}$$

This is a function whose amplitude decreases exponentially with z, see Fig. 5.4. According to (5.64) the probability $|\varphi_{\mathrm{II}}(z)|^2$ of finding the particle at a point

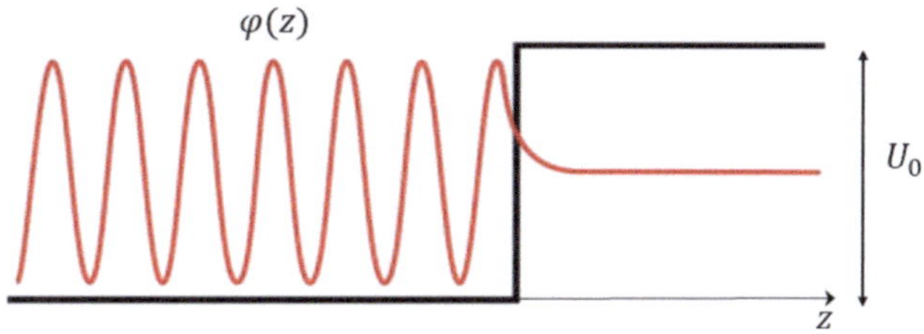

Fig. 5.4 The wave function to the left and right of the discontinuity point, for energies less than U_0. In the quantum description we have an exponentially decreasing probability of finding the quantum particle on the right side of the discontinuity, even though the particle energy is less than the height of the barrier

z inside the step is proportional to e^{-2qz}, i.e., a non-zero probability. A result like this does not exist in classical physics. Unlike classical particles, quantum particles penetrate the step potential and the probability of finding them to the right of the discontinuity decreases with z. The penetration depth at the step potential is proportional to $1/q \approx \hbar/\sqrt{2m_2^*(U_o - E)}$. The penetration depth becomes infinite when $E = U_0$.

To conclude this part, It is important to emphasize that in the step potential problem the energy is not quantized and the wave functions are propagating functions in the region I and exponentially decreasing in the region II. Let us now look at the other case.

2. *When* $E > U_o$. Since in both regions of the step potential the energy is greater than the potential energy, the differential equations on the left and right of the discontinuity are similar and have oscillating solutions. For region I we have the same equation as in (5.39). In region II, the equation is different from the one we had in (5.40). Now the equation is

$$-\frac{\hbar^2}{2m_2^*}\frac{d^2\varphi_{\text{II}}(z)}{dz^2} = (E - U_o)\,\varphi_{\text{II}}(z), \quad \text{with} \quad z > 0, \tag{5.65}$$

and $(E - U_o) > 0$. The solutions in regions I and II are, respectively,

$$\varphi_{\text{I}}(z) = a_1 e^{ik_1 z} + b_1 e^{-ik_1 z}, \quad \phi_{\text{I}}(z) = \begin{pmatrix} a_1 e^{ik_1 z} \\ b_1 e^{-ik_2 z} \end{pmatrix}, \quad \text{for} \quad z \leq 0, \tag{5.66}$$

with $k_1 = \sqrt{2m_1^* E/\hbar^2}$, and

$$\varphi_{\text{II}}(z) = a_2 e^{ik_2 z} + b_2 e^{-ik_2 z}, \quad \phi_{\text{II}}(z) = \begin{pmatrix} a_2 e^{ik_2 z} \\ b_2 e^{-ik_2 z} \end{pmatrix} \quad \text{for} \quad z > 0, \tag{5.67}$$

where $k_2 = \sqrt{2m_2^*\,(E - U_o)\,/\hbar^2}$. With these solutions there is no risk of divergence and it makes no sense to impose the condition of finiteness. However, it is clear that if the particles come only from the left or only from the right, it is not physically possible to have on the other side of the discontinuity particles moving in the opposite direction to that of the incidence. Let us assume that the particles come only from the left. In this case, the second term in the first of (5.67) must be cancelled out. This means that b_2 must be zero. Before setting this coefficients to zero, we will write the continuity conditions as

$$\begin{pmatrix} a_2 \\ b_2 \end{pmatrix} = \frac{m_2^*}{2k_2} \begin{pmatrix} \frac{k_2}{m_2^*} + \frac{k_1}{m_1^*} & \frac{k_2}{m_2^*} - \frac{k_1}{m_1^*} \\ \frac{k_2}{m_2^*} - \frac{k_1}{m_1^*} & \frac{k_2}{m_2^*} + \frac{k_1}{m_1^*} \end{pmatrix} \begin{pmatrix} a_1 \\ b_1 \end{pmatrix}. \tag{5.68}$$

This relation differs slightly from the continuity conditions in (5.52). Here the wave number q is replaced by ik_2. If we now set $b_2 = 0$, we have

$$\begin{pmatrix} a_2 \\ 0 \end{pmatrix} = \frac{m_2^*}{2k_2} \begin{pmatrix} \frac{k_2}{m_2^*} + \frac{k_1}{m_1^*} & \frac{k_2}{m_2^*} - \frac{k_1}{m_1^*} \\ \frac{k_2}{m_2^*} - \frac{k_1}{m_1^*} & \frac{k_2}{m_2^*} + \frac{k_1}{m_1^*} \end{pmatrix} \begin{pmatrix} a_1 \\ b_1 \end{pmatrix}. \tag{5.69}$$

This is the equation that we must use for the specific case in which the incidence is from the left-hand side. If we multiply the matrices on the right, and equate term by term with the vector on the left, we have, on the one hand, the equality

$$a_2 = \frac{1}{2}\left(1 + \frac{k_1 m_2^*}{k_2 m_1^*}\right) a_1 + \frac{1}{2}\left(1 - \frac{k_1 m_2^*}{k_2 m_1^*}\right) b_1. \tag{5.70}$$

and, on the other hand, the equation

$$\frac{1}{2}\left(1 - \frac{k_1 m_2^*}{k_2 m_1^*}\right) a_1 + \frac{1}{2}\left(1 + \frac{k_1 m_2^*}{k_2 m_1^*}\right) b_1 = 0. \tag{5.71}$$

This means that

$$b_1 = \frac{k_1 m_2^* - k_2 m_1^*}{k_1 m_2^* + k_2 m_1^*} a_1, \quad \text{and} \quad a_2 = \frac{2k_1 m_2^*}{k_1 m_2^* + k_2 m_1^*} a_1. \tag{5.72}$$

With these coefficients, the solutions in regions I and II become, respectively:

$$\varphi_{\mathrm{I}}(z) = a_1 e^{ik_1 z} + a_1 \frac{k_1 m_2^* - k_2 m_1^*}{k_1 m_2^* + k_2 m_1^*} e^{-ik_1 z}; \tag{5.73}$$

and

$$\varphi_{\mathrm{II}}(z) = a_1 \frac{2k_1 m_2^*}{k_1 m_2^* + k_2 m_1^*} e^{ik_2 z}. \tag{5.74}$$

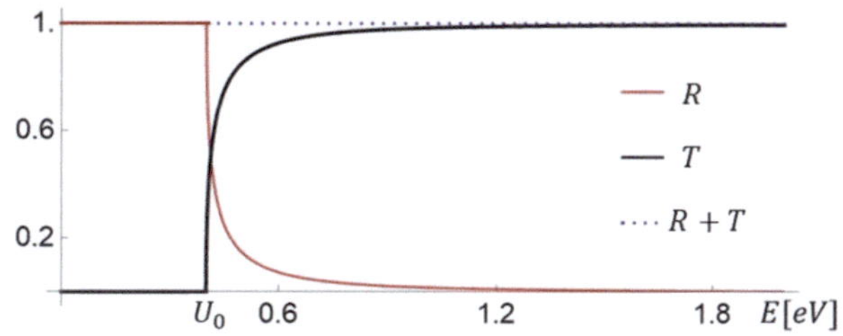

Fig. 5.5 The reflection and transmission coefficients of the step potential for $E > U_0$. When the energy exceeds the height of the barrier, the transmission increases rapidly to 1 and the reflection tends to zero

The second term of φ_{I} is the function that describes the fraction of particles reflected by the step. It can be easily verified that the incident, reflected and transmitted currents are respectively:

$$\begin{aligned}
\mathbf{j}_{\mathrm{inc}} &= \frac{\hbar k_1 |a_1|^2}{m_1^*}\,\widehat{\mathbf{z}},\\
\mathbf{j}_{\mathrm{refl}} &= -\frac{\hbar k_1 |a_1|^2}{m_1^*}\left(\frac{k_1 m_2^* - k_2 m_1^*}{k_1 m_2^* + k_2 m_1^*}\right)^2 \widehat{\mathbf{z}},\\
\mathbf{j}_{\mathrm{trans}} &= \frac{\hbar k_2 |a_1|^2}{m_2^*}\left(\frac{2k_1 m_2^*}{k_1 m_2^* + k_2 m_1^*}\right)^2 \widehat{\mathbf{z}}.
\end{aligned} \tag{5.75}$$

In this class of systems, particles can eventually move from $-\infty$ to $+\infty$ or in the opposite direction. To interpret and understand the behavior of these systems, it is useful to define the reflection and transmission coefficients. One way to define them is as the fractions of the incident current density that are reflected and transmitted, i.e.

$$R = \frac{|\mathbf{j}_{\mathrm{refl}}|}{|\mathbf{j}_{\mathrm{inc}}|} = \left(\frac{k_1 m_2^* - k_2 m_1^*}{k_1 m_2^* + k_2 m_1^*}\right)^2 \quad \text{and} \quad T = \frac{|\mathbf{j}_{\mathrm{trans}}|}{|\mathbf{j}_{\mathrm{inc}}|} = \frac{4k_1 k_2 m_1^* m_2^*}{(k_1 m_2^* + k_2 m_1^*)^2}. \tag{5.76}$$

These coefficients are plotted in Fig. 5.5 as functions of the particle energy. The transmission probability is zero when the particle energy is less than the step height, and increases rapidly when the energy is greater.

The flux of particles must be conserved; therefore, the sum $R + T$ must satisfy the following equality

$$R + T = 1. \tag{5.77}$$

5.4 Reflection and Transmission Amplitudes and Coefficients

In the last section the transmission and reflection coefficients were determined in terms of the particle current densities. These can also be obtained knowing the *reflection and transmission amplitudes*.[3]

It turns out that, using the transfer matrix method to represent the continuity and boundary conditions, it is much easier to obtain these amplitudes, and therefore the reflection and transmission coefficients. To establish the relationship between the elements of a transfer matrix and the reflection and transmission amplitudes, we are going to briefly introduce the dispersion or *scattering* formalism.

An alternative description of the transport phenomenon is the scattering approach. In this framework, interaction between particles and potentials or fields are treated as a scattering process. The wave functions that describe the particles either transmitted through or reflected from the interaction region, are characterized by transmission t and reflection r amplitudes. As illustrated in Fig. 5.6, particles approaching the scattering region, from the left and right are described by the functions $\varphi_{il}(z_1)$ and $\varphi_{ir}(z_2)$, respectively. After interacting with the potential, these waves are partially reflected and partially transmitted. r and t are the reflection and transmission amplitudes when the incidence is from the left, and r' and t' the amplitudes when the incidence is from the right. In terms of these amplitudes, the outgoing wave functions $\varphi_{ol}(z_1)$ and $\varphi_{or}(z_2)$, can be written as:

$$\varphi_{ol} = r\,\varphi_{il} + t'\varphi_{ir}; \tag{5.78}$$

$$\varphi_{or} = t\,\varphi_{il} + r'\varphi_{ir}. \tag{5.79}$$

These relations can be expressed in matrix form as

$$\begin{pmatrix} \varphi_{ol}(z_1) \\ \varphi_{or}(z_2) \end{pmatrix} = \begin{pmatrix} r & t' \\ t & r' \end{pmatrix} \begin{pmatrix} \varphi_{il}(z_1) \\ \varphi_{ir}(z_2) \end{pmatrix} = S \begin{pmatrix} \varphi_{il}(z_1) \\ \varphi_{ir}(z_2) \end{pmatrix}. \tag{5.80}$$

The matrix S is the *scattering matrix*, whose elements are the scattering amplitudes. Many physical systems are analyzed by sending particles into an interaction region and observing the scattered states. The core object in this analysis is the *scattering* matrix (S-matrix). This matrix encapsulates all the information about the interaction with the potential.

The relation between incoming functions $\varphi_{il}(z_1)$ and outgoing $\varphi_{ir}(z_2)$, is one of the possible relations. We have seen, in the last section that we can define state vectors and a matrix, the transfer matrix M that connects the state vectors. Indeed,

[3] Here it is assumed that the amplitudes t and r are complex scalar functions, in which case $T = tt^*$ and $R = rr^*$. This is the case in systems with only one propagating mode (i.e. one in the positive sense and one in the opposite sense), but in systems of higher dimension and with more propagating modes, the amplitudes are complex matrix functions. In that case, $T = tt^\dagger$ and $R = rr^\dagger$.

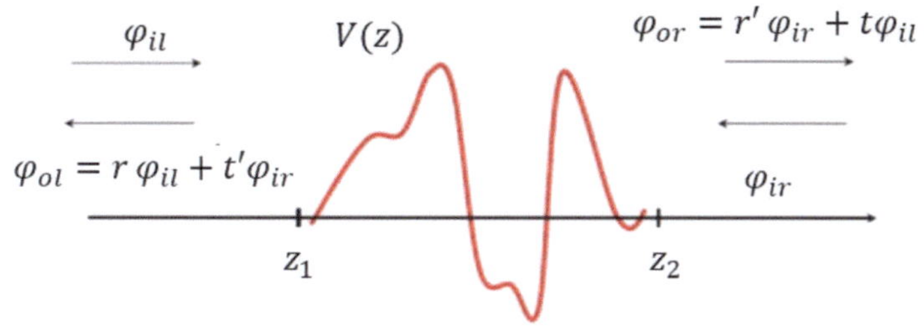

Fig. 5.6 The wave functions incident, reflected and transmitted by an arbitrary scattering potential $V(z)$

we can define the state vectors

$$\phi(z_1) = \begin{pmatrix} \varphi_{il}(z_1) \\ \varphi_{ol}(z_1) \end{pmatrix} \quad \text{and} \quad \phi(z_2) = \begin{pmatrix} \varphi_{or}(z_2) \\ \varphi_{ir}(z_2) \end{pmatrix} \tag{5.81}$$

at the left and right hand sides of the potential in Fig. 5.6, and find the transfer matrix $M(z_2, z_1)$ that connects these state vectors as follows

$$\begin{pmatrix} \varphi_{or}(z_2) \\ \varphi_{ir}(z_2) \end{pmatrix} = \begin{pmatrix} \alpha & \beta \\ \gamma & \delta \end{pmatrix} \begin{pmatrix} \varphi_{il}(z_1) \\ \varphi_{ol}(z_1) \end{pmatrix} = M(z_2, z_1) \begin{pmatrix} \varphi_{il}(z_1) \\ \varphi_{ol}(z_1) \end{pmatrix}. \tag{5.82}$$

An important relation in the scattering approach is the relation between the S matrix and the transfer matrix M. Our purpose is to establish this relation. If we combine (5.78) and (5.82), it is possible to obtain the following equations, where only the incident wave functions appear, that is:

$$\begin{aligned} (\alpha + \beta r - t)\varphi_{il}(z_1) &= (r' - \beta t')\varphi_{ir}(z_2); \\ (\gamma + \delta r)\varphi_{il}(z_1) &= (1 - \delta t')\varphi_{ir}(z_2). \end{aligned} \tag{5.83}$$

Since $\varphi_{il}(z_1)$ and $\varphi_{ir}(z_2)$ are linearly independent functions, all coefficients in these equations must cancel out. Given the set of conditions

$$\begin{aligned} \alpha + \beta r - t = 0 \quad r' - \beta t' = 0; \\ \gamma + \delta r = 0 \quad 1 - \delta t' = 0, \end{aligned} \tag{5.84}$$

it is easy to obtain the very important and useful relations

$$t = \alpha - \beta\frac{1}{\delta}\gamma, \quad t' = \frac{1}{\delta}, \quad r = -\frac{1}{\delta}\gamma, \quad \text{and} \quad r' = \beta\frac{1}{\delta}. \tag{5.85}$$

One of the properties that must be invoked when dealing with scattering and transfer matrices is the conservation of flux. As we have already mention, this property corresponds to the conservation of current density of particles, reflected in the unitarity condition $S^\dagger S = I$ that the S matrices must satisfy. For the scattering amplitudes, this condition implies the following relations

$$t^\dagger t + r^\dagger r = 1, \quad t'^\dagger t' + r'^\dagger r' = 1, \quad t'^\dagger r + r'^\dagger t = 0 = r^\dagger t' + t^\dagger r'. \tag{5.86}$$

What implication does the conservation of flux has for transfer matrices? Imposing this condition of conservation on the current density leads to the conclusion [5] that the transfer matrix must satisfy the requirement

$$M \begin{pmatrix} I & 0 \\ 0 & -I \end{pmatrix} M^\dagger = \begin{pmatrix} I & 0 \\ 0 & -I \end{pmatrix} \tag{5.87}$$

where I is the unit matrix of dimension $N \times N$, being N the number of propagating modes. The flux conservation condition in terms of the elements of the transfer matrix is expressed by the fulfillment of the following relations

$$\alpha\alpha^\dagger - \beta\beta^\dagger = 1, \quad \delta\delta^\dagger - \gamma\gamma^\dagger = 1, \quad \alpha\gamma^\dagger - \beta\delta^\dagger = 0 = \gamma\alpha^\dagger - \delta\beta^\dagger. \tag{5.88}$$

A physical symmetry that characterize most of the interactions is the time reversal invariance (TRI) symmetry, i.e. interactions that are invariant under the transformation $t \to -t$ (here t is the time), in which case (see Appendix E) it can be shown that $S^{-1} = S^*$, which means that the scattering matrix is symmetric, $S = S^T$. An important consequence of this symmetry is that the reflection amplitudes are symmetric and

$$t^T = t' \quad \text{and} \quad t = \frac{1}{\delta^T}. \tag{5.89}$$

For the transfer matrix, this TRI symmetry implies that

$$\begin{pmatrix} 0 & I \\ I & 0 \end{pmatrix} M \begin{pmatrix} 0 & I \\ I & 0 \end{pmatrix} = M^*. \tag{5.90}$$

5.4.1 Reflection and Transmission Amplitudes in the 1D Approximation

We shall now see how the previous relations appear in the one-dimensional and one propagating mode approximation,[4] where the transfer matrix elements and scattering amplitudes are scalar. In this case the † must be replaced by $*$ and the flux conservation conditions for the scattering and transfer matrices are

$$t^* t + r^* r = 1, \quad t'^* t' + r'^* r' = 1, \quad t'^* r + r'^* t = 0 = r^* t' + t^* r', \tag{5.91}$$

[4] In the one propagating mode case the state vectors have two components, one in the positive direction and one in the negative direction, and the transfer matrices are 2×2. If the interactions are spin dependent, these matrices are 4×4.

and

$$\alpha\alpha^* - \beta\beta^* = 1, \quad \delta\delta^* - \gamma\gamma^* = 1, \quad \alpha\gamma^* - \beta\delta^* = 0 = \gamma\alpha^* - \delta\beta^*. \tag{5.92}$$

It can be further shown that if the transmission amplitude t is given by

$$t = \alpha - \frac{\beta\gamma}{\delta} = \frac{\det M}{\delta}, \tag{5.93}$$

the flux conservation condition

$$t^*t + r^*r = \frac{|\det M|^2}{|\delta|^2} + \frac{|\gamma|^2}{|\delta|^2} = 1, \tag{5.94}$$

together with the second condition in the (5.92), requires that

$$\det M = 1. \tag{5.95}$$

Therefore, when the transfer matrix determinant is 1, the flux conservation principle is fulfilled and the relation between the transfer matrix elements and the scattering amplitudes can be written as

$$t = \frac{1}{\delta} \quad \text{and} \quad r = -\frac{\gamma}{\delta}. \tag{5.96}$$

The transmission and reflection coefficients are generally defined as

$$T = tt^* = \frac{1}{|\delta|^2} \quad \text{and} \quad R = rr^* = \frac{|\gamma|^2}{|\delta|^2}. \tag{5.97}$$

What happens when the determinant of the transfer matrix is different from 1? If $\det M \neq 1$, the matrix M should be redefined as follows

$$M \quad \Rightarrow \quad M_u = \frac{1}{\sqrt{\det M}} M. \tag{5.98}$$

In this case, the matrix

$$M_u = \begin{pmatrix} \alpha_u & \beta_u \\ \gamma_u & \delta_u \end{pmatrix}, \tag{5.99}$$

fulfills the flux conservation requirement, $\det M_u = 1$, and its matrix elements satisfy (5.92). The transmission amplitude t, written in terms of the matrix elements of M_u, becomes

$$t = \alpha_u - \frac{\beta_u\gamma_u}{\delta_u} = \frac{\det M_u}{\delta_u} = \frac{1}{\delta_u} = \frac{\sqrt{\det M}}{\delta}. \tag{5.100}$$

The last relation shows that in the case that $\det M = 1$, the transmission amplitude reduces to (5.96), otherwise, by writing it as

$$t = \frac{\sqrt{\det M}}{\delta}, \tag{5.101}$$

the fulfillment of the flux conservation requirement is guaranteed. Thus, $t = t'$, and the transfer matrix, as shown in Appendix E, has the following symmetry

$$M = \begin{pmatrix} \alpha & \beta \\ \beta^* & \alpha^* \end{pmatrix}. \tag{5.102}$$

Since time reversal invariance implies that $\delta = \alpha^*$ and $\gamma = \beta^*$, the transmission and reflection coefficient amplitudes are written as

$$t = \frac{1}{\alpha^*}, \quad T = tt^* = \frac{1}{|\alpha|^2}, \quad r = -\frac{\beta^*}{\alpha^*}, \quad \text{and} \quad R = rr^* = \frac{|\beta|^2}{|\alpha|^2}. \tag{5.103}$$

Generally, when the potential energy is symmetric, the determinant is 1. The transfer matrix of the rectangular barrier that we will study in the next section is unimodular, however the transfer matrix of the step potential is not. In any case, if we know the transfer matrix, and its determinant, we can rescale as mentioned before, and we can obtain immediately the transmission and reflection *amplitudes*, and with these quantities the *coefficients* of transmission and reflection.

For the step potential studied in the previous section, the potential is asymmetric. When $E > U_0$, the determinant of the transfer matrix is k_1/k_2. In this case, to fulfill the flux conservation principle, the transfer matrix must be redefined as mentioned before. If

$$\alpha_u = \delta_u = \frac{1}{\sqrt{\det M}} \frac{k_2 m_1^* + k_1 m_2^*}{2k_2 m_1^*} = \frac{k_2 m_1^* + k_1 m_2^*}{2\sqrt{k_1 k_2 m_1^* m_2^*}}, \tag{5.104}$$

and

$$\beta_u = \gamma_u = \frac{1}{\sqrt{\det M}} \frac{k_2 m_1^* - k_1 m_2^*}{2k_2 m_1^*} = \frac{k_2 m_1^* - k_1 m_2^*}{2\sqrt{k_1 k_2 m_1^* m_2^*}}, \tag{5.105}$$

the transmission and reflection amplitudes are then

$$t = \frac{2\sqrt{k_1 k_2 m_1^* m_2^*}}{k_2 m_1^* + k_1 m_2^*} = t' \quad \text{and} \quad r = -\frac{\gamma_u}{\delta_u} = \frac{k_1 m_2^* - k_2 m_1^*}{k_2 m_1^* + k_1 m_2^*} = -r'. \tag{5.106}$$

The transmission and reflection coefficients become

$$T = tt^* = \frac{4k_1 k_2 m_1^* m_2^*}{\left(k_2 m_1^* + k_1 m_2^*\right)^2} \quad \text{and} \quad R = rr^* = \frac{\left(k_2 m_1^* - k_1 m_2^*\right)^2}{\left(k_2 m_1^* + k_1 m_2^*\right)^2}. \tag{5.107}$$

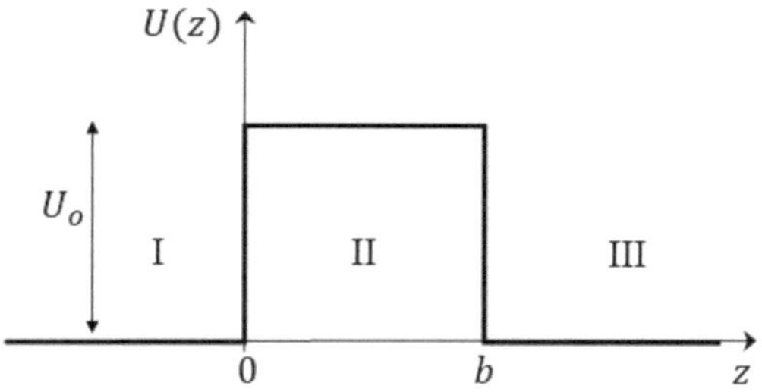

Fig. 5.7 Rectangular potential barrier, the parameters, and the regions in which the Schrödinger equation is solved

These are precisely the coefficients that were obtained in the previous section for the step potential. The ease with which coefficients are determined once the transfer matrix is known, is greater when systems are more complex. This is one of the advantages of the transfer matrix method and one of the reasons why it is convenient to work with this method. Transfer matrices have also other advantages. Physically, the condition $\det M = 1$ ensures the preservation of flux, meaning that no probability is lost or created during transmission or reflection, and therefore the transfer matrix correctly represents a physically admisible quantum process.[5]

5.5 Rectangular Potential Barrier

The rectangular potential barrier has been studied since the early years of the development of quantum mechanics because it is easy to solve and provides qualitative results valid for real barriers. It especially allows us to understand one of the most striking phenomena in the behavior of quantum systems: *the tunneling effect.* The potential profile like the one shown in Fig. 5.7 arises naturally when a semiconductor structure with three semiconductor layers is grown, such that the energy gap in the first and third layers is smaller than in the second. With the development of modern experimental techniques, it has been possible to grow semiconductor heterostructures in which the electrons in the conduction band do indeed experience abrupt changes in the potential when they move from one layer to another. If on top of a $GaAs$ substrate, a layer of $Al_xGa_{1-x}As$ with thickness b is grown, and on top of this layer a layer of $GaAs$, the barrier $GaAs/Al_xGa_{1-x}As/GaAs$ is formed. In this system the barrier potential height U_o is related to the concentration x. If in this system the concentration is 30%, the height is $U_o \simeq 0.23\,\text{eV}$.

Let us now consider the rectangular potential shown in the Fig. 5.7 defined as

$$U(z) = \begin{cases} 0 & z \leq 0, \ \ z \geq b, \\ U_o & 0 < z < b. \end{cases} \tag{5.108}$$

Our purpose is to solve the Schrödinger equation of this rectangular barrier when the energy E of the incident particles (which we will assume to come from the left)

[5] See [7].

is: (a) less than U_o and (b) greater than U_o. We will study these cases separately. The first in more detail. In none of them will we have quantization.

1. If $E < U_o$, the Schrödinger equations in regions I and III, in the effective mass approximation, are

$$-\frac{\hbar^2}{2m_1^*}\frac{d^2\varphi_{\mathrm{I,III}}(z)}{dz^2} = E\,\varphi_{\mathrm{I,III}}(z), \quad \text{if} \quad z \leq 0 \quad \text{and if} \quad z \geq 0. \tag{5.109}$$

As in the previous examples of free particles and step potential, the functions that satisfy these equations are:

$$\varphi_{\mathrm{I}}(z) = a_1 e^{ikz} + b_1 e^{-ikz}, \quad \text{if} \quad z \leq 0, \tag{5.110}$$

and

$$\varphi_{\mathrm{III}}(z) = a_3 e^{ikz} + b_3 e^{-ikz}, \quad \text{if} \quad z \geq b, \tag{5.111}$$

with $k = \sqrt{2m_1^* E/\hbar^2}$. In region II the Schrödinger equation is

$$\frac{\hbar^2}{2m_2^*}\frac{d^2\varphi_{\mathrm{II}}(z)}{dz^2} = (U_o - E)\,\varphi_{\mathrm{II}}(z), \quad \text{if} \quad z > 0, \tag{5.112}$$

and, as in the step potential problem, the solution is

$$\varphi_{\mathrm{II}}(z) = a_2 e^{qz} + b_2 e^{-qz}, \quad \text{if} \quad 0 < x < b, \tag{5.113}$$

with $q = \sqrt{2m_2^*(U_o - E)/\hbar^2}$. The functions (5.110), (5.111) and (5.113) are formally solutions of the differential equations. There still remains the problem of determining all the coefficients a_i and b_i.

The continuity conditions at points $z = 0$ and $z = b$ give rise to the following equations. From continuity at $z = 0$, we have[6]:

$$\begin{aligned} a_1 + b_1 &= a_2 + b_2, \\ ik(a_1 - b_1) &= q(a_2 - b_2). \end{aligned} \tag{5.114}$$

From the continuity in $z = b$, we have:

$$\begin{aligned} a_2 e^{qb} + b_2 e^{-qb} &= a_3 e^{ikb} + b_3 e^{-ikb}, \\ \frac{q}{m_2^*}(a_2 e^{qb} - b_2 e^{-qb}) &= i\frac{k}{m_1^*}(a_3 e^{ikb} - b_3 e^{-ikb}). \end{aligned} \tag{5.115}$$

[6] Although in this case the effective masses m_1^* and m_2^* are different, due to the potential symmetry the effect of this difference is negligible.

These equations allow us to obtain four of the six coefficients. In an open system like this, functions with coefficients a_1 and b_3 describe particles incident from the left and right, respectively. These amplitudes (a_1 and b_3) are arbitrary and are fixed by the experiment. For this reason, it is convenient to leave the coefficients a_1 and b_3 as independent parameters, and the other coefficients expressed in terms of a_1 and b_3. The algebra required to solve for the coefficients grows with the number of equations and tends to become cumbersome. In the step potential problem, the matrix representation was introduced to express some coefficients in terms of the others and in the previous section we saw that the elements of the transfer matrix have a simple relationship with the transmission and reflection amplitudes. This is one of the advantages of using the transfer matrix method, it facilitates the multiple substitutions and works in an orderly and simple manner.

The continuity conditions at $z = 0$ and $z = b$, can be written in the form

$$\begin{pmatrix} a_2 \\ b_2 \end{pmatrix} = \frac{m_2^*}{2q} \begin{pmatrix} \dfrac{q}{m_2^*} + i\dfrac{k}{m_1^*} & \dfrac{q}{m_2^*} - i\dfrac{k}{m_1^*} \\ \dfrac{q}{m_2^*} - i\dfrac{k}{m_1^*} & \dfrac{q}{m_2^*} + i\dfrac{k}{m_1^*} \end{pmatrix} \begin{pmatrix} a_1 \\ b_1 \end{pmatrix}; \tag{5.116}$$

$$\begin{pmatrix} a_3 e^{ikb} \\ b_3 e^{-ikb} \end{pmatrix} = \frac{m_1^*}{2k} \begin{pmatrix} \dfrac{k}{m_1^*} - i\dfrac{q}{m_2^*} & \dfrac{k}{m_1^*} + i\dfrac{q}{m_2^*} \\ \dfrac{k}{m_1^*} + i\dfrac{q}{m_2^*} & \dfrac{k}{m_1^*} - i\dfrac{q}{m_2^*} \end{pmatrix} \begin{pmatrix} a_2 e^{qb} \\ b_2 e^{-qb} \end{pmatrix}. \tag{5.117}$$

In this problem, we can also obtain *bona fide* transfer matrices that relate state vectors at distant points. The transfer matrix $M(z_2, z_1)$, that relates the state vectors at points z_1 and z_2, behaves as a propagator transferring the physical information from one point to another. We have seen before that when the points z_1 and z_2 belong to a region of constant potential, the vectors differ only in their phases. For example, if $z_1 = 0^+$ and $z_2 = b^-$ are points inside the region II, the state vectors $\phi_{\text{II}}(0^+)$ and $\phi_{\text{II}}(b^-)$ are related as follows

$$\begin{pmatrix} a_2 e^{qb} \\ b_2 e^{-qb} \end{pmatrix} = \begin{pmatrix} e^{qb} & 0 \\ 0 & e^{-qb} \end{pmatrix} \begin{pmatrix} a_2 \\ b_2 \end{pmatrix} = M(b^-, 0^+) \begin{pmatrix} a_2 \\ b_2 \end{pmatrix}. \tag{5.118}$$

The transfer matrix $M(b^-, 0^+)$ is diagonal, introduces a phase factor and propagates the state vector from $z = 0^+$ to $z = b^-$. Combining the transition matrices in (5.116) and (5.117) with the transfer matrix in (5.118), we can establish the relation

$$\begin{pmatrix} a_3 e^{ikb^+} \\ b_3 e^{-ikb^+} \end{pmatrix} = M_b(b^+, 0^-) \begin{pmatrix} a_1 \\ b_1 \end{pmatrix}, \tag{5.119}$$

where

$$M_b(b^+,0^-)$$
$$= \frac{m_1^* m_2^*}{4qk} \begin{pmatrix} \frac{k}{m_1^*} - i\frac{q}{m_2^*} & \frac{k}{m_1^*} + i\frac{q}{m_2^*} \\ \frac{k}{m_1^*} + i\frac{q}{m_2^*} & \frac{k}{m_1^*} - i\frac{q}{m_2^*} \end{pmatrix} \begin{pmatrix} e^{qb} & 0 \\ 0 & e^{-qb} \end{pmatrix} \begin{pmatrix} \frac{q}{m_2^*} + i\frac{k}{m_1^*} & \frac{q}{m_2^*} - i\frac{k}{m_1^*} \\ \frac{q}{m_2^*} - i\frac{k}{m_1^*} & \frac{q}{m_2^*} + i\frac{k}{m_1^*} \end{pmatrix}, \tag{5.120}$$

is the barrier transfer matrix that transfers the state vector from the left to the right of the barrier (from $z = 0^-$ to $z = b^+$). To simplify the notation we shall define

$$\kappa = k m_2^* \quad \text{and} \quad \upsilon = q m_1^* \tag{5.121}$$

Multiplying the matrices in (5.120) we have *the transfer matrix of the barrier*:

$$M_b(b^+,0^-) = \begin{pmatrix} \cosh qb + i\frac{\kappa^2 - \upsilon^2}{2\upsilon\kappa}\sinh qb & -i\frac{\kappa^2 + \upsilon^2}{2\upsilon\kappa}\sinh qb \\ i\frac{\kappa^2 + \upsilon^2}{2\upsilon\kappa}\sinh qb & \cosh qb - i\frac{\kappa^2 - \upsilon^2}{2\upsilon\kappa}\sinh qb \end{pmatrix}. \tag{5.122}$$

This is an important result and can be used as a *building block* for more complex systems that contain more than one barrier. For convenience we will denote the transfer matrix as

$$M_b(b^+,0^-) = \begin{pmatrix} \alpha_b & \beta_b \\ \gamma_b & \delta_b \end{pmatrix}, \tag{5.123}$$

with $\gamma_b = \beta_b^*$ and with $\delta_b = \alpha_b^*$. These relations follow from the fact that the rectangular barrier is a time reversal invariant potential. In this case, the (5.119) is written in the form

$$\begin{pmatrix} a_3 e^{ikb} \\ b_3 e^{-ikb} \end{pmatrix} = \begin{pmatrix} \alpha_b & \beta_b \\ \beta_b^* & \alpha_b^* \end{pmatrix} \begin{pmatrix} a_1 \\ b_1 \end{pmatrix}, \tag{5.124}$$

with

$$\begin{aligned} \alpha_b &= \cosh qb + i\frac{\kappa^2 - \upsilon^2}{2\kappa\upsilon}\sinh qb, \\ \beta_b &= -i\frac{\kappa^2 + \upsilon^2}{2\kappa\upsilon}\sinh qb. \end{aligned} \tag{5.125}$$

As said before, the transfer matrix contains all the physical information of the interaction between the particle and the potential. By using the transfer matrix properly, we can not only determine the wave functions but also, when possible, the eigenvalues, and in systems such as the rectangular barrier, the coefficients T and R.

5.5.1 Wave Functions for the Rectangular Barrier

Let us now see how we obtain the wave functions in the three regions of the barrier. Without losing generality, we will assume that the particles are incident only from the left, in this case the coefficient b_3 in (5.111) must vanish. We have then

$$\begin{pmatrix} a_3 e^{ikb} \\ 0 \end{pmatrix} = \begin{pmatrix} \alpha_b & \beta_b \\ \beta_b^* & \alpha_b^* \end{pmatrix} \begin{pmatrix} a_1 \\ b_1 \end{pmatrix}, \tag{5.126}$$

The transfer matrix remains the same. By multiplying the second row of M_b by the state vector to its right, we have

$$\beta_b^* a_1 + \alpha_b^* b_1 = 0, \tag{5.127}$$

and consequently

$$b_1 = -\frac{\beta_b^*}{\alpha_b^*} a_1. \tag{5.128}$$

The coefficient b_1 is the coefficient of the reflected wave function. And, according to (5.103), it is also proportional to the reflection amplitude r. If we substitute this coefficient in (5.110) we have

$$\varphi_{\mathrm{I}}(z) = a_1 e^{ikz} - \frac{\beta_b^*}{\alpha_b^*} a_1 e^{-ikz}, \tag{5.129}$$

while the wave function in region III takes the form

$$\varphi_{\mathrm{III}}(z) = a_3 e^{ikb} = \left(\alpha_b - \beta_b \frac{\beta_b^*}{\alpha_b^*} \right) a_1 e^{ik(z-b)}. \tag{5.130}$$

It is not difficult to show that the transfer matrix (5.122) is unimodular with determinant 1. For this reason

$$\varphi_{\mathrm{III}}(z) = \frac{1}{\alpha_b^*} a_1 e^{ik(z-b)}, \quad \text{with} \quad z \geq b. \tag{5.131}$$

To obtain the wave function in region II, we use the relation (5.116), in addition to

$$\begin{pmatrix} a_2 e^{qz} \\ b_2 e^{-qz} \end{pmatrix} = \begin{pmatrix} e^{qz} & 0 \\ 0 & e^{-qz} \end{pmatrix} \begin{pmatrix} a_2 \\ b_2 \end{pmatrix}, \tag{5.132}$$

with which we easily obtain

$$\varphi_{\mathrm{II}}(z) = a_1 \left[\left(1 - \frac{\beta_b^*}{\alpha_b^*} \right) \cosh qz + i \frac{\kappa}{v} \left(1 + \frac{\beta_b^*}{\alpha_b^*} \right) \sinh qz \right]. \tag{5.133}$$

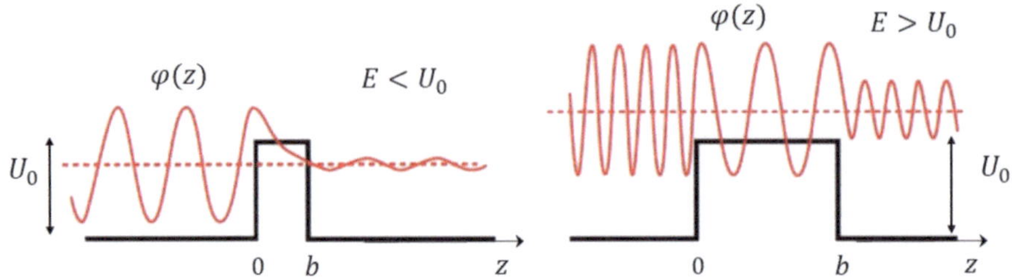

Fig. 5.8 The wave functions inside and outside the potential barrier, and the tunnel effect. In the figure on the left, the wave function is drawn for $E = 0.18\,\text{eV}$ (less than $U_o = 0.23\,\text{eV}$) and in the figure on the right for $E = 0.28\,\text{eV}$ (greater than U_o). The widths of the potential on the left and right are $b = 6\,\text{nm}$ and $b = 20\,\text{nm}$, respectively. Owing to the tunnel effect, the flux, in the figure on the left, crosses the potential barrier, even when the energy is less than the height of the barrier

In Fig. 5.8 the absolute values of the functions defined in (5.129), (5.131) and (5.133) are plotted for two different values of the energy. In the figure on the left, the energy is less than U_o; It is clear that even when an important fraction is reflected, there is a non-zero transmitted portion on the other side of the barrier. This is the *tunnel effect*. One of the most important facts in quantum phenomenology. The wave function in the barrier region drops exponentially. In the figure on the right, the energy is greater than the height of the barrier; in this case, we have important fractions of transmitted and reflected flux. In the barrier zone the wave function retains its wavelike character, but with a lower frequency.

5.5.2 Reflection and Transmission Coefficients for the Rectangular Barrier

Before obtaining the transmission coefficients we will remember some of the results obtained in this problem and their relationship with the transmission and reflection amplitudes.

In (5.128) and (5.131) we obtained the coefficients $b_1 = -(\beta_b^*/\alpha_b^*)a_1$ and $a_3 = (1/\alpha_b^*)a_1 e^{-ikb}$. Comparing these expressions with the ones we had in (5.103), it is clear that the coefficients b_1 and a_3, can also be expressed as

$$b_1 = r a_1 \quad \text{and} \quad a_3 = t a_1 e^{-ikb}, \tag{5.134}$$

with which the functions $\varphi_{\text{I}}(z)$ and $\varphi_{\text{III}}(z)$ in (5.129) and (5.131) take the form (for simplicity we choose[7] $a_1 = 1$):

$$\begin{aligned} \varphi_{\text{I}}(z) &= e^{ikz} + r\, e^{-ikz} \quad \text{if} \quad z \le 0, \\ \varphi_{\text{III}}(z) &= t\, e^{ik(z-b)} \quad \text{if} \quad z \ge b. \end{aligned} \tag{5.135}$$

[7] This choice of incident amplitude has no effect on the transfer matrix or the results obtained here.

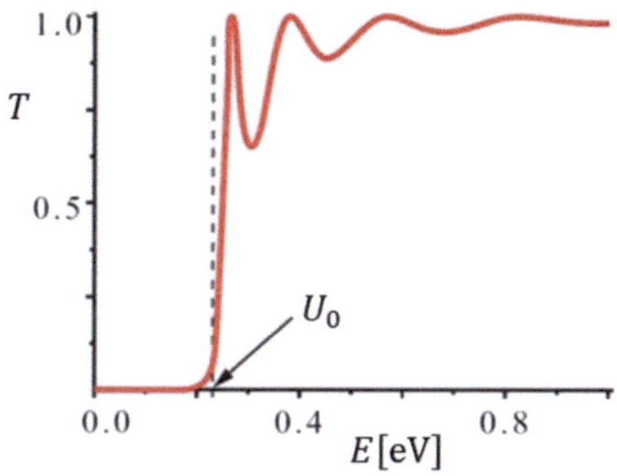

Fig. 5.9 Transmission coefficient T as a function of energy for a rectangular potential barrier with $U_o = 0.23\,\text{eV}$ and $b = 6\,\text{nm}$. The transmission coefficient does not cancel out when the energy is less than the barrier height U_o. For energies greater than U_o the transmission is resonant when it equals to 1 only for certain values of the energy, the resonant energies

In these expressions the role played by the reflection and transmission amplitudes in the wave functions is evident. It can be verified that the reflection and transmission coefficients of the barrier are, respectively,

$$T = tt^* = \frac{1}{|\alpha_b|^2} = \frac{1}{\left|\cosh qb + i\dfrac{\kappa^2 - \upsilon^2}{2\kappa\upsilon}\sinh qb\right|^2}, \tag{5.136}$$

and

$$R = rr^* = \frac{|\beta_b^*|^2}{|\alpha_b^*|^2} = \left(-\frac{\kappa^2 + \upsilon^2}{2\kappa\upsilon}\sinh qb\right)^2 T. \tag{5.137}$$

The latest results, and especially the solution $\varphi_{\text{III}}(z)$ in the (5.131), deserve additional comment. While in the classical description the probability that a particle with energy $E < U_o$ crosses the potential barrier is zero, in the quantum theory that probability is given by

$$|\varphi_{\text{III}}(z)|^2 = \frac{|a_1|^2}{|\alpha_b^*|^2} = |a_1|^2\, T, \tag{5.138}$$

which is proportional to the transmission coefficient T and, of course, to the probability $|a_1|^2$ for the particle to hit the barrier.

From the exact expression (5.136), for the transmission coefficient, it is possible to show that in the limit $qb \gg 1$ (corresponding to high and/or wide barriers) the approximate expression for the transmission coefficient is

$$T \cong e^{-\frac{2b}{\hbar}\sqrt{2m_2^*(U_o - E)}}. \tag{5.139}$$

It is evident here that the transmission tends to 1 when the energy is close to U_o and decreases exponentially with the width b and the height U_o. In Fig. 5.9, we plot the transmission coefficient T of (5.136) as a function of the energy, assuming that $U_o = 0.23$ eV and $b = 6$ nm. For small energies the transmission probability is small but not zero. When the energy is greater than U_o the transmission coefficient has a resonant behavior and tends to 1.

Let us now see the behavior of physical quantities when the energy E is greater than the barrier height.

2. If $E > U_o$, the solutions to the Schrödinger equation are:

$$\varphi_{\rm I}(z) = a_1 e^{ik_1 z} + b_1 e^{-ik_1 z} \quad \text{for} \quad z \leq 0; \tag{5.140}$$

$$\varphi_{\rm II}(z) = a_2 e^{ik_2 z} + b_2 e^{-ik_2 z}, \quad \text{for} \quad 0 < z < b; \tag{5.141}$$

$$\varphi_{\rm III}(z) = a_3 e^{ik_1 z} + b_3 e^{-ik_1 z}, \quad \text{for} \quad z \geq b. \tag{5.142}$$

Here $k_1 = \sqrt{2m_1^* E/\hbar^2}$ and $k_2 = \sqrt{2m_2^*(E - U_o)/\hbar^2}$. As in the step potential problem, the solution $\varphi_{\rm II}(z)$ is now oscillatory. All expressions derived in the case $E < U_o$, are also obtained for $E > U_o$, but with the wave number q replaced by ik_2. Taking into account this change, and defining

$$\kappa_1 = k_1 m_2^* \quad \text{and} \quad \kappa_2 = k_2 m_1^*, \tag{5.143}$$

the transfer matrix of (5.122) becomes

$$M_b(b^+, 0^-) = \begin{pmatrix} \cos k_2 b + i\dfrac{\kappa_1^2 + \kappa_2^2}{2\kappa_1\kappa_2}\sin k_2 b & -i\dfrac{\kappa_1^2 - \kappa_2^2}{2\kappa_1\kappa_2}\sin k_2 b \\ i\dfrac{\kappa_1^2 - \kappa_2^2}{2\kappa_1\kappa_2}\sin k_2 b & \cos k_2 b - i\dfrac{\kappa_1^2 + \kappa_2^2}{2\kappa_1\kappa_2}\sin k_2 b \end{pmatrix}.$$

The hyperbolic functions are changed by trigonometric functions, and using the basic relations (5.103) between scattering amplitudes and transfer matrix, the transmission coefficient for these energies is

$$T = \frac{1}{\left| \cos k_2 b + i\dfrac{\kappa_1^2 + \kappa_2^2}{2\kappa_1\kappa_2}\sin k_2 b \right|^2}, \quad \text{for} \quad E > U_o. \tag{5.144}$$

This function can be written also in the form

$$T = \frac{\rm I}{1 + \left(\dfrac{\kappa_1^2 - \kappa_2^2}{2\kappa_1\kappa_2}\right)^2 \sin^2 k_2 b}. \tag{5.145}$$

The oscillating behavior of the transmission coefficient in Fig. 5.9 is due precisely to the trigonometric term of the denominator. It is clear that when the oscillating function $\sin k_2 b$ cancels out, the denominator is reduced to 1 and the transmission coefficient T is 1. Indeed, if

$$\sin k_2 b = 0 \quad \text{then} \quad k_2 b = l\pi \quad \text{with} \quad l = 0, \pm 1, \pm 2, ... \tag{5.146}$$

The energies in which this resonant phenomenon occurs are

$$E_l = \hbar^2 l^2 \pi^2 / 2m_2^* b + U_o. \tag{5.147}$$

To conclude this section we must notice that in the barrier problem the energy spectrum is continuous; there is no quantization.

5.6 Currents Through Polarized Potentials

When heterostructures are used in electronic transport devices, and these are connected to a battery during operation, potential differences and electric currents across the structure are established. In this section we focus on heterostructures commonly found in real transport devices, specifically, the step potential and the potential barrier. When these systems are connected to a battery, their profiles modify as shown in Figs. 5.10b and 5.14b. For these potential profiles, the Schrödinger equations are still solvable. When the Schrödinger equation is solvable, we can derive a transfer matrix and compute the corresponding transport coefficients. For the specific cases considered here, we will first obtain the transfer matrices, and taking into account the relationships between the matrix elements and the scattering amplitudes, we will be able to determine important electronic transport properties of the system.

From the general definition of the current density in quantum systems, (5.10), the incident currents are written, in general, as

$$j_{ie} = \frac{1}{L^2} e \frac{\hbar k_i}{m_1^*} |\varphi_i(z_L)|^2 = \frac{1}{L^2} e v_i |\varphi_i(z_L)|^2. \tag{5.148}$$

Here and in the following, we assume that the particles move in the three-dimensional space primarily along the z-axis, and that the systems are invariant under translation in the transverse (x-y) plane. This is the one-dimensional approximation.[8] The factor $1/L^2$ in (5.148) ensures that the current density is expressed per unit volume. We also assume that $|\varphi_i(z_L)|^2$ contains the normalization constant $|A|^2 \propto 1/L$. As previously discussed, the reflection and transmission coefficients are directly related

[8] The 1D approximation of 3D systems, whose dimensions in the transverse directions are generally large, must not be confused with quasi-1D systems whose transverse dimensions are of the order of a few nanometers, and much smaller than their lengths. These systems are part of the low-dimensional systems, briefly referred to in Chap. 13.

to the reflected and transmitted currents. Using these definitions, the reflected and transmitted currents can be written as

$$j_{re} = Rj_{ie} = \frac{1}{L^2} Rev_i |\varphi_i(z_L)|^2 \quad \text{and} \quad j_{te} = Tj_{ie} = \frac{1}{L^2} Tev_i |\varphi_i(z_L)|^2. \tag{5.149}$$

From these expressions, it is clear that the currents can be determined once the coefficients R and T are calculated. This is not always possible. It is also evident that most of the dependence on the electron energy E and the applied potential V is contained in R and T. In these equations the factor $|\varphi_i(z_L)|^2$ comes from the amplitude of the incident electron. For simplicity, we can set this factor to 1, without affecting the physical results. To distinguish from the most general case involving a current of many particles, we refer to the current in (5.149), as the *current per particle*. If multiple electrons move with energy E, the factor $|\varphi_i(z_L)|^2$ should be taken proportional to the transverse degeneracy of the energy level E, or more generally, to the number of electrons moving with the same energy E. Since this factor acts merely as a scaling term, we can also choose it, for simplicity, equal to 1.

Due to flux conservation, the total electric current per particle $j_e(E, V)$ on one side of the system, given by the sum of incident and reflected currents, must be equal to the transmitted current, that is:

$$j_e(E, V) = j_{ie} - j_{re} = j_{te} = \frac{1}{L^2} eT(E, V)v_i |\varphi_i(z_L)|^2. \tag{5.150}$$

When electrons move with different energies, the occupation probabilities factor is given by the Fermi distribution

$$f_{\mu_L}(E) = \frac{1}{e^{(E-\mu_L)/k_B T} + 1}, \tag{5.151}$$

where μ_L is the Fermi level of the injected electrons. To determine the total current, the density of energy levels should be considered with energies varying from the edge of the conduction band to $E = \infty$. Thus, the total current is obtained after evaluating the integral

$$\begin{aligned} J_e(V) &= \int_{E_c}^{\infty} j_e(E, V) g_{3d}(E) f_{\mu_L}(E) dE \\ &= \int_{E_c}^{\infty} f_{\mu_L}(E) T(E, V) e v_i(E) g_{3d}(E) dE. \end{aligned} \tag{5.152}$$

Generally the integrand is peaked at the Fermi energy. In the one-dimensional approximation, we replace the density of energy levels $g_{3d}(E)$ by $g_{1d}(E) = L/\pi\hbar v_i$. If we choose the zero of the energy at the edge of the conduction band, that is, if $E_c = 0$, the electric current $I = J_e L^2$ will be

$$I(V) = \frac{e}{\pi\hbar} \int_0^{\infty} f_{\mu_L}(E) T(E, V) dE. \tag{5.153}$$

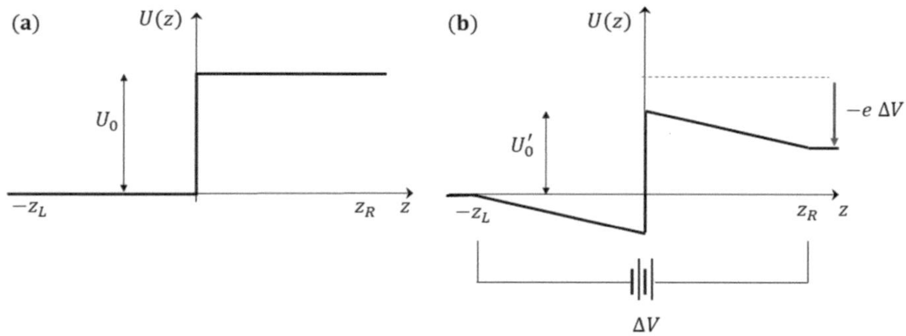

Fig. 5.10 Step potential with forward polarization. The effective height of the potential decreases to $U_o^{'}$, electrons with energies $U_0 - eV < E < U_o^{'}$ will have a non-zero probability of tunneling the barrier, and reaching the point z_R (see Fig. 5.11)

Let us now calculate the specific currents for the polarized step potential and the polarized barrier potential, separately.

5.6.1 Current Through a Polarized Step Potential

The step potential is a barrier qualitatively similar to the one formed at an np junction, but with an abrupt change at the interface. In real semiconductor systems, there are factors that make the change of the potential energy not so abrupt, such as atomic diffusion and partial filling of the atomic layers. We have seen in the previous section that, for an unpolarized step potential, particles moving from left to right require an energy E greater than the barrier height U_0 to be transmitted. In this case, tunneling is not possible because the right-hand side layer is assumed infinitely thick. Even though the wave function for $E < U_o$ can slightly penetrate the barrier, it decays exponentially and effectively vanishes. The situation changes when the step potential is polarized. If the step potential is connected to a battery as shown in Fig. 5.10, a constant electric field is established between the contact points z_L and z_R. This creates a potential difference δV across the barrier. In this system, conduction band electrons accelerate under the field,[9] and effectively their kinetic energy increases as a function of z. While the total energy remains constant, the distribution between potential and kinetic energies varies. As shown in Fig. 5.10b the potential energy gradually decreases along the field direction, except at the interface where the potential energy suddenly grows. This means that, when applying a potential difference between the points z_L and z_R, the electrons see, just as in the polarized np junction, that the height $U_0^{'}$ of the barrier in front of them gets lower in direct polarization and higher for reverse polarization. As a consequence, in forward polarization the transmission probability, even at energies lower than $U_o^{'}$, can be different from zero.

[9] Notice that in this model we are, by definition, excluding all inelastic interactions.

To obtain the currents through the polarized step potential, we consider first with the incident electric current density given by

$$j_{ie} = \frac{1}{L^2} e \frac{\hbar k_i}{m_1^*} |\varphi_i(z_L)|^2 = \frac{1}{L^2} e v_i |\varphi_i(z_L)|^2. \tag{5.154}$$

The total current density per particle at z_L is equal to the incident current plus the reflected current moving in the opposite direction. This current, due to flux conservation, is also equal to the transmitted current, that is:

$$j_e(z_L) = \frac{1}{L^2} e \frac{\hbar k_i}{m_1^*} (1 - |r|^2) ||\varphi_i(z_L)|^2 = \frac{1}{L^2} e \frac{\hbar k_i}{m_1^*} |t|^2 |\varphi_i(z_L)|^2. \tag{5.155}$$

Here $|\varphi_i(z_L)|^2$ represents the probability density of finding a quantum particle with charge e and energy E at the position z_L. As mentioned above, in a system with many particles, this factor is proportional to the concentration of incident particles with energy E, and when particles with different energies are present, we have to integrate upon the energy in order to include all the injected particles. However, before performing the integration, we must first determine the transmission coefficient $T(E, V)$ for a particle (or particles) with energy E interacting with the polarized potential.

Our objective is to calculate $T(E, V)$ exactly. If the electric potential in a point z, between the contact points at $-z_L$ and z_R, is

$$\phi(z) = V \frac{z + z_L}{z_L + z_R}, \tag{5.156}$$

the potential energy is

$$U(z) = \begin{cases} -eV \dfrac{z + z_L}{z_L + z_R} & z \leq 0 \\[2ex] U_o - eV \dfrac{z + z_L}{z_L + z_R} & z > 0. \end{cases} \tag{5.157}$$

It is usual and convenient to define the electric force $F = eV/L$, with $L = z_L + z_R$. Using this force, the Schrödinger equations on the left and right sides of the discontinuity are

$$-\frac{\hbar^2}{2m_1^*} \frac{d^2 \varphi_{\text{I}}(z)}{dz^2} - (z_L + z) F \varphi_{\text{I}}(z) = E\, \varphi_{\text{I}}(z), \quad \text{for} \quad -z_L < z \leq 0 \tag{5.158}$$

and

$$-\frac{\hbar^2}{2m_2^*} \frac{d^2 \varphi_{\text{II}}(z)}{dz^2} - (z_L + z) F \varphi_{\text{II}}(z) = (E - U_0)\, \varphi_{\text{II}}(z), \quad \text{for} \quad 0 < z < z_R. \tag{5.159}$$

An alternative method to solve these equations is the WKB approximation, but in this case, in which the potential energy is linear, it is possible to solve the differential equations exactly, transforming them into the Airy equations. If we make the well known changes of variable

$$\xi_L = \left(\frac{2m_1^* F}{\hbar^2}\right)^{1/3} \left(\frac{E}{F} + z_L + z\right) \quad \text{and} \quad \xi_R = \left(\frac{2m_2^* F}{\hbar^2}\right)^{1/3} \left(\frac{E - U_0}{F} + z_L + z\right), \tag{5.160}$$

we obtain the Airy equations

$$\frac{d^2 \varphi(\xi_i)}{d\xi_i^2} + \xi_i \varphi(\xi_i) = 0 \quad \text{with} \quad i = L, R, \tag{5.161}$$

whose general solutions are

$$\varphi(\xi_i) = a_i \mathrm{Ai}(-\xi_i) + b_i \mathrm{Bi}(-\xi_i), \tag{5.162}$$

where $\mathrm{Ai}(-\xi)$ and $\mathrm{Bi}(-\xi)$ are the Airy functions. In this case, it is convenient to define the transfer matrices $W(z_2, z_1)$, which relate vectors of wave functions and their derivatives[10] at points z_1 and z_2, as follows:

$$\begin{pmatrix} \varphi(z_2) \\ \varphi'(z_2) \end{pmatrix} = W(z_2, z_1) \begin{pmatrix} \varphi(z_1) \\ \varphi'(z_1) \end{pmatrix}. \tag{5.163}$$

For simplicity we denote $\mathrm{Ai}(-\xi(z))$ as $A(z)$ and the same for $\mathrm{Bi}(-\xi(z))$ and their derivatives. It is easy to see that

$$W(z_2, z_1) = \frac{1}{\mathscr{W}} \begin{pmatrix} A'(z_1)B(z_2) - B'(z_1)A(z_2) & B(z_1)A(z_2) - A(z_1)B(z_2) \\ -B'(z_1)A'(z_2) + A'(z_1)B'(z_2) & B(z_1)A'(z_2) - A(z_1)B'(z_2) \end{pmatrix}, \tag{5.164}$$

with $\mathscr{W} = A'(z_1)B(z_1) - A(z_1)B'(z_1)$. The transfer matrix, connecting the functions and their derivatives at points $-z_L$ and z_R, is

$$W(z_R, -z_L) = W(z_R, 0)W(0, -z_L). \tag{5.165}$$

Given the transfer matrix $W(z_R, -z_L)$, we can obtain the transfer matrix $M(z_R, -z_L)$ through the similarity transformations

$$M_{sp}(z_R, -z_L) = \begin{pmatrix} 1 & 1 \\ q & -q \end{pmatrix}^{-1} W(z_R, -z_L) \begin{pmatrix} 1 & 1 \\ ik_L & -ik_L \end{pmatrix} = \begin{pmatrix} \alpha_{sp} & \beta_{sp} \\ \gamma_{sp} & \delta_{sp} \end{pmatrix}, \tag{5.166}$$

[10] For the definition of this matrix and its relationship with the matrix M, see P. Pereyra *Physica Status Solidi b*, **259**, 210405 (2022).

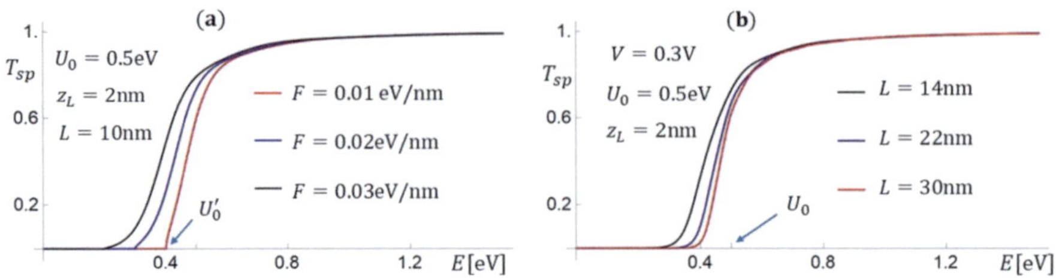

Fig. 5.11 Transmission coefficient of polarized step potential, as a function of the particle energy. In the left panel, for different values of the electric force F, and, in the right, for different lengths $L = z_L + z_R$. The higher F, the lower the height of the barrier, and the transmission begins at lower energies. In the right panel, the applied potential is $V = 0.3$ V. The probability of transmission, for a given value of E, is lower for longer lengths

with $k_L(-z_L) = \sqrt{2m_1^*/\hbar^2}\sqrt{E}$ and $q(z_R) = \sqrt{2m_2^*/\hbar^2}\sqrt{U_o - E - LF}$. Now that we have the transfer matrix $M(z_R, -z_L)$, and knowing the relation with the scattering amplitudes, we can calculate the reflection and transmission coefficients and we can see how they behave, especially as functions of the incident particle energy and also as functions of the applied potential V.

In Fig. 5.11a, b, we plot the transmission coefficient

$$T_{sp}(E, V) = \frac{|\det M_{sp}|}{|\delta_{sp}|^2}, \tag{5.167}$$

as a function of the energy. In the left panel, for different values of the electric force. In the right panel, for different lengths $L = z_L + z_R$ and a fixed value of the applied potential V. Notice that the greater the electric force F is (i.e. the greater the potential difference V), the lower the height of the barrier and the transmission begins at lower energies. On the right-side panel, the applied potential is $V = 0.3$ V, but the strengths of electric forces are different. In all cases, the probability of transmission, for the same value of E, is greater when the slope $F = eV/L$ of the potential $eV(z + z_L)(z_L + z)$ increases, either because V increases or because L decreases.

To evaluate the electric current per particle j_{et} we need the speed v_i given by

$$v_i = \frac{\hbar k_i}{m^*}\bigg|_{z=-z_L}, \tag{5.168}$$

which can explicitly be written as

$$v_i = \frac{\hbar}{m_1^*}\sqrt{\frac{2m_1^*}{\hbar^2}\left(E + eV\frac{z + z_L}{L}\right)}\bigg|_{z=-z_L} = \sqrt{\frac{2E}{m_1^*}}. \tag{5.169}$$

With this speed and the transmission coefficient, the transmitted current density, per particle, takes the form

$$j_{et}(E) = \frac{1}{L^3}eT(E, V)\sqrt{\frac{2E}{m_1^*}} = e\sqrt{\frac{2E}{m_1^*}}\frac{|\det M_{sp}|}{|\delta_{sp}|^2}. \tag{5.170}$$

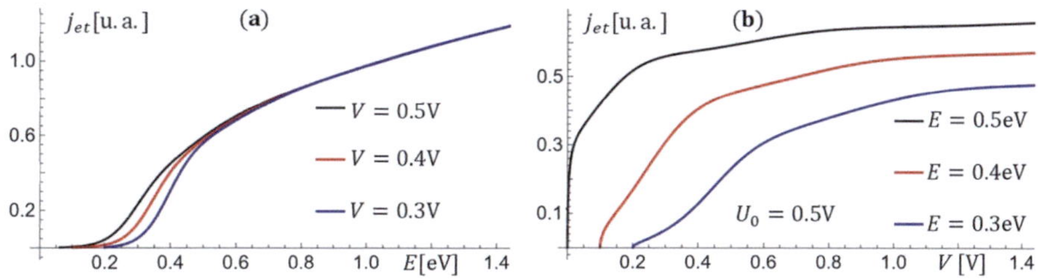

Fig. 5.12 Current per particle in the polarized step potential. In **a** as a function of the energy of the particle, for different values of the potential V. When the energy of the incident-particle is small, the currents depend strongly on V, but when the energies are large, the currents are practically independent of V. In **b** the current as a function of the potential V for different values of the incident-particle energy. The currents tend to limiting values, which depend on the incident energy

In Fig. 5.12 this current is plotted as a function of the energy of the incident particle and also as a function of the applied potential V. In (a) we have the current as a function of the energy, for different values of the applied potential V. For small energies, the currents vary strongly with V, but when the energy of the incident particle increases, the currents become practically independent of V. In (b) we have the current as a function of the potential V, for different values of the incident particle energy. The current tends to limiting values, which depends on the energy of the incident particle.

To conclude this part, we will plot the total electric current

$$I_{sp} = \frac{e}{\pi\hbar}\int_0^\infty f_{\mu_L}(E)T_{sp}(E,V)dE, \tag{5.171}$$

as a function of V and of the Fermi energy E_F of the incident electrons. In Fig. 5.13 the current I_{sp} through the step potential is plotted as a function of the applied potential V, for different values of E_F. When the potential V is small, the behavior of the current is almost linear and tends to a saturation current for potentials of the order of $2U_0/e$. For these graphs, we assumed that $U_0 = 0.5\,\text{eV}$.

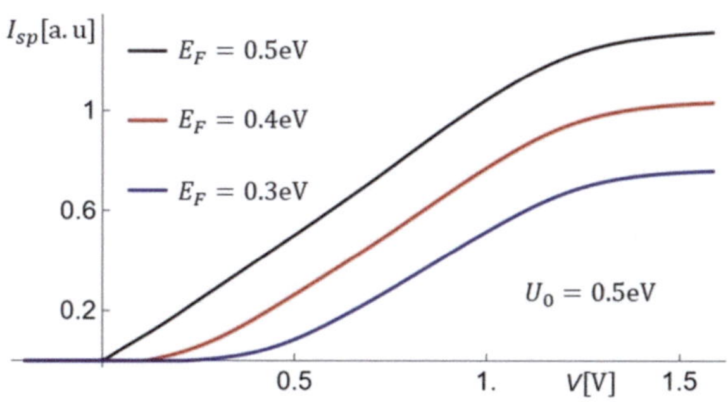

Fig. 5.13 Electric current I_{sp} through the step potential as a function of the applied potential V, for different values of the Fermi energy. For this graph it was assumed that $U_0 = 0.5\,\text{eV}$. For low potentials the behavior is almost linear and tends to a saturation current for potentials of the order of $2U_0/e$

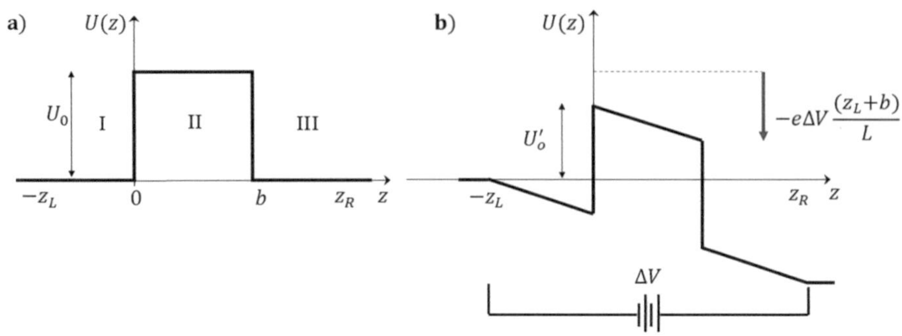

Fig. 5.14 A rectangular barrier connected to a battery. The potential energy varies linearly with z and the initial barrier height U_0 decreases with V

5.6.2 Current Through a Polarized Rectangular Barrier

When a potential barrier is connected to a battery and a potential difference V is established, the change in the potential profile is as shown in the Fig. 5.14. The Schrödinger equation of each region transforms into an Airy equation, as for the step potential in the last section. Thus, the solutions of regions I and III are

$$\varphi_{\text{I,III}}(\xi_0) = a\text{Ai}(-\xi_0) + b\text{Bi}(-\xi_0) \quad \text{with} \quad \xi_0 = \left(\frac{2m_1^* F}{\hbar^2}\right)^{1/3} \left(\frac{E}{F} + z_L + z\right), \tag{5.172}$$

and the solution in the barrier is

$$\varphi_{\text{II}}(\xi_b) = a\text{Ai}(-\xi_b) + b\text{Bi}(-\xi_b) \quad \text{with} \quad \xi_b = \left(\frac{2m^* F}{\hbar^2}\right)^{1/3} \left(\frac{E - U_o}{F} + z_L + z\right). \tag{5.173}$$

The transfer matrix that connects the wave functions and their derivatives at points $-z_L$ and z_R is similar to the one that we had for the step potential. In the rectangular barrier there is one more layer, hence one more matrix, i.e.

$$W(z_R, -z_L) = W(z_R, b)W(b, 0)W(0, -z_L). \tag{5.174}$$

Again, it is convenient to transform the matrix $W(z_R, -z_L)$ into the transfer matrix

$$M_{rb}(z_R, -z_L) = \begin{pmatrix} 1 & 1 \\ ik_R & -ik_R \end{pmatrix}^{-1} W(z_R, -z_L) \begin{pmatrix} 1 & 1 \\ ik_L & -ik_L \end{pmatrix} = \begin{pmatrix} \alpha_{rb} & \beta_{rb} \\ \gamma_{rb} & \delta_{rb} \end{pmatrix}, \tag{5.175}$$

whose matrix elements have a simple relation with the scattering amplitudes. Therefore, the transmission coefficient is given by

$$T_{rb}(E, V) = \frac{|\det M_{rb}|}{|\alpha_{rb}|^2}. \tag{5.176}$$

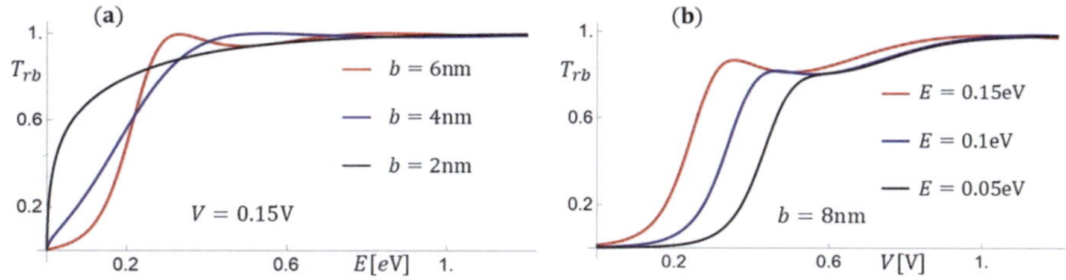

Fig. 5.15 Transmission coefficients through the potential barrier connected to a battery. In the left hand side panel the coefficient is plotted as a function of the energy, for different values of the barrier width b. In the right hand side panel the coefficient is plotted as a function of the potential V. The initial height of the barrier is $U_0 = 0.23\,\text{eV}$

Replacing this coefficient, we can evaluate the current density per incident particle as

$$j_{et}(E) = \frac{1}{L^3} eT(E, V)\sqrt{\frac{2E}{m_1^*}} = \frac{1}{L^3} e\sqrt{\frac{2E}{m_1^*}} \frac{|\det M_{rb}|}{|\delta_{rb}|^2}. \tag{5.177}$$

In Fig. 5.15, we plot the transmission coefficients through the polarized barrier as functions of the energy and as functions of the applied potential V. To set the potential parameters, we assume that the rectangular barrier is made of $GaAs$ and $Al_xGa_{1-x}As$ with $x = 0.3$. This means $U_0 = 0.23$ eV. In Fig. 5.15a the transmission coefficients as functions of the incident energy, are plotted for different values of the barrier width b. It is clear that, in the low energy region, the transmission is greater when barriers are narrower, and lower when they are wider. In Fig. 5.15b we plot the transmission coefficients as functions of the potential V, for different values of the incident energy. In this case the width of the barrier is set as $b = 10$ nm. Although the transmission coefficients depend strongly on the applied potential. It is evident that the greater are the incident energy and the applied potential, the greater is the transmission.

In Fig. 5.16 the current densities are plotted using the formulas in the (5.176) and (5.177). In panel (a), the current is plotted as a function of the energy, for different values of V. When the energy of the incident particle is small, the current is sensitive to the applied potential. As the energy increases, the current tends to become independent of the potential. In the right-hand side panel the currents are plotted as functions of V. When the energy is kept fixed, the current density behavior is similar to that of the transmission coefficient shown in Fig. 5.15b. For Fig. 5.16a, b, we assumed that $U_0 = 0.23$ eV and $b = 10$ nm.

The total current through the polarized rectangular barrier can also be evaluated. In this case the current is given by

$$I_{rb} = \frac{e}{\pi\hbar}\int_0^\infty f_{\mu_L}(E) T_{rb}(E, V) dE. \tag{5.178}$$

In Fig. 5.17 we plot this current as a function of V, and for different values of the Fermi energy. For this plot, Fermi energies much lower than the height of the barrier

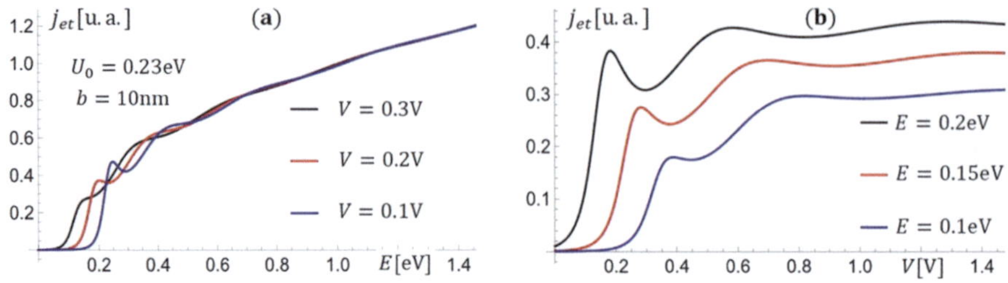

Fig. 5.16 Current density per particle through the potential barrier connected to a battery. In **a** the current j_{et} as a function of the energy, and in **b** as a function of the potential V. In all cases $U_0 = 0.23$ eV and $b = 10$ nm

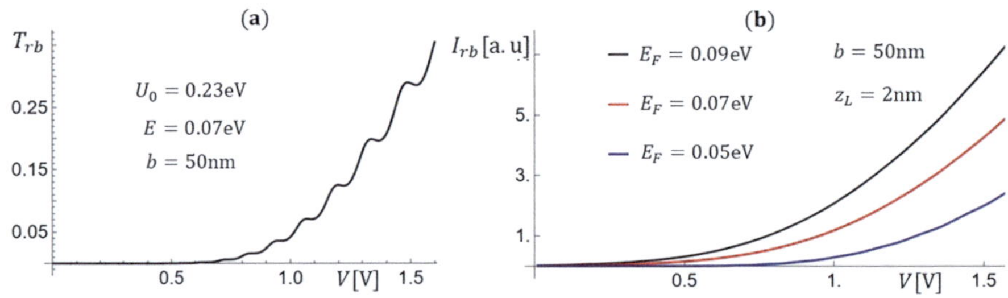

Fig. 5.17 Transmission and current I_{rb} as functions of the potential V. In the left panel a resonant transmission coefficient for a particle with energy $E = 0.07$ eV, as a function of V, and in the right panel the currents I_{rb}, as functions of V, for incident electrons with different Fermi energies. For these plots we assumed that $U_0 = 0.23$ eV, the contact point at $z_L = 2$ nm, and the barrier width $b = 50$ nm

were considered, and barriers with relatively large widths. The left panel shows the transmission coefficient of a particle with energy $E = 0.07$ eV, which is roughly of the order of the Fermi energies considered for the right-hand side panel. From the non linear and resonant behavior of the transmission coefficient, for a given energy E, we see that after integration only the non linear behavior of the current, as a function of the applied potential V, remains. In real systems where inelastic interactions occur, this effect is even more natural. This shows that with a simple model such as the polarized rectangular barrier through which particles can pass with different energies but distributed with the Fermi-Dirac occupation function, one can simulate (qualitatively and even quantitatively well) the effects in a real system that inelastic interactions would produce.

This formalism can, in principle, be applied to any quantum structure. The main problem lies, generally, in the calculation of the transmission coefficients as functions of energy and also as functions of the potential V.

Among the first proposals for calculating quantum transport properties in terms of transmission coefficients, that to some extent takes into account the loss of quantum coherence due to inelastic interactions, is the Landauer formalism that we will summarize in the next section.

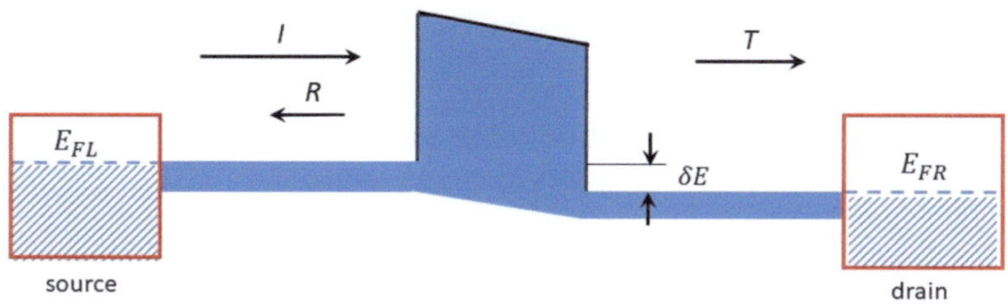

Fig. 5.18 A "realistic" system connected by conductors in thermodynamic equilibrium with reservoirs, whose Fermi energies are E_{FL} and E_{FR}

5.6.3 Landauer Conductances

It is a well established fact in classical electronic transport theory that the current density and the electric current, in polarized systems, are proportional to the applied electric potential V. Being the proportionality constants, the conductance and resistance, defined as

$$G = \frac{I}{V} \quad \text{and} \quad \rho = \frac{V}{I}. \tag{5.179}$$

As in classical systems, conductance and resistance are essential quantities that characterize electronic transport in quantum systems. In the previous sections, we analyzed specific quantum structures for which the transmission and reflection coefficients could be calculated explicitly, allowing us to determine transmitted currents. A natural and important question arises: Do these theoretical quantities correspond to those measured experimentally? This question leads us to the issue of the measurement configuration, particularly the distinction between two-probe and four-probe measurements, and the corresponding difference in the Landauer formulas. The Landauer conductance formulas provide elegant expressions in which the conductance depends directly on the transmission and reflection coefficients.

We shall now derive the Landauer conductance formulas, and we will comment later on the relation of these formulas with the measurement configuration and the underlying physical assumptions. In Fig. 5.18, the system of interest, through which the current circulates and whose conductance we aim to calculate, is connected to two electron reservoirs via ideal (perfect) conductors, or leads. These reservoirs serve as a source and drain of electrical charge, and are assumed to maintain thermodynamic equilibrium, with Fermi levels μ_L and μ_R, respectively. As mentioned earlier, the calculation of the conductance in such systems has been extensively discussed in the literature.[11] Here we will present a summary of Landauer's formalism in the one-dimensional approximation of one propagating mode. The extension to multiple propagating modes is considered in the Appendix F.

To derive the first Landauer formula, we will express the carrier concentration in terms of the number of occupied states to the left and right of the scattering region.

[11] See for example [8–10].

On the left, the number of particles is the sum of the incident plus the reflected particles, thus

$$\langle n_L \rangle = \langle n_i \rangle + \langle n_r \rangle, \tag{5.180}$$

with

$$\langle n_i \rangle \propto |\varphi_i|^2 \quad \text{and} \quad \langle n_r \rangle \propto \langle rr^* \rangle |\varphi_i|^2, \tag{5.181}$$

we have

$$\langle n_L \rangle \propto \big(1 + \langle rr^* \rangle\big) |\varphi_i|^2. \tag{5.182}$$

On the right-hand side of the scatterer, the concentration of the transmitted particles is

$$\langle n_R \rangle \propto |\varphi_t|^2 = \langle tt^* \rangle |\varphi_i|^2, \tag{5.183}$$

and the density of current is

$$j_{tL} = \frac{1}{L^2} e \langle v \rangle \langle n_R \rangle \propto \frac{1}{L^2} e \langle v \rangle \langle tt^* \rangle |\varphi_i|^2. \tag{5.184}$$

Since $\langle n_L \rangle \neq \langle n_R \rangle$, we can define the difference

$$\delta n = \langle n_L \rangle - \langle n_R \rangle = 2R|\varphi_i|^2. \tag{5.185}$$

Here R is written for $\langle rr^* \rangle$, and the relation $R = 1 - T$ was considered. It is clear that whenever the scatterer region has some degree of disorder, R and T represent averages. If, on the other hand, the difference δn is written as

$$\delta n = \frac{\partial n}{\partial E} \delta E = \frac{1}{L} g_{1d}(E) \delta E, \tag{5.186}$$

with $\delta E \simeq e\delta V$, it is easy to verify that using the density of energy levels $g_{1d}(E) = L/(\pi\hbar\langle v \rangle)$, and combining (5.185) and (5.186), we have

$$|\varphi_i|^2 \simeq \frac{e}{2R\pi\hbar\langle v \rangle} \delta V. \tag{5.187}$$

This allows us to write the electric current of electrons with energy between E and $E + \delta E$, as

$$\delta I_e = j_{tL} L^2 \simeq \frac{e^2}{2\pi\hbar} \frac{T}{R} \delta V = G\delta V, \tag{5.188}$$

where

$$G = \frac{e^2}{h} \frac{T}{R}. \tag{5.189}$$

Since the physical quantities are averages, this is also written as

$$G = \frac{e^2}{h}\frac{\langle T\rangle}{\langle R\rangle}. \tag{5.190}$$

This is, chronologically, the first of the Landauer conductance formulas. To derive the second formula we follow Landauer's argument. When the system is polarized [4], Landauer assumes that the number of particles per unit area moving to the right, in the transmitted region, corresponds to 1/2 of the density of states per unit volume and energy, $\partial n/\partial E$, multiplied by the difference of the chemical potential of the reservoirs that fed the current, and by the transmission coefficient T of the sample. Hence, the transmitted Landauer current density j_{Lt} is written as

$$j_{tL} = \frac{1}{L^3} e \frac{1}{2\hbar}\frac{\partial E}{\partial k_t}\frac{\partial n}{\partial E} T\Delta\mu. \tag{5.191}$$

In this case, an important assumption is that all transmitted particles, with energies in the interval $\Delta\mu = eV$, have the same transmission coefficient T. Since the particles move in only one direction, we have

$$\frac{\partial E}{\partial k_t}\frac{\partial n}{\partial E} = \frac{\partial n}{\partial k_t} = g_{1d}(k). \tag{5.192}$$

In the second equation of (1.36) we had the density $g_{1d}(k) = L/\pi$. Replacing this density, we obtain

$$j_{tL} = \frac{e^2}{L^2 h} T V, \tag{5.193}$$

thus, the conductance G defined as $j_e L^2/V = I/V$ becomes

$$G = \frac{L^2 j_{te}}{V} = \frac{e^2}{h} T. \tag{5.194}$$

For spin degenerate systems the unit of conductance e^2/h must be multiplied by a factor 2. In the first formula, the conductance is proportional to T/R, while in the second it is to T. It has been argued that the conductance $G \propto T/R$ results from two-probe measurements when the potential difference is measured between points in the ideal conductors, and $G \propto T$ results from four-probe measurements, when the contacts are in the reservoirs. The reader interested in this topic can consult the literature [11].

5.6.3.1 On the Scattering Coefficients in Diffusive Systems

Electronic transport is in general, in the diffusive regime, when the length L of the system is in the range

$$\ell << L << N\ell \approx \xi_{1D}, \tag{5.195}$$

where ℓ, N and ξ_{1D} are the mean free path, the number of propagating modes and the quasi-1D localization length, respectively.

It has been shown in the theory of disordered conductors of [5,6] that, in the diffusive regime, the average transmission and reflection coefficients, for time reversal invariant systems, are given as

$$\langle T \rangle \simeq \frac{N\ell}{\ell + L} - \frac{1}{3}\frac{L^3}{(\ell + L)^3} \quad \text{and} \quad \left\langle \frac{R}{T} \right\rangle \leq \frac{1}{2}\left(e^{2L/\alpha\ell} - 1\right), \tag{5.196}$$

with α a number.

5.7 The Effects of the Diffusion of Impurities at Interfaces

As mentioned in the introduction to this chapter, the practical growth of heterostructures involves a wide range of real-world challenges that experimental physicists have to control and take into account. Among these are impurity diffusion at the interface and incomplete filling of atomic layers. Although various techniques have been developed to mitigate these issues, one important phenomenon that deserves attention is the position-dependent effective mass.

When impurities diffuse across the interface or the terminating atomic layers are not fully occupied, the resulting change in the potential profile is no longer abrupt. In such cases, some of the standard modeling assumptions must be revised. The first consequence is that the effective mass, typically constant within each semiconductor layer, now varies smoothly near the interface. This gives rise to what is known as a position-dependent effective mass.

There are several approaches to incorporate this effect into the theoretical models. If the heterostructure grows along the z-direction and the one-dimensional approximation is applicable, a way to express the Schrödinger equation with a position-dependent effective mass $\mu(z) \equiv m^*(z)$ while maintaining the Hermitian nature of the Hamiltonian is

$$-\frac{\hbar^2}{2\mu^{1/2}(z)}\frac{d}{dz}\frac{\hbar^2}{2\mu^{1/2}(z)}\frac{d}{dz}\varphi(z) + (V - E)\varphi(z) = 0. \tag{5.197}$$

If in the vicinity of the A/B interface the mass is modeled as

$$\mu(z) = \frac{m_B^* - m_A^*}{1 + e^{az}} + m_A^*, \tag{5.198}$$

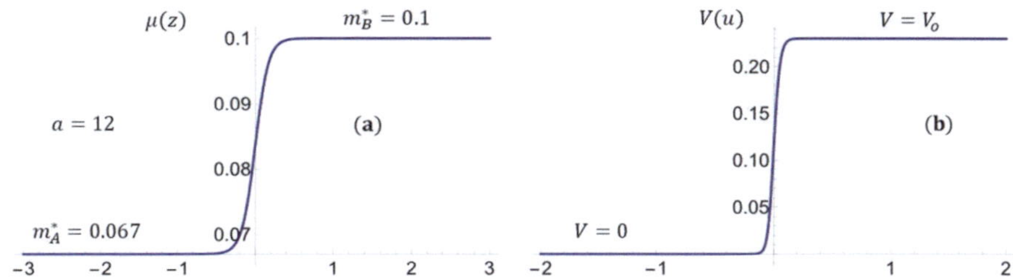

Fig. 5.19 Continuous variation of the effective mass and potential at an interface A/B. In panel **a** the variation of the effective mass $m_A^* = 0.067$ in the $GaAs$, up to $m_B^* = 0.1$ in a layer of $Al_xGa_{1-x}As$, with $x = 0.3$. In panel **b** the continuous variation of the potential, across the interface, as a function of the new variable u

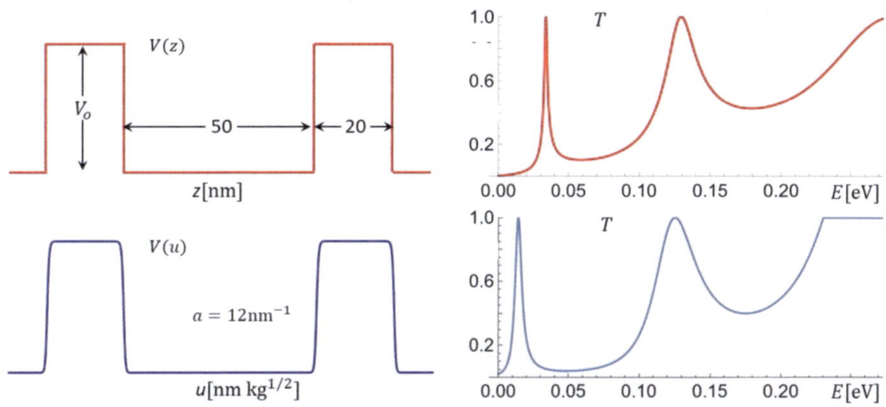

Fig. 5.20 The transmission coefficients for effective masses that change abruptly, top panels, and gradually, bottom panels

which describes the gradual change observed in Fig. 5.19a, it is possible, with a change of variable, to transfer the dependence on z from the mass to the potential [12]. The constant a in (5.198) is determined by the specific system being modeled. With the change of variable, $du/dz = \mu^{1/2}(z)$, the mass returns to its constant values, but now the potential that was sectionally constant, becomes the function

$$V_{l,r} = \frac{V_o}{2} \text{ Exp}\left[a\left(2\sqrt{m_A^*}u - (m_B^* + m_A^*)z]\right)/(m_B^* - m_A^*)\right]. \tag{5.199}$$

This function has a gradual variation with z, as seen in panel (b) of Fig. 5.19. The Schrödinger equation with these potentials can be solved, for example, in the WKB approximation. Figure 5.20 shows the transmission coefficients when the potentials, for a double barrier, are sectionally constant and when they vary continuously as described by the potentials in (5.199). It is evident that the position-dependent effective mass in the vicinity of the interface has an effect especially on the position of the lowest energy level, which may lead to red shifts in optical devices.

5.8 Problems

1. Obtain the numerical values of the current density (of electrons) at points $x_1 = -5$ nm and $x_2 = 5$ nm, located one on the left and one on the right of the discontinuity of the step potential whose height is $U_0 = 0.3$ eV, when the energy is: (a) $E = 0.2$ eV and (b) $E = 0.4$ eV.
2. Obtain the incident, reflected and transmitted currents from (5.75)
3. Given the incident, reflected and transmitted currents of the step potential, (5.75), write them in terms of the reflection and transmission coefficients, and show that $R + T = 1$.
4. Obtain the Landauer conductance as a function of the height U_0 of the step potential when (a) the energy has a definite value, less than U_0 and (b) a definite value greater than U_0.
5. Complete the derivation of (5.5).
6. Obtain the current density on the left side of the step potential studied in Sect. 5.1 when $E < U_0$.
7. Derive the relations (5.83) and obtain (5.122).
8. Show that (5.120) can indeed be written in the form of (5.122).
9. Show that the coefficients in (5.136) and (5.5.2) satisfy the relation $R + T = 1$. Obtain the reflection coefficient of the barrier when $E > U_0$.
10. Obtain the expression for the transmission coefficient from (5.139).
11. Using the definition in (5.162) and the solutions (5.163), obtain the transfer matrix in (5.164).
12. Show that the relationship in (5.166) between the matrices W and M is correct.
13. If the semiconductors that combine to produce a step potential and a barrier are $CdSe$ and $ZnSe$, plot the transmission and reflection potentials.
14. Same as in the previous problem but with GaN and $Al_xGa_{1-x}N$ semiconductors.
15. Obtain the currents I_{sp} in Fig. 5.13 when $U_0 = 0.3$ eV and $b = 40$ nm.

References

1. C. Weisbuch, B. Vinter, *Quantum Semiconductor Structures, Fundamentals and Applications* (Academic Press Inc., Boston, 1991)
2. R. Kubo, J. Phys. Soc. Jpn. **12**, 570 (1957)
3. B.L. Altshuler, A.G. Aronov, D.E. Khmelnitskii, A.I. Larkin, in *Quantum Theory of Solids*, ed. by I.M. Lifshits (MIR, URSS, 1982)
4. R. Landauer, Phys. Letts. A **85**, 91 (1981)
5. P.A. Mello, P. Pereyra, N. Kumar, Ann. Phys. **181**, 290 (1988)
6. P.A. Mello, N. Kumar, *Quantum Transport in Mesoscopic Systems. Complexity and Statistical Fluctuations* (Oxford University Press, Oxford, 2004)
7. P. Pereyra, *Fundamentals of Quantum Physics* (Springer, Berlin, 2011)
8. D.J. Thouless, Phys. Rev. Lett. **47**, 972 (1981)
9. E.N. Economou, C.M. Soukoulis, Phys. Rev. Lett. **46**, 618 (1981)
10. M.Ya. Azbel, J. Phys. C **14**, L225 (1981)
11. M. Büttiker, Y. Imry, R. Landauer, S. Pinhas, Phys. Rev. B **31**, 6207 (1985)
12. De la Cruz Leonel, *Transporte en heterostructuras con masas efectivas dependiente de la posición PTI* (UAM, Mexico, 2020)

Theory of Finite Periodic Systems

6

Abstract

This chapter provides a concise review of the theory of finite periodic systems, with a focus on semiconductor superlattices. We summarize the general solution applicable to systems with an arbitrary number of propagating modes, unit cells, and potential profiles. Closed-form expressions are derived for the transmission and reflection coefficients, enabling the determination of resonant energies in open systems and energy eigenvalues in confined systems, along with their corresponding resonant states and eigenfunctions. We also examine the symmetry properties of the eigenfunctions and introduce novel selection rules governing optical transitions between conduction and valence bands in these periodic structures.

6.1 Introduction

A great variety of natural and artificial systems have a periodic structure. When the classical theories of physics were developed, other branches of natural sciences, such as metallurgy and crystallography, made enormous contributions classifying solids and crystals by their symmetries and physical, chemical, and other kind of properties. Later, when the quantum theory was developed, the attempts to apply the new and more accurate formalism to crystals and solids was a natural aim. The complexity of these systems, with many particles but evident regularities, found in Floquet's [1] and Bloch's theorems [2] useful and important tools to deal with these systems. Despite the fact that the theorems are valid for infinite systems, solid-state physics explains many properties of solids, but at the same time, as the energy spectra become continuous and the physical systems reach mesoscopic and even nanoscopic dimensions, the theory becomes a kind of semi-classical theory. In fact, when systems are implicitly assumed of infinite size, the discretization of the energy, which is a basic feature of quantum theory, is lost. The energy bands become continuous. An important quantum property that solid-state physics did not abandon in its description, is the band structure, specifically the existence of valence and conduction bands.

P. Pereyra, *Contemporary Theory of Semiconductors and Devices*,
Graduate Texts in Physics, https://doi.org/10.1007/978-3-032-00296-9_6

The bandstructure is an important consequence of the phase coherence in periodic systems.

Currently there are two theoretical formalisms to study periodic systems. Solid State Physics, developed for over 90 years, based on the Bloch-Floquet theorem, and the Theory of Finite Periodic Systems (TFPS), where the transfer matrix method and its properties are applied to obtain solutions of n unit-cells systems, based on the single-cell solution. The latter is appropriate for systems with periodicity in one direction, such as semiconductor superlattices. Both theories predict a band structure for valence electron energies. Bloch Theory predicts continuous bands[1] and the TFPS predicts discrete bands with n energy levels per band, being n equal to the number of atoms or quantum wells in the periodic sequence. The density of levels predicted by the TFPS is discrete and, as n increases, it tends to the density of levels of the Kronig-Penney model. Unlike Bloch's theory, where the wave function is a phase multiplied by a practically unknown function that must have the periodic potential periodicity, in the TFPS, general analytical expressions are derived to explicitly evaluate energy eigenvalues, eigenfunctions and scattering amplitudes.

To calculate the optical response of superlattices, we will refer to the TFPS. We will present a summary of the general results. Although this theory is valid for an arbitrary number of propagating modes, through an arbitrary number of unit cells with arbitrary potential profiles (or refractive indices) [4], we will focus on the one-dimensional version of the formalism, i.e. on the TFPS for one propagating mode that will be applied to superlattices with periodicity in one direction, the direction of growth. It is important to make clear that when we say only one mode, we are actually considering the mode that moves in one direction and in the opposite direction. In few cases, as in point contact, quantum Hall effect, spintronics, and graphene, we will refer to more than one propagating mode (multichannel) systems.

The application of transfer matrices to one-dimensional periodic systems has been so natural that one of the most important results, the n cell transfer matrix, first reported by Jones [5] in 1941, and later by Abelés [6,7] in 1950, was rediscovered at least a dozen times later [8–18].

As we saw in the previous chapter, one of the important properties of the transfer matrix is the simple relationship that its elements have with the transmission and reflection amplitudes. These physical variables are suitable for describing electronic or wave transport in open periodic systems. In these systems, the transmission coefficients are resonant, the eigenvalues of the energy are the resonant energies, and the solutions of the Schrödinger equation are the resonant functions [4,19]. In lasers the superlattices of the active zone are limited by semiconductor layers known as cladding layers, that is, by finite repulsive potentials that confine the quantum particles. The TFPS has been extended to also obtain analytical expressions for the calculation of eigenvalues and eigenfunctions in superlattices limited by infinite and

[1] The most explicit calculation based on Bloch's theorem can be seen in the Kronig-Penney model. See [3].

finite potential barriers. We will summarize that part of the TFPS [19], as well as the symmetries analysis related to selection rules for optical transitions [20].

Griffiths and Steinke [21] used the transfer matrix approach to study wave propagation in different types of 1D, locally periodic media. Sprung et al. [22] studied the relationship between bound states and surface states in finite periodic systems. The variety of applications of this formalism is very large [23–35]. In addition to the optical response discussed in Chap. 8, we will summarize in Chaps. 12 and 14 the propagation of Gaussian packets of electrons and photons through semiconductor superlattices and optical superlattices [36]. We will also, in Chap. 15, summarize the theory of spin transport through magnetic superlattices [37–40].

6.2 Propagating Modes and Transfer Matrix

Most of the formulas written here are valid for electromagnetic systems and quantum systems. To be specific, we will be referring to semiconductors of electronic systems, but they can be applied to electromagnetic systems [36,41,42].

When we deal with transport through a system of length $l = z_R - z_L$, and cross section $w_x w_y$ connected to perfect conductors of equal cross section (see Fig. 6.1), the transverse potential energy $V_C(x, y)$ that confines the charge carriers in the transverse plane, can be modeled as the potential for an infinite well. To this potential we add a potential $V_P(x, y, z)$, which we will then require to be periodic, at least in the direction of growth z, between z_R and z_L. For simplicity, we will consider that $V_P(x, y, z)$ is a sectionally constant function of z, with discontinuities at some points $z = \ell_r$ (with $r = 0, ..., m_\ell$), and infinity outside of the region defined by $\{0 \leq x \leq w_x,\ 0 \leq y \leq w_y\}$. The coordinates ℓ_r represent the points at which the semiconductor layers begin and end, and can coincide with the end points of the unit cells. For periodic systems with n unit cells along the z axis, the extreme points of the unit cells will be at $z = z_j$ (with $j = 0, ..., n$, $z_0 = z_L$, and $z_n = z_R$). When a unit cell contains two discontinuity points, $m_\ell = 2n$. To solve the Schrödinger equation

$$-\frac{\hbar^2}{2m}\nabla^2\Psi(\mathbf{r}) + \big(V_C(x, y) + V_P(\mathbf{r})\big)\Psi(\mathbf{r}) = E\Psi(\mathbf{r}), \tag{6.1}$$

we follow the procedure of Bagwell [43]. We first solve the Schrödinger equation in the perfect conductors described by the equation

$$-\frac{\hbar^2}{2m}\left(\frac{\partial^2}{\partial x^2} + \frac{\partial^2}{\partial y^2}\right)\phi_{n_x n_y} + V_C(x, y)\,\phi_{n_x n_y} = E_{n_x n_y}\phi_{n_x n_y}, \tag{6.2}$$

plus the hard walls at $x = y = 0$, $x = w_x$ and $y = w_y$. The solutions for this system are the functions $A \sin n_x\pi x/w_x \sin n_y\pi y/w_y$, similar to those of a two-dimensional infinite well. These solutions constitute a set of orthogonal and normalized functions $\phi_{n_x n_y}(x, y)$. The quantum numbers n_x, n_y and spin projections define the physical channels $i = \{n_x, n_y, s\} = 1, 2, ..., s\mathcal{N}$. For a given Fermi energy E, *the (open)*

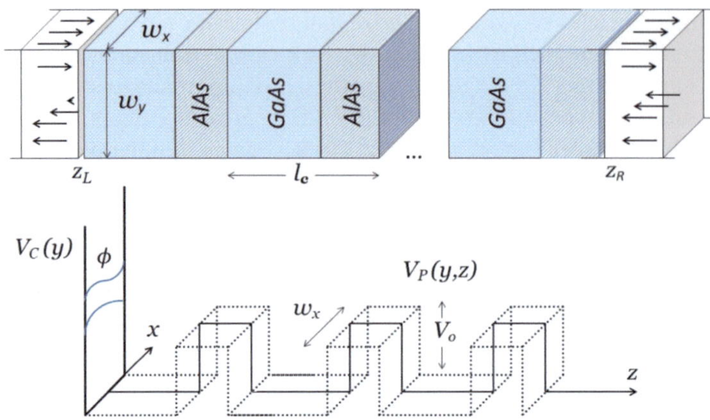

Fig. 6.1 A quasi-1D periodic system (upper panel), the transverse confinement potential $V_C(y)$ and a periodic potential $V_P(y, z)$. Reproduced from [44], with permission of Wiley

channels or propagation modes are those states whose energies are greater than the transverse energies defined by

$$E_{Ti} = \frac{\hbar^2\pi^2}{2m}\left(\frac{n_x^2}{w_x^2} + \frac{n_y^2}{w_y^2}\right) \leq E. \tag{6.3}$$

The electron energy is divided, in general, into the transverse energy E_{Ti} and the longitudinal energy. Consequently, the longitudinal wave number is

$$k_{iz}^2 = \frac{2m}{\hbar^2}E - \pi^2\left(\frac{n_x^2}{w_x^2} + \frac{n_y^2}{w_y^2}\right) \geq 0, \tag{6.4}$$

If the energies E_{Ti} are the threshold energies for the opening of a new channel or propagating mode, we can call the wave numbers $k_{Ti} = \sqrt{2mE_{Ti}}/\hbar$, the threshold wave numbers.

Following Bagwell [43], we can use the set of functions $\{\phi_i(x, y)\}$ as a basis and express the wave function $\Psi(\mathbf{r})$ in the periodic system as the combination

$$\Psi^r(x, y, z) = \sum_i \phi_i(x, y)\,\varphi_i^r(z) \quad \text{for} \quad \ell_r < z < \ell_{r+1}. \tag{6.5}$$

If we introduce this function into the Schrödinger equation (6.1), multiply by $\phi_i^*(x, y)$ and integrate, we obtain the system of coupled equations

$$\frac{d^2}{dz^2}\varphi_i^r(z) + \left(k^2 - k_{Ti}^2\right)\varphi_i^r(z) = \sum_j K_{ij}^r(z)\,\varphi_j^r(z) \quad i = 1, 2, ... \tag{6.6}$$

where $k = \sqrt{2mE}/\hbar$, $k_{Ti} = \sqrt{2mE_{Ti}}/\hbar$ and $K^r_{ij}(z)$ are the elements of the coupling matrix

$$K^r_{ij}(z) = \frac{2m}{\hbar^2}\int \phi^{r*}_i(x,y)\, V_P(x,y,z)\, \phi^r_j(x,y)\, dxdy \quad \text{for} \quad \ell_r < z < \ell_{r+1}. \tag{6.7}$$

The set of coupled equations (6.6) is infinite, therefore impossible to solve in general. Supported by physical conditions, which justify approximations such as the one-dimensional approximation, we can cut to a finite number $N = s\mathcal{N}$, called the number of physical channels, which, depending on the Fermi energy, may include only open channels or open channels plus some closed (evanescent) channels.[2] If for $i \neq j$, $K^r_{ij}(z) = 0$, the channels are said to be decoupled.

6.3 Transfer Matrix for a System with n Unit Cells

The multiplicative property of transfer matrices, noted in the previous chapter, makes them especially appropriate for dealing with layered systems. Owing to this property, when we joint two identical semiconductor structures (the unit cells) of length L/n and transfer matrix M each, the transfer matrix M_2 of the two-cell system, of length $2L/n$, is $M_2 = MM = M^2$. It is now well established that knowing the transfer matrix of a unit cell, we have all the information about the wave functions in that cell. In the same way, knowing the transfer matrix M_2, we have all the information about the wave functions of the system of two unit cells, and so on. Knowing the transfer matrix of a system with two or more cells, we have the possibility of determining other quantities such as the eigenvalues or scattering amplitudes, whose formal relations with the transfer matrix elements were obtained in the last chapter. Here we will derive closed and general relations for n-cell systems. By applying the multiplicative property over and over again, we can express the global transfer matrix (for the system with n cells) as

$$M_n = M^n = \begin{pmatrix} \alpha & \beta \\ \gamma & \delta \end{pmatrix}^n \equiv \begin{pmatrix} \alpha_n & \beta_n \\ \gamma_n & \delta_n \end{pmatrix}. \tag{6.8}$$

An important advantage of the transfer matrix method is that one can analytically determine the matrices α_n, β_n, etc., and therefore deduce analytical expressions for global physical quantities of systems with n-cells. It is true that if the objective is to work numerically, it will be sufficient to diagonalize M as $U\Lambda U^\dagger$ and write the transfer matrix of n cells as $U\Lambda^n U^\dagger$. However, doing so one loses much of the descriptive power of the transfer matrix method, which lies in deriving new analytical expressions for fundamental physical quantities.

As we saw in the previous chapter, an important property of transfer matrices is their direct relation with the scattering amplitudes. The relations that were obtained

[2] For a more detailed analysis of the alternatives for choosing transversal solutions in the leads or within the system, see [45,46].

in Sect. 5.4, are valid also for the transfer matrix elements of the whole system. This means that

$$\begin{pmatrix} \alpha_n & \beta_n \\ \gamma_n & \delta_n \end{pmatrix} = \begin{pmatrix} \left(t_n^\dagger\right)^{-1} & r_n' \left(t_n'\right)^{-1} \\ -\left(t_n'\right)^{-1} r_n & \left(t_n'\right)^{-1} \end{pmatrix}. \tag{6.9}$$

In the particular and rather amply used one-dimensional, one propagating mode approach the transfer matrices are 2×2 and their matrix elements α, ... δ are complex functions of the potential parameters. The same applies to the scattering amplitudes. If, in general, one has N propagating modes, the matrix is of dimension $2N \times 2N$ and α, β, γ and δ are matrix-blocks of dimension $N \times N$. Independently of the number of propagating modes, the transfer matrices are compatible with the physical symmetries of the system. They must satisfy conservation principles and satisfy the corresponding requirements. The time reversal symmetry is analyzed in the Appendix E, and we referred to flux conservation in Sect. 5.4.

We now outline some general results that were published in [4, 17], valid for an arbitrary number of propagating modes, and in the following section we will focus on those valid when only one propagating mode is present.

6.3.1 Basic Results for Multichannel Systems

Since

$$M_n = M M_{n-1}, \tag{6.10}$$

it is clear that, independent of the matrix dimension, we have

$$\alpha_n = \alpha\, \alpha_{n-1} + \beta\, \gamma_{n-1} \tag{6.11}$$

$$\beta_n = \alpha\, \beta_{n-1} + \beta\, \delta_{n-1} \tag{6.12}$$

$$\gamma_n = \gamma\, \alpha_{n-1} + \delta\, \gamma_{n-1} \tag{6.13}$$

$$\delta_n = \gamma\, \beta_{n-1} + \delta\, \delta_{n-1}, \tag{6.14}$$

with $\alpha_0 = \delta_0 = I_N$ and $\beta_0 = \gamma_0 = 0$. Starting from these relations, it is not difficult to obtain the recurrence relation

$$\beta_n = \left(\alpha + \beta\delta\beta^{-1}\right) \beta_{n-1} + \left(\beta\gamma - \beta\delta\beta^{-1}\alpha\right) \beta_{n-2}, \tag{6.15}$$

and similar ones for α_n, γ_n and δ_n. All these relations are three-term recurrence relations with matrix coefficients of dimension $N \times N$. If we define the matrix function

$$p_{N,n-1} = \beta^{-1}\beta_n, \tag{6.16}$$

equations (6.15) becomes the noncommutative polynomial recurrence relation (NCPRR)

$$p_{N,n} + \zeta_i p_{N,n-1} + \eta_i p_{N,n-2} = 0. \tag{6.17}$$

with $\zeta_i = -\beta^{-1}\alpha\beta - \delta$ and $\eta_i = \delta\beta^{-1}\alpha\beta - \gamma\beta$. The solution to this recurrence relation is (see [4, 17]) the matrix polynomial

$$p_{N,n} = \sum_{k=0}^{K} q_{n-k} \sum_{l=0}^{k} p_{N,l} g_{k-l} \quad \text{with} \quad K = \begin{cases} n & \text{for} \quad n < 2N \\ 2N-1 & \text{for} \quad n \geq 2N-1 \end{cases} \tag{6.18}$$

where

$$g_j = (-1)^j \sum_{l_1 < l_2 < \ldots < l_j}^{2N} \lambda_{l_1}\lambda_{l_2}\ldots\lambda_{l_j} \quad \text{with} \quad g_0 = 1, \tag{6.19}$$

and

$$q_m = \sum_{i=1}^{2N} \frac{\lambda_i^{2N+m-1}}{\prod_{j\neq i}^{2N}(\lambda_i - \lambda_j)}. \tag{6.20}$$

The initial polynomials $p_{N,l}$ can be obtained from the recurrence relation. In terms of this polynomials, the transfer matrix of the n-cell system can be written as (for lighter notation we drop the subindex N)

$$M_n = \begin{pmatrix} p_n - \gamma^{-1}\delta\gamma p_{n-1} & \beta p_{n-1} \\ \gamma p_{n-1} & p_n - \beta^{-1}\alpha\beta p_{n-1} \end{pmatrix}. \tag{6.21}$$

The relation with the scattering amplitudes was already given in (6.9). In terms of the polynomials and the unit-cell matrix elements, the scattering amplitudes for particles coming from the left are

$$t_n = \frac{1}{p_n - p_{n-1}(\gamma^{-1}\delta\gamma)^\dagger} \quad \text{and} \quad r_n = \frac{1}{p_n - \beta^{-1}\alpha\beta p_{n-1}}\gamma p_{n-1},$$

and for particles coming from the right

$$t'_n = \frac{1}{p_n - \beta^{-1}\alpha\beta p_{n-1}} \quad \text{and} \quad r'_n = \beta p_{n-1}\frac{1}{p_n - \beta^{-1}\alpha\beta p_{n-1}}.$$

6.3.2 Important Relations for 1D, One Propagating Mode Systems

In the case of one propagating mode, the matrix elements in (6.11)–(6.14), obtained from $M_n = M M_{n-1}$, together with the initial conditions $\alpha_0 = \delta_0 = 1$ and $\beta_0 = \gamma_0 = 0$, lead easily to the *recurrence relation*

$$\beta_n = (\alpha + \delta)\, \beta_{n-1} + (\beta\gamma - \delta\alpha)\, \beta_{n-2}\,, \tag{6.22}$$

and similar ones for α_n, γ_n and δ_n.

When time-reversal invariance, ($\alpha = \delta^*$ and $\beta = \gamma^*$) and flux conservation ($\det M = \delta\alpha - \gamma\beta = 1$) are taken into account, the recurrence relations simplify. An additional reduction can be obtained by defining the function p_n such that

$$p_{n-1} = \beta^{-1}\beta_n \tag{6.23}$$

In this case, the recurrence relation becomes

$$p_n + 2\alpha_R\, p_{n-1} + p_{n-2} = 0 \quad \text{for} \quad n \geq 1, \tag{6.24}$$

with $\alpha_R = \Re e\alpha$ (i.e., the real part of α) with the initial conditions $p_{-1} = 0$ and $p_0 = 1$. This is precisely the recurrence relation of the second-class Chebyshev polynomials U_n evaluated at α_R. With these relations, the transfer matrix M_n, in the one propagation mode limit, is the Abelès transfer matrix

$$M_n = \begin{pmatrix} U_n - \alpha^* U_{n-1} & \beta U_{n-1} \\ \beta^* U_{n-1} & U_n - \alpha U_{n-1} \end{pmatrix}. \tag{6.25}$$

Although the one propagating mode approach represents the simplest version of the TFPS, it has been applied with success to calculate transmission coefficients for different types of systems in the 1D approximation. These results can be used for systems with more than one propagating mode and weak channel coupling, as well as for first-order approximations.

It is clear that the matrix polynomials obtained in the previous section reduce to the scalar relations and Chebyshev polynomials obtained in this section.

6.3.3 Scattering Amplitudes and Transport Properties

In the particular but widely used case of a 1-D system with one propagating mode, the transmission amplitude

$$t_n = \frac{t^\dagger}{p_n t^\dagger - p_{n-1}}, \tag{6.26}$$

takes the form

$$t_n = \frac{t^*}{t^* U_n - U_{n-1}}. \tag{6.27}$$

This is an extremely simple function of Chebyshev polynomials of second class, $U_n(\alpha_R)$ and $U_{n-1}(\alpha_R)$, and the transmission amplitude of a unit cell t. Using the identity $U_n U_{n-2} = U_{n-1}^2 - 1$, or alternatively that $|\alpha_n|^2 = 1 + |\beta_n|^2$, it is easy to show that the transmission coefficient $T_n = |t_n|^2$ can be written as [19]

$$T_n = \frac{T}{T + U_{n-1}^2(1-T)}, \tag{6.28}$$

with an evident resonant behavior. Here $T = |t|^2$ is the transmission coefficient of a unit cell. Transmission resonances occur precisely when the polynomial U_{n-1} becomes zero. Therefore, the ν-th resonant energy $E^r_{\mu,\nu}$ is the solution of

$$(\alpha_R)_\nu = \cos\frac{\big(\nu + (\mu-1)n\big)\pi}{n}, \tag{6.29}$$

with $\nu = 1, 2, 3..., n-1$. The index μ labels the bands and ν the intraband states. If the periodic system is formed in the conduction or valence band, μ labels subbands. Equation (6.29) is a dispersion relation for resonant states, a discrete dispersion relation, which differs from the continuous dispersion relations that result in hybrid approaches that combine transfer matrices with Bloch functions.

6.3.4 Landauer Conductance in the 1D TFPS

A physical quantity that acquires a particularly clear representation in the TFPS is the Landauer conductance of a system composed of n unit cells. The Landauer conductance introduced in Sect. 5.6.3, with T_n and R_n of the 1D TFPS, takes the form

$$G_n = \frac{1}{(U_{n-1})^2}\, G. \tag{6.30}$$

where G is the conductance of a single unit cell. This expression carries information about the transmission and tunneling effect, while the inverse of U_{n-1} has information about the quantum coherence responsible for the resonant character of the conductance in a periodic system. The conductance G_n resonates when the energy E coincides with one of the resonant energies $E_{\mu,\nu}$. At resonances the transmission is 1 and the resistance $R_n = 0$.

6.3.5 Quantum Coherence and Band Formation

The most important property in periodic systems is the band structure, and yet the heuristic and semiclassical arguments used in textbooks to convince the student of this phenomenon, obscure its quantum nature. In the TFPS the formation of bands is natural and easy to understand. One only needs to plot the coefficient T_n for some

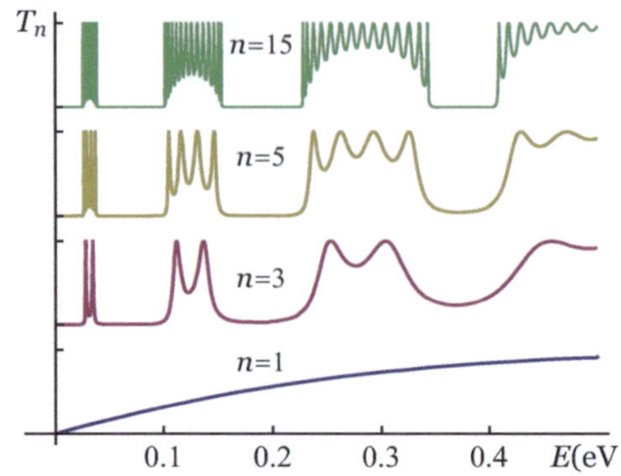

Fig. 6.2 The transmission coefficient T_n of a periodic potential composed of rectangular wells and barriers, for different numbers of unit cells n. We plot here for $n = 1, 3, 5$ and 15, and the potential parameters $a = 2$ nm, $b = 10$ nm and $V_o = 0.23$ eV. As n grows beyond, say $n = 5$, a clear band structure is seen. Reprinted from [47] Springer

values of n, to visualize how, by increasing the number of unit cells, the bands start to form. This phenomenon results from the coherent superposition of wave functions that undergo multiple reflections within the periodic system. The phases of these wave functions, regardless of where they are reflected, interfere and the bands appear. In the theory of periodic systems, this phase information is carried by the matrix polynomials p_n, in the one-dimensional case, by the Chebyshev polynomials.

In Fig. 6.2 the transmission coefficients T_n of (6.28) are plotted as functions of the electrons incident energy E through the superlattice $(GaAs/Al_xGa_{1-x}As)^n$ at the edge of the conduction band of the heterostructure. For this example, $x = 0.3$. Using the information given in the previous chapter, the reader can verify that the matrix element α of the unit-cell transfer matrix, in a sequence of wells (of width a) and rectangular barriers (of width b), is

$$\alpha = e^{ika}\left(\cosh qb + i\frac{k^2 - q^2}{2qk}\sinh qb\right), \tag{6.31}$$

with $k = \sqrt{2m_e^* E/\hbar^2}$ and $q = \sqrt{2m_e^*(V_o - E)/\hbar^2}$. The series of graphs of the transmission coefficient T_n in Fig. 6.2 shows that by increasing n a band structure is gradually built. It is also evident that when n is of order 5, the band structure begins to be better defined.

6.4 Eigenvalues, Eigenfunctions and Dispersion Relations

The calculation of eigenvalues and eigenfunctions is one of the most important objectives in solving differential equations of dynamic systems and also of solving the Schrödinger equation. The calculation of the energy eigenvalues and the corresponding eigenfunctions in bounded quasi-1D periodic systems is as important as the calculation of transmission coefficients and resonant states in open systems. Many

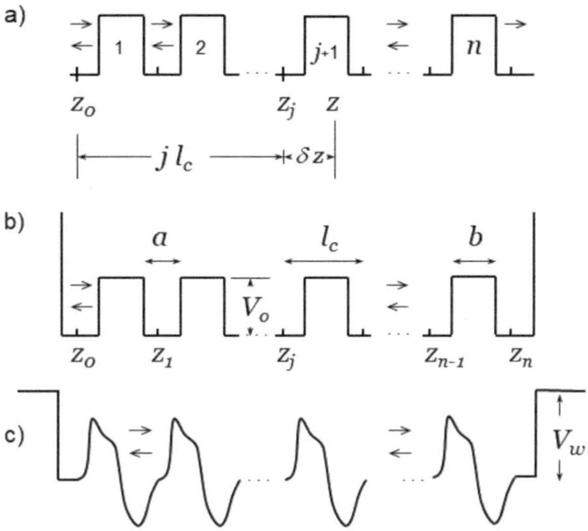

Fig. 6.3 Open, bounded, and quasi-bounded superlattices, and a sketch for determining the transfer matrix and wave function at an arbitrary point z inside the superlattice (SL). In **a** the SL is open. In this system, dispersion amplitudes can be defined, as well as energies and resonant wave functions. In **b** the SL is limited by infinite barriers, while in **c** the SL is limited by finite barriers. In the last two cases, the energy eigenvalues, eigenfunctions and dispersion relations can be obtained. To determine the wave function at any point within the SLs, say at point z of cell $j+1$ (with $j = 0, 1, 2, ..., n-1$) we need the transfer matrix $M(z, z_o)$, which is the product of $M_j = M(z_j, z_o)$ and $M_p = M(z, z_j)$, as shown below panel (**a**). Reprinted from [19], with permission from Elsevier

properties of periodic systems, such as transition probabilities in the optical response of semiconductor superlattices, depend on these quantities. In Bloch theory the possibility of finding eigenvalues is excluded by definition, but not in the theory of finite periodic systems. The TFPS was initially oriented to calculate scattering amplitudes and, in some cases, resonant energies. Subsequently, it was extended, and rigorous and compact analytical expressions were obtained for the calculation of energy eigenvalues, $E_{\mu,\nu}$ and the corresponding eigenfunctions $\Psi_{\mu,\nu}(z)$ in bounded superlattices [19]. We will present here only the main formulas for the two configurations of interest. Details can be consulted in [19,20]. All results obtained below are exact, and based on the transfer matrix method.

We will first consider open superlattices as in Fig. 6.3a, we will then outline the derivation of analytical expressions for confined superlattices as in Fig. 6.3c. We will also see that the eigenfunctions of symmetric SLs have well-defined parity symmetries, and we will present the new selection rules for the inter- and intra-subband transition probabilities.[3]

[3] For a detailed analysis of this topic, see [20,23].

6.4.1 Density of Levels, Energies and Resonant Functions in Open Superlattices

We will assume that the system of n cells is connected to ideal conductors. Although the results that will be obtained here are valid in general, that is, for any shape of the potential profile, for specific calculations we will consider that the potential of the superlattice is sectionally constant, as shown in Fig. 6.3a, also known as the Kronig-Penney model.

In open superlattices, the resonant behavior of the transmission coefficient has always been recognized as a natural characteristic in these systems and, also, almost the entire scientific community assumes that the energy spectrum of the bands must be continuous and that the Bloch functions are characteristic properties of periodic systems. In this section we will present the formulas that allow an exact calculation of the true energies and wave functions for open SLs. Since the characteristics of these results contrast eloquently with well-established and widely accepted theory, we will present specific results.

In the previous sections, we have seen that resonant transmission occurs when the energy is such that α_R is a zero of the Chebyshev polynomial $U_{n-1}(\alpha_R)$, and the relation

$$(\alpha_R)_{\mu,\nu} = \cos\frac{\nu + (\mu - 1)n}{n}\pi. \tag{6.32}$$

is satisfied. It is found that, for each value of the integer $\mu = 1, 2, ...,$ the quantum number ν can take the values $\nu = 1, 2, ..., n-1$, which correspond to the $n-1$ zeros of the Chebyshev polynomial U_{n-1}, per band. The resonant energies for a periodic system with n unit cells (and $n-1$ confining wells) are solutions of (6.32). Solving this equation, we have the entire set of resonant energies $E^r_{\mu,\nu}$. These energies are characterized by the two quantum numbers mentioned before: the quantum number μ that labels the subbands (or cycles on the unit circle) and the quantum number ν, which labels the intrasubband resonant energies.

With the resonant energies, one can determine the density of resonant levels for any number of unit cells n. We will consider a specific example, and we will show that as n increases, the density of levels approaches the level density that Kronig-Penney [3] obtained for $n = \infty$, which is given by

$$\rho(E) = \frac{1}{\pi}\frac{d}{dE}\cos^{-1}[\alpha_R]. \tag{6.33}$$

If the potential profile consists of rectangular wells and barriers, (6.32) has the explicit form

$$\cos k_\nu a \cosh q_\nu b - \frac{k_\nu^2 - q_\nu^2}{2k_\nu q_\nu}\sin k_\nu a \sinh q_\nu b = \cos\frac{\nu + (\mu - 1)n}{n}\pi, \tag{6.34}$$

where $k_\nu^2 = 2m_v^* E^r_{\mu,\nu}/\hbar^2$ and $q_\nu^2 = 2m_b^*(V_o - E^r_{\mu,\nu})/\hbar^2$.

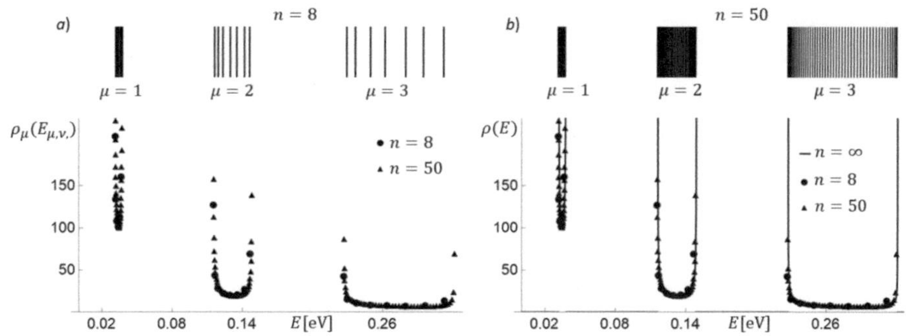

Fig. 6.4 In **a** and **b** the discrete spectra and level densities $\rho_\mu(E_{\mu,\nu})$ of resonant states are shown. In the upper panels of (**a**) and (**b**) are the spectra for superlattices with $n = 8$ and $n = 50$, squares and triangles, respectively. In the lower panels are the levels densities $\rho_\mu(E_{\mu,\nu})$. In (**a**) the densities for $n = 8$ and $n = 50$ and in (**b**) the same densities as in (**a**) but together with the Kronig-Penney levels density $\rho(E)$ that is valid for the continuous limit ($n = \infty$). The levels densities shown here are for the first three subbands of the $GaAs\ (Al_{0.3}Ga_{0.7}As/GaAs)^n$ superlattice, with $a = 10\,\text{nm}$, $b = 3\,\text{nm}$, $V_o = 0.23\,\text{eV}$. Reproduced from [44], with permission of Wiley Publisher

An example of a rectangular SL is the $GaAs\ (Al_xGa_{1-x}As/GaAs)^n$, where x defines the impurity concentration of Al and n is the number of unit cells. In Fig. 6.4a, b, we show the resonant energies and the level density of the resonant states $\rho_\mu(E_{\mu,\nu})$, for the superlattice in which the width of the $GaAs$ well is equal to 10 nm, and the width of the $Al_xGa_{1-x}As$ barrier is 3 nm. The barrier height for $x = 0.3$ is $V_o = 0.23\,\text{eV}$.

The upper panels (a) and (b) show the resonant spectra within the first three conduction subbands, for $n = 8$ in (a) and for $n = 50$ in (b). In the lower panel, the subband level densities, for $n = 8$ and $n = 50$ (squares and triangles, respectively). As expected, and shown in (b), the continuous levels density $\rho(E)$, predicted by Kronig-Penney, is reached when the number of cells $n \to \infty$.

The resonant functions are easily obtained using the transfer matrix definition. The state vector at any point z of the superlattice, say within cell $j+1$, is determined by the relation

$$\phi(z) = \begin{pmatrix} \overrightarrow{\varphi}(z) \\ \overleftarrow{\varphi}(z) \end{pmatrix} = M_p M_j \begin{pmatrix} \overrightarrow{\varphi}(z_o) \\ \overleftarrow{\varphi}(z_o) \end{pmatrix} = M_j M_p \begin{pmatrix} \overrightarrow{\varphi}(z_o) \\ \overleftarrow{\varphi}(z_o) \end{pmatrix}. \tag{6.35}$$

Here $\overrightarrow{\varphi}(z_o)$ and $\overleftarrow{\varphi}(z_o)$ are the wave functions moving to the right and to the left, respectively, at z_o. $M_j = M(z_j, z_o)$ is the transfer matrix for j complete cells, and M_p is the transfer matrix $M(z, z_j)$ connecting the state vectors at points z_j and z, see Fig. 6.3. Assuming that the incidence is only from the left side, we have

$$\overleftarrow{\varphi}(z_o) = -\frac{\beta_n^*}{\alpha_n^*}\overrightarrow{\varphi}(z_o) = r_n \overrightarrow{\varphi}(z_o). \tag{6.36}$$

Here α_n and β_n are the elements of the transfer matrix of n cell system, and r_n the reflection amplitude. Thus, the wave function at z, for any value of the energy E, is given by

$$\Psi(z, E) = \overrightarrow{\varphi}(z_o)\left[\left(\alpha_j - \beta_j\frac{\beta_n^*}{\alpha_n^*}\right)(\alpha_p + \gamma_p) + \left(\beta_j^* - \alpha_j^*\frac{\beta_n^*}{\alpha_n^*}\right)(\beta_p + \delta_p)\right]. \tag{6.37}$$

Evaluating this function for $E = E^r_{\mu,\nu}$, we have the ν-th resonant wave function in the μ-th subband, that is:

$$\Psi^r_{\mu,\nu}(z) = \Psi(z, E^r_{\mu,\nu}). \tag{6.38}$$

In one-dimensional periodic systems, the resonant wave function $\Psi^{(*)}_{\mu,\nu}(z)$ is a simple but non-trivial combination of Chebyshev polynomials. It is easy to verify that (6.37) implies

$$\Psi(z_n, E) = \frac{1}{\alpha_n^*}\overrightarrow{\varphi}(z_o) = t_n\overrightarrow{\varphi}(z_o) = \overrightarrow{\varphi}(z_n),$$

and

$$\begin{aligned}\Psi(z_o, E) &= (1 - \frac{\beta_n^*}{\alpha_n^*})\overrightarrow{\varphi}(z_o) = (1 + r_n)\overrightarrow{\varphi}(z_o)\\ &= \overrightarrow{\varphi}(z_o) + \overleftarrow{\varphi}(z_o).\end{aligned}$$

In these equations t_n and r_n are the transmission and reflection amplitudes of the n cells, respectively.

To make the difference with the standard approach even more convincing, we show in Fig. 6.5 the resonant wave functions and a gap function. It is evident that, unlike Bloch functions which are periodic, the exact resonant functions are not periodic.

6.4.2 The Eigenvalues and Eigenfunctions for Bounded Superlattices

The superlattices bounded by symmetric or asymmetric cladding layers represent a class of structures known as multiple quantum wells (MQW), widely used in optical devices. When the barrier height inside the superlattice is V_o, the barrier heights in the left and right cladding layers are V_{lb} and V_{rb}, and the energy is $E < V_o$, V_{lb}, V_{rb}, the TFPS allows us to obtain general formulas for the calculation of the energy eigenvalues $E_{\mu,\nu}$ and the corresponding eigenfunctions $\Psi^{qb}_{\mu,\nu}(z)$. If the superlattice is symmetric of length $L = nl_c$, that is, of exactly n unit cells, the eigenvalues equation is

$$h_w U_n + f_w U_{n-1} = 0, \tag{6.39}$$

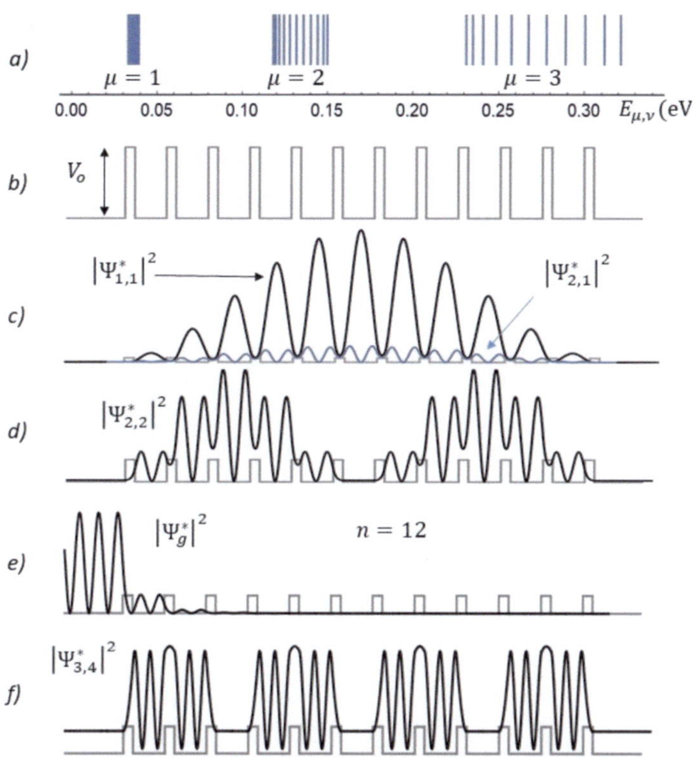

Fig. 6.5 Resonant wave functions for the $GaAs\ (Al_xGa_{1-x}As/GaAs)^n$ superlattice. In all cases we plot the amplitude squared, in arbitrary units. In **c** the resonant functions $\left|\Psi^r_{1,1}(z)\right|^2$ and $\left|\Psi^r_{2,1}(z)\right|^2$ plotted with the same scale. In **d** and **f** the resonant functions $\left|\Psi^r_{2,2}(z)\right|^2$ and $\left|\Psi^r_{3,4}(z)\right|^2$, corresponding to the energies $E^r_{2,2} = 0.11761667862$ eV and $E^r_{3,4} = 0.24557249944$ eV, respectively. In **e** is the wave function for an energy in the gap, stationary outside and exponentially decreasing within the superlattice. Reproduced from [19], with permission from Elsevier

where

$$h_w = 1, \quad \text{and} \quad f_w = \alpha_I \frac{q_w^2 - k^2}{2q_w k} - \alpha_R - \beta_I \frac{q_w^2 + k^2}{2q_w k}, \tag{6.40}$$

when $V_{lb} = V_{rb} \equiv V_w$, $q_w^2 = 2m(V_w - E)/\hbar^2$. k is the wave vector at points z_o^+, and z_n^-, and α_R and α_I are the real and imaginary parts of the element α of the transfer matrix of the unit cell.

A similar but slightly more realistic structure is the confined superlattice with all wells of equal width, and length $L = nl_c + a$. The eigenvalue equation for this system is also (6.39) but the functions h_w and f_w change. For the superlattice shown in Fig. 6.6 the functions h_w and f_w are

$$h_w = \frac{q_w^2 - k^2}{2q_w k} \sin ka + \cos ka, \tag{6.41}$$

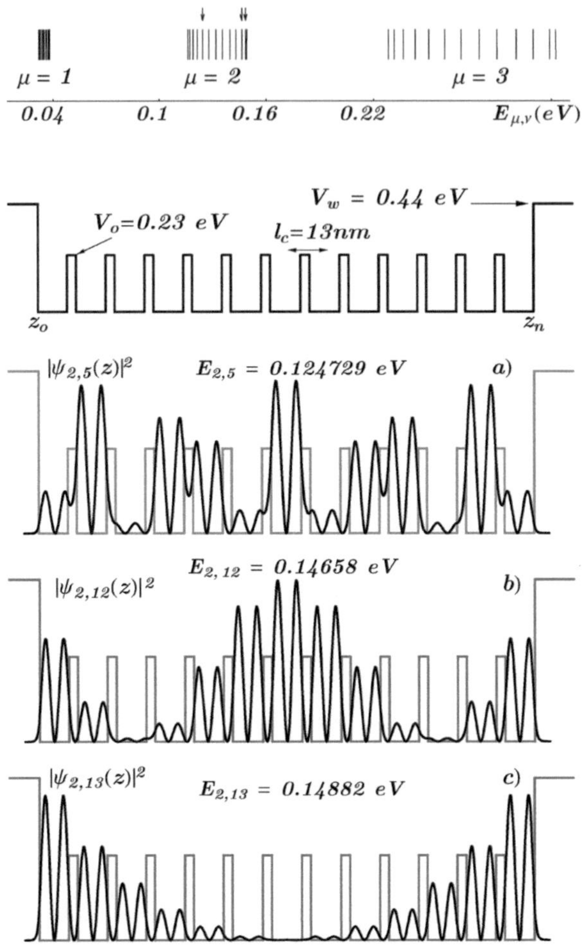

Fig. 6.6 Eigenfunctions $|\Psi_{\mu,\nu}(z)|^2$ in the second subband of the $GaAs\,(Al_xGa_{1-x}As/GaAs)^n$ superlattice bounded by finite potentials of $AlAs$ cladding layers, for $x = 0.3$ and $n = 12$. The height of the confinement potential V_w is 0.44 eV. The eigenfunctions correspond to the energy levels indicated in the top panel. The eigenfunction $|\Psi_{2,13}(z)|^2$ is a surface state, with maximum amplitude at the surface that decreases exponentially towards the interior of the superlattice. Reproduced from [19], with permission from Elsevier

and

$$f_w = \frac{q_w^2 - k^2}{2q_wk}(\alpha_I \cos ka - \alpha_R \sin ka) - \alpha_R \cos ka - \alpha_I \sin ka - \beta_I \frac{q_w^2 + k^2}{2q_wk}. \tag{6.42}$$

With a procedure similar to that followed for the resonant functions, we show that the wave function at a point z inside cell $j+1$ is given by

$$\Psi^{qb}(z,E) = \frac{A}{g_n}\left\{\left[(\alpha_p+\gamma_p)\alpha_j+(\beta_p+\delta_p)\beta_j^*\right]\left(1-i\frac{q_w}{k}\right)\right.$$
$$\left.+\left[(\alpha_p+\gamma_p)\beta_j+(\beta_p+\delta_p)\alpha_j^*\right]\left(1+i\frac{q_w}{k}\right)\right\}. \quad (6.43)$$

Here A is a normalization constant and

$$g_n = \alpha_{ni}\frac{q_w^2+k^2}{2q_w k} - \beta_{ni}\frac{q_w^2-k^2}{2q_w k} - \beta_{nr}, \quad (6.44)$$

with $\alpha_{ni} = (\alpha_n - \alpha_n^*)/2$, $\beta_{nr} = (\beta_n + \beta_n^*)/2$ and $\beta_{ni} = (\beta_n - \beta_n^*)/2$.

Again, by evaluating the wave function at $E = E_{\mu,\nu}$, we have the corresponding eigenfunction

$$\Psi^{qb}_{\mu,\nu}(z) = \Psi^{qb}(z, E_{\mu,\nu}). \quad (6.45)$$

The above equations allow us to determine not only the rigorous solutions of the Schrödinger equation, but also of the Maxwell equation if the particles that propagate are photons represented by fields through one-dimensional finite periodic optical structures. In Fig. 6.6, we plot the eigenfunctions $\Psi^{qb}_{2,5}(z)$, $\Psi^{qb}_{2,12}(z)$ and $\Psi^{qb}_{2.13}(z)$. In 6.6b–c the *surface functions* begin to appear with maximum amplitude at the extremes and with exponential decay towards the interior of the superlattice. As in previous figures, two main characteristics can be distinguished: (i) symmetry with respect to the center of the superlattice, and (ii) fast oscillations modulated by envelope functions. When the $n+1$ eigenfunctions of a subband are plotted, another energy-related symmetry is discovered, symmetry with respect to the center of the subband. The effects of these symmetries are discussed with more detail in [20,23].

6.5 Parity Symmetries and Selection Rules

The parity of resonant functions and in particular of eigenfunctions is an important property. We will analyze the parity symmetries for the systems considered in the previous section. Since the eigenfunctions depend on the Chebyshev polynomials, their symmetry properties are closely related to the symmetries of the Chebyshev polynomials. It is worth remembering that all Chebyshev polynomials, which appear in the physical expressions derived here for 1D superlattices, are evaluated at α_R, the real part of α.

6.5.1 Parity Symmetry of Resonant Functions

In the last section, the resonant wave functions were expressed as

$$\Psi^r_{\mu,\nu}(z) = \vec{\varphi}\,(z_o)\left[\left(\alpha_j - \beta_j\frac{\beta_n^*}{\alpha_n^*}\right)\left(\alpha_p + \gamma_p\right) + \left(\beta_j^* - \alpha_j^*\frac{\beta_n^*}{\alpha_n^*}\right)\left(\beta_p + \delta_p\right)\right]\Bigg|_{E^r_{\mu,\nu}}. \tag{6.46}$$

To determine the spatial inversion symmetries of these functions, we evaluate the resonant functions at two points, symmetric with respect to the superlattice center. If we consider the points $z_n = L/2$ and $z_o = -L/2$, it is easily found that the real part has the symmetry

$$\Re\mathrm{e}\big[\Psi^r_{\mu,\nu}(z_{n-1})\big] = \frac{1}{U_n}\Re\mathrm{e}\big[\Psi^r_{\mu,\nu}(z_1)\big]. \tag{6.47}$$

while the imaginary parts of $\Psi_{\mu,\nu}(z_1)$ and $\Psi_{\mu,\nu}(z_{n-1})$ satisfy the relation (Figs. 6.7 and 6.8)

$$\Im\mathrm{m}\big[\Psi^r_{\mu,\nu}(z_{n-1})\big] = -\frac{1}{U_n}\Im\mathrm{m}\big[\Psi^r_{\mu,\nu}(z_1)\big] \tag{6.48}$$

These relations together imply that (here * means complex conjugate)

$$\Psi^r_{\mu,\nu}(L/2) = U_n\Psi^{r*}_{\mu,\nu}(-L/2), \tag{6.49}$$

which also depends on the symmetries of the Chebyshev polynomial U_n, evaluated at the resonant energies. From the symmetries of the Chebyshev polynomials,[4]

$$U_n\Big|_{E_{\mu,\nu}} = \begin{cases} (-1)^{\nu} & \text{for } n \text{ even} \\ (-1)^{\nu+\mu+1} & \text{for } n \text{ odd.} \end{cases} \tag{6.50}$$

we conclude that

$$\Psi^r_{\mu,\nu}(L/2) = \begin{cases} (-1)^{\nu}\Psi^{r*}_{\mu,\nu}(-L/2) & \text{for } n \text{ even} \\ (-1)^{\nu+\mu+1}\Psi^*_{\mu,\nu}(-L/2) & \text{for } n \text{ odd.} \end{cases} \tag{6.51}$$

[4] For a more detailed presentation of this topic see Pereyra [20].

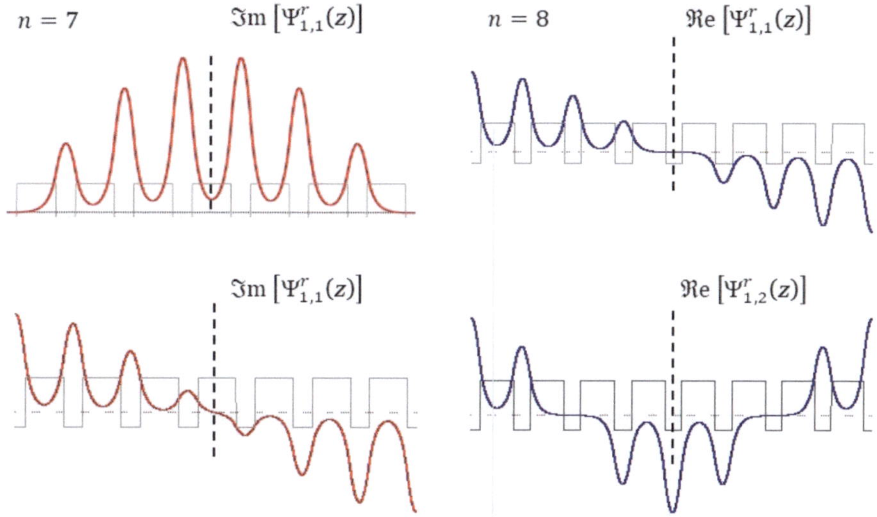

Fig. 6.7 Parity symmetries of resonant wave functions. In the left column the imaginary parts of eigenfunctions in a superlattice with $n = 7$, and in the right column the real parts of eigenfunctions with $n = 8$. As the symmetry relations in (6.48) and (6.47) establish, the function $\Im\mathrm{m}\Psi^r_{1,1}(z)$ in the left column is symmetric and $\Im\mathrm{m}\Psi^r_{1,2}(z)$ is antisymmetric. In the right column, $\Re\mathrm{e}\Psi^r_{1,1}(z)$ is antisymmetric and $\Re\mathrm{e}\Psi^r_{1,2}(z)$ is symmetric. Reproduced from [44], with permission of Wiley Publisher

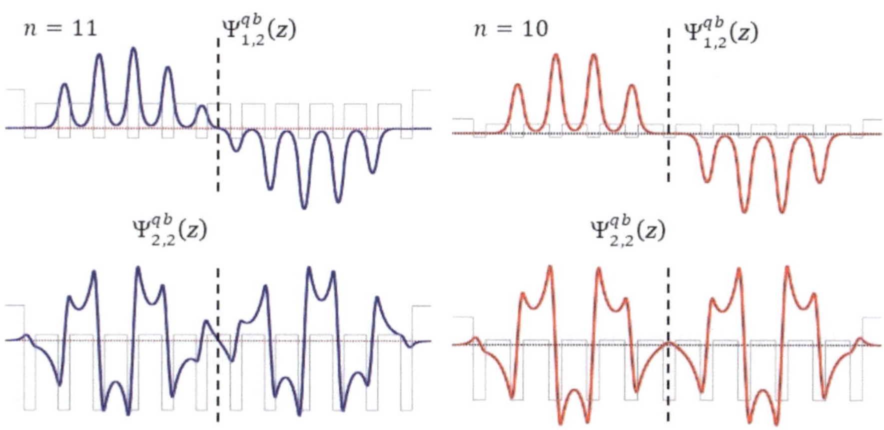

Fig. 6.8 Eigenfunctions of superlattices with length $L = nl_c + a$, bounded by finite walls. These graphs show that certainly, as predicted by the parity symmetry relation (6.52), the eigenfunctions $\Psi^{qb}_{1,2}(z)$ and $\Psi^{qb}_{2,2}(z)$ in the left column (for $n = 11$) are antisymmetric, while the eigenfunctions $\Psi^{qb}_{1,2}(z)$ and $\Psi^{qb}_{2,2}(z)$ in the right column (for $n = 10$) are antisymmetric and symmetric, respectively. Reproduced from [44], with permission of Wiley Publisher

Table 6.1 Parity symmetry sign $s^c_{\mu,\nu}$ for resonant (r), bounded (b) and quasi-bounded (qb) SLs, the parity of n (the number of unit cells), and the quantum numbers μ and ν

$s^c_{\mu,\nu}$	SL length	n even	n odd
$s^r_{\mu,\nu}$	nl_c	$(-1)^{\nu}$	$(-1)^{\nu+\mu+1}$
$s^b_{\mu,\nu}$	nl_c	$(-1)^{\nu+1}$	$(-1)^{\nu+\mu}$
$s^b_{\mu,\nu}$	nl_c+a	$(-1)^{\nu+\mu}$	$(-1)^{\nu+1}$
$s^{qb}_{\mu,\nu}$	nl_c+a	$(-1)^{\nu+\mu}$	$(-1)^{\nu+1}$

6.5.2 Parity Symmetry of the Eigenfunctions in Bounded Superlattices

It was shown in [20] that when the SLs are bounded by finite lateral barriers, the symmetries of the eigenfunctions, for a superlattice of length $L = nl_c + a$, are

$$\Psi^{qb}_{\mu,\nu}(z) = \begin{cases} (-1)^{\nu+1}\Psi^{qb}_{\mu,\nu}(-z) & \text{for } \; n \; \text{ odd} \\ \\ (-1)^{\nu+\mu}\Psi^{qb}_{\mu,\nu}(-z) & \text{for } \; n \; \text{ even.} \end{cases} \tag{6.52}$$

Table 6.1 summarizes the parity symmetries of the eigenfunctions of the different types of superlattices. The parity symmetries are written as

$$\Psi^c_{\mu,\nu}(z) = s^c_{\mu,\nu}\Psi^c_{\mu,\nu}(-z), \tag{6.53}$$

where $s^c_{\mu,\nu}$ is a sign factor, whose values depend on the quantum numbers μ and ν, the parity of n (the number of unit cells), and depend also on whether is an open superlattice $c = o$, confined by infinite walls $c = b$, or finite walls $c = qb$.

6.5.3 Selection Rules for Transitions in Superlattices

The parity relations of the wave function lead to the following selection rules for the transition probabilities from levels in the conduction band to levels in the valence band.[5] When the number of unit cells n is even, the symmetry selection rules are:

$$\int dz \Psi^{q,v}_{\mu',\nu'}(z)\frac{\partial}{\partial z}\Psi^{q,c}_{\mu,\nu}(z) \begin{cases} = 0 & \text{when} \quad P[\mu'+\nu'] = P[\mu+\nu] \\ \neq 0 & \text{when} \quad P[\mu'+\nu'] = P[\mu+\nu+1] \end{cases}. \tag{6.54}$$

[5] In transitions with infrared emission, the functions $\Psi^{q,v}_{\mu',\nu'}(z)$ are replaced by the functions $\Psi^{q,c'}_{\mu',\nu'}(z)$ of the subband c' in the conduction band. See Pereyra [23].

Here $P[\mu + \nu]$ means parity of $\mu + \nu$, c refers to the conduction band, v refers to the valence band and q represents a quasi-bounded superlattice. When n is odd, the symmetry selection rules are

$$\int dz \Psi^{q,v}_{\mu',\nu'}(z) \frac{\partial}{\partial z} \Psi^{q,c}_{\mu,\nu}(z) \begin{cases} = 0 & \text{when} \quad P[\nu'] = P[\nu] \\ \neq 0 & \text{when} \quad P[\nu'] = P[\nu + 1] \end{cases} . \tag{6.55}$$

Similar relations are valid for infrared (intraband) transitions, but with additional constraints $\mu \geq \mu'$ and, when $\mu = \mu'$ we must also have $\nu > \nu'$.

References

1. G. Floquet, Ann. Sci. de E.N.S. **12**, 47 (1983)
2. F. Bloch, Z. Physik **52**, 555 (1928)
3. R.L. Kronig, W.G. Penney, Proc. R. Soc. A 499 (1931)
4. P. Pereyra, E. Castillo, Phys. Rev. B **65**, 205120 (2002)
5. R.C. Jones, J. Opt. Soc. Am. **31**, 500 (1941)
6. F. Abelès, Ann. Phys. **3**, 504 (1948)
7. M. Born, E. Wolf, *Principles of Optics: Electromagnetic Theory of Propagation, Interference and Diffraction of Light*, 7th edn. (Cambridge University Press, 2020)
8. H. Jones, J. Phys. F Metal Phys. **3**, 2075 (1973)
9. M. Cvetic, L. Picman, J. Phys. A Math. Gen. **14**, 379 (1981)
10. M. Pacheco, F. Claro Phys. Status Solidi (b) **114**, 399 (1982)
11. D.J. Vezzetti, M. Cahay, J. Phys. D **19**, L53 (1986)
12. R. Pérez-Alvarez, H. Rodriguez-Coppola, Phys. Status Solidi (b) **145**, 493 (1988)
13. T.H. Kalotas, A.R. Lee, Eur. J. Phys. **12**, 275 (1991)
14. D.J. Griffiths, N.F. Taussing, Am. J. Phys. **60**, 883 (1992)
15. D.W. Sprung, H. Wu, J. Martorell, Am. J. Phys. **61**, 1118 (1993)
16. M.G. Rozman, P. Reineker, R. Tehver, Phys. Lett. A **187**, 127 (1994)
17. P. Pereyra, Phys. Rev. Lett. **80**, 2677 (1998); P. Pereyra, J. Phys. A **31**, 4521 (1998)
18. P. Pereyra, Phys. Status Solidi B **259**, 2100405 (2022)
19. P. Pereyra, Ann. Phys. **320**, 1 (2005)
20. P. Pereyra, Ann. Phys. **378**, 264 (2017)
21. D.J. Griffiths, C.A. Steinke, Am. J. Phys. **69**, 137 (2001)
22. D.W.L. Sprung, P. Jagiello, J.D. Sigetich, J. Martorel. Phys. Rev. B **67**, 085318 (2003)
23. P. Pereyra, Ann. Phys. **397**, 159 (2018)
24. A. Kunold, P. Pereyra, J. Appl. Phys. **93**, 5018 (2003)
25. L. Zhang, Phys. Status Solidi (b) **248**, 2120 (2011)
26. T.H. Hanley, B.J. Gallacher, H.T.D. Grigg, Mech. Syst. Sig. Process. **89**, 27 (2017)
27. C. Pacher, E. Gornik, Phys. Rev. B **68**, 155319 (2003)
28. H. Li, J.M. Shao, H.B. Zhang, G.W. Yang, Nanoscale **6**, 3127 (2014)
29. M. Paulescu, E. Tulcan-Paulescu, P. Gravila, Int. J. Mod. Phys. B **24**, 2121 (2010)
30. C. Pacher, M. Peev, Eur. Phys. J. B **59**, 519 (2007)
31. G. Cattapan, P. Lotti, Eur. Phys. J. B **66**, 517 (2008)
32. H.-S. Wei, G.Y. Wu, J. Phys. Condens. Matter **20**, 055226 (2008)
33. M. Szelag, M. Szopa, J. Phys. Conf. Ser. **104**, 012006 (2008)
34. A. Robledo-Martinez, J.C. Sandoval, P. Pereyra, Microelectron. J. **40**, 788 (2009)
35. P. Pereyra, Adv. Cond. Matter Phys. 4824921 (2017)
36. H.P. Simanjuntak, P. Pereyra, Phys. Rev. B **67**, 045301 (2003)
37. J.L. Cardoso, P. Pereyra, A. Anzaldo-Meneses, Phys. Rev. B **63**, 153301 (2001)

38. W.-T. Lu, W. Li, Y.-L. Wang et al., J. Appl. Phys. **112**, 083712 (2012)
39. C.-Z. Ye, R. Xue, Y.-H. Nie, J.-Q. Liang, Phys. Lett. A **373**, 1290 (2009)
40. P. Pereyra, V.G. Ibarra-Sierra, J.L. Cardoso, Microelectron. J. **40**, 779 (2009)
41. P. Pereyra, Photonics **7**, 29 (2020)
42. P. Pereyra, Photonics **8**, 86 (2021)
43. P.F. Bagwell, T.P.E. Broekaert, T.P. Orlando, C.G. Fonstad, J. Appl. Phys. **68**, 4634 (1990)
44. P. Pereyra, Semicond. Sci. Technol. **39**, 095006 (2024)
45. A. Anzaldo-Meneses, P. Pereyra, Ann. Phys. **322**, 2114 (2007)
46. P.A. Mello, N. Kumar, *Quantum Transport in Mesoscopic Systems. Complexity and Statistical Fluctuations* (Oxford University Press, Oxford, 2004)
47. P. Pereyra, *Fundamentals of Quantum Physics* (Springer, Berlin, 2011)

Optical Transitions, Quantum Description

7

Abstract

In this chapter, we begin with a brief overview of transition rates expressed through the Einstein coefficients for spontaneous and stimulated processes, along with the densities of states for electrons and photons, and the joint density of states in quantum structures. We then review the Hamiltonian describing particle-field interactions and the transition amplitudes derived from time-dependent perturbation theory. We proceed to calculate the transition probabilities per unit frequency and per unit time, considering both the idealized case and the case where a finite mean lifetime of the initial states is included. When this lifetime is taken into account, we derive expressions for the asymptotic transition rate, the optical susceptibility, and the Einstein coefficients, all written in terms of the momentum matrix elements of the quantum structure. These quantities are further extended to include explicit time dependence by deriving a universal time-distribution factor that modulates their evolution in time. This factor, analogous in form to Planck's spectral distribution, governs the temporal behavior of transition probabilities, optical response, and emission coefficients.

7.1 Rates of Spontaneous and Induced Transitions

Transitions of electrons from one energy level E_i to another E_f involve the absorption or emission of energy. Most intraband transitions (within the same band) and especially interband transitions (between different bands), are associated with the absorption or emission of photons. In this chapter, we will be concerned with these types of transitions and derive general expressions for calculating the probabilities of both induced and spontaneous transitions. Following the spirit of Einstein's original model, we define the rates of emission and absorption, per unit time and per unit energy, as

P. Pereyra, *Contemporary Theory of Semiconductors and Devices*,
Graduate Texts in Physics, https://doi.org/10.1007/978-3-032-00296-9_7

$$\begin{aligned} R_{sp} &= A_{sp} g_{fi}(E) f_{if}(E,T), \\ R^i_f &= B^i_f g_{if}(E) f_{if}(E,T) \wp_{ph}(\omega,T), \\ R^f_i &= B^f_i g_{if}(E) f_{if}(E,T) \wp_{ph}(\omega,T). \end{aligned} \tag{7.1}$$

The Einstein coefficients for spontaneous emission A_{sp} and stimulated emission (or absorption) B^a_b (B^b_a) are essential quantities for predicting transition rates, as they depend on the semiconductor-field interaction. In this chapter, we will derive these coefficients. The function $g_{if}(E)$ represents the joint density of states and depends on the initial and final energy levels E_i and E_f. The functions $f_{if}(E,T)$ describe the probabilities that the initial state is occupied and the final state unoccupied. The probability of emitting or absorbing a photon with energy $E = \hbar\omega_t = |E_f - E_i|$, induced by the interaction with the electromagnetic field, also depends on the density of photonic energy levels, which is proportional to the spectral distribution $\wp_{ph}(E,T)$. Some of these quantities were introduced in Chap. 1. Before discussing the Einstein coefficients, we briefly review these contributing factors.

For interband transitions, when the semiconductor layers are large in all three spatial dimensions, the appropriate density of levels to use in computing transition rates is the joint density of states per unit of energy, as derived in Chap. 1. Including spin degeneracy, this density is given by

$$g^{3d}_{if}(E) = \frac{L_x L_y L_z}{2\pi^2} \left(\frac{2m_r}{\hbar^2}\right)^{3/2} \sqrt{\hbar\omega_t - E_g}. \tag{7.2}$$

Here ω_t is the frequency of the emitted or absorbed photon and m_r is the reduced mass. In this case, $m_r = m_e m_h/(m_e + m_h)$.

For intraband transitions the analysis is formally similar, with appropriate adjustments. When the emitting system is a heterostructure, its dimensions are typically not equal in all three directions. The transverse dimensions are usually much larger while the growth direction z is much smaller. In quantum wells, multiple quantum wells or superlattices, the layer thickness along the growth direction is typically on the order of nanometers. As a result, the energy levels within the confining wells are quantized, and the densities of energy levels becomes effectively two-dimensional. In Chap. 1, we have seen that the joint density of levels for such two-dimensional systems is

$$g^{qw}_{\mu'\mu}(E) = \frac{L_x L_y}{2\pi} \frac{m_r}{\hbar^2}, \tag{7.3}$$

This density is constant with respect to energy, but is given per unit of energy. For transverse lengths on the order of 0.1μ and effective masses around $0.1m_e$, the resulting density of levels is typically on the order of $10^2\,(\text{eV})^{-1}$.

In the definition of the recombination rates we also have the occupation probabilities

$$f_{if}(E,T) = f_i(E_i,T)\big(1 - f_f(E_f,T)\big) \\ = \frac{1}{1+e^{(E_i-\mu_F)/k_BT}}\Big(1 - \frac{1}{1+e^{(E_f-\mu_F)/k_BT}}\Big). \tag{7.4}$$

While the thermal energy k_BT is very small, typically of the order of 0.026 meV at room temperature, the energy differences $E_i - \mu_F$ and $E_f - \mu_F$ are much larger in magnitude. This difference is generally used to justify certain simplifications for $f_{if}(E,T)$. In the emission process $E_f < \mu_F < E_i$, and $E_i < \mu_F < E_f$ in absorption.

The transition probability also depends on the number of photon modes available for emission or absorption within the cavity volume $\mathscr{V} = \mathscr{L}^3$. The density of these accessible modes can be determined using the same approach employed in Chap. 1 to calculate the density of levels in a semiconductor. If the components of the photon's wave vector $\mathbf{k}$ are given by $k_x = n_x\pi/\mathscr{L}$, $k_y = n_y\pi/\mathscr{L}$ and $k_z = n_z\pi/\mathscr{L}$, then the number of modes with wave vector magnitude less than or equal to k is

$$N_k = \frac{4\pi}{3}\left(\frac{k\mathscr{L}}{2\pi}\right)^3 = \frac{\mathscr{L}^3\omega^3}{6\pi^2c^3}. \tag{7.5}$$

The derivative of this number with respect to k is the photonic density of levels

$$g(k) = \frac{\mathscr{L}^3}{2\pi^2}k^2. \tag{7.6}$$

If we take into account that $\omega = kc$ and $E = \hbar\omega$, and also that $g(k)dk = g(\omega)d\omega$ and $g(\omega)d\omega = g(E)dE$, the densities per unit frequency and per unit energy are, respectively, the following

$$g(\omega) = \frac{\mathscr{L}^3\omega^2}{2\pi^2c^3}, \quad \text{and} \quad g(E) = \frac{\mathscr{L}^3E^2}{2\pi^2\hbar^2c^3}. \tag{7.7}$$

These functions are consistent with Planck's spectral distribution for black-body radiation. In the black-body radiation problem, it was known that the total energy density u of the radiation field (i.e., energy per unit volume) is proportional to T^4. The challenge was to determine the spectral energy distribution $\wp_{ph}(\omega,T)$, which gives the energy per unit volume and per unit frequency, such that

$$du = \wp_{ph}(\omega,T)d\omega \quad \text{and} \quad u = \int_0^\infty \wp_{ph}(\omega,T)d\omega = \sigma T^4. \tag{7.8}$$

The frequency distribution that satisfies this condition along with other experimental observations is the well-known Planck spectral distribution, given by [1]

$$\wp_{ph}(\omega,T) = \frac{\omega^2}{\pi^2c^3}\frac{\hbar\omega}{e^{\hbar\omega/k_BT}-1}. \tag{7.9}$$

The relation between this frequency distribution and the photonic density of levels is simple. In fact, if we remember that

$$\langle E\rangle = \frac{\hbar\omega}{e^{\hbar\omega/k_BT}-1} = \hbar\omega n_\phi, \tag{7.10}$$

where n_ϕ is the photonic occupation number, introduced in (3.109), the spectral distribution can be written as

$$\wp_{ph}(\omega,T) = \frac{\omega^2}{\pi^2c^3}\langle E\rangle = 2g(\omega)\frac{\langle E\rangle}{\mathscr{L}^3} = g_{ph}(\omega)\frac{\langle E\rangle}{\mathscr{L}^3} \tag{7.11}$$

This is the density of levels per unit of frequency, obtained in (7.7), multiplied by the average energy per unit of volume. The factor $g_{ph}(\omega)$ is the density of levels accessible to photons, defined as

$$g_{ph}(\omega) = 2g(\omega) = \frac{\mathscr{L}^3\omega^2}{\pi^2c^3}. \tag{7.12}$$

This density includes the two possible photon polarizations. In certain applications, it is necessary to determine the number of electromagnetic modes in a wave number range from k to $k+dk$, and confined to a specific solid angle $d\Omega$. In that case, the number of accessible modes is given by

$$dN_{k,\Omega} = \frac{\mathscr{L}^3}{(2\pi)^3}dk_xdk_ydk_z = \frac{\mathscr{L}^3}{(2\pi)^3}k^2dkd\Omega = g_{k,\Omega}dkd\Omega. \tag{7.13}$$

However, when the wave vectors of the incident or emitted radiation are uniformly distributed in all directions, the transition probabilities become independent of the angles θ and ϕ. In such case, we integrate over all directions, and the relevant densities per unit of solid angle and per unit of wave number, frequency or energy, are, respectively:

$$g_{k,\Omega} = \frac{\mathscr{L}^3}{(2\pi)^3}k^2; \quad g_{\omega,\Omega} = \frac{\mathscr{L}^3\omega^2}{(2\pi)^3c^3}; \quad \text{and} \quad g_{E,\Omega} = \frac{\mathscr{L}^3E^2}{(2\pi)^3\hbar^2c^3}. \tag{7.14}$$

In spontaneous emissions, all frequencies are possible. On the other hand, when the photonic field is that of the resonant cavity, the field tends to be monochromatic and the function $\wp_{ph}(E,T) \propto \delta(\omega-\omega_t)$.

7.2 Transition Coefficients in Quantum Structures

The calculation of transition coefficients for spontaneous and induced processes, can be carried out within the framework of the time-dependent perturbation theory of quantum physics. We will present here some general expressions and results for absorption and emission probabilities adapted to heterostructures with quantized energy in the growth direction z.

To describe the electron-field system we consider the Hamiltonian

$$\widehat{H} = \widehat{H}_o + \mathscr{H} + \widehat{H}_I \tag{7.15}$$

where $\widehat{H}_o$ describes the particle in the absence of an electromagnetic field, $\mathscr{H}$ describes the field and $\widehat{H}_I$ represents the particle-field interaction, i.e.

$$\widehat{H} = \frac{1}{2m}\mathbf{p}^2 + V(\mathbf{r}) + \sum_{\alpha}\sum_{\mathbf{k}} \hbar\omega_{\mathbf{k}\alpha}\left(\widehat{b}^{\dagger}_{\mathbf{k}\alpha}\widehat{b}_{\mathbf{k}\alpha} + \frac{1}{2}\right) + \widehat{H}_I, \tag{7.16}$$

Here $V(\mathbf{r})$ is the potential energy of the electron in the semiconductor structure, $\mathbf{k}$ the propagation vector, and α the polarization. The operator $\widehat{b}^{\dagger}_{\mathbf{k}\alpha}\widehat{b}_{\mathbf{k}\alpha}$ is the photon number operator with $\widehat{b}^{\dagger}_{\phi_\alpha}$ and $\widehat{b}_{\phi_\alpha}$ the photon creation and annihilation operators acting on the photonic states. The Hamiltonian $\widehat{H}_I$, which describes the electron-field interaction, is sometimes expressed in terms of the electric field, others in terms of the magnetic field. Expressed in terms of the magnetic field, and assuming the Coulomb gauge ($\nabla \cdot \mathbf{A} = 0$) and zero scalar potential, we have[1]

$$\widehat{H}_I = \frac{e}{mc}\mathbf{A}(\mathbf{r}, t) \cdot \mathbf{p} + \frac{e^2}{2mc^2}\mathbf{A}^2(\mathbf{r}, t), \tag{7.17}$$

with

$$\mathbf{A}(\mathbf{r}, t) = \sum_{\alpha=1}^{2}\sum_{\mathbf{k}} \sqrt{\frac{\hbar}{\mathscr{V}2\epsilon_0\omega_{\mathbf{k}}}}\left[\mathbf{e}_\alpha \widehat{b}_{\mathbf{k}\alpha} e^{i\mathbf{k}\cdot\mathbf{r} - i\omega_{\mathbf{k}}t} + \mathbf{e}^*_\alpha \widehat{b}^{\dagger}_{\mathbf{k}\alpha} e^{-i\mathbf{k}\cdot\mathbf{r} + i\omega_{\mathbf{k}}t}\right]. \tag{7.18}$$

Here $\mathscr{V}$ is the normalization volume. The polarization vectors $\mathbf{e}_\alpha$ are perpendicular to the propagation vector $\mathbf{k}$.

Before obtaining the transition probabilities for this specific case, we will recall some general relations of time-dependent perturbations theory that are useful in the calculation of transition probabilities, which will allow us to derive, first, the Wentzel-Fermi golden rule and, subsequently, the transition probabilities in more realistic situations, when the finite lifetime of the initial states describing the photonic states, is taken into account.

[1] An comprehensive coverage of light-matter interaction can be seen in Y. B. Band's book [2].

7.2.1 Transition Probabilities and the Wentzel-Fermi Golden Rule

To lighten the notation slightly, we will consider only the Hamiltonians $\widehat{H}_o$ and $\widehat{H}_I$, and, just for this section, we will use $\widehat{U}(t)$ instead of $\widehat{H}_I(\mathbf{r}, t)$, to avoid using too many subscripts.

If the wave function $|\psi_n\rangle_{\mathrm{o}} = e^{-iE_nt/\hbar}\varphi_n(\mathbf{r})$ is a solution of

$$\widehat{H}_o|\psi_n\rangle_{\mathrm{o}} = ih\frac{\partial|\psi_n\rangle_{\mathrm{o}}}{\partial t}, \tag{7.19}$$

then the solution of

$$(\widehat{H}_o + \widehat{U}(t))|\psi\rangle = ih\frac{\partial|\psi\rangle}{\partial t}. \tag{7.20}$$

is written, in the time dependent perturbation theory (see Appendix D), as the combination

$$|\psi\rangle = \sum_n a_n(t)|\psi_n\rangle_{\mathrm{o}}, \tag{7.21}$$

where the time-dependent coefficients $a_n(t)$, are determined from

$$a_l(t) = -\frac{i}{\hbar}\int_0^t U_{l,n}(t')e^{i\omega_{ln}t'}dt', \quad \text{for} \quad l \neq n, \quad \text{and} \quad \omega_{ln} = \frac{E_l - E_n}{\hbar} \tag{7.22}$$

These coefficients represent the probability amplitudes that after a time t, the system has made a transition to the state $|\psi_l\rangle_{\mathrm{o}}$. Consequently, the transition probability from the state $|\psi_n\rangle_{\mathrm{o}}$ to the state $|\psi_l\rangle_{\mathrm{o}}$, is

$$|a_l(t)|^2 = \frac{1}{\hbar^2}\left|\int_0^t U_{l,n}(t')e^{i\omega_{ln}t'}dt'\right|^2. \tag{7.23}$$

This also means that the probability to remain in the state $|\psi_n\rangle_{\mathrm{o}}$ is

$$|a_n(t)|^2 = 1 - \sum_{l\neq n}|a_l(t)|^2 = 1 - \frac{1}{\hbar^2}\sum_{l\neq n}\left|\int_0^t U_{l,n}(t')e^{i\omega_{ln}t'}dt'\right|^2. \tag{7.24}$$

If the potential changes slowly over time, the matrix element $U_{l,n}$ is nearly constant and can be taken outside the integral. After integrating the rest, we obtain the transition amplitude

$$a_l(t) = -\frac{{}_{\mathrm{o}}\langle\psi_l|U|\psi_n\rangle_{\mathrm{o}}}{E_l - E_n}\left(1 - e^{i\omega_{ln}t}\right), \quad \text{for} \quad l \neq n, \tag{7.25}$$

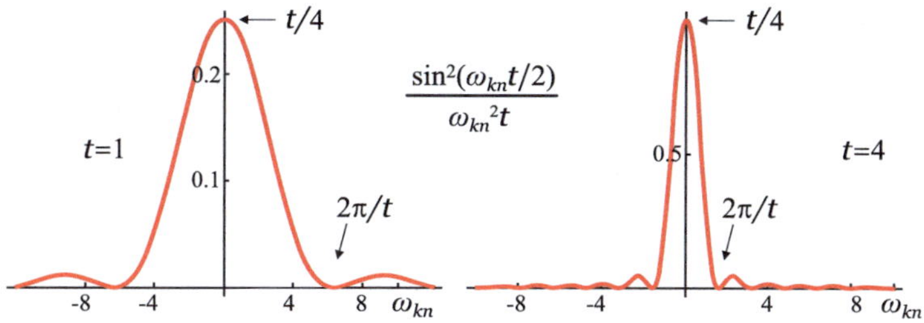

Fig. 7.1 The behavior of the function $\sin^2(\omega_{kn}t/2)/\omega_{kn}^2 t$ for $t = 1$ and for $t = 4$. The maximum is equal to $t/4$ and the region of greatest amplitudes extends from $-2\pi/t$ to $2\pi/t$. The integral of this function from $-\infty$ to ∞ is $\pi/2$, for any value of t

and the transition probability per unit of time, of the state $|\psi_n\rangle_\mathrm{o}$ to the state $|\psi_l\rangle_\mathrm{o}$ will be

$$\mathrm{W}_{ln} = \frac{|a_l(t)|^2}{t} = \frac{4\left|\,{}_\mathrm{o}\langle\psi_l\,|U|\psi_n\rangle_\mathrm{o}\right|^2}{\hbar^2}\frac{\sin^2(\omega_{ln}t/2)}{\omega_{ln}^2 t}. \tag{7.26}$$

The function $\sin^2(\omega_{ln}t/2)/\omega_{ln}^2 t$ reaches its maximum when $\omega_{ln} = 0$. The maximum value increases with time but the region that contributes the most becomes narrower, as can be seen in Fig. 7.1. The frequency range of greatest contribution is basically the region in which $|\omega_{ln}| \leq 2\pi/t \equiv \epsilon/\hbar$. In the limit $t \to \infty$ the function tends to $(\pi/2)\delta(E_n - E_l)$.

A normal procedure is to replace the function $\sin^2(\omega_{ln}t/2)/\omega_{ln}^2 t$ by a function such that the integral in the frequency interval $(-\pi/t, \pi/t)$, is also $\pi/2$. That function is $t/4$ and the sum is restricted to the states $|\psi_l\rangle_\mathrm{o}$ of the energy range $(-\pi\hbar/t, \pi\hbar/t)$. If $\rho(E_l)$ is the density of states, the number of states in this interval is

$$\Delta N = \rho(E_l)\frac{2\pi\hbar}{t}. \tag{7.27}$$

These approximations imply the transformation

$$\frac{4\left|\,{}_\mathrm{o}\langle\psi_l\,|U|\psi_n\rangle_\mathrm{o}\right|^2}{\hbar^2}\frac{\sin^2(\omega_{ln}t/2)}{\omega_{ln}^2 t} \quad\to\quad \frac{4\left|\,{}_\mathrm{o}\langle\psi_l\,|U|\psi_n\rangle_\mathrm{o}\right|^2}{\hbar^2}\rho(E_l)\frac{2\pi\hbar}{t}\frac{t}{4}. \tag{7.28}$$

In this way, we have that the probability of transition per unit of time can also be written as

$$\mathrm{W}_{ln} = \frac{2\pi\left|\,{}_\mathrm{o}\langle\psi_l\,|U|\psi_n\rangle_\mathrm{o}\right|^2}{\hbar}\rho(E_l). \tag{7.29}$$

This expression is known as the *Wentzel-Fermi golden rule*, and it has been extensively used in the study of quantum transitions. However for the analysis of the optical response of semiconductor structures, this model is insufficient. We need to extend the model to include in the optical transitions, a property that turned out to be significant, the effect of the finite lifetime of the excited states.

7.3 Transition from States with Finite Half-Lives

In the previous section, the finiteness of the half-life times (or mean lifetimes) of the excited energy levels was ignored and the explicit form of the interaction Hamiltonian $\widehat{H}_I$ and the operators $\widehat{b}^{\dagger}_{\mathbf{k}_j}$ and $\widehat{b}_{\mathbf{k}_j}$ action on the photonic states $|n_\phi\rangle$ were also not taken into account. When the initial state $|\psi_i\rangle_{\mathrm{o}}|n_\phi\rangle_i$ has a half-life time τ_i, the probability of finding the system in the initial state decreases as $e^{-t/2\tau_i}$.

The coefficient $a_{if}(t)$, which represents the probability amplitude of a transition from the state $|\psi_i\rangle_{\mathrm{o}}|n_\phi\rangle_i$, at $t = 0$ to the state $|\psi_f\rangle_{\mathrm{o}}|n_\phi\rangle_f = |\varphi_i\rangle_{\mathrm{o}}|n_\phi\rangle_i e^{-iE_it/\hbar}$, at time t, is now expressed as

$$a_{if}(t) = -\frac{i}{\hbar}\int_0^t {}_{\mathrm{o}}\langle\Phi_f\,|H_I(t')|\Phi_i\rangle_{\mathrm{o}} e^{i(E_f-E_i+i\hbar/2\tau_i)t'/\hbar}dt', \quad \text{for} \quad f \neq i. \tag{7.30}$$

Here $|\Phi_i\rangle_{\mathrm{or}} = |\varphi_i\rangle_{\mathrm{o}}|n_\phi\rangle_i$, where $|\varphi_i\rangle_{\mathrm{o}}$ describes the electrons in the semiconductor and $|n_\phi\rangle_i$ the photonic state with $n_\phi = n_{\mathbf{k}\alpha}$ the number of photons. $n_{\mathbf{k}\alpha}$ represents the number of photons whose propagation vector is $\mathbf{k}$ and polarization α.

The *energy of the electromagnetic field* is defined by the Hamiltonian

$$\mathscr{H} = \sum_\phi \hbar\omega_\phi\left(\widehat{b}^{\dagger}_\phi\widehat{b}_\phi + \frac{1}{2}\right) = \sum_\alpha\sum_{\mathbf{k}} \hbar\omega_\alpha\left(\widehat{b}^{\dagger}_{\mathbf{k}\alpha}\widehat{b}_{\mathbf{k}\alpha} + \frac{1}{2}\right). \tag{7.31}$$

Here and in what follows we will write ω_α instead of $\omega_{\mathbf{k}\alpha}$. The operators $\widehat{b}^{\dagger}_\phi$ and $\widehat{b}_\phi$, acting on photonic states, create or annihilate photons and have the following properties

$$\widehat{b}_{\phi_j}|n_{\phi_j}\rangle = \sqrt{n_{\phi_j}}\;|n_{\phi_1}, n_{\phi_2}, ..., n_{\phi_j}-1, ...\rangle \tag{7.32}$$

$$\widehat{b}^{\dagger}_{\phi_j}|n_{\phi_j}\rangle = \sqrt{n_{\phi_j}+1}\;|n_{\phi_1}, n_{\phi_2}, ..., n_{\phi_j}+1, ...\rangle \tag{7.33}$$

and the number operator $\widehat{b}^{\dagger}_{\mathbf{k}_j\alpha}\widehat{b}_{\mathbf{k}_j\alpha}$ has the property

$$\widehat{b}^{\dagger}_{\mathbf{k}_j\alpha}\widehat{b}_{\mathbf{k}_j\alpha}|n_{\phi_j}\rangle = n_{\phi_j}|n_{\phi_j}\rangle. \tag{7.34}$$

If we substitute the interaction Hamiltonian into (7.30), the transition amplitude in the International System of Units (SI), takes the form

$$a_{if}(t) = -\frac{ie}{\hbar m}\int_0^t \sum_{\alpha=1}^{2}\sum_{\mathbf{k}} \sqrt{h/4\pi\epsilon_o\mathscr{V}\omega_\alpha}\;{}_{\mathrm{o}}\!\left\langle\Phi_f\,|\widehat{b}^{\dagger}_{\mathbf{k}\alpha}e^{-i\mathbf{k}\cdot\mathbf{r}}\mathbf{p}|\Phi_i\right\rangle_{\mathrm{o}}$$

$$\cdot\,\mathbf{e}_\alpha e^{i\left(E_f-E_i+\hbar\omega_\alpha+i\hbar/2\tau_i\right)t'/\hbar}dt'. \tag{7.35}$$

To simplify the evaluation of the matrix element ${}_{o}\langle\Phi_f\,|\widehat{b}^{\dagger}_{\mathbf{k}\alpha}e^{-i\mathbf{k}\cdot\,bfr}\mathbf{p}|\Phi_i\rangle_o\cdot\mathbf{e}_\alpha$, it is common to use the dipolar approximation, in which the wavelengths are assumed to be in the optical region, that is, the region in which the wave number $k=2\pi/\lambda$ of the radiated field is very small.[2] In this approximation, the exponential $\exp(i\mathbf{k}\cdot\mathbf{r})\simeq 1+i\mathbf{k}\cdot\mathbf{r}$ is practically 1, and we have

$$a_{if}(t)=-\frac{i}{\hbar}\sum_{\alpha=1}^{2}\frac{e}{m}\sqrt{\frac{h(n_\phi+1)}{4\pi\epsilon_o\mathcal{V}\omega_\alpha}}\;{}_{o}\langle\varphi_f\,|\mathbf{e}_\alpha\cdot\mathbf{p}|\varphi_i\rangle_o\int_0^t e^{i(E_f-E_i+\hbar\omega_\alpha+i\hbar/2\tau_i)t'/\hbar}dt'. \tag{7.36}$$

When absorption processes are considered, the operator $\widehat{b}_{\mathbf{k}\alpha}$ should be considered instead of $\widehat{b}^{\dagger}_{\mathbf{k}\alpha}$, and as a consequence we have n_ϕ instead of $n_\phi+1$. In the absence of an external field, $n_\phi=0$. In this case there is no absorption but emission is still possible. That is the spontaneous emission. Evaluating the integral in (7.36) results in

$$a_{if}(t)=\sum_{\alpha=1}^{2}\frac{e}{m}\sqrt{\frac{h(n_\phi+1)}{4\pi\epsilon_o\mathcal{V}\omega_\alpha}}\;{}_{o}\langle\varphi_f\,|\mathbf{p}\cdot\mathbf{e}_\alpha|\varphi_i\rangle_o\frac{e^{i(E_f-E_i+\hbar\omega_\alpha+i\hbar/2\tau_i)t/\hbar}-1}{(E_f-E_i+\hbar\omega_\alpha+i\hbar/2\tau_i)}, \tag{7.37}$$

whose square modulus gives the transition probability

$$|a_{if}(t)|^2=\frac{\hbar e^2}{m^2 2\epsilon_o\mathcal{V}}\left|\sum_{\alpha=1}^{2}\sqrt{\frac{n_\phi+1}{\omega_\alpha}}\frac{{}_{o}\langle\varphi_f\,|\mathbf{p}\cdot\mathbf{e}_\alpha|\varphi_i\rangle_o}{(E_f-E_i+\hbar\omega_\alpha+i\hbar/2\tau_i)}\right.$$
$$\left.\left(e^{i(E_f-E_i+\hbar\omega_\alpha+i\hbar/2\tau_i)t/\hbar}-1\right)\right|^2. \tag{7.38}$$

When the functions φ_i and φ_f depend only on the variable z, as in semiconductor structures with quantization along the z direction, only the matrix elements of the z component of the linear momentum remain. In this case

$${}_{o}\langle\varphi_f\,|\mathbf{p}\cdot\mathbf{e}_\alpha|\varphi_i\rangle_o\quad\rightarrow\quad\text{either}\quad{}_{o}\langle\varphi_f\,|\mathbf{p}_z\cdot\mathbf{e}_\alpha|\varphi_i\rangle_o\quad\text{or}\quad{}_{o}\langle\varphi_f\,|p_z\hat{\mathbf{u}}_z\cdot\mathbf{e}_\alpha|\varphi_i\rangle_o. \tag{7.39}$$

These systems are quasi-two-dimensional, and taking into account the parameters shown in Fig. 7.2, the dot products with the polarization vectors are expressed in the form

$$\hat{\mathbf{u}}_z\cdot\mathbf{e}_1=\sin\theta\cos\phi\;\;\text{y}\;\;\hat{\mathbf{u}}_z\cdot\mathbf{e}_2=\sin\theta\sin\phi. \tag{7.40}$$

---

[2] This corresponds to wavelengths of the radiation field much larger than the dimensions of the system.

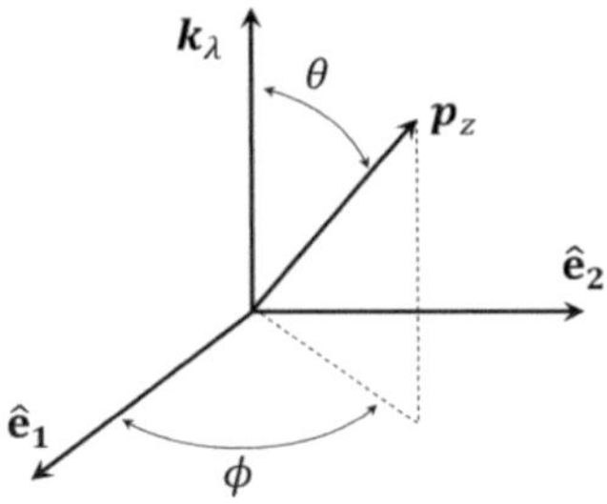

Fig. 7.2 The wave vector $\mathbf{k}_\alpha$, the polarization $\mathbf{e}_1$ and $\mathbf{e}_2$ of the radiation, and the moment $\mathbf{p}_z$

If we substitute (7.37) and (7.39) we obtain a transition coefficient that depends on the direction of emission (θ, ϕ), i.e.

$$|a_{if}(t)|^2 = \frac{e^2}{\hbar m^2 2\epsilon_o \mathscr{V}} |_o\langle\varphi_f|p_z|\varphi_i\rangle_o|^2 \left| \sum_{\alpha=1}^{2} \sqrt{\frac{n_\phi+1}{\omega_\alpha}} \mathbf{u}_z \right.$$
$$\left. \cdot \mathbf{e}_\alpha \frac{e^{i(E_f - E_i + \hbar\omega_\alpha + i\hbar/2\tau_i)t/\hbar} - 1}{(E_f - E_i + \hbar\omega_\alpha + i\hbar/2\tau_i)} \right|^2, \tag{7.41}$$

Since emissions occur in all directions, randomly, one can multiply by the solid angle element $\sin\theta d\theta d\phi/4\pi$ and integrate. This gives

$$\frac{1}{4\pi}\int_0^{2\pi}\int_0^{\pi} \mathbf{u}_z \cdot \mathbf{e}_\alpha \mathbf{u}_z \cdot \mathbf{e}_{\alpha'} \sin\theta d\theta d\phi = \frac{1}{3}\delta_{\alpha\alpha'}, \tag{7.42}$$

and the emission probability takes the form

$$|a_{if}(t)|^2 = \frac{\hbar e^2}{6m^2\epsilon_o\mathscr{V}} |_o\langle\varphi_f|p_z|\varphi_i\rangle_o|^2 \sum_{\alpha=1}^{2} \frac{n_\phi+1}{\omega_\alpha} \frac{\left[1 + e^{-2t\gamma_i} - 2e^{-t\gamma_i}\cos\Omega t\right]}{\left|E_f - E_i + \hbar\omega_\alpha + i\hbar\gamma_i\right|^2}. \tag{7.43}$$

Here we have defined the frequency

$$\Omega = \omega_\alpha - \frac{E_i - E_f}{\hbar} = \omega_\alpha - \omega_t \quad \text{and} \quad \gamma_i = \frac{1}{2\tau_i}. \tag{7.44}$$

Before we discuss the calculation of the transition probability per unit time (or transition rate), we shall conclude this section considering the density of photonic states per unit frequency.

7.3.1 Probability of Emission per Unit of Frequency

If we multiply the transition coefficient $|a_{if}^{in}(t)|^2(\omega, t)$ of (7.43) by the density of modes of the photonic field $g(\omega) = \mathscr{L}^3\omega_\alpha^2/2\pi^2c^3$ we have the probability of emission per unit frequency

$$\mathscr{P}_{if}(\omega, t) = |a_{if}^{in}(t)|^2 \frac{\mathscr{L}^3\omega_\alpha^2}{2\pi^2c^3} \tag{7.45}$$

which will become a dimensionless quantity if we multiply by $d\omega_\alpha$, to integrate. If we also take into account that light has two polarization modes, we can write (recall that $\mathscr{V} = \mathscr{L}^3$)

$$\mathscr{P}_{if}(\omega, t) = \frac{\hbar e^2}{6m^2\epsilon_o\mathscr{V}} |\,{}_o\langle\varphi_f\,|p_z|\varphi_i\rangle_o|^2 \frac{2(n_\phi+1)}{\omega} \frac{\left[1 + e^{-2t\gamma_i} - 2e^{-t\gamma_i}\cos\Omega t\right]}{(E_f - E_i + \hbar\omega)^2 + \hbar^2\gamma_i^2} \frac{\mathscr{L}^3\omega^2}{2\pi^2c^3}, \tag{7.46}$$

which, after simplifying some factors takes the form

$$\mathscr{P}_{if}(\omega, t) = \frac{e^2\hbar\omega(n_\phi+1)}{6m^2\pi^2\epsilon_o c^3} \frac{|\,{}_o\langle\varphi_f\,|p_z|\varphi_i\rangle_o|^2}{(E_f - E_i + \hbar\omega)^2 + \hbar^2\gamma_i^2} \left[1 + e^{-2t\gamma_i} - 2e^{-t\gamma_i}\cos\Omega t\right]. \tag{7.47}$$

This is an important intermediate result. An key quantity in this result is the matrix element ${}_o\langle\varphi_f\,|p_z|\varphi_i\rangle_{or}$, whose evaluation generally requires approximate methods. This is due to the wide range of specific characteristics that the emitting system may exhibit.[3]

We will now follow two procedures that yield two conceptually different expressions of the transition probability. In the first approach, we integrate over all frequencies and examine the behavior of the time dependent factors. This analysis shows that, for short times, the transition probability grows linearly with t, while for longer times it approaches an asymptotic value proportional to the mean-lifetime τ_i of the initial state. This justifies dividing the total transition probability by either the elapsed time t or by τ_i (in the large-time limit), yielding a meaningful expression for the transition probability rate, similar in spirit to how average velocity is defined. This method allows us to derive the optical response, the Einstein coefficients and the transition rates defined at the beginning of this chapter. In the second approach,

[3] We explicitly evaluate matrix elements for specific cases in the solved problems section at the end of the chapter.

which will be treated mainly in the Appendix G.1, we take the time derivative before integrating over the frequency. This yields the time resolved transition probability, very much as the instantaneous velocity. Interestingly, the time-dependent factor that governs this distribution has a shape reminiscent of the Planck frequency distribution for thermal radiation. In this formulation, the optical response, Einstein coefficients and transition rates are all time-dependent, and vanish in the limit $t \rightarrow \infty$.

7.4 Time-Independent Coefficients and Transition Probability

As mentioned in the previous section, we can follow two approaches to derive the physical quantities relevant to the analysis of the optical properties. In this section, we will derive well-known quantities associated to long-time and asymptotic processes. In the following section, we will obtain the same quantities using a time-resolved approach, which is more appropriate for analyzing fast processes. In both cases, the starting point is the transition probability per unit frequency

$$\mathscr{P}_{if}(\omega, t) = \frac{e^2 \hbar\omega(n_\phi + 1)}{6m^2\pi^2\epsilon_o c^3} \frac{|{}_o\langle\varphi_f|p_z|\varphi_i\rangle_o|^2}{(E_f - E_i + \hbar\omega)^2 + \hbar^2\gamma_i^2}\left[1 + e^{-2t\gamma_i} - 2e^{-t\gamma_i}\cos\Omega t\right]. \tag{7.48}$$

This probability has two important factors: the energy-dependent factor $|(E_f - E_i \pm \hbar\omega + i\hbar\gamma_i)|^2$, with sign $+$ for emission and $-$ for absorption (not shown in (7.48) because it corresponds to emission), and the time-dependent factor. The energy-dependent component essentially forms a Lorentzian distribution

$$\rho_L(E) = \frac{\hbar\gamma_i/\pi}{|E_f - E_i \pm \hbar\omega + i\hbar\gamma_i|^2} \sim \delta(E_f - E_i \pm \hbar\omega), \tag{7.49}$$

and gives a resonant character to the probability function $\mathscr{P}_{if}(\omega, t)$. Due to this resonance, the primary interaction between the semiconductor structure and the radiation field occurs with the components of the field whose frequencies match $\omega_\alpha = |E_f - E_i|/\hbar = \omega_{if}$, corresponding to the frequencies of the emitted or absorbed photons.

The resonance width at half maximum is related to the parameter $\hbar\gamma_i$, which is associated with the broadening of the energy level E_i. Since $\gamma_i = (2\tau_i)^{-1}$, the relation $\Gamma_i = \hbar\gamma_i$ is consistent with the Heisenberg uncertainty relation $\Gamma_i\tau_i \geq \hbar/2$. The time-dependent factor in $\mathscr{P}_{if}(\omega, t)$ has a relatively complex behavior, depending primarily on whether the observation time t is smaller or larger than the mean lifetime τ_i. In the following sections, we will analyze these limits to derive both the probability per unit time as well as the Einstein coefficients.

Before determining the transition probability per unit time and the Einstein coefficients, we can introduce a physical quantity that is very important in the analysis of the optical response of light-emitting device, *the optical susceptibility*.

7.4.1 Susceptibility and Optical Response at the Limit $t \to \infty$

The transition probability is the first result that we can use to analyze experimental results associated with optical transitions. In the limit $t \to \infty$ the probability acquires a form feasible to evaluate and to compare with experimental results. In general optical spectra are obtained by accumulating data over more or less large time intervals, so the spectra can be compared with $\mathscr{P}_{if}(\omega, \infty)$. In this limit, the probability $\mathscr{P}_{if}(\omega, t)$ in (7.48), reduces to

$$\mathscr{P}_{if}(\omega, \infty) = \frac{e^2\hbar\omega(n_\phi + 1)}{6m^2\pi^2\epsilon_o c^3}\frac{|\,{}_o\langle\varphi_f\,|p_z|\varphi_i\rangle_o|^2}{(E_f - E_i + \hbar\omega)^2 + \hbar^2\gamma_i^2}. \tag{7.50}$$

An important characteristic of emitting systems, such as quantum wells, superlattices, etc., is that several energy levels participate in the emission and excitation processes, i.e. there is more than one possible initial state and more than one possible final state. In evaluating the response of these systems, the contribution of all possible transitions must be considered, and the transition probability takes the form

$$\mathscr{P}(\omega, \infty) = \sum_{i,f}\frac{e^2\hbar\omega(n_\phi + 1)}{6m^2\pi^2\epsilon_o c^3}\frac{|\,{}_o\langle\varphi_f\,|p_z|\varphi_i\rangle_o|^2}{(E_f - E_i + \hbar\omega)^2 + \hbar^2\gamma_i^2}. \tag{7.51}$$

This function describes the optical response or susceptibility $\chi(\omega)$ of the optical device, as a function of frequency. A simplified way of defining the optical response, for a qualitative description (in arbitrary units) is

$$\chi(\omega) = a_0\sum_{i,f}\frac{|\,{}_o\langle\varphi_f\,|p_z|\varphi_i\rangle_o|^2}{(E_f - E_i + \hbar\omega)^2 + \hbar^2\gamma_i^2}. \tag{7.52}$$

This is an important and useful formula and can be applied to a wide variety of light emitting systems.

7.4.2 Energy and Time Factors Behavior, the Transition Probability W

As discussed above, the transition probability $\mathscr{P}_{if}(\omega, t)$, has a frequency and a time-dependent factors. Our purpose is to derive the transition probability per unit time. We will first analyze the behavior of these factors, and we will then integrate over the frequency. From the resulting time dependence we will define the transition probability per unit time. Both factors in $\mathscr{P}_{if}(\omega, t)$ define the function

$$F(\omega, t) = \frac{\omega}{\Omega^2 + \gamma_i^2}\left[1 + e^{-2t\gamma_i} - 2e^{-t\gamma_i}\cos\Omega t\right] = \frac{\omega}{\Omega^2 + \gamma_i^2}\, f(\omega, t), \tag{7.53}$$

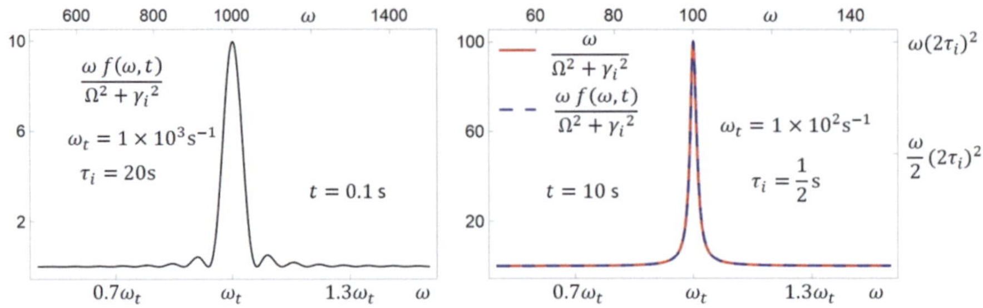

Fig. 7.3 Behavior of the function $F(\omega, t)$. In the left panel the oscillatory behavior is seen when $t \ll \tau_i$ and in the right panel when $t \gg \tau_i$

In Fig. 7.3 we plot the function

$$F(\omega, t) = \frac{\omega f(\omega, t)}{\Omega^2 + \gamma_i^2} = \omega \frac{1 + e^{-2t\gamma_i} - 2e^{-t\gamma_i} \cos \Omega t}{(\omega - \omega_t)^2 + \gamma_i^2}, \tag{7.54}$$

for small times (on the left) and for large times (on the right). The concentration of the function around the frequency ω_t is a consequence of the denominator which, as mentioned before, is resonant when $\Omega = 0$, that is, for $\omega = \omega_t$. This function behaves almost as a delta function $\delta(\omega - \omega_t)$. For small times, it behaves as shown in the left panel of the Fig. 7.3. The maximum value, at the peak, is of the order of $\omega_t t^2$, and the average width of $2\pi/t$. This means that the area under the curve, as we will see later, when integrated into the variable ω, is of the order of $2\pi\omega_t t$. If, on the other hand, the time is large or tends to infinity, the function $f(\omega, t) \to 1$ and $F(\omega, t)$ reduces to the Lorentzian

$$F(\omega, t \to \infty) = f_L(\omega) = \frac{\omega}{\Omega^2 + \gamma_i^2}. \tag{7.55}$$

In the right panel of Fig. 7.3 the functions $F(\omega, t)$ and $f_L(\omega)$ are plotted as functions of ω. The coincidence is perfect. We can also see that the height of the peak is $\omega_t/\gamma_i^2 = \omega_t(2\tau_i)^2$, while the average width at half maximum is $\pi\gamma_i$. The area is again of the order of $2\pi\omega_t\tau_i$. If $F(\omega, t)$ is integrated over ω, the result as a function of time has the behavior shown in Fig. 7.4. For $t < \tau_i$, it corresponds to the function $2\pi\omega_t t$, as mentioned above. As time increases, the integral asymptotically approaches the constant $2\pi\omega_t\tau_i$. These results are important and confirm the approximate evaluations of the areas in the graphs of Fig. 7.3. With these results we can also divide by t or by τ_i, as in the usual procedure, and define an approximate expression for the transition probability:

$$W_{if} = \frac{1}{\tau_i} \int \mathscr{P}_{if}(\omega, t) d\omega_\alpha \tag{7.56}$$

Thus

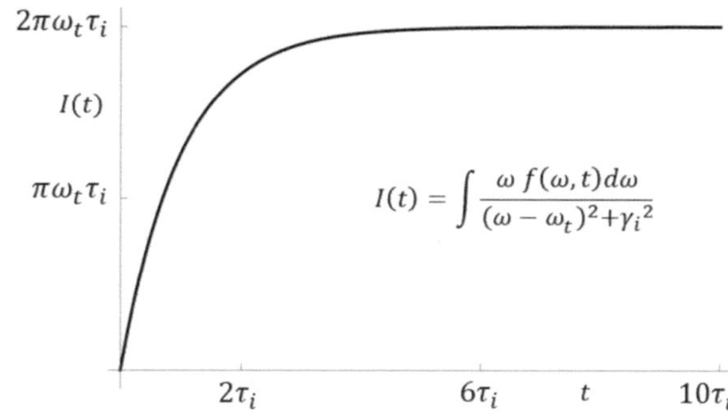

Fig. 7.4 Integral of the function $F(\omega, t)$. Linear behavior equal to $2\pi\omega_t t$ for small t, and constant equal to $2\pi\omega_t\tau_i$ for large times

$$W_{if} \simeq \frac{2\pi}{\hbar} \mid {}_o\langle\varphi_f\,|p_z|\varphi_i\rangle_o|^2 \frac{\omega_t e^2(n_\phi+1)}{6m^2\pi^2\epsilon_o c^3} \tag{7.57}$$

which is, essentially, the Fermi golden rule.

7.4.3 Induced and Spontaneous Emission Coefficients

The derivation of W_{if}, in (7.56), brings us closer to the determination of the Einstein coefficients B^i_f, and A_{sp}. In Einstein's model the probability of emission per unit of time is defined as

$$W_{if} = B^i_f\rho(\omega) + A_{sp}. \tag{7.58}$$

The first term describes the transitions induced by the interaction with the electromagnetic field, whose spectral density is given by $\rho(\omega)$, while the second term describes emissions due to spontaneous decay. Let us return to the transition probability W^{in}_{if} that was written in (7.57) as

$$W_{if} \simeq \frac{2\pi}{\hbar} \mid {}_o\langle\varphi_f\,|p_z|\varphi_i\rangle_o|^2 \frac{\omega_t e^2(n_\phi+1)}{6m^2\pi^2\epsilon_o c^3} \tag{7.59}$$

It is evident that in the absence of a photonic field, i.e., when $n_\phi = 0$, we have the probability of spontaneous emission given by

$$W^{sp}_{if} \simeq \frac{2\pi}{\hbar} \mid {}_o\langle\varphi_f\,|p_z|\varphi_i\rangle_o|^2 \frac{\omega_t e^2}{6m^2\pi^2\epsilon_o c^3} \tag{7.60}$$

while the probability of induced emission can be identified with

$$W^{in}_{if} \simeq \frac{2\pi}{\hbar} \mid {}_o\langle\varphi_f\,|p_z|\varphi_i\rangle_o|^2 \frac{\omega_t e^2 n_\phi}{6m^2\pi^2\epsilon_o c^3} \tag{7.61}$$

such that the total emission probability is given as

$$W_{if} = W^{in}_{if} + W^{sp}_{if}. \tag{7.62}$$

If we consider the relationship between n_ϕ in (7.10) and the Planck distribution $\wp_{ph}(\omega, T)$ in (7.9) the probability of induced emission can be written as

$$W_{if}^{in} = \frac{\pi e^2}{3\epsilon_o \hbar^2 m^2 \omega_t^2} \mid {}_o\langle \varphi_f \mid p_z \mid \varphi_i \rangle_o \mid^2 \wp_{ph}(\omega, T) = B_f^i \wp_{ph}(\omega, T), \tag{7.63}$$

with the emission coefficient

$$B_f^i = \frac{\pi e^2}{3\epsilon_o \hbar^2 m^2 \omega_t^2} \mid {}_o\langle \varphi_f \mid p_z \mid \varphi_i \rangle_{or} \mid^2. \tag{7.64}$$

The spontaneous emission coefficient, on the other hand, takes the form

$$W_{if}^{sp} = \frac{\omega_t e^2}{3\epsilon_o \hbar m^2 \pi c^3} \mid {}_o\langle \varphi_f \mid p_z \mid \varphi_i \rangle_o \mid^2 = A_f^i. \tag{7.65}$$

From these results we have the well-known relationship

$$A_f^i = \frac{\hbar \omega_t^3}{\pi^2 c^3} B_f^i \tag{7.66}$$

between the spontaneous emission coefficient and the induced emission coefficient. Both coefficients in the above equations are in terms of the matrix elements ${}_o\langle \varphi_f \mid p_z \mid \varphi_i \rangle_{or}$. It is common to write the coefficients B_f^i and A_f^i in terms of the matrix elements of the dipolar moment. To write the coefficients in terms of these matrix elements we can use the identity between expected values in the Heisenberg representation

$$\langle \varphi_f \mid \frac{d}{dt} \hat{f}(t) \mid \varphi_i \rangle = i \frac{E_f - E_i}{\hbar} \langle \phi_f \mid \hat{f}(t) \mid \phi_i \rangle. \tag{7.67}$$

Here $\hat{f}(t) = e^{iE_f t/\hbar} \hat{f} e^{-iE_i t/\hbar}$. Applied to z and $p_z = m dz/dt$, we have the relationship

$$e^2 \mid {}_o\langle \varphi_f \mid p_z \mid \varphi_i \rangle_o \mid^2 = m^2 \omega_t^2 \mid {}_o\langle \varphi_f \mid ez \mid \varphi_i \rangle_{or} \mid^2. \tag{7.68}$$

Using this identity the induced transition probability can be written in the form

$$W_{if}^{in} = \frac{\pi}{3\hbar^2 \epsilon_o} \mid {}_o\langle \varphi_f \mid ez \mid \varphi_i \rangle_o \mid^2 \wp_{ph}(\omega, T) = B_f^i \wp_{ph}(\omega, T), \tag{7.69}$$

with

$$B_f^i = \frac{\pi}{3\hbar^2 \epsilon_o} \mid {}_o\langle \varphi_f \mid ez \mid \varphi_i \rangle_{or} \mid^2 \tag{7.70}$$

and the probability of spontaneous transition as

$$W_{if}^{sp} = \frac{\omega_t^3}{3\hbar \epsilon_o \pi c^3} \mid {}_o\langle \varphi_f \mid ez \mid \varphi_i \rangle_o \mid^2 = A_f^i. \tag{7.71}$$

It is evident from these relationships that the three Einstein coefficients are related and can be determined once one of them is known.

7.4.4 Transition Rates

With the results obtained so far, the transition rates defined in Sect. 7.1 can now be written as

$$R_{sp} = \frac{\omega_t^3}{3\hbar\epsilon_o\pi c^3} |\, {}_{\mathrm{o}}\langle\varphi_f\, |ez|\varphi_i\rangle_{\mathrm{o}}|^2 f_{if}(E,T) g_{if}(E), \tag{7.72}$$

$$R^i_f = \frac{\pi}{3\hbar^2\epsilon_o} |\, {}_{\mathrm{o}}\langle\varphi_f\, |ez|\varphi_i\rangle_{\mathrm{o}}|^2 f_{if}(E,T) g_{if}(E) \wp_{ph}(\omega,T), \tag{7.73}$$

$$R^f_i = \frac{\pi}{3\hbar^2\epsilon_o} |\, {}_{\mathrm{o}}\langle\varphi_f\, |ez|\varphi_i\rangle_{\mathrm{o}}|^2 f_{if}(E,T) g_{if}(E) \wp_{ph}(\omega,T). \tag{7.74}$$

In all these equations, the density $\wp_{ph}(\omega, T)$ is the Planck density and $g_{if}(E)$ is the joint density of levels defined above.

7.5 The Temporal Distribution of Transition Probabilities

Let us once again consider the transition probability per unit of frequency

$$\begin{aligned}\mathscr{P}_{if}(\omega,t) =& \frac{e^2\hbar\omega(n_\phi+1)}{6m^2\pi^2\epsilon_o c^3} \frac{|{}_{\mathrm{o}}\langle\varphi_f\, |p_z|\varphi_i\rangle_{\mathrm{o}}|^2}{(E_f - E_i + \hbar\omega)^2 + \hbar^2\gamma_i^2} \\ &\times \left[1 + e^{-2t\gamma_i} - 2e^{-t\gamma_i}\cos\Omega t\right].\end{aligned} \tag{7.75}$$

As already mentioned, this probability contains both a time-dependent factor, and a resonant energy-dependent factor. Since our goal is to determine the transition probability per unit time, we must eventually integrate over the frequency variable. In the previous section, we consider the function

$$F(\omega,t) = \omega\frac{1 + e^{-2t\gamma_i} - 2e^{-t\gamma_i}\cos\Omega t}{(\omega-\omega_t)^2 + \gamma_i^2}, \tag{7.76}$$

and studied how its behavior evolves with time. When we integrate $F(\omega, t)$ over the frequency ω, the result grows linearly with t and for long times reaches a maximum equal to τ_i. This dependence is not assumed but emerges from the integration process itself, and provides a natural way to define the transition probability per unit time, by dividing the integrated $\mathscr{P}_{if}(\omega, t)$ over the frequency by t.

In this section we follow an alternative but equivalent procedure: instead of integrating first and dividing by t afterward, we begin by differentiating with respect to t and then integrate over the frequency to directly obtain the time-resolved transition

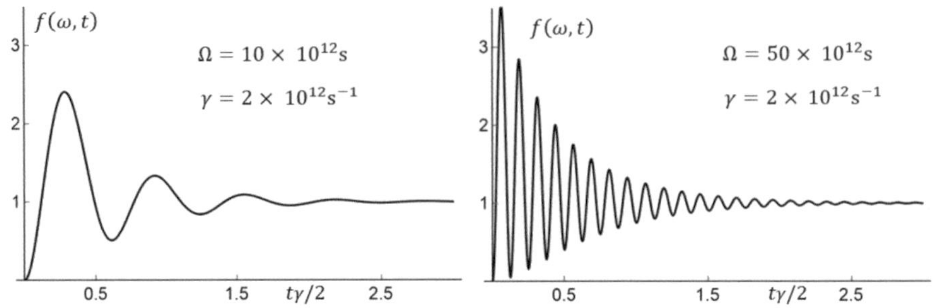

Fig. 7.5 Behavior of the function $f(\omega, t)$. In the left panel the oscillatory behavior of the function $f(\omega, t)$ when $\Omega\tau_i \sim 1$ and in the right panel when $\Omega\tau_i \gg 1$

probability per unit time, making it possible to analyze its temporal evolution in more detail.

The time-dependent factor in $\mathscr{P}_{if}(\omega, t)$ is given by

$$f(\omega, t) = 1 + e^{-2t\gamma_i} - 2e^{-t\gamma_i} \cos \Omega t, \tag{7.77}$$

whose behavior is illustrated in Fig. 7.5. For higher frequencies, the oscillations become more rapid as can be seen in plots where the frequency differs by a factor of 5. Moreover, as shown in Fig. 7.3, the function becomes concentrated around the resonant frequency ω_t.

In addition to the behavior of the function $f(\omega, t)$, as a part of the function $F(\omega, t)$, we are interested in the time-dependent behavior of the transition probability obtained by differentiation.

The time dependence of $F(\omega, t)$ provides valuable insight into how the transition develops in time. It is shown in the Appendix G that the derivative of function $F(\omega, t)$, given by

$$G(\omega, t) = \frac{\partial F(\omega, t)}{\partial t} = \frac{2\omega e^{-2t\gamma_i}(-\gamma_i + e^{t\gamma_i}(\gamma_i \cos \Omega t + \Omega \sin \Omega t))}{(E_f - E_i + \hbar\omega)^2 + \Gamma_i^2}, \tag{7.78}$$

can accurately be approximated, as seen in Fig. G.1, by the function

$$D(\omega, t) = \frac{2\omega\gamma_i e^{-2t\gamma_i}(-1 + e^{t\gamma_i})}{(E_f - E_i + \hbar\omega)^2 + \Gamma_i^2}, \tag{7.79}$$

which is the product of two distributions, one for frequencies and the other one for time, which we define as

$$\rho_w(\omega, \tau_i) = \frac{\omega}{(E_f - E_i + \hbar\omega)^2 + \Gamma_i^2} \quad \text{and} \quad \rho_t(t, \tau_i) = 2\gamma_i e^{-2t\gamma_i}(-1 + e^{t\gamma_i}). \tag{7.80}$$

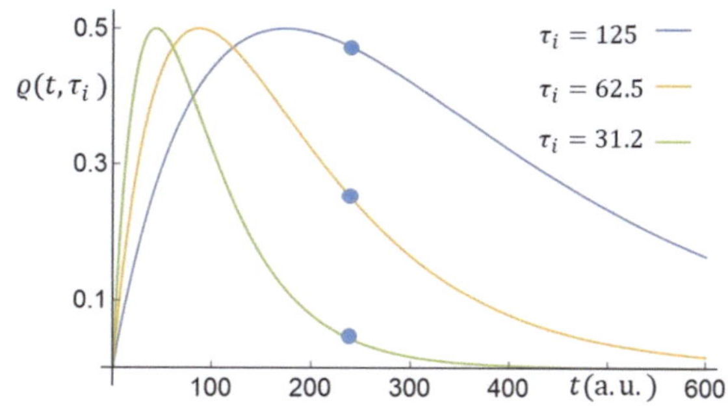

Fig. 7.6 Temporal distribution $\varrho(t, \tau_i)$ for three values of τ_i. If t, is greater than the times in which $\varrho(t, \tau_i)$ is maximum, the values of the function $\varrho(t, \tau_i)$ are different (see blue dots). The transition probability tends to zero quickly, the faster the smaller τ_i is

In terms of these distributions, the function $D(\omega, t)$ can be written as

$$D(\omega, t) = \rho_w(\omega, \tau_i)\rho_t(t, \tau_i). \tag{7.81}$$

In the Appendix G, the time-dependent transition probability per unit time $W_{if}(t)$ is derived. The function obtained there is

$$W_{if}(t) = \frac{\pi}{\hbar} \mid {}_o\langle\varphi_f \mid p_z \mid \varphi_i\rangle_o\mid^2 \frac{\omega_t e^2 \langle n_\phi\rangle}{6m^2\pi^2\epsilon_o c^3}\, \varrho(t, \tau_i) \tag{7.82}$$

where $\varrho(t, \tau_i)$ is the time distribution defined as

$$\varrho(t, \tau_i) = 2e^{-2t\gamma_i}(-1 + e^{t\gamma_i}) = \frac{\rho_t(t, \tau_i)}{\gamma_i}, \tag{7.83}$$

whose behavior over time is shown in Fig. 7.6, for different mean lifetimes of the initial state. Note that the maximum value of $\varrho(t, \tau_i)$ is 1/2 and occurs when $t = 2\tau_i \ln 2$.

The transition probability in (7.82) is a time-dependent version of Fermi's golden rule.

The temporal distribution $\varrho(t, \tau_i)$ also affects the optical response that was introduced in Sect. 7.4.1. Taking into account this time distribution, the optical susceptibility of (7.52) takes the form

$$\chi(\omega, t) = a_0 \sum_{i,f} \frac{\mid {}_o\langle\varphi_f \mid p_z \mid \varphi_i\rangle_o\mid^2}{(E_f - E_i + \hbar\omega)^2 + \Gamma_i^2}\, \varrho(t, \tau_i). \tag{7.84}$$

7.6 Solved Problems

1. **Problem 1**. Obtain the Einstein coefficients as functions of time by following a procedure where the derivation and integration orders are reversed.

   Solution. An alternative to obtain the Einstein coefficients is to use the relationship (7.49)

$$\frac{1}{|E_f - E_i + \hbar\omega_\alpha + i\hbar\gamma_i|^2} \simeq \frac{\pi}{\hbar\gamma_i}\delta(E_f - E_i + \hbar\omega_\alpha) = \frac{\pi}{\hbar^2\gamma_i}\delta(\omega_\alpha - \omega_t) \tag{7.85}$$

   substitute it into

$$\begin{aligned}\int \mathscr{P}_{if}(\omega_\alpha, t)d\omega_\alpha &= \int d\omega_\alpha |a_{if}^{in}(t)|^2(\omega, t)\frac{L^3\omega_\alpha^2}{2\pi^2c^3} \qquad (7.86)\\ &= \frac{e^2}{12m^2\pi^2\hbar\epsilon_o c^3}\,|_o\langle\varphi_f|p_z|\varphi_i\rangle_o|^2 2n_\phi \\ &\quad \int d\omega_\alpha\frac{\pi\omega_\alpha}{2\gamma_i}\delta(\omega_\alpha - \omega_t)f(\omega_\alpha, t)\end{aligned}$$

   and integrate. The integration is direct and results in

$$\int \mathscr{P}_{if}(\omega_\alpha, t)d\omega_\alpha = \frac{2e^2|_o\langle\varphi_f|p_z|\varphi_i\rangle_o|^2}{6m^2\pi^2\hbar\epsilon_o c^3}\frac{\pi\omega_t n_\phi}{\gamma_i}\left[1 - e^{-t\gamma_i}\right]^2. \tag{7.87}$$

   If we derive with respect to time, we obtain the probability of emission per unit of time

$$W_{if}^{in}(t) = \frac{2e^2\pi|\,_o\langle\varphi_f|p_z|\varphi_i\rangle_o|^2}{6\hbar^2m^2\epsilon_o\omega_t^2\gamma_i}\,\wp_{ph}(\omega, T)\;2\gamma_i e^{-2t\gamma_i}(e^{t\gamma_i} - 1). \tag{7.88}$$

   This means that

$$B_f^i(t) = \frac{e^2\pi|\,_o\langle\varphi_f|p_z|\varphi_i\rangle_o|^2}{3\hbar^2m^2\epsilon_o\omega_t^2}\,2e^{-2t\gamma_i}(e^{t\gamma_i} - 1). \tag{7.89}$$

   Both $A_{sp}(t)$ and $B_f^i(t)$ vary with time and, as shown in Appendix G, have a temporal distribution similar to the Planck spectrum, with a maximum at $t = 2\tau_i \ln 2$.

2. **Problem 2**. To illustrate the use of transition probability, we will consider two specific examples. In both cases we are going to evaluate the matrix element $|\,_o\langle\varphi_f|p_z|\varphi_i\rangle_o|^2$, and assume that $t = t_{max}$. We saw in the previous section that at this time $\varrho(t, \tau_i) = 1/2$. Determine:

   (a) the transition rates in a homogeneous GaAs semiconductor layer, of thickness L,
   (b) obtain the probabilities and transition times in a quantum well when the GaAs layer is in between two layers of $Ga_{1-x}Al_xAs$.

Solution.

(a) We will assume that the GaAs semiconductor layer, of thickness L in the z direction and wide in the x-y plane, is between hard walls with infinite potential for $z < 0$ and $z > L$. Electrons and holes in the conduction and valence bands are confined along the z axes by infinite rectangular wells of width L. In principle, the atomic potential that electrons feel in the well is periodic, with the periodicity of the crystalline structure. But, in the effective mass approximation, we assume a constant potential. It is then acceptable to calculate the matrix elements using the stationary eigenfunctions $\sin k_{ei} z$ and $\cos k_{hi} z$, with $i = 1, 2, ...$

Given the normalized eigenfunctions

$$|\varphi_i^b\rangle_{\rm o} = \sqrt{\frac{2}{L}} \sin \frac{n_i \pi z}{L} \quad \text{and eigenvalues} \quad E_i^b = \frac{\hbar^2 n_i^2 \pi^2}{2m_e L^2}, \tag{7.90}$$

where $b = v, c$ for valence or conduction band, the objective is to obtain the matrix elements

$$_{\rm o}\langle\varphi_f^v|p_z|\varphi_i^c\rangle_{\rm o} = \int_0^L \varphi_f^v(z) \frac{\partial}{\partial z} \varphi_i^c(z) dz. \tag{7.91}$$

It can be verified that the non-zero elements are

$$_{\rm o}\langle\varphi_f^v|p_z|\varphi_i^c\rangle_{\rm o} = \frac{16\hbar^2 k_i^2 k_f^2}{\left(k_i^2 - k_f^2\right)^2 L^2} \quad \text{for} \quad n_i + n_f = \text{odd}. \tag{7.92}$$

In the calculation of the matrix elements $|\,_{\rm o}\langle\varphi_f|p_z|\varphi_i\rangle_{\rm o}|^2$ it is necessary to refer the energies to the same zero of the energy, which we will place at the upper edge of the valence band. Furthermore, we have to consider the energy densities in the initial and final states. As explained in the first chapter, because the transitions involve two energy levels, it is the joint density of levels which must be considered. In two-dimensional systems, this density is

$$g_\mu(E_e) = \frac{L_x L_y m_e^*}{2\pi\hbar^2}. \tag{7.93}$$

It is common to associate the matrix element $_{\rm o}\langle\varphi_f^v|p_z|\varphi_i^c\rangle_{\rm o}$ with an energy parameter E_p defined as

$$E_p = \frac{2m_e}{\hbar^2} P^2 = \frac{2m_e}{\hbar^2} \frac{3\hbar^2}{m_e^2} \langle p_{cv}\rangle^2 = \frac{6}{m_e} \langle p_{cv}\rangle^2 \tag{7.94}$$

with values that, depending on the semiconductor, fall between 18 and 25 eV. In Fig. 7.7 E_p is plotted for

$$\langle p_{cv}\rangle^2 = \left(\frac{V}{V_c}\right)^3 \frac{n_i^2 n_f^2}{e^3} \pi^2 \frac{16\hbar^2 k_{ei}^2 k_{hf}^2}{\left(k_{ei}^2 - k_{hf}^2\right)^2 L^2}. \tag{7.95}$$

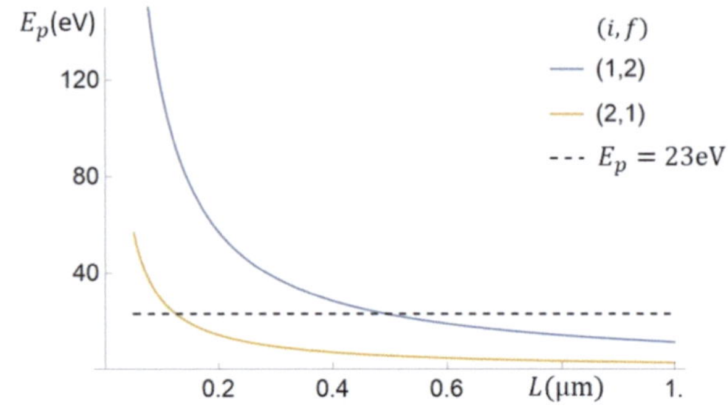

Fig. 7.7 The energy parameter E_p for $(n_i, n_f) = (1, 2)$ and $(2, 1)$, as a function of L. The constant value that is usually reported for this parameter, calculated by using other methods, is also shown

Here V/V_c is the ratio between the volumes of the semiconductor and that of a unit cell. In addition

$$k_{ei}^2 = \frac{n_i^2\pi^2}{L^2} + \frac{2m_e E_g}{\hbar^2} \quad \text{and} \quad \frac{\hbar^2 k_{hf}^2}{2m_h} = -\frac{\hbar^2 k_f^2}{2m_{ev}} \tag{7.96}$$

We also consider that $m_{ec}^* = 0.067m_e$, $V = L^3$ and $V_c = l_c^3$ with $l_c = 0.3$ nm. In Fig. 7.7, we show the energy E_p defined in (7.94), for the energy levels $(n_i, n_f) = (1, 2)$ and $(2, 1)$. Those closest to the band edge, are of the order of a few eVs, as reported in the literature.
Figure 7.8 shows the transition probabilities W_{if} with $|\,_o\langle\varphi_f|p_z|\varphi_i\rangle_o|^2$ multiplied by $n_i^2 n_f^2 \pi^2$ as a function of the thickness of the GaAs layer when the transitions start at level $n_i = 1$, left panel, and when they start at level $n_i = 6$ right panel. We see that the higher the initial energy level, the greater the probability of recombination. The probabilities decrease rapidly with the growth of L. In short: lower initial energies are more stable.
The level index of final state also has an effect. The probabilities increase as n_f increases. The recombination rate, or mean-life times, τ_{if} are plotted in Fig. 7.9. Notice that the initial levels with lower energy are more stable. Also, given an initial level n_i, the minimum transition rate occurs at levels with the largest level index n_f.

(b) The semiconductor layers of $Ga_{1-x}Al_xAs$ have a larger gap, thus the electrons and holes are confined in quantum wells (QWs) of width a and depth V_o. The depth of the wells depends on the concentration x and the depth of the well in the conduction band is approximately twice the depth in the valence band. This system is studied in the Appendix C and the eigenvalue equation obtained there is

$$\cot ka = \frac{k^2 - q^2}{2kq} \quad \text{which implies that} \quad q = \begin{cases} k \tan ka/2 & \text{even functions} \\ -k \cot ka/2 & \text{odd functions} \end{cases} \tag{7.97}$$

On the other hand, the eigenfunctions are defined in (C.22). The eigenfunctions penetrate the lateral barriers and decay exponentially. If these functions

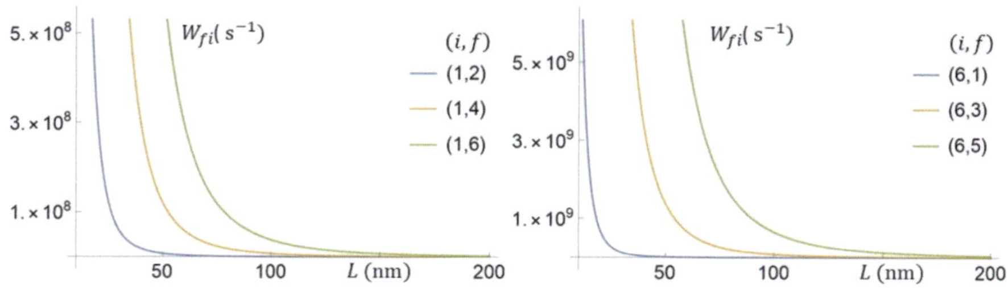

Fig. 7.8 Transition probabilities in the infinite quantum well, as a function of the indices n_i and n_f and of L. The transition probability decreases as L increases. It is greater if the energy of the initial state is greater. It is also sensitive to the relation order between the indices n_i and n_f. If $n_f > n_i$ the probability decreases as n_f increases, but if $n_f < n_i$ the probability increases as n_f increases

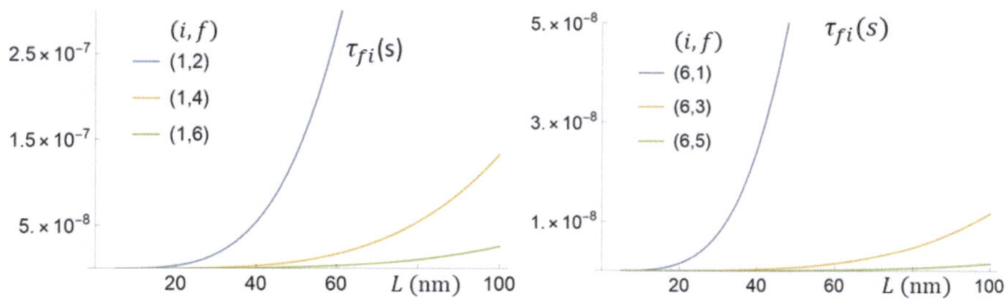

Fig. 7.9 The behavior of transition times is compatible with that of transition probabilities. In general, transition times increase with L. Lower energy states are more stable

(normalized and evaluated for each of the eigenvalues in the conduction and valence band) are considered, and one calculates the matrix elements

$$|\,{}_{o}\langle\varphi_f\,|p_z|\varphi_i\rangle_{o}|^2 = \hbar^2\left[\int_{-\infty}^{0}\varphi_{1f}^{v}\frac{\partial}{\partial z}\varphi_{1i}^{c}dz + \int_{0}^{a}\varphi_{2f}^{v}\frac{\partial}{\partial z}\varphi_{2i}^{c}dz + \int_{a}^{\infty}\varphi_{3f}^{v}\frac{\partial}{\partial z}\varphi_{3i}^{c}dz\right]^2, \tag{7.98}$$

only those integrals with functions whose parities (in the indices i and f) are different, differ from zero. Figure 7.10 shows transition times, for $a = 10\,\text{nm}$ and for $a = 20\,\text{nm}$, along with the times evaluated for the system of the previous problem. In general, the transition times in wider and deeper wells are greater than in thinner and shallow wells.

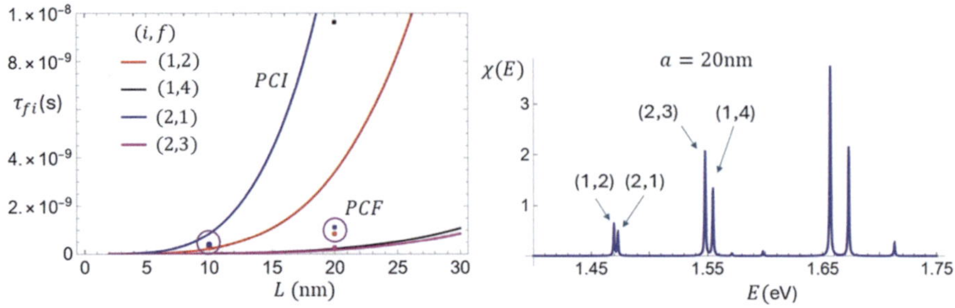

Fig. 7.10 The behavior of the transition rate in infinite and finite quantum wells, and the optical susceptibility ξ. In the left panel the transition times in homogeneous layer bounded by infinite walls (continuous curves) and transition times in finite quantum wells (dots) for widths $a = L = 10$ nm and $a = L = 20$ nm. The susceptibility in the right panel is for $a = L = 20$ nm. The low energy transitions (1, 2) and (2, 1), are indicated with dots or solid curves in red and blue, respectively. The transition times in quantum wells with $a = L = 20$ nm are greater than for $a = L = 10$ nm. This means that the transition probabilities are lower in wells of greater width. From the continuous curves, it can be concluded that deeper wells imply longer transition times and lower transition probabilities

7.7 Problems

1. Show that integration over the angles results in (7.44).
2. Show that the integral over ω of the function $F(\omega, t)$ behaves as indicated in the Fig. 7.4
3. Show that the value of t at which the time distribution $\rho_t(t, \tau_i)$ is maximum is $t = t_{max} = 2\tau_i \ln 2$, and that the maximum value of $\rho_t(t, \tau_i)$ is $1/4\tau_i$.
4. Obtain the values of $\varrho(t, \tau_i)$ for the half-life times indicated in the Fig. 7.6, when $t = 300$ s.
5. Prove that $\rho_t(t < \tau_i, \tau_i) \simeq 2\gamma_i^2 t$.
6. Show that in the infinite well (solved Problem 1) the expected value is the one indicated in (7.92).
7. Obtain the expected value of z, for the infinite well, and using (7.92) show that the following relation is satisfied

$$|\,{}_{\mathrm{o}}\langle\varphi_f\,|p_z|\varphi_i\rangle_{\mathrm{o}}|^2 = m^2\omega_{if}^2|\,{}_{\mathrm{o}}\langle\varphi_f\,|z|\varphi_i\rangle_{\mathrm{or}}|^2. \tag{7.99}$$

8. Obtain the transition probabilities W_{if} and the corresponding transition times for the infinite well when $i = 4$, $f = 1, 2, 3, 4, 5$, and 6, and $L = 15$ nm.
9. Obtain the transition probabilities W_{if} and the corresponding transition times τ_{if} for a quantum well formed with a layer of $GaAs$, between two layers of $Ga_{1-x}Al_xAs$, with x such that $V_o = 0.23$ eV in the conduction band and $V_o = 0.115$ eV in the valence band. Obtain W_{if} and τ_{if} for $i = 4$, $f = 1, 2, 3, 4, 5$, and 6, and $a = 15$ nm.

References

1. M. Planck, Verh. Deut. Phys. Ges. **2**, 202 (1900); M. Planck, Verh. Deut. Phys. Ges. **2**, 237 (1900)
2. Y.B. Band, *Light and Matter* (Wiley, Chichester, 2006)

Part II

Optical and Electronic Devices, Semiclassical and Quantum

LED np, QWs Laser Diodes

8

Abstract

In this chapter, we focus on light-emitting semiconductor structures, including *pn* diodes and quantum heterostructures. Building on the generation-recombination Boltzmann equations introduced in Chap. 4, we derive analytical expressions for the generation-recombination rates, illumination-induced response, and generation-recombination currents as functions of carrier concentration and applied bias voltage. We explore the mechanisms of carrier generation, recombination, optical gain, and pumping within the depletion region of *pn* diodes. Using the spontaneous transition probability formulas developed in Chap. 7, we carry out explicit calculations for two-level systems involving the conduction and valence bands, and analyze how optical gain varies with the applied bias. The optical response formalism introduced earlier is then applied to quantum wells and multiple quantum well structures. A detailed case study is presented for GaN-based superlattices, where we apply the Theory of Finite Periodic Systems (developed in Chap. 6) to analyze their response. This theoretical framework enables us to accurately reproduce the complete optical spectrum observed experimentally by Nakamura [1], with improved agreement achieved by incorporating charge polarization effects. Moreover, to our knowledge, this is the only theoretical approach that fully recovers the discrete energy level structure of the conduction and valence bands-features that are otherwise lost in the continuous band description of conventional Bloch theory for infinite systems. This makes the present treatment not only precise but also conceptually distinctive in its ability to bridge the gap between finite and infinite system models in semiconductor optics.

8.1 Introduction

LED stands for light-emitting device. The first semiconductor LEDs were developed based on the generation-recombination mechanism occurring in the transition (or depletion) region of a biased *pn* junction.

P. Pereyra, *Contemporary Theory of Semiconductors and Devices*,
Graduate Texts in Physics, https://doi.org/10.1007/978-3-032-00296-9_8

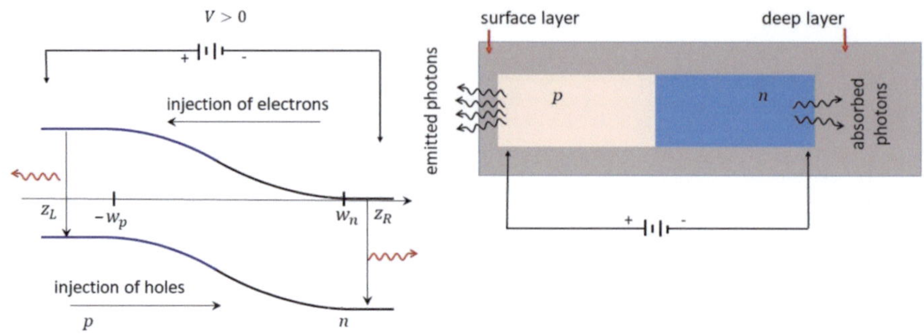

Fig. 8.1 The pn junction is a light-emitting device. If the current of electrons towards the p side is greater than that of holes in the opposite direction ($j_n > j_p$) and the p side is covered by a thin layer, the photons emitted in that area leave the device, while those emitted in the opposite end, are absorbed

As explained in Chap. 4, carrier generation and recombination occur in pn junctions. These processes are governed by various capture and emission mechanisms. These include the increase of minority carrier populations due to an applied external voltage across the junction, excitation and emission processes driven by electromagnetic fields, and non-radiative emissions mediated by energy states within the bandgap-particularly deep-level trap states (Fig. 8.1).

We have seen in Chap. 4 that the generation-recombination rates (4.94) and (4.95) are closely related to the injection currents in the forward-biased pn junction, satisfying the following relations:

$$-\frac{1}{e}\big(j_n(0) - j_n(-w_p)\big) = w_p\, r^n_{gr}(V) \quad \text{and} \quad \frac{1}{e}\big(j_p(w_n) - j_p(0)\big) = w_n\, r^p_{gr}(V). \tag{8.1}$$

The generation-recombination coefficients $r^n_{gr}(V)$ and $r^p_{gr}(V)$ were obtained by using plausibility arguments, see (4.86)–(4.91). We will determine here the generation-recombination rates starting from (Fig. 8.2)

$$r^n_{gr}(z, V) = \left.\frac{\partial n}{\partial t}\right|_g - \left.\frac{\partial n}{\partial t}\right|_r = -\frac{\delta n(z, V)}{\tau_n} \tag{8.2}$$

and

$$r^p_{gr}(z, V) = \left.\frac{\partial p}{\partial t}\right|_g - \left.\frac{\partial p}{\partial t}\right|_r = -\frac{\delta p(z, V)}{\tau_p}, \tag{8.3}$$

and by taking into account the concentrations n and p in the transition zone of a polarized junction. In Chap. 4, the *minority* concentrations for polarized and non-polarized junctions were obtained. On the p side, as

$$n(z) = n_p\, e^{e\phi_n(z)/k_B T} \quad \text{and} \quad n_V(z) = n(z) e^{e\phi_{VL}(z)/k_B T} \qquad -w_p < z \le 0, \tag{8.4}$$

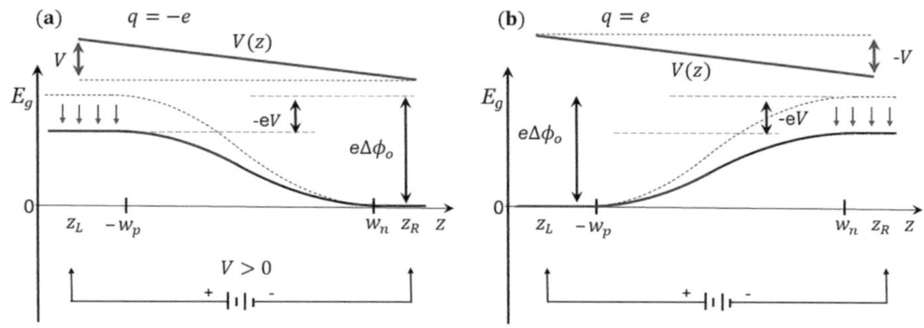

Fig. 8.2 The height of the contact barrier $e\Delta\phi_o$ for electrons and holes changes to $e\Delta\phi_o - eV$ in the presence of a potential V. When the polarization is direct $V > 0$, the contact barrier decreases. In this, also known as forward polarization, the potential energy of electrons at $z = z_L$ is $-eV(z)$, while at $z = z_R$ it is zero. In reverse bias, the potential energy of the holes in $z = z_R$ is lower than in $z = z_L$

and on the n side, like

$$p(z) = p_n\, e^{e\phi_p(z)/k_BT} \qquad \text{and} \qquad p_V(z) = p(z)e^{e\phi_{VR}(z)/k_BT} \qquad 0 \leq z < w_n. \tag{8.5}$$

In these expressions the potentials $\phi_n(z)$, $\phi_p(z)$, $\phi_{VL}(z)$ and $\phi_{VR}(z)$ are

$$\phi_{VL}(z) = V\frac{z_R - z}{z_L + z_R}, \qquad \phi_n(z) = \frac{eN_A}{2\epsilon\epsilon_o}(z + w_p)^2, \tag{8.6}$$

and

$$\phi_{VR}(z) = -V\frac{z + z_L}{z_L + z_R}, \qquad \phi_p(z) = -\frac{eN_D}{2\epsilon\epsilon_o}(z - w_n)^2. \tag{8.7}$$

The variations of the minority concentrations in the transition zone, as functions of z and V, are then

$$\delta n_p(z, V) = n_p\, e^{e\phi_n(z)/k_BT}\left(e^{e\phi_{VL}(z)/k_BT} - 1\right) \tag{8.8}$$

$$\delta p_n(z, V) = p_n\, e^{-e\phi_p(z)/k_BT}\left(e^{-e\phi_{VR}(z)/k_BT} - 1\right). \tag{8.9}$$

To simplify these results, we will assume that $z_L \simeq w_p$, $z_R \simeq w_n$ and $z_R + z_L = L$, and evaluate the generation-recombination rates at $z = 0$. Furthermore, if $N_A \simeq N_D$, we have $z_L/L \simeq z_R/L \simeq 1/2$, and the generation-recombination rates take the form

$$r_{gr}^n(0, V) = \frac{n_p e^{e^2 N_A w_p^2/2\epsilon\epsilon_o k_BT}}{\tau_n}\left(e^{eV/2k_BT} - 1\right) \tag{8.10}$$

and

$$r^p_{gr}(0, V) = \frac{p_n e^{e^2 N_D w_n^2/2\epsilon\epsilon_o k_B T}}{\tau_p}\left(e^{eV/2k_BT} - 1\right). \tag{8.11}$$

If we remember that

$$\Delta\phi_o = \frac{e^2 N_D w_n^2}{2\epsilon\epsilon_o} + \frac{e^2 N_A w_p^2}{2\epsilon\epsilon_o}, \tag{8.12}$$

the above expressions can be written in terms of the intrinsic concentrations as

$$r^n_{gr}(0, V) \simeq \frac{n_i}{\tau_n}\left(e^{eV/2k_BT} - 1\right) \quad \text{and} \quad r^p_{gr}(0, V) \simeq \frac{p_i}{\tau_p}\left(e^{eV/2k_BT} - 1\right), \tag{8.13}$$

since $\tau_n = \tau_p = \tau$ and $n_i = p_i$ we have $r^n_{gr}(0, V) = r^p_{gr}(0, V)$. If the recombination-generation rate is the same at all points in the transition zone, and is multiplied by the cross-sectional area A of the diode and the effective length L, we can obtain the number of recombinations per unit of time induced by the potential difference V. With this number we will evaluate, on the one hand, the *illumination intensity* $\mathtt{I}$, and on the other, the *generation-recombination current.*

The electrons that recombine emit photons whose energy is $\hbar\omega = E_g$. In a previous chapter, the light intensity was defined as $\mathtt{I} = \hbar\omega c\phi$, where c is the speed of light and ϕ the number of photons per unit of volume, that is, as the energy per unit of surface and time. The number of photons that are generated in the transition zone (of length L) per unit of volume and per unit of time is r_{gr}, thus, the number of photons emitted or absorbed per unit of surface and unit of time is given by the product $\sim r_{gr}L$. Multiplying this number by the energy of the photon we have the illumination intensity produced in the pn junction

$$\mathtt{I} = \hbar\omega r_{gr} L = \frac{\hbar\omega L n_i}{\tau_n}\left(e^{eV/2k_BT} - 1\right). \tag{8.14}$$

If we multiply by the area A and by the charge e, we obtain the charge that contributes to this process per unit of time. This has the same units as the electric current and it is defined as the generation-recombination current

$$I_{gr}(V) \sim r_{gr} e A L = \frac{eAL}{\tau} n_i\left(e^{eV/2k_BT} - 1\right) = I_{r0}\left(e^{eV/2k_BT} - 1\right), \tag{8.15}$$

which is similar to the generation-recombination current in (4.98) and to the current in (B.18) obtained within the Shockley, Read and Hall model.[1]

[1] This is an alternative model in which the generation-recombination rate is determined based on the deep-level mechanism. It is a relatively complex model and a summary is presented in the Appendix B.

8.2 Stimulated Emission, Coherent Light

Let us consider that the LED is in a resonant cavity. As mentioned before, the objective of the cavity is to filter the frequencies of the electromagnetic field, and the interaction with the cavity field induces transitions with the resonant frequency. A well-known device that does this frequency selection is the Fabry-Perot resonator. This device is made up of a pair of mirrors that reflect (from one mirror to the other) the emitted photons. If the distance between the mirrors is $L = n\lambda/2$ with n integer and

$$\lambda = \frac{c}{\nu} = \frac{2\pi c}{\omega} \tag{8.16}$$

the cavity is resonant for photons with frequency $\omega = \pi c n/L$ and, in the ideal case, only photons with this frequency interfere constructively and persist within the cavity. Photons with frequencies different from ω, including its submultiples, undergo destructive interference and are suppressed. At the end of Sect. 3.5, we referred to the semiconductor structure np as a two-level system with concentrations n_1 and n_2, and wrote the relation (3.106) that we repeat here

$$\frac{dn_2}{dt} = \Big(n_1\alpha(\omega) - n_2\gamma(\omega)\Big)S\frac{\mathtt{I}}{\hbar\omega} - \frac{n_2}{\tau_{21}} + W_p n_1. \tag{8.17}$$

Here $\gamma(\omega)$ and $\alpha(\omega)$ are the emission and absorption coefficients, induced by the photonic field, and $\mathtt{I}$ is the field intensity or illumination intensity. In the previous chapter, we introduced the Einstein coefficients B_b^a in the context of emission and absorption processes. These coefficients are also related to $\gamma(\omega)$ and $\alpha(\omega)$. In this section, we will express (8.17) in terms of the electron and hole concentrations in the conduction and valence bands. The first two terms on the right-hand side of this equation define a key quantity in the operation of light-emitting devices: the *gain*, denoted by $\mathscr{G}$. We will examine this quantity with more detail in the following section.

If charge carriers on the p side of the conduction band are the minority carriers increased by the external potential V, we can replace the concentration n_2 in (8.17) by n_{pV}. Let us write this concentration as n_c, and for n_1 we will write n_v. The electrons concentration in the valence band that interests us is that of excitable electrons. In general these concentrations depend on several factors and on the occupation probability. If g_n and g_p represent the degeneracy of levels in these bands, and the relation $\alpha(\omega)/\gamma(\omega) = g_n/g_p$ is satisfied, the rate of change of n_c can be expressed as

$$\frac{dn_c}{dt} = -\left(n_c - n_v\frac{g_n}{g_p}\right)\frac{\mathtt{I}}{\hbar\omega}\gamma(\omega) - \frac{n_c}{\tau_{np}} + W_p n_v, \tag{8.18}$$

The penultimate term represents losses due to spontaneous emission, while the final term accounts for an increase in population resulting from a mechanism generically referred to as *pumping*.

The light emitted by a diode can be coherent or incoherent. When the diode is confined in a resonant cavity, only photons with wavelength congruent with the length of the resonant cavity survive. After some time, the population of photons has the resonant frequency, and the absorption and emission is predominantly of photons with the resonant frequency, hence the light is coherent. But, if electrons are injected or excited with a non-monochromatic field, the recombining electrons come from different energy levels, their frequencies are not exactly the same, and the emitted light will be incoherent.

Intra-band transitions (of electrons or holes) are of very small frequencies with wavelengths of the order of 2000 nm. These transitions are responsible for infrared (IR) radiation; if the wavelengths are shorter, say around 1000 nm, the radiation is the near infrared radiation (NIR). Inter-band transitions are generally in the range of visible radiation wavelengths, between 400 and 700 nm. Radiation with higher frequencies or shorter wavelengths is the ultraviolet radiation.

8.2.1 Spontaneous Emission in a Two-Level System

Suppose there is no field-confining cavity. In this case, the rate of change of the carrier concentration n, described by (8.18), will consist of only two components: spontaneous emission and pumping. That is,

$$\frac{dn_c}{dt} = -\frac{n_c}{\tau_{np}} + W_p n_v. \tag{8.19}$$

The intensity of the emitted light depends on both the bias potential V, through $n_c = n_{pV}$, and the recombination time τ_{np}, which is the inverse of the spontaneous transition probability W^{sp}, given in (7.71). As observed before, the spontaneous transition is modeled as a process in which the electron from the state $|\varphi_i\rangle$, in the conduction band, interacting with the electromagnetic vacuum, recombines with a hole in the final state $|\varphi_f\rangle$, in the valence band, and emits a photon. The probability of spontaneous transition from the (7.71), has the form

$$W^{sp}_{fi} = \frac{\omega_t^3}{3\hbar\epsilon_o\pi c^3}|{}_o\langle\varphi_f\, |ez|\varphi_i\rangle_{or}|^2. \tag{8.20}$$

To determine this probability it is required, as we saw in the first of the solved examples in the previous chapter, to obtain the matrix elements, or alternatively to follow the standard procedure which consists of expressing the transition probability as

$$W_{sp} = \frac{\omega^3}{3\hbar\epsilon_o\pi c^3} O_{cv}, \tag{8.21}$$

where O_{cv} is the matrix element of the electric dipolar moment

$$O_{cv} = |\ \langle f_v|e\ r|i_c\rangle\ |^2 = \frac{e^2}{m_o^2\omega^2}\ |\ \langle f_v|\hat{p}|i_c\rangle\ |^2, \tag{8.22}$$

This is calculated using a classical analog that models the spontaneous emission of atomic electrons subjected to harmonic forces, an external electromagnetic field, and friction-like forces that account for decay. The resulting quantities, known as oscillator strengths, represent the probability of transition and quantify the number of electrons per molecule resonant with a specific frequency ω_{0i} characterized by a decay constant γ_i. In the quantum approach the evaluation of the matrix elements is made with the exact or approximate wave functions of the electrons in the conduction and valence bands. This is not always possible. In the following sections we will show important cases in which it is possible. It turns out that the dipolar moment transition probability obtained in terms of oscillator intensities is two orders of magnitude larger than that obtained when the transition between states is calculated using functions of the one-dimensional infinite rectangular well. The actual physical systems are three dimensional (3D) and, generally, the physical symmetries are such that the one-dimensional approximation[2] can be assumed. In this approximation, to each energy level with wave number k_{zi} corresponds a degeneracy ΔN of levels in the transverse direction which is proportional to the two-dimensional density of levels g_{2D}. For an energy level width ΔE_i, the degeneracy is

$$\Delta N = g_{2D} \Delta E_i = L^2/(4\pi)\big(2m_e^*/\hbar^2\big)\Delta E_i = 0.1399\, L^2 \Delta E_i. \tag{8.23}$$

In the previous chapter we saw that the transition times for $L_z \simeq 100\,\text{nm}$ are of the order of 10^{-6}s, which means that $\Delta E_i \simeq \hbar/\tau_i \simeq 6 \times 10^{-10}$eV. For $L \simeq 10^6\,\text{nm}$, the degeneracy will be of the order of 0.84×10^2, that is, of the order of 100. Taking this into account, it is found that in most semiconductors

$$\frac{2\mid p_{cv}\mid^2}{m_o} \cong 23\text{eV}. \tag{8.24}$$

If the semiconductor is GaAs and the *hole*-energy levels are near the band edge, the emitted photon has an energy $\hbar\omega \simeq 1.52\,\text{eV}$ and the transition probability is

$$W^{sp} = \frac{e^2\hbar\omega}{3\pi\epsilon_0 m_o^2 c^3 \hbar^2} \mid p_{cv}\mid^2 = 5.4\ 10^7\text{s}^{-1}. \tag{8.25}$$

Thus, the recombination time will be

$$\tau_{sp} = \frac{1}{W^{sp}} = 0.185 ns. \tag{8.26}$$

For this recombination time, the power necessary to emit a photon is

$$W = \frac{\hbar\omega}{\tau_{sp}} = W^{sp}\hbar\omega = 5.4\ 10^7 \hbar\omega\ \text{eV/s} = 8.21\ 10^7\ \text{eV/s}. \tag{8.27}$$

[2] We have earlier mention that in heterostructures, the transverse dimensions in the x-y plane are much larger than in the growth direction z, the system is studied as if it were one-dimensional.

With this information we can ask: what potential difference must be established to develop the power W? To keep the emission process stable, the pumping power $W_b = e\Delta V/\tau_{sp}$, must be equal to W, i.e,

$$e\Delta V/\tau_{sp} = W, \tag{8.28}$$

therefore

$$\Delta V = \frac{\tau_{sp}}{e}W = \frac{\hbar\omega}{e} = 1.519V. \tag{8.29}$$

Finally, it is appropriate to note that to keep constant the spontaneous emission rate, it is necessary to maintain the minority population n_{pV} stable. This requires that the pumping rate satisfies the ratio

$$W_p = \frac{n_{pV}}{n_v}W^{sp} = \frac{n_{pV}}{n_v}\frac{e^2\hbar\omega}{3\pi\epsilon_0 m_o^2 c^3\hbar^2} \mid p_{cv} \mid^2 . \tag{8.30}$$

8.3 Population Inversion and Gain

In Chap. 4, we discussed that maintaining a steady-state generation-recombination process requires a continuous injection of charge carriers. In a resonant cavity, however, the presence of an electromagnetic field adds an additional layer of complexity. This field not only contributes to generation by exciting electrons into the conduction band, but also stimulates recombination processes that emit photons. As a result, it plays a crucial role in sustaining and enhancing the optical gain $\mathscr{G}$, defined by the first two terms in the second member of (8.18). In the definition of the gain, the correlation between the populations n_c, in conduction band, and the concentration n_v in valence band, is fundamental. For a device to favor light emission induced by a monochromatic field, we need a large population n_c and large populations of

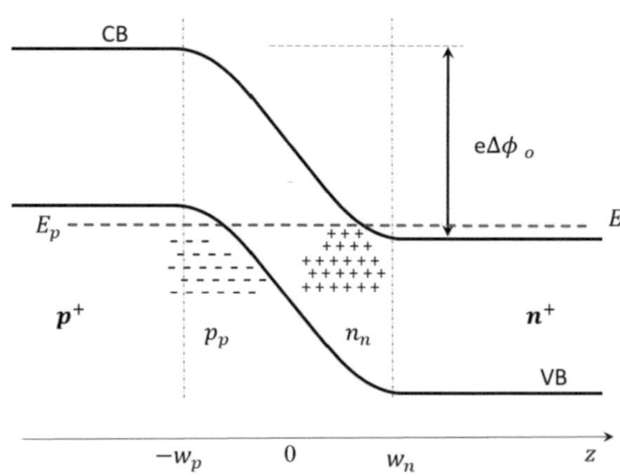

Fig. 8.3 The p^+n^+ junction with heavily doped semiconductors. In these systems the quasi-Fermi levels move inside the conduction and valence bands. In the transition zone, electrons penetrate to p side with concentration p_p, equal to that of the majority holes. The same with the holes on the n side. In these semiconductors $n_n \simeq p_p$

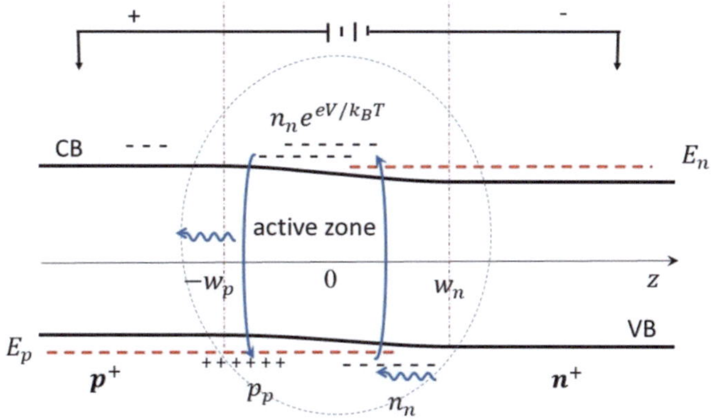

Fig. 8.4 When a potential difference V is applied in direct polarization, the bands tend to align. Simultaneously, the charges of the transition region return to their corresponding semiconductors. The VB in the vicinity of the interface has, on the p side, a concentration $\simeq p_p$ of holes, and on the n side, a concentration $\simeq n_n$ of electrons. In the transition zone there is also a concentration n_{pV}. That is the active zone of light emitting systems

electrons and holes in the valence band. This is achieved when the semiconductors are heavily doped. If we have concentrations n^+ and p^+, we can treat the electrons and holes separately with their corresponding quasi-Fermi levels E_n and E_p which, as shown in Fig. 8.3, are practically within the bands. In addition to this, the populations of charges traversing the interface are large. We saw in the first chapters that the concentration of electrons entering the p side is proportional to p_p and the concentration of holes on the n side is proportional to n_n. If a potential difference $V > 0$ is applied, the minority population of electrons grows by a factor e^{eV/k_BT}. When the potential is large enough, the bands tend to the flat band configuration, the concentration $n_c \to n_{pV} = n_p e^{eV/k_BT}$ and, in the transition zone we have a large population of electrons with concentration $n_v \simeq p_p$ that absorb photons and are injected into the CB, and a large population of holes that favor the recombination of electrons in the CB. As indicated in Fig. 8.4, on either side of the interface, there are concentrations p_p and n_n in the VB, and a concentration n_{eV} in the CB. In terms of these concentrations, the rate of variation of n, taking into account that $B_i^f = B_f^i$, is

$$\begin{aligned}\frac{dn}{dt} &= -(n_{pV} - p_p)B_f^i \rho(\omega_{if}) - n_{pV}\frac{1}{\tau_{sp}} + W_p p_p \\ &= -\mathscr{G}(V,T)B_f^i \rho(\omega_{if}) - n_{pV}\frac{1}{\tau_{sp}} + W_p p_p. \end{aligned} \tag{8.31}$$

The first term of these expressions defines the *gain*, which, as we saw before, depends crucially on the correlation of majority p_p and minority n_p populations on the p side. This correlation of populations can be modified by applying a polarization potential. If we use the relations defined earlier for n_{pV} and p_p, the gain can be written as

$$\mathscr{G} = \Delta n = \frac{n_p^2 e^{eV/k_B T} - n_i^2}{n_p} = N_c e^{-(E_p+E_g)}\left(e^{2(E_p+eV)/k_B T} - e^{2E_n/k_B T}\right). \tag{8.32}$$

Figure 8.5 shows the behavior of the population inversion and the gain as functions of V. For small values of V, the gain is negative, but when V is greater than V_u, known as the threshold potential, the CB's electron population grows rapidly. This population inversion is the condition to maintain the emission regime stimulated by the photonic field. From the previous equation, we can conclude that one has population inversion if

$$e^{(2E_p+eV)/k_B T} \geq e^{2E_n/k_B T} \qquad \text{i.e., if} \qquad 2E_p + 2eV \geq 2E_n. \tag{8.33}$$

Thus, the threshold potential is given by the difference of the quasi-Fermi levels

$$eV_u = (E_n - E_p) \sim E_g. \tag{8.34}$$

The diode keeps emitting under steady conditions when the potential difference is equal to or greater than V_u.

8.4 Optical Devices in Heterostructures

In Ref. [2] H. Kroemer suggested the possibility of creating quantum confinements when semiconductor structures are grown in such a way that, instead of different dopings, the semiconductors differ in their fundamental *gaps*. In Chap. 5 it was discussed how this procedure produces wells and barriers by alternating layers of semiconductors with different gaps. In addition to the gaps, an important property is the alignment and relative position of the bands in the heterojunction. Different types of band alignment are shown in Fig. 8.6. If the heterojunctions of the semiconductors are type I, and the gap in the central layer is greater than in the lateral layers, a barrier is formed, but if in the central layer it is smaller than in the lateral layers, a potential well is formed. In type I structures, the gap of the central layer is completely

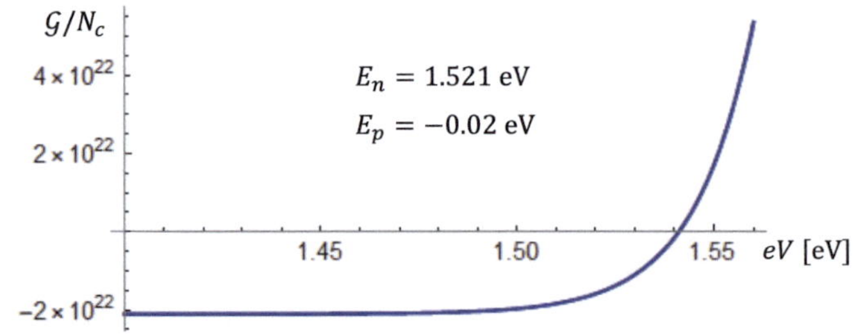

Fig. 8.5 Inversion population. The gain of (8.32), in units of the effective concentration N_c, as a function of the external potential V

enclosed by the gap of the lateral layers. The bandgap difference ΔE_g distributes between the conduction and the valence bands. We will, in general, assume that the discontinuities in the conduction and valence bands are $2/3\Delta E_g$ and $1/3\Delta E_g$, respectively. The original idea was to grow lateral layers to inject charges into the central layer.

In the 1970s, began the semiconductor heterostructures era. Since those years, methods for growing semiconductor films were developed with an accuracy that, after two decades, reached the atomic layer precision. The improvement of layer growth methods made it possible to obtain increasingly thinner layers, create narrow quantum wells, and to grow an endless variety of semiconductor structures with diverse band profiles, for diverse purposes. The emergence of heterostructures represented an important change in the development of optical and electronic transport devices, and also in theoretical approaches.

From a theoretical point of view there is an important change in the description of these systems. The semiclassical theory presented in the first chapters loses the relevance that it had in the first era of semiconductor technology, in which the concepts introduced in the previous chapters were essential. The nanoscopic dimensions of heterostructures and the effective mass approximation allow us to include other properties and phenomena. Important new results are obtained when the Schrödinger equation (1.5) for heterostructures is solved, mainly in the effective mass approximation. Although in heterostructures the diffusion of impurities that defined the potential profile in the np junctions is small, the interfaces are not, in general, ideal. There are reasons, some of which were discussed at the end of Chap. 5, of why the gap energy does not change abruptly at the interfaces, as assumed in a first approximation.

To analyze the physics of a quantum well we will rely on a specific structure of $(Al, Ga)As$. A common quantum well is the $Al_xGa_{1-x}As/GaAs/Al_xGa_{1-x}As$ structure where a layer of $GaAs$ is grown between two of $Al_xGa_{1-x}As$ in which a fraction x of Ga atoms have been replaced by Al atoms. In $Al_xGa_{1-x}As$ the gap

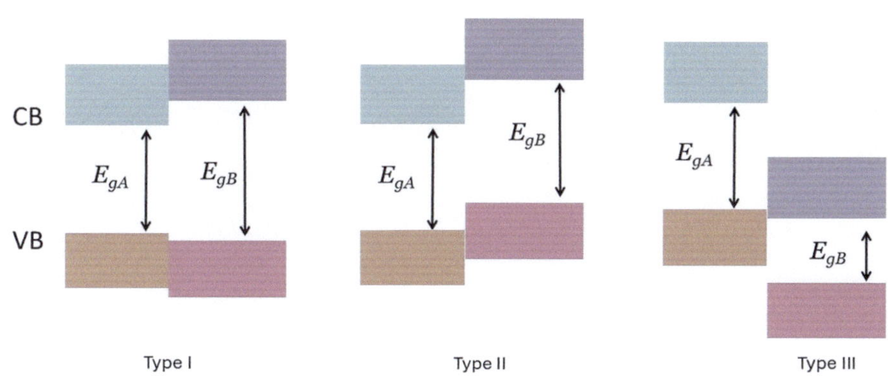

Fig. 8.6 Different types of band alignment in semiconductor heterojunctions

changes with the concentration x, according to the empirical formula

$$E_g(x) = 1.519 + 1.155x + 0.37x^2. \tag{8.35}$$

Thus the gap that for $x = 0$ is $E_g(0) = 1.519\,\text{eV}$ changes to $E_g = 1.8988\,\text{eV}$ when $x = 0.3$. As a consequence, if the interfaces are ideal, the edges of the conduction and valence bands form a sharply changing profile as shown in Fig. 8.7. If electrons and holes are injected into the A layers of $Al_xGa_{1-x}As$, they will fall to lower energy levels in the electron and hole wells of the B layer of $GaAs$. The discretization of the energy levels is one of the important features of the quantum wells. It is in some ways similar to the frequency filter achieved with the resonant cavity. Another important consequence of these systems is that the initial and final quantum states can be determined with sufficient precision and it is not necessary to rely on the quasi-empirical evaluation of oscillator strengths $\langle p_{cv}\rangle^2$. In the following sections we will briefly refer to the optical responses of a quantum well and of a superlattice.

8.4.1 Optical Transitions in a Quantum Well

In the previous chapter it was shown that the transition probability per unit time from one state $\phi_i = |i\rangle$ to another $\phi_f = |f\rangle$ could be written in the form

$$W_{if} = \frac{e^2\hbar\omega}{3\epsilon_o m_o^2 c^3} \frac{\left|\int dz[\varphi_f^h(z)]^* \frac{\partial}{\partial z}\varphi_i^e(z)\right|^2}{(\hbar\omega - E_i^e + E_f^h)^2 + \Gamma_i^2}\langle n_\omega\rangle = \frac{1}{\tau_{if}}. \tag{8.36}$$

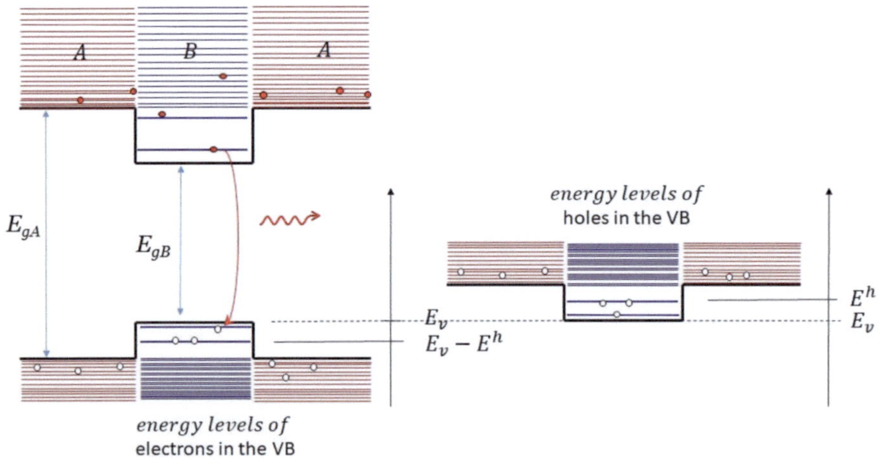

Fig. 8.7 Heterostructure of three semiconductor layers that form a quantum well. When $E_{gA} > E_{gB}$, wells are formed for electrons in the CB and for holes in the VB. The right panel shows the positive hole levels in the VB that correspond to the negative electron levels in the VB on the left

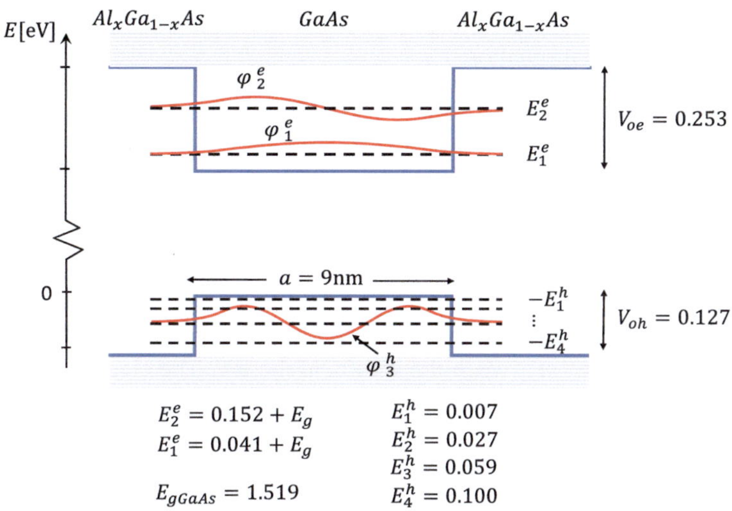

Fig. 8.8 Eigenvalues and eigenfunctions in the CB and VB wells for $x = 0.3$. The depths of the wells are $V_{oe} = 2V_{oh} = 0.2532\,\text{eV}$ The effective masses (see Table 1.4) are $m_e^* = (0.67 + 0.083\,x)m_o$ and $m_{hh}^* = (0.51 + 0.25\,x)m_o$. With these parameters, 2 eigenvalues result in the CB and 4 in the VB as indicated in the figure. Some of the eigenfunctions are also shown. The eigenfunctions have well-defined parity with respect to the center of the well

If there is no external field or resonant cavity in which the emitted photons are contained, the factor $\langle n_\omega \rangle$ can be removed. When the physical system that absorbs or emits photons has several energy levels from which the transition could begin and several levels to which the particles that make the transition could reach, we must add over the initial and final states, i.e.

$$W(\omega) = \sum_{i,f} W_{if} = \frac{e^2\hbar\omega}{3\epsilon_o m_o^2 c^3} \sum_{i,f} \frac{\left| \int dz \left[\varphi_f^h(z)\right]^* \frac{\partial}{\partial z} \varphi_i^e(z) \right|^2}{\left(\hbar\omega - E_i^e + E_f^h\right)^2 + \Gamma_i^2}. \tag{8.37}$$

In quantum wells whose width is a few tens of atomic monolayers, there are two or three quantum levels and the selection rules reduce further the allowed transitions, hence the spectrum has one or two peaks only.

Consider a one-dimensional potential well of depth V_{ob} and width a, as illustrated in Fig. 8.8. The Schrödinger equation for this system yields a discrete set of eigenfunctions $\varphi_\mu^b(z)$ and corresponding energy eigenvalues E_μ^b, where the index $b = e, h$ refers to electrons in the conduction band (CB) and holes in the valence band (BV), respectively. Since this is a well-known textbook problem, we consider here only the final results and include details of its solution in the Appendix C for completeness. There, in the Appendix, the rectangular potential well is treated using the transfer matrix method introduced in Chap. 5 which provides a systematic approach for piecewise-constant potentials. The energy eigenvalues are obtained by solving the standard transcendental equation

$$\cot ka = \frac{k^2 - q^2}{2kq}. \tag{8.38}$$

Here

$$k = \sqrt{\frac{2m_b^*}{\hbar^2}E} \qquad \text{and} \qquad q = \sqrt{\frac{2m_b^*}{\hbar^2}(V_o - E)}. \tag{8.39}$$

The mass m_b^* is the effective mass of electrons m_e^*, if the well is in the CB, or the effective mass of holes m_h^*, when the well is in the BV. The eigenfunctions $\varphi_\mu(z)$ are obtained when the function

$$\varphi(z, E) = A \begin{cases} e^{qz} & \text{for } z < 0, \\ \cos kz + \dfrac{q}{k}\sin kz & \text{for } 0 < z < a, \\ \dfrac{k^2 + q^2}{2kq}\sin ka e^{-q(z-a)} & \text{for } z < a, \end{cases} \tag{8.40}$$

is evaluated for $E = E_\mu^b$, that is: $\varphi_\mu(z) = \varphi(z, E)|_{E=E_\mu^b}$. A is a normalization constant. In the heterostructure $Al_xGa_{1-x}As/GaAs/Al_xGa_{1-x}As$ with $x = 0.3$ the band gap difference is $\Delta E_g \sim 0.38$ eV, which under the 2/3 and 1/3 distribution, imply that the depths of the wells in the conduction and valence bands are $V_{oe} = 2V_{oh}$ $= 0.2532$ eV and the effective masses (see Table 1.4) are $m_e^* = (0.67 + 0.083\,x)m_o$ and $m_{hh}^* = (0.51 + 0.25\,x)m_o$. If the width of the well is $a = 9$ nm (equivalent to 16 unit cells), the eigenvalues in the CB band and in the VB are those indicated in Fig. 8.8. This figure also shows some of the eigenfunctions. An important characteristic of the eigenfunctions is that they have well defined parity $P[\varphi_\mu(z)]$, related to the quantum number μ. In the quantum well the parities, in the conduction and valence bands, are:

$$P[\varphi_\mu^e(z)] = P[\mu + 1] \qquad \text{and} \qquad P[\varphi_{\mu'}^h(z)] = P[\mu' + 1]. \tag{8.41}$$

With the eigenfunctions in (8.40), the matrix elements in (8.36) and (8.37) can be calculated. Taking into account the parity symmetries of the eigenfunctions, it is expected that the matrix elements

$$\int_{-a/2}^{a/2} dz[\varphi_{\mu'}^h(z)]^* \frac{\partial}{\partial z}\varphi_\mu^e(z) \quad \text{be:} \quad \begin{cases} = 0 \text{ if} & \mu + \mu' \quad \text{es even} \\ \neq 0 \text{ if} & \mu + \mu' \quad \text{is odd} \end{cases}. \tag{8.42}$$

These selection rules predict the possible and the prohibited transitions. In our example, the quantum numbers are $\mu = 1, 2$ and $\mu' = 1, 2, 3, 4$. Therefore, only four transitions are possible. So if

$$\omega_{if} = \frac{E_i^e + E_f^h}{\hbar}, \tag{8.43}$$

there will be transitions for the frequencies $\omega_{21}, \omega_{41}, \omega_{12}$ and ω_{32}. Figure 8.9 shows the optical spectra calculated for the optical response

$$\chi_{if} = \frac{\left|\int dz[\varphi_f^h(z)]^* \frac{\partial}{\partial z}\varphi_i^e(z)\right|^2}{\left(\hbar\omega - E_i^e - E_f^h\right)^2 + \Gamma_i^2} \propto W_{if} \tag{8.44}$$

For the plots in Fig. 8.9, we considered different values of the parameter Γ_i, which due to the uncertainty relation $\Gamma_i \simeq \hbar/\tau_i$ are related to different decay times τ_i of the initial state. Γ_i is also related to the experimental precision. If the decay times are very large, the states are quasi-stable, the Gamma parameter is small, and the resonances in the optical response are narrow. It is clear that if the experimental precision is low, it does not distinguish one resonance from another. For the optical responses in Fig. 8.9 different Γs are considered, however, in each case, the decay times of all initial states are taken equal. For the graph in panel (*b*) it is assumed that $\Gamma \simeq 0.01$ eV that is, smaller than the distance between the energy eigenvalues. With this theoretical or experimental precision, the spectrum that will be observed (or calculated) is like the one shown in that panel. If the precision is a factor of 10 less, i.e., if $\Gamma \simeq 0.1$ eV a spectrum like that in panel (*a*) is obtained, in which case, the individual resonances can not be distinguished. If the calculation or experimental precision is increased, say up to $\Gamma \simeq 0.003$ eV we have an optical spectrum like the one shown in panel (*c*), where all transitions allowed by the selection rules are clearly distinguished. In this graph, the frequencies ω_{21}, ω_{41}, ω_{12} and ω_{32} of the corresponding transitions are indicated above the peaks of the resonances. Using the energy eigenvalues, the reader can verify if the assignments are correct. In many experimental results, it is normal that not all peaks, predicted in the theoretical optical susceptibility, are seen. To improve the prediction of susceptibility, it can be multiplied by the occupation probability of the energy levels, and also include the temporal distribution of the transition probabilities, if the corresponding decay times τ_i are known. The inclusion of the transition probability factor will generally reduce the relative height of the resonances at higher frequencies. In general, the lasting transition is $E_1^e \to E_2^h$.

8.4.2 Emission in Multiple-Quantum-Wells Superlattices

Structures that emit visible light include multiple quantum wells (MQW) and superlattices. One particularly important superlattice is $(In_xGa_{1-x}N/In_yGa_{1-y}N)^n$ that emits blue light.

Superlattices are periodic structures in which a basic structure, for example a well followed by a barrier, is repeated a finite number of times. In a superlattice system, finiteness of the number of unit cells is an eloquent characteristic. In superlattices produced for light emission purposes the number of unit cells is on the order of a dozen. As already explained in Chap. 6, periodicity and, consequently, quantum coherence are responsible for the quantum phenomenon known as splitting of energy levels. For this effect to be observable, it is necessary that the separation of the quantum wells

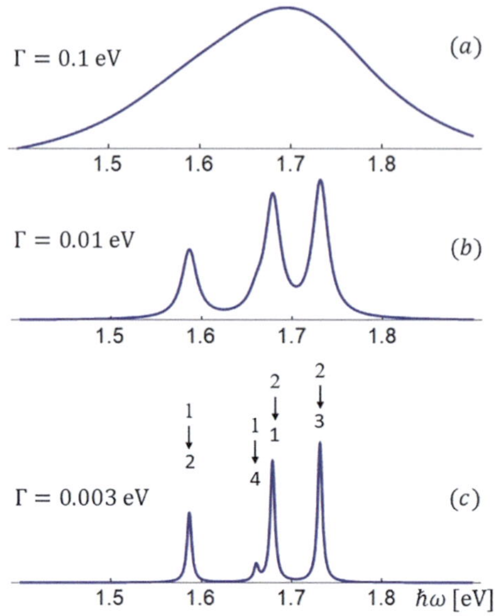

Fig. 8.9 Optical response of a potential well, with the eigenvalues and eigenfunctions determined by the (8.38) and (8.40), for the same system but different values of the precision parameter Γ. In panel (**a**), $\Gamma = 0.1$ eV, of the order of the separation of the eigenvalues. In panel (**b**) $\Gamma = 0.01$ eV and in the (**c**) $\Gamma = 0.003$ eV. In the last panel we can see all peaks allowed by the selection rules (8.42). A pair of quantum numbers that correspond to initial and final states is attached to each peak

is not so large. It was already noticed that in periodic systems, the number of energy levels is equal to the number of wells in the crystalline direction characterized by a periodic potential. The symmetric double quantum well (DQW) is a superlattice with one confined unit cell, and two confining barriers. The energy levels in the DQW appear in pairs, slightly separated (split) from each other, and practically with the same values as the energy levels in the single well. This characteristic, among others, motivated a lot of experimental work and many applications. When we have a superlattice, the Schrödinger equation is solved with the formalism of the theory of finite periodic systems. We have seen in Chap. 6 that in this formalism analytical expressions for the eigenvalues and the eigenfunctions can be obtained. In a periodic heterostructure the electrons of the CB and also the holes of the VB feel, at the corresponding band edges, a periodic modulation of the potential. As in any periodic system, a band structure appears related to this periodic profile, and since these are within the conduction and valence bands, they are called subbands. The number of levels in these subbands is the same as the number of wells in the superlattice. In the effective mass approximation, the potential profiles in the conduction and valence bands are like the ones shown in Fig. 8.10.

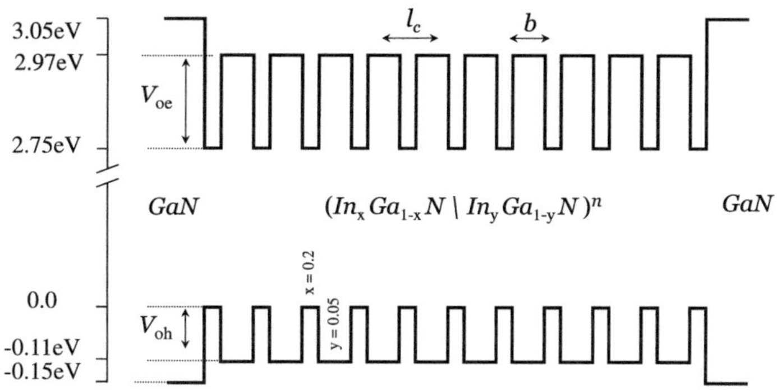

Fig. 8.10 Superlattice $(In_xGa_{1-x}N/In_yGa_{1-y}N)^n$ with x =0.2 and, y =0.05 and $n = 10$. For the spectrum in Figs. 8.12-8.14, the width of the wells is a = 2.6 nm and of the barriers is b =5.2 nm. This system emits blue light. The superlattice is contained between two layers of GaN. Reproduced from [3], with permission from Elsevier

If we use the TFPS, the first quantity that we have to obtain is the transfer matrix M of a unit cell,[3] say cell $j + 1$, which satisfies the relation

$$\phi(z_{j+1}) = M\phi(z_j) = \begin{pmatrix} \alpha & \beta \\ \gamma & \delta \end{pmatrix}\phi(z_j). \tag{8.45}$$

The matrix elements α, β, etc., are potential-dependent functions. When the number of propagating modes is N (N moving in the positive direction and N in the opposite direction), these elements are matrices of dimension $N \times N$. For the purposes of the superlattice that we are interested here, we consider only one propagating mode (one in the growing direction and one in the opposite) and the matrix elements are scalars. In the theory of periodic systems it is shown that, knowing the unit-cell transfer matrix, the elements of the transfer matrix M_n, of the entire system, i.e. of the n unit cells, are expressed in the form

$$\alpha_n = U_n - \gamma^{-1}\delta\gamma U_{n-1}, \qquad \text{and} \qquad \beta_n = \beta U_{n-1}. \tag{8.46}$$

For systems invariant under time inversion, $\gamma = \beta^*$ and $\delta = \alpha^*$ with

$$\begin{aligned} \alpha &= e^{ika}\Big(\cosh qb + i\frac{k^2 - q^2}{2qk}\sinh qb\Big), \\ \beta &= -i\frac{k^2 + q^2}{2qk}\sinh qb. \end{aligned} \tag{8.47}$$

Here $k = \sqrt{2m_1^* E/\hbar^2}$ and $q = \sqrt{2m_2^*(V_o - E)/\hbar^2}$.

[3] To better visualize, in this part the functions that come from the matrix are written in blue of a cell.

To calculate the transitions from the subbands in the CB to the subbands in the VB, it is necessary to first obtain the eigenvalues and eigenfunctions of electrons and holes. Before referring specifically to these energy levels in the subbands, it is important to notice that in Bloch's Theory, specifically in the Kronig-Penney model, which is an infinite sequence of rectangular wells and barriers, the continuous subbands shown in the left panels of Fig. 8.12 are obtained. We do not use that model here but we show that the subbands predicted in the Kronig-Penney model and those obtained in the TFPS have some features in common. The eigenvalues of the superlattice stretching from z_L to z_R, with $z_0 - z_L = z_R - z_n = a/2$, see Fig. 8.11, are obtained from equation

$$\Re\mathrm{e}\left[\alpha_n e^{ikd}\right] - \frac{k^2 - q_w^2}{2q_w k}\Im\mathrm{m}\left[\alpha_n e^{ikd}\right] - \frac{k^2 + q_w^2}{2q_w k}\beta_{nI} = 0. \qquad (8.48)$$

where q_w and k are the wave numbers to the left (right) and right (left) of the discontinuity point z_L (z_R). Figure 8.12b shows the discrete subbands calculated by solving this equation. For each set of parameters we have $n+1$ eigenvalues $E_{\mu,\nu}$ per subband. As explained in Chap. 6, the eigenvalues are characterized by two quantum numbers: μ that labels the subbands and ν that labels the intra-subband levels. It was also mentioned before that $\nu = 1, 2, 3, ..., n+1$. When the height of the lateral potentials in the 'cladding layers' is greater than the height of the internal barriers, i.e. if $V_w > V_o$, two energy levels separate from the subbands and their corresponding eigenfunctions are localized at the ends of the superlattice. These are the surface states.

The right panel of Fig. 8.12 shows the subbands for the superlattice $In_yGa_{1-y}N/(In_xGa_{1-x}N\ /In_yGa_{1-y}N)^n$ with $x = 0.2$, $y = 0.05$, $n = 10$, $a = 2.6$ nm and $b = 5.2$ nm. For comparison, these (discrete) subbands were calculated with the same potential parameters that were used to obtain the continuous subbands of the Kronig-Penney model.

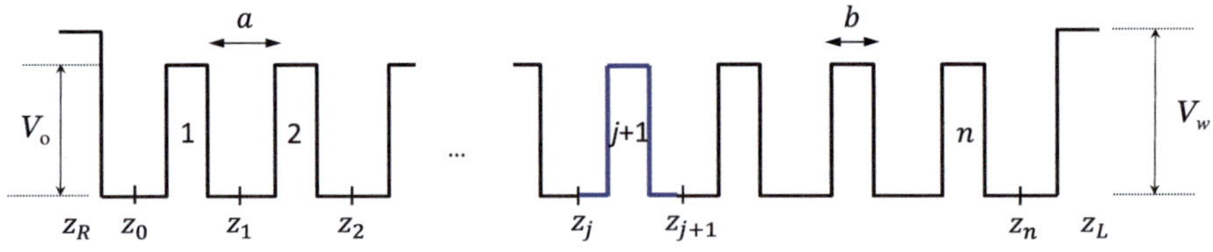

Fig. 8.11 Periodic, sectionally constant potential, that results from changes in the band edges, both in the CB and in the VB, due to the energy gap difference ($\Delta E = E_{gA} - E_{gB} = V_{oe} + V_{oh}$), in the alternating layers

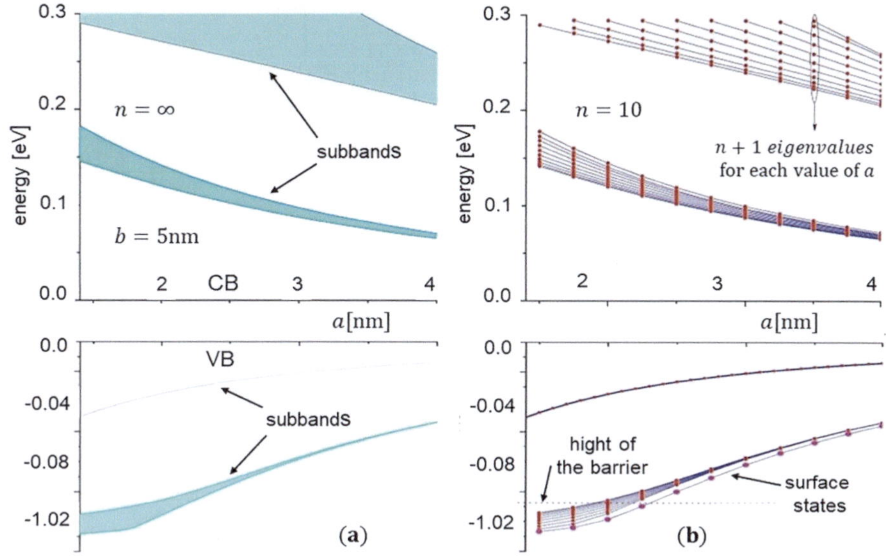

Fig. 8.12 Subbands in the conduction and valence bands. **a** Continuous subbands calculated with the Kronig-Penney model. **b** Discrete subbands calculated with the TFPS. The energies are plot as functions of the width of the wells a, and for $In_yGa_{1-y}N/(In_xGa_{1-x}N/In_yGa_{1-y}N)^n$ with $x = 0.2$, $y = 0.05$, $b = 5.2$ nm, and $n = \infty$ in (**a**), $n = 10$ in (**b**). Reprinted from [3], with Permission of Elsevier

The eigenfunctions are obtained by substituting the eigenvalues $E_{\mu,\nu}$ into the equation

$$\begin{aligned}\Psi_{\mu,\nu}(z) = \frac{A_o e^{ika/2}}{2k}\Big[&\Big(\big(\alpha_p(z)+\gamma_p(z)\big)\alpha_j + \big(\beta_p(z)+\delta_p(z)\big)\beta_j^*\Big)(k - iq_w)\\ &+\Big(\big(\alpha_p(z)+\gamma_p(z)\big)\beta_j + \big(\beta_p(z)+\delta_p(z)\big)\alpha_j^*\Big)e^{-ika}(k+iq_w)\Big]_{E=E_{\mu,\nu}}\end{aligned} \tag{8.49}$$

A_o is a normalization constant and z any point in the $j+1$ unit cell. α_j, β_j,... are elements of the transfer matrix $M_j(z_j, z_0)$ that connects state vectors at points separated by exactly j unit cells. $\alpha_p(z)$, $\beta_p(z)$ ... are the elements of the transfer matrix $M_p(z, z_j)$ that connects the state vectors $\Phi(z_j)$ and $\Phi(z)$, for $z_j \leq z \leq z_{j+1}$.

Figure 8.13 also shows the eigenvalues $E_{\mu,\nu}$ that result from solving (8.48). This figure shows one subband in the CB and two in the VB of the $(In, Ga)N$ superlattice. On the right of the figure the subbands are amplified. In each subband μ, we denote the two levels related to the surface eigenfunctions as $s\mu$, and the rest of the levels as $g\mu$. This figure also indicates the transitions from the CB to the second subband of the VB. As in the quantum well, the eigenvalues and eigenfunctions allow us to evaluate the susceptibility

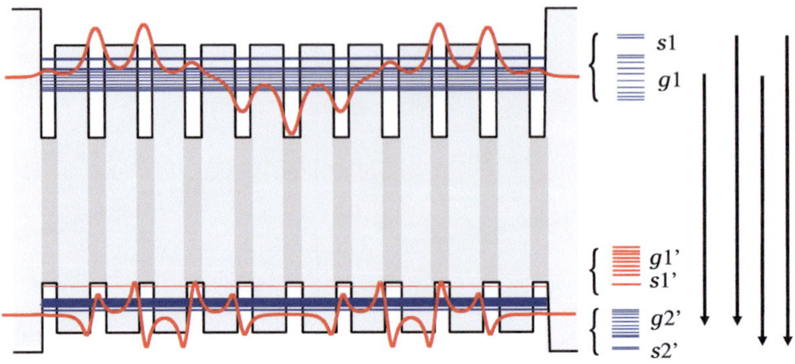

Fig. 8.13 If (8.48) is solved, subbands shown here and in the Fig. 8.12b, with $n + 1$ eigenvalues $E_{\mu,\ nu}$ per subband, are obtained. Reprinted from [3], with Permission of Elsevier

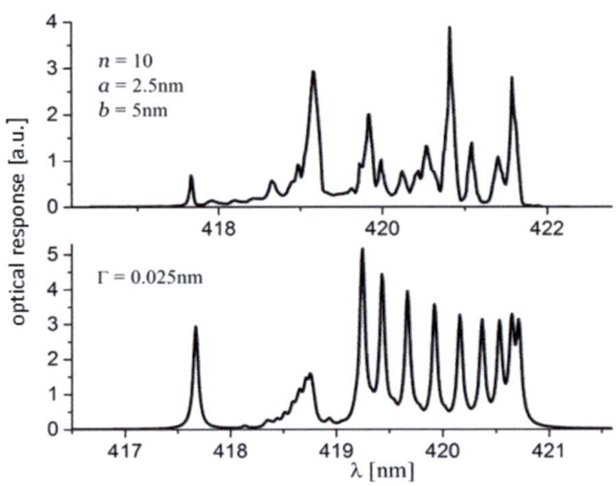

Fig. 8.14 Optical response of a $(In, Ga)N$ superlattice with characteristics indicated in Fig. 8.10. In the upper panel the experimental results and in the lower panel the theoretical results obtained with the formalism outlined in the text. Reprinted from [3], with Permission of Elsevier

$$\chi = \sum_{\nu,\nu',\mu,\mu'} f_{eh} \frac{\left| \int dz \left[\Psi^{v}_{\mu',\nu'}(z)\right]^{*} \frac{\partial}{\partial z} \Psi^{c}_{\mu,\nu}(z) \right|^2}{\left(\hbar\omega - E^{c}_{\mu,\nu} + E^{v}_{\mu',\nu'} + E_B\right)^2 + \Gamma^2} \tag{8.50}$$

where E_B is the exciton binding energy, which is of the order of 10 meV. Figure 8.14 shows the experimental results of Nakamura [1] for the superlattice sketched in Fig. 8.10, and the spectrum obtained with the TFPS formalism. The agreement is good, there is however a small difference. The theoretical spectrum extends from $\simeq 417.6$ nm to $\simeq 420.6$ nm, 1 nm less that the experimental spectrum. There are two reasons for this difference. One is the small asymmetry in the lateral confining cladding layers, the second is that we omitted the local Stark effect characteristic of wurzite structures. We will consider this correction in the next section.

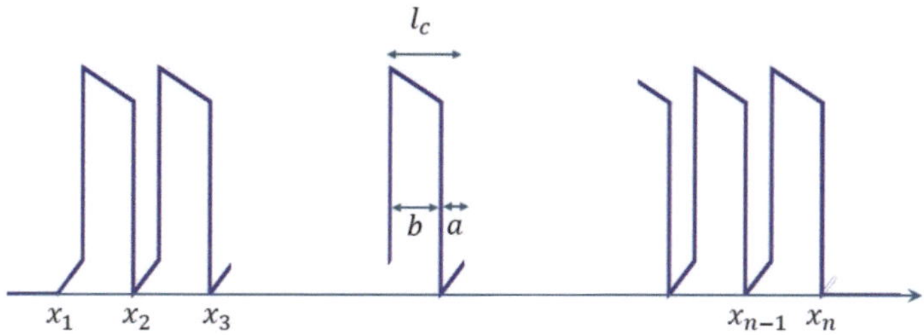

Fig. 8.15 GaN superlattice and the local Stark effect. The polarization of charges in the wells leads to a potential profile as the one drawn in this figure

8.4.3 Charge Polarization Effect in the Blue Laser Spectrum

As mentioned in Chap. 1, the lack of inversion symmetry in wurzite configurations leads to charge polarization in layered structures, hence to local electric fields. An example of this kind of systems is the GaN superlattice considered in the last section. Due to charge polarization, the superlattice profile is as shown in Fig. 8.15. Since our goal is to correct the small difference noticed in the calculation of the spectrum studied in previous section, the potential parameters that we will consider here are the same as in the last section, i.e. wells width $a = 2.78\,\text{nm}$, barriers width b=5.56nm, $V_o = 0.2252\,\text{eV}$, effective masses $m_w^* = 0.128m_e$, $m_b^* = 0.114m_e$[4] and the number of unit cells $n = 10$. Due to the spontaneous dielectric polarization the localized quasi-two-dimensional electron and hole distributions on the opposite sides of the quantum wells, produce local electric forces F_w in the well and barriers, such that $F_b = -F_w a/b$. Despite the changes in signs and magnitudes of the electric forces, we can, using transfer matrices as those obtained in Sect. 5.6 for biased step potential and rectangular barrier, with the appropriate adjustments, solve the Schrödinger equations. The Schrödinger equations in the well and barrier of the j-th unit cell, are

$$\frac{d^2\varphi_l(x)}{dx^2} - \frac{2m_w^* F_w}{\hbar^2}\left(x - \frac{E}{F_w}\right)\varphi_l(x) = 0, \qquad x_j < x < x_j + a, \qquad l = 2j \tag{8.51}$$

and

$$\frac{d^2\varphi_{l+1}(x)}{dx^2} - \frac{2m_b^* F_b}{\hbar^2}\left(\frac{V_o - E}{F_b} - x\right)\varphi_{l+1}(x) = 0, \qquad x_j + a < x < x_j + a + b. \tag{8.52}$$

respectively. In the last equation we leave the Airy equation as a function of $-x$. Instead of writing the Airy equation solutions as $Ai(-\eta)$ and $Bi(-\eta)$. This is an

[4] These parameters correspond to the conduction band profile in the semiconductor superlattices $(In_xGa_{1-x}N/In_yGa_{1-y}N)^n In_xGa_{1-x}N$.

alternative, however one must be careful with the sign when derivatives with respect to x are performed. Introducing the changes of variable

$$\xi = \eta_w \left(x - \frac{E}{F_w} \right) \quad \text{and} \quad \zeta = \eta_b \left(\frac{V_o - E}{F_b} - x \right), \tag{8.53}$$

where

$$F_b = F_w \frac{a}{b}, \quad \eta_w = \left(\frac{2m_w^* F_w}{\hbar^2} \right)^{1/3} \quad \text{and} \quad \eta_b = \left(\frac{2m_b^* F_b}{\hbar^2} \right)^{1/3}, \tag{8.54}$$

we end up with two Airy equations, whose solutions, respectively, are

$$\varphi_l(x, E) = a_l \,\mathrm{Ai}(\xi) + b_l \,\mathrm{Bi}(\xi), \qquad x_j < x < x_j + a, \qquad l = 2j \tag{8.55}$$

and

$$\varphi_{l+1}(x, E) = a_{l+1} \,\mathrm{Ai}(\zeta) + b_{l+1} \,\mathrm{Bi}(\zeta) \qquad x_j + a < x < x_j + l_c. \quad l_c = a + b. \tag{8.56}$$

As long as the potential periodicity is kept, the calculation of the physical quantities can easily be done using the transfer matrices approach and the compact formulas of the theory of finite periodic systems, briefly introduced in Chap. 6. As mentioned there, all we need, to determine the n-cells transfer matrix, is the unit-cell transfer matrix defined in (8.62), which we write here again as

$$W_{l_c j}(x_j, x_{j-1}) = W_b(x_j, x_{j-1} + a) W_w(x_{j-1} + a, x_{j-1}) \equiv \begin{pmatrix} \theta_{l_c} & \mu_{l_c} \\ \nu_{l_c} & \chi_{l_c} \end{pmatrix} \tag{8.57}$$

where

$$W_{aj} = W_w(x_j + a, x_j) = \begin{pmatrix} \theta(\xi_{j0}, \xi_{ja}) & \mu(\xi_{j0}, \xi_{ja}) \\ \nu(\xi_{j0}, \xi_{ja}) & \chi(\xi_{j0}, \xi_{ja}) \end{pmatrix}. \tag{8.58}$$

and

$$W_{bj} = W_b(x_j, x_{j-1} + a) = \begin{pmatrix} \theta(\zeta_{j0}, \zeta_{jb}) & -\mu(\zeta_{j0}, \zeta_{jb}) \\ -\nu(\zeta_{j0}, \zeta_{jb}) & \chi(\zeta_{j0}, \zeta_{jb}) \end{pmatrix} \tag{8.59}$$

The functions $\theta(z_1, z_2)$, $\mu(z_1, z_2)$, $\nu(z_1, z_2)$ and $\chi(z_1, z_2)$ are defined as

$$\begin{aligned}
\theta(z_1, z_2) &= \frac{1}{\mathrm{W}_0} \Big(\mathrm{Ai}(z_2)\mathrm{Bi}'(z_1) - \mathrm{Ai}'(z_1)\mathrm{Bi}(z_2) \Big), \\
\mu(z_1, z_2) &= \frac{1}{\mathrm{W}_0} \Big(\mathrm{Ai}(z_1)\mathrm{Bi}(z_2) - \mathrm{Ai}(z_2)\mathrm{Bi}(z_1) \Big), \\
\nu(z_1, z_2) &= -\frac{\eta_s}{\mathrm{W}_0} \Big(\mathrm{Ai}'(z_1)\mathrm{Bi}'(z_2) - \mathrm{Ai}'(z_2)\mathrm{Bi}'(z_1) \Big),
\end{aligned}$$

$$\chi(z_1, z_2) = \frac{1}{\mathrm{W}_0}\Big(\mathrm{Ai}(z_1)\mathrm{Bi}'(z_2) - \mathrm{Ai}'(z_2)\mathrm{Bi}(z_1)\Big). \tag{8.60}$$

Here W_0 is the Wroskian

$$\mathscr{W}_0 = \mathrm{Ai}(z_1)\mathrm{Bi}'(z_1) - \mathrm{Ai}'(z_1)\mathrm{Bi}(z_1). \tag{8.61}$$

The minus signs in (8.59) are because of the derivative with respect to x. It has been shown in the theory of finite periodic systems that the n-cells transfer matrix

$$W_{nl_c} \equiv \begin{pmatrix} \theta_{nl_c} & \mu_{nl_c} \\ \nu_{nl_c} & \chi_{nl_c} \end{pmatrix}, \tag{8.62}$$

can be transformed into the transfer matrix M_{nl_c} and derive then the scattering amplitudes of the whole superlattice of length $L = nl_c$. To obtain the matrix M_{nl_c} we use the similarity transformation

$$M_{nl_c} = \begin{pmatrix} 1 & 1 \\ ik & -ik \end{pmatrix}^{-1} W_{nl_c} \begin{pmatrix} 1 & 1 \\ ik & -ik \end{pmatrix} = \begin{pmatrix} \alpha_n & \beta_n \\ \gamma_n & \delta_n \end{pmatrix} \tag{8.63}$$

or, alternatively, transform the single cell transfer matrix W_{l_c} into the transfer matrix M_{l_c} through the similarity transformation

$$M_{l_c} = \begin{pmatrix} 1 & 1 \\ ik & -ik \end{pmatrix}^{-1} W_{l_c} \begin{pmatrix} 1 & 1 \\ ik & -ik \end{pmatrix} = \begin{pmatrix} \alpha & \beta \\ \gamma & \delta \end{pmatrix} \tag{8.64}$$

and use the closed formulas of the theory to derive the physical quantities of interest. If we have an open superlattice as the one shown in (8.15) we can calculate the transmission coefficient and the resonant enrgies. If the purpose is to obtain the optical response, we have to confine the superlattice and calculate the energy eigenvalues and eigenfunctions. Let us transform the unit cell transfer matrix W_{l_c} into the transfer matrix M_{l_c}. From the last equation we have

$$\alpha = \frac{1}{2}(\theta_{lc} + \chi_{lc}) + \frac{i}{2k}(k^2\mu_{lc} - \nu_{lc}) = \delta^* \tag{8.65}$$

and

$$\beta = \frac{1}{2}(\theta_{lc} - \chi_{lc}) - \frac{i}{2k}(k^2\mu_{lc} + \nu_{lc}) = \gamma^*. \tag{8.66}$$

These quantities are enough to determine the n-cell transfer matrix M_n, whose matrix elements are

$$\begin{aligned} \alpha_n &= U_n - \delta\, U_{n-1}, & \beta_n &= \beta\, U_{n-1} \\ \delta_n &= U_n - \alpha\, U_{n-1}, & \gamma_n &= \gamma\, U_{n-1}, \end{aligned} \tag{8.67}$$

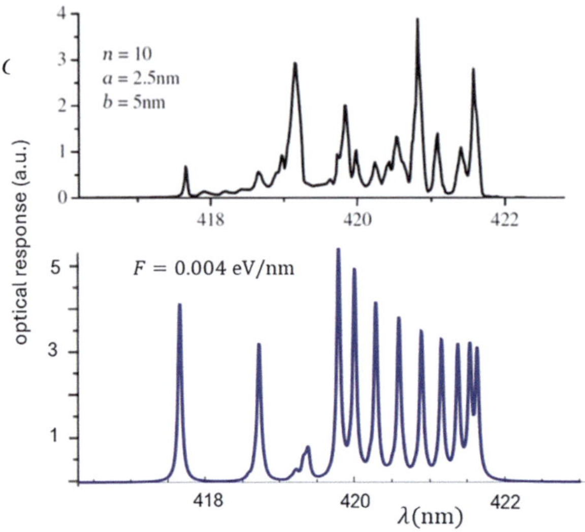

Fig. 8.16 Optical response of the superlattice $(In_xGa_{1-x}N/In_yGa_{1-y}N)^nIn_x($ when the local Stark field is taken into account

where U_n is the Chebyshev polynomial of the second kind, of order n, defined by the well known recurrence relation

$$U_n - (\alpha + \alpha^*)\, U_{n-1} + U_{n-2} = 0. \tag{8.68}$$

It is clear from this relation that the Chebyshev polynomials depend on α_R, the real part of α. As explained in Chap. 6, the transfer matrices M and M_n that satisfy the flux conservation principle are unimodular. If in addition, the system is invariant under time-reversal, i.e. that $\delta = \alpha^*$ and $\gamma = \beta^*$, we have

$$|\alpha|^2 - |\beta|^2 = 1 \quad \text{and} \quad |\alpha_n|^2 - |\beta_n|^2 = 1. \tag{8.69}$$

Before we present the optical response, we recall that having the matrix element α_n we can calculate the transmission coefficient

$$T_n = \frac{1}{|\alpha_n|^2} = \frac{1}{|U_n - \delta\, U_{n-1}|^2}. \tag{8.70}$$

To calculate the optical response, lateral cladding layers confine the superlattice, and using equations (8.48) and (8.49) we evaluate the eigenvalues and eigenfunctions. With these results, and the formalism developed in [4], that was also used in the last section, we obtain the optical response shown in Fig. 8.16, when the local field is $F = 0.0004$ eV/nm. The agreement with the experimental result is as expected, good. Now the width of the theretical spectrum coincides with the width of the experimental spectrum. The only difference that remains for a perfect account is the difference in the peaks hight.

A hint to the solution is the fact that the shorter resonances in the experimental spectrum are those of higher energy, or smaller wavelength. Generally, the mean lifetime is smaller in the higher energy levels. It is known for a harmonic oscillator that the energy-levels lifetime τ_i in the excited state ϕ_{n_i} is given by

$$\tau_i = 6\pi\epsilon_0 \frac{mc^3}{e^2\omega^2}\frac{1}{n_i}. \tag{8.71}$$

We have seen in Chap. 7 that the time dependent optical response must be written as

$$\chi(\omega, t) = a_0 \sum_{i,f} \frac{|_o\langle\varphi_f|p_z|\varphi_i\rangle_o|^2}{(E_f - E_i + \hbar\omega)^2 + \Gamma_i^2}\varrho(t, \tau_i). \tag{8.72}$$

with

$$\varrho(t, \tau_i) = 2e^{-t/\tau_i}\left(-1 + e^{t/2\tau_i}\right) \tag{8.73}$$

being t the time elapsed, from the time in which the energy levels were occupied, until the measurement is performed. In other words, if the spectrum was measured at t_{exp} and the energy levels were occupied at $t_0 = 0$, the elapsed time is $t = t_{exp} - t_0 = t_{exp}$. It is of course difficult to know this time. As far as we know, no experimental result gives this parameter. If we assume that the mean lifetime of the energy levels in the subband of the superlattice behaves as the lifetime of the energy levels of the harmonic oscillator and are given by (8.71) with $1/\nu$, instead of $1/n_i$, and we choose t_{exp} larger than τ_1, τ_2 and τ_3, but smaller than τ_4, τ_5,..., the theoretical spectrum will look closer to the experimental one.

8.5 Problems

1. Assuming that the quasi-Fermi levels are given by

$$\begin{aligned} E_n &= E_c + k_B T \ln \frac{n}{N_c} \\ E_p &= E_v - k_B T \ln \frac{p}{N_v}, \end{aligned} \tag{8.74}$$

and knowing that when a potential V is applied the minority concentration n changes into n_{pV} and the majority p remains as p_p, show that the population inversion condition can be expressed as

$$V \geq \frac{k_B T}{e} \ln \frac{N_c N_v}{n_i^2}. \tag{8.75}$$

And also that

$$V \geq \frac{E_g}{e}. \tag{8.76}$$

2. For the quantum well parameters given right after (8.40), show that the eigenvalues are those indicated in Fig. 8.8.
3. For the QW of the previous problem, obtain the eigenvalues when the width of the well is $a = 6.75$ nm, corresponding to 12 unit cells.
4. Calculate numerically the transition matrix elements for the quantum well defined after (8.40). Verify that they satisfy the selection rule in (8.42).
5. Evaluate the optical response (8.44) for the resonance broadenings Γ considered in Fig. 8.9.
6. Repeat the calculation of the concentration n from the exercise in 3.8 section, but assuming that the quasi-levels are given by the expressions in (8.74), instead of those of the Joyce-Dyxon approach.
7. Obtain the gain $\mathcal{G}/N_c$ for a CdSe pn junction.
8. If a potential well is made of $In_yGa_{1-y}N/In_xGa_{1-x}N/In_yGa_{1-y}N/$ with $x = 0.18$ and $y = 0.04$, obtain its optical spectrum.
9. Obtain the optical response if the superlattice of Sect. 8.4.2 consists of 3 unit cells.

References

1. S. Nakamura, M. Senoh, S. Nagahama, N. Iwasa, T. Yamada, T. Matsushita, H. Kiyoku, Y. Sugimoto, Appl. Phys. Lett. **68**, 3269 (1996)
2. H. Kroemer, Proc. IEEE **51**, 1782 (1963)
3. P. Pereyra, Ann. Phys. **397**, 159 (2018)
4. P. Pereyra, Ann. Phys. **320**, 1 (2005)

Solar Cells

9

Abstract

In this chapter, we build upon the general semiconductor properties and their response to local and external fields introduced in Chap. 4. We begin with an overview of the solar radiation power density, the efficiency of electron-hole pair generation, optical absorption, and the resulting photocurrent. We then derive the rate of pair generation per unit area over a wavelength window for crystalline silicon, given a specific incident power density. We then present a detailed analysis of three-layer semiconductor solar cell structures, focusing on light-induced electron-hole generation and the device mechanisms that convert light into photocurrent. Explicit expressions for electron and hole currents are derived to describe carrier transport. To understand key features of solar cell operation, particularly the relationship between dissipated power, dark current, and photocurrent, we analyze both open-circuit and short-circuit conditions, deriving the induced voltage in terms of temperature, photocurrent, and dark current. We also explore nanowire solar cells with axial heterostructures, deriving simplified expressions for optical gain and responsivity. Special attention is given to the potential of GaN-based superlattices, where spontaneous polarization arising from crystal asymmetry enhances carrier separation. Finally, we compute the responsivity as a function of the number of unit cells and evaluate the resulting photocurrent in such devices.

9.1 Introduction

An important device based on three semiconductor layers, typically made of the same material, such as crystalline silicon (c-Si) but with different doping levels, is the solar cell. Its operation can be readily understood using the physical principles introduced in the previous chapters.

Solar cells are semiconductor structures designed to convert solar radiation into electrical energy. They consist of at least three semiconductor layers (see Fig. 9.1),

P. Pereyra, *Contemporary Theory of Semiconductors and Devices*,
Graduate Texts in Physics, https://doi.org/10.1007/978-3-032-00296-9_9

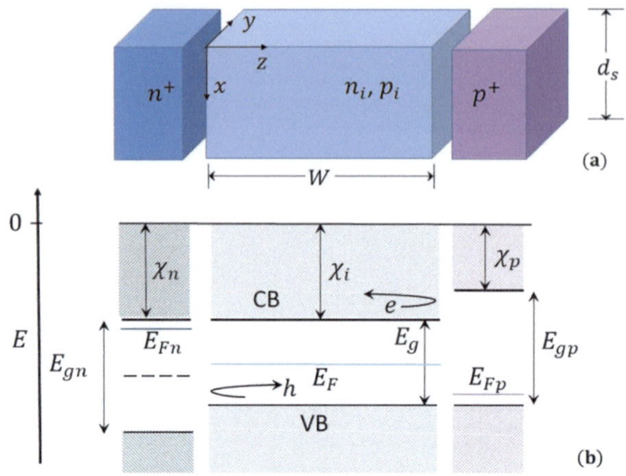

Fig. 9.1 Semiconductor layers of a solar cell, and profiles of the conduction and valence bands when the layers are isolated. For the side layers to act as selective barriers for electrons and holes, it is required that $E_{gn}, E_{gp} > E_g$ and, furthermore, that the electrical affinities satisfy the relations $\chi_p < \chi_i$ and $\chi_n \simeq \chi_i$

each serving a distinct function. Two of these layers are heavily doped, one n^+, the other p^+. The central layer generally has moderate doping, and is referred to, for simplicity, as the absorption layer, characterized by carrier concentrations n_a, p_a, or by its intrinsic concentrations n_i, p_i. In the absorption layer, incoming photons are absorbed, promoting electrons to the conduction band and leaving behind holes in the valence band. The role of the outer n^+ and p^+ layers is to create electric fields at the n^+/p_i and n_i/p^+ interfaces. These fields drive the charge-separation process of the generated electron-hole (*e*-*h*) pairs, propelling electrons and holes in opposite directions and thereby inducing current. Moreover, it is desirable for the outer layers to function as selective barriers, permitting the flow of one type of charge carrier while blocking the other. To achieve this selective behavior, the energy gaps of n^+ and p^+ layers must be larger than that of the central absorption layer, with appropriate alignment of the valence and conductance band edges, as illustrated in Fig. 9.1b.

9.2 Induced Potentials at the Interfaces

When the semiconductor layers shown in Fig. 9.1 are brought into contact, charge migrations occurs, similar to the observed in a standard *n*-*p* junction. Carriers move from the highly doped n^+ and p^+ layers toward the central absorption layer, which has lower impurity concentration. As a direct consequence, the conduction and valence band profiles are modified, as illustrated in Fig. 9.2. This band bending gives rise to internal electric fields that facilitate the separation of photo-generated electron-hole (*e*-*h*) pairs, ideally before recombination can occur.

If the parameters and charge distributions at the junctions are as indicated in Fig. 9.3, and we follow the procedure used to obtain the potential difference in (4.9), we obtain the contact potentials

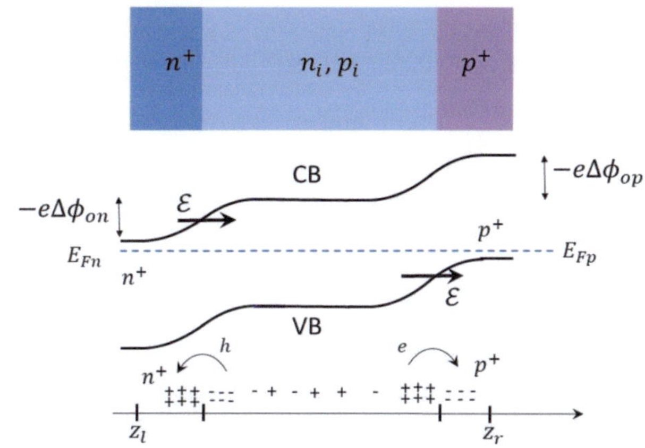

Fig. 9.2 Profile of conduction and valence bands in electrostatic equilibrium. When the layers come into contact, the charges in majority concentrations (of the doped layers) move and establish the electric fields indicated in the figure. When the radiation excites *e*-*h* pairs, the electron and hole separate by the action of the fields

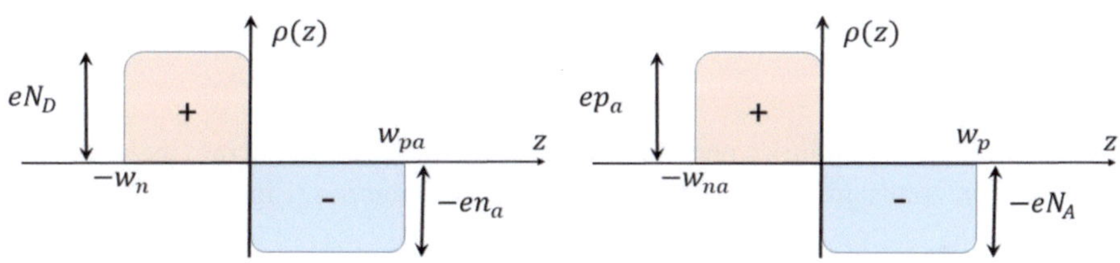

Fig. 9.3 Charge densities at the joints n^+/p_i and n_i/p^+

$$\Delta\phi_{on} = -\frac{e}{2\epsilon\epsilon_o}\left(N_D w_n^2 + n_a w_{pa}^2\right) \quad \text{and} \quad \Delta\phi_{op} = -\frac{e}{2\epsilon\epsilon_o}\left(N_A w_p^2 + p_a w_{na}^2\right), \tag{9.1}$$

which yield a total potential difference

$$\Delta\phi_o = \Delta\phi_{on} + \Delta\phi_{op} = -\frac{e}{2\epsilon\epsilon_o}\left(N_D w_n^2 + n_a w_{pa}^2 + N_A w_p^2 + p_a w_{na}^2\right). \tag{9.2}$$

If $N_D w_n = n_a w_{pa}$ and $N_A w_p = p_a w_{na}$, we obtain

$$w_n = \sqrt{\frac{2\epsilon\epsilon_o n_a|\Delta\phi_{on}|}{eN_D(n_a + N_D)}} \quad \text{and} \quad w_p = \sqrt{\frac{2\epsilon\epsilon_o p_a|\Delta\phi_{op}|}{eN_A(p_a + N_A)}}. \tag{9.3}$$

On the other hand, the contact potential can also be expressed in terms of the Fermi energies, and the following relations can be written

$$-e\Delta\Phi_{on}| = E_{Fn} - E_F \quad \text{and} \quad -e\Delta\Phi_{op} = E_F - E_{Fp}. \tag{9.4}$$

In Sect. 2.1 it was shown that $E_F \sim E_g/2$. In electrostatic equilibrium, the potential-energy difference between points z_r and z_l, in Fig. 9.2, is

$$\Delta U_o = -e\Delta\phi_{op} - e\Delta\phi_{on} = E_{Fn} - E_{Fp}, \tag{9.5}$$

with E_{Fn} and E_{Fp} the Fermi energies in layers with donor and acceptor impurities, respectively. As seen before, these energy levels are very close to the conduction and valence bands. Consequently, the potential-energy difference is of the order of the energy gap E_g, which in crystalline silicon is ~ 1.2 eV. This means that $\Delta U_o \sim 1$ eV. We can now obtain the order of magnitude of the widths of the transition zones from

$$\Delta U_o = -e\Delta\phi_{op} - e\Delta\phi_{on} = \frac{e^2}{2\epsilon\epsilon_o}\left(N_D w_n^2 + n_a w_{pa}^2 + N_A w_p^2 + p_a w_{na}^2\right). \tag{9.6}$$

If $e/2\epsilon\epsilon_o = 7.736 \times 10^{-10}$ Vm, $N_A \sim N_D \sim 10^{23} m^{-3}$, and $w_n \sim w_p$, we have

$$w_n \simeq 0.57 \times 10^{-7} m = 0.057 \mu m. \tag{9.7}$$

This shows that for impurity concentrations of the order of 10^{17}/cm^3, the widths of the transition zones are of the order of one tenth of a micron (Fig. 9.4).

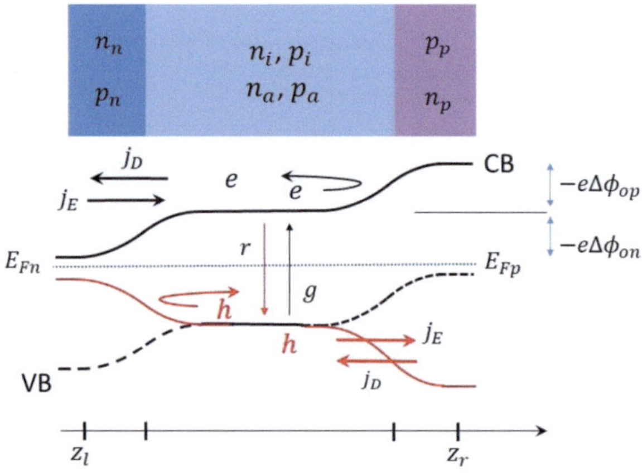

Fig. 9.4 Under conditions of electrostatic equilibrium. Electrons in the conduction band feel an attractive potential on the left and a repulsive potential on the right. Holes in the valence band feel the potential shown by the red line, repulsive on the left and attractive on the right. The drift and diffusion currents at each interface are balanced, and so are the generation and recombination rates of *e-h* pairs

9.3 Minority Concentrations and Solar Radiation

Before analyzing the effect of radiation on the solar cell, it is convenient to express the minority concentrations, on the n^+ and p^+ layers, as functions of the contact potentials in (9.4), build by the carrier migrations at the interface of n^+ and p_a, and at the interface of n_a and $p+$. As a consequence minority concentrations (see the relations (4.23) and (4.24) given by the expressions

$$p_n = p_p\, e^{-e\Delta\phi_{op}/k_B T} \qquad \text{and} \qquad n_p = n_n\, e^{-e\Delta\phi_{on}/k_B T}, \tag{9.8}$$

are established. We have also seen in Chap. 4 that, if in addition to these potentials, an external potential is applied, in direct or reverse polarization, the band profiles and minority concentrations modify. In direct polarization, see Fig. 9.5a, the height of the barriers decreases and the minority populations increase exponentially, whereas in reverse polarization, see Fig. 9.5b, the minority concentrations decrease exponentially. This means that

$$n_p \quad \rightarrow \quad n_{pV} = n_n\, e^{(-e\Delta\phi_{op}+eV)/k_B T} = n_p\, e^{eV/k_B T} \tag{9.9}$$

and

$$p_n \quad \rightarrow \quad p_{nV} = p_p\, e^{(-e\Delta\phi_{on}+eV)/k_B T} = p_n\, e^{eV/k_B T}. \tag{9.10}$$

Important consequences of the increase or decrease of minority concentrations in the presence of an external potential can be seen when the circuits of Fig. 9.5 are closed. In these circuits electron and hole currents, proportional to the minority concentrations, are established. As was shown in Chap. 4 (see (4.59), (4.60) and (4.61)), the currents

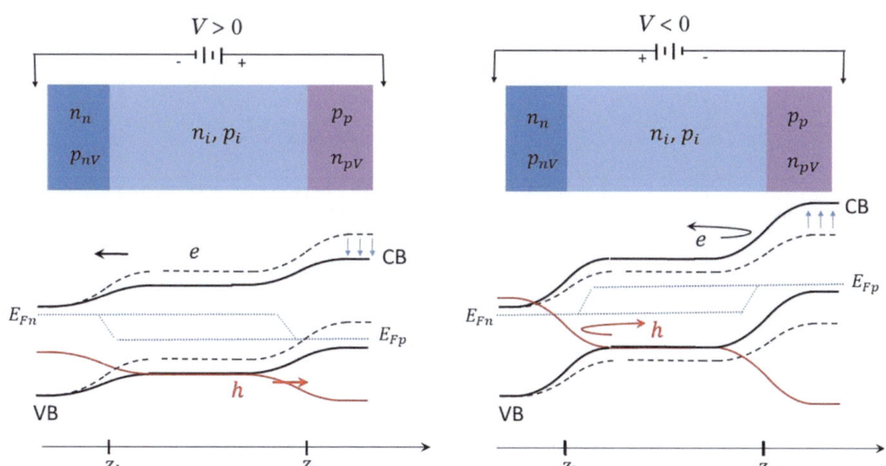

Fig. 9.5 A solar cell with direct polarization (**a**) and reverse polarization (**b**)

at the edges of the transition zones, associated with minority populations sum a total current

$$I(V) = I_0\left(e^{eV/k_BT} - 1\right) \tag{9.11}$$

where

$$I_0 = \left(\frac{eD_e n_p A}{L_e} + \frac{eD_h p_n A}{L_h}\right). \tag{9.12}$$

Here D_e and D_h are the diffusion constants, L_e and L_h the diffusion lengths and A the area of the interfaces. We will see in the next section that when the cell is exposed to sunlight and the *e*-*h* pairs are excited, and move to the n^+ and p^+ layers, they behave as if an external potential was applied, establishing, in addition to the excited-electron and hole currents, minority-charge-concentration currents. These currents are known in the solar cell field as *dark currents*. The magnitude of these currents is important to determine some of the characteristic parameters of solar cells. We will return to these currents in Sect. 9.5, where we will consider again open and closed circuits.

9.4 Solar Radiation. Spectrum and Characteristic Variables

The solar radiation, as shown in Fig. 9.6, covers practically the entire electromagnetic spectrum. The incident radiation is typically measured in watts per unit area per unit of wavelength. This physical quantity is known as power density and denoted as $P(\lambda)$. Figure 9.6 shows three power density spectra: one corresponds to the radiation at the surface of the Sun (AM0 radiation). This spectrum peaks at the wavelength expected for black body radiation at T=6000K. The dotted spectrum represents the black body radiation spectrum at this temperature, and the other corresponds to the

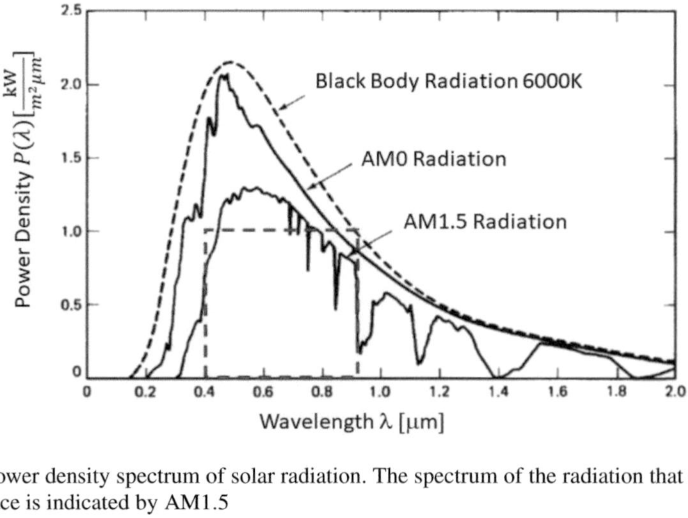

Fig. 9.6 Power density spectrum of solar radiation. The spectrum of the radiation that reaches the Earth surface is indicated by AM1.5

radiation that reaches the Earth's surface, labeled AM1.5 radiation. The AM stands for *air mass* and the number 1.5 indicates that the incidence angle θ_i with respect to the zenith is such that $1/\cos\theta_i=1.5$, i.e., that $\theta_i = 48.2°$ from the zenith.

In the AM1.5 spectrum, the ultraviolet radiation is largely absorbed by atmospheric ozone, while infrared radiation near $\lambda = 0.95\mu$m, primarily by CO_2. The power density in the green region of the visible spectrum is a little more than 1 kW/m$^2\mu$m.

The radiation that reaches the surface of a solar cell is partially reflected and partially absorbed. If the refractive indices of the air and of the semiconductor in the absorption zone are n_0 and n_s, respectively, the reflection coefficient of solar radiation is given approximately by the relation

$$R_s = \left|\frac{n_0 - n_s}{n_0 + n_s}\right|^2. \tag{9.13}$$

On the other hand, the energy of a photon is $E_f = h\nu = hc/\lambda$, the number of photons of wavelength λ that enter into the absorbtion zone per second, per m^2 and per unit wavelength (measured in μm), is given by

$$\mathcal{N}(\lambda) = \frac{P(\lambda)\lambda}{hc}(1 - R_s)\left[\frac{\text{photons}}{\text{s m}^2\mu\text{m}}\right]. \tag{9.14}$$

These photons excite *e-h* pairs in the conduction and valence bands. These electrons and holes move, dragged by the local electric fields, in opposite directions, as indicated by arrows in panel a) of Fig. 9.5. The intensity of the electron and hole currents depends strongly on the density of incident photons $\mathcal{N}(\lambda)$ on the absorption layer.

The solar radiation that reaches the semiconductor, is progressively absorbed as it penetrates the absorption layer. The absorption depends on an absorption coefficient $\alpha(\lambda)$, and the efficiency with which the absorbed photons generate electron-hole pairs, which is denoted by η_g (the pairs generation efficiency). With these quantities, the density of incident photon, the absorption coefficient and the efficiency, we can define the generation rate of pairs per unit volume and per unit wavelength, as a function of the penetration depth x. One of the expressions to determine the generation rate of pairs in a solar cell is the Lambert-Beer law:

$$g(\lambda, x) = \eta_g \mathcal{N}(\lambda)\alpha(\lambda)e^{-\alpha(\lambda)x}. \tag{9.15}$$

If we integrate this rate upon λ, for an interval $\Delta\lambda$ of the spectrum, assuming that in that interval $P \sim P_m$, and α and R_s are almost constant, we obtain the rate of pair generation per unit of volume (evaluated for the interval $\Delta\lambda = \lambda_2 - \lambda_1$), at a depth x, given by

$$G(x) = \int_{\lambda_1}^{\lambda_2} \eta_g \mathcal{N}(\lambda)\alpha(\lambda)e^{-\alpha(\lambda)x}d\lambda \tag{9.16}$$

$$G(x) = \eta_g \frac{P_m}{2hc}(\lambda_2^2 - \lambda_1^2)(1 - R_s)\alpha e^{-\alpha x}. \tag{9.17}$$

Likewise, if the factor $\alpha e^{-\alpha x}$ is integrated over the variable x from $x = 0$ to $x = d_s$ we have the fraction of incident radiation that is absorbed by the absorption layer. This fraction is

$$\delta_d = \int_0^{d_s} \alpha e^{-\alpha x} dx = 1 - e^{-\alpha d_s}. \tag{9.18}$$

For $\alpha = 1.11 \times 10^4 \text{cm}^{-1}$ and thickness d_s =10μm, the absorbed fraction δ_d is

$$\delta_d = 1 - e^{-1.11 \ 10^4 \times 10 \ 10^{-4}} = 1 - 0.0000167017 \simeq 1, \tag{9.19}$$

practically 100%. It is clear that if the thickness of the absorption layer is smaller, part of the radiation will be transmitted. Generally, the thicknesses are hundreds of microns, and the pair generation efficiency η_g is assumed to be ~1. So, a good approximation for the rate of pair generation per unit area is

$$\Gamma = \frac{P_m}{2hc}(\lambda_2^2 - \lambda_1^2)(1 - R_s). \tag{9.20}$$

Let us evaluate this rate for solar radiation incident on a c-Si layer, with refractive index $n_s = 4.293 + 0.045i$ and reflection coefficient $R_s \sim 0.38$. We will assume also that the solar radiation excites pairs *e*-*h* with wavelengths in the interval $\Delta\lambda$, ranging from $\lambda = 0.4$μm to $\lambda = 0.92$μm with a medium power density $P_m = 1\,\text{kW/m}^2\mu\text{m}$, as shown in red box of Fig. 9.6, Replacing these quantities, the number of photons per m^2 and unit of time entering the absorption layer of the solar cell is

$$\Gamma_f = \frac{P_m}{2hc}(\lambda_2^2 - \lambda_1^2)(1 - R_s) = 1.07123 \times 10^{21} \left[\frac{\text{photons}}{\text{s m}^2}\right] \tag{9.21}$$

This corresponds, see solved problem 2, to an approximate total-current density of 343 A/m^2.

9.4.1 Alternative Derivation of the Pair Current

The electron-hole pairs generated by solar radiation modify the concentrations of charge carriers within the cell. The generation rate of pairs per unit time, denoted by $G = G_e = G_h$, is proportional to the density $\mathcal{N}$ of photons that strike and are absorbed. If L_y is the transverse length of the cell, and L_n, L_p and W are the widths of the n^+, p^+ and absorption layers, the effective absorption area is $(L_n + L_p + W)L_y$. The products $G_e\tau_n(L_n + W/2)L_y$ and $G_h\tau_p(L_p + W/2)L_y$ represent, respectively, the total number of electrons and holes, generated and sustained per unit time during their lifetimes τ_n and τ_p. Multiplying each term by the elementary charge e and the respective drift velocities given by, $-L_e/\tau_n$ for electrons and L_h/τ_p for holes, yields the following expression for the electron and hole currents

$$I_{ph} = eG_e\tau_n\Big(L_n + \frac{W}{2}\Big)L_yL_e/\tau_n + eG_h\tau_p\Big(L_p + \frac{W}{2}\Big)L_yL_h/\tau_p. \tag{9.22}$$

Substituting $G \simeq P_m(1 - R_s)\lambda_g/hc$ and assuming that $A \simeq L_yL_e \simeq L_yL_h$, the current density associated with the pairs generated by the solar radiation can be written as

$$j_{ph} = eG(L_n + L_p + W) = eP_m(1 - R_s)(L_n + L_p + W)\lambda_g/hc. \tag{9.23}$$

For $P_m = 1\,\text{kW/m}^2\mu\text{m}$, $\lambda_g = 0.55\mu\text{m}$ (the wavelength of green light), the pair generation rate (or number of photons that penetrate) will be $G = 1.71663 \times 10^{24}$ photons/$sm^2\mu$m. If the total length of the cell is $L_T = L_p + L_n + W = 1.2\mu\text{m}$, we obtain a current $j_{ph} = 330\,\text{A/m}^2$, which is of the order of the current that was obtained above.

9.5 The Solar Cell in Open and Closed Circuits

Let us now examine how this device operates when exposed to radiation and connected to an external circuit, allowing the radiation-induced current to flow. As explained earlier, the absorbed energy promotes electrons to the conduction band while, simultaneously, holes are created in the valence band. Figure 9.7 illustrates the generation of *e-h* pairs under illumination, along with the potential barriers encountered by electrons in the conduction band (black line) and holes in the valence band (red line). To analyze the effect of incident radiation on the solar cell, we will consider two scenarios: one where the ends of the structure are electrically insulated (open

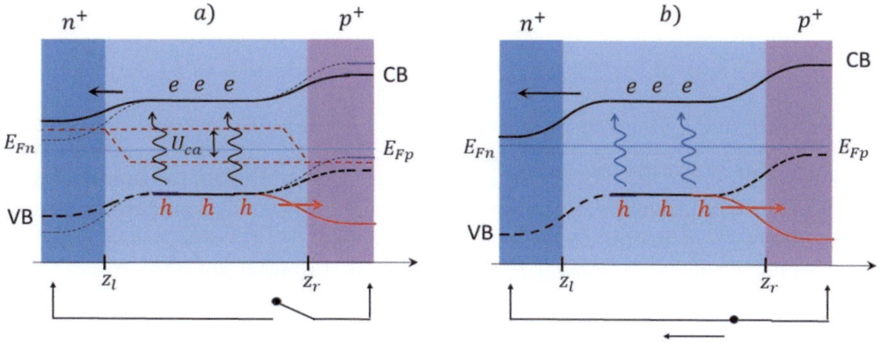

Fig. 9.7 Solar cell in open circuit (**a**) and in closed circuit (**b**). In an open circuit the excited electrons in the CB move towards the n^+ layer. The electric potential ϕ^+ becomes more negative and the potential energy $-e\phi^+$ more positive. The potential difference between the ends of the cell is $V_{ca} = U_{ca}/e$. In a closed circuit, the bands remain as before illuminating

circuit) and another where they are connected through an external load, as shown in Fig. 9.7a and b, respectively. When the circuit is open, the radiation-induced displacement of electrons toward the n^+ layer and holes toward the p^+ layer, modifies the contact potentials and the barrier heights. The accumulated charges, negative on the left and positive on the right of 9.7a), act similarly to an externally applied forward bias. As previously discussed, this leads to charges in minority carrier concentrations. If the change in the contact potential in the n^+ layer, due to radiation, is denoted by $-\Delta\phi_n^+$, the electrochemical potential becomes $U_n^+ = E_{Fn} + e\Delta\phi_n^+$. Similarly, for the p^+ layer, where the electrochemical potential becomes $U_p^+ = E_{Fp} - e\Delta\phi_p^+$. Assuming that before illumination the Fermi levels were aligned, i.e., that $E_{Fn} = E_{Fp}$, the open-circuit voltage V_{ca} (the potential difference between the two ends of the solar cell) is given by

$$V_{ca} = \frac{U_p^+ - U_n^+}{e} = \Delta\phi_n^+ + \Delta\phi_p^+. \tag{9.24}$$

If instead the circuit is closed externally, as shown in Fig. 9.7b, the open-circuit voltage V_{ca} vanishes. In this configuration, the excited electrons and holes move through the external circuit, producing a photocurrent with a current density j_{ph}, previously calculated to be approximately 343 A/m^2 for crystalline silicon (c-Si). To estimate the magnitude of the open-circuit voltage V_{ca}, the potential established when the circuit is open, but which disappears when the circuit is closed, we assume that the photocurrent j_{ph} behaves analogously to a forward-bias diode current. Thus, we can use the ideal diode equation:

$$j_{ph} = J_0\big(e^{eV_{ca}/k_BT} - 1\big) \tag{9.25}$$

where the saturation current is given by

$$J_0 = en_i^2\left(\frac{D_e}{N_D L_e} + \frac{D_h}{N_A L_h}\right). \tag{9.26}$$

Solving for the open-circuit voltage V_{ca} yields:

$$V_{ca} = \frac{k_B T}{e}\ln\left(\frac{j_{ph}}{J_0} + 1\right). \tag{9.27}$$

For a c-Si cell where the electron diffusion length is L_e=500μm, the hole diffusion length L_h=10μm, acceptor concentration $N_A = 2 \times 10^{24}\text{m}^{-3}$, donor concentration $N_D = 10^{25}\text{m}^{-3}$, diffusion coefficients $D_e = D_h = 2.5810^{-3}\text{m}^2s^{-1}$ and temperature T =300K, the saturation current, or total drift current, will be

$$J_0 = 4.67 \times 10^{-9}\text{A/m}^2 \tag{9.28}$$

substituting into the equation for V_{ca} gives

$$V_{ca} = 0.0258\ln\left(\frac{340}{J_0} + 1\right) = 0.65\text{V}. \tag{9.29}$$

9.6 The Solar Cell in Operation

Now that we have defined the magnitude of the electric current generated by solar radiation, we turn to analyzing how the total current behaves during operation, specifically, how it is affected by the contribution of minority carrier currents when the cell is exposed to light.

As discussed in the previous section, illumination induces a photocurrent density j_{ph}. We also saw in Sect. 9.5 that charge separation caused by radiation establishes a potential difference V_{ca}. If instead of a short circuit, the solar cell is connected to an external load with effective resistance R, a voltage drop RI_T occurs across the device. The resulting potential across the cell $V_{ca} - Rj_T A$, where A is the cross sectional area and j_T the total current density, can be interpreted as an applied potential $V = V_{ca} - Rj_T A$. It was shown in Sect. 4.6, that in the presence of an applied voltage, the minority carrier concentrations are modified according to (4.55) and (4.56). For the solar cell configuration under consideration, the minority carrier concentrations take the following forms:

$$\delta n_p(z) = n_p\big(e^{eV/k_BT} - 1\big)e^{-(z-w_p)/L_e} \qquad \text{for} \qquad z > w_p, \tag{9.30}$$

to the right of w_p, and

$$\delta p_n(z) = p_n\big(e^{eV/k_BT} - 1\big)e^{(z-w_n)/L_h} \qquad \text{for} \qquad z < w_n, \tag{9.31}$$

to the left of w_n. With these concentrations and the relationships $\mathbf{j}_n = eD_e\nabla\delta n_p$ and $\mathbf{j}_p = -eD_h\nabla\delta p_n$, we obtain the currents

$$j_n(w_p) = -\frac{eD_e n_p}{L_e}\big(e^{eV/k_BT} - 1\big)e^{-(z-w_p)/L_e}|_{z=w_p} = -\frac{eD_e n_p}{L_e}\big(e^{eV/k_BT} - 1\big) \tag{9.32}$$

$$j_p(w_n) = -\frac{eD_h p_n A}{L_h}\big(e^{eV/k_BT} - 1\big)e^{(z+w_n)/L_h}|_{z=w_n} = -\frac{eD_h p_n A}{L_h}\big(e^{eV/k_BT} - 1\big). \tag{9.33}$$

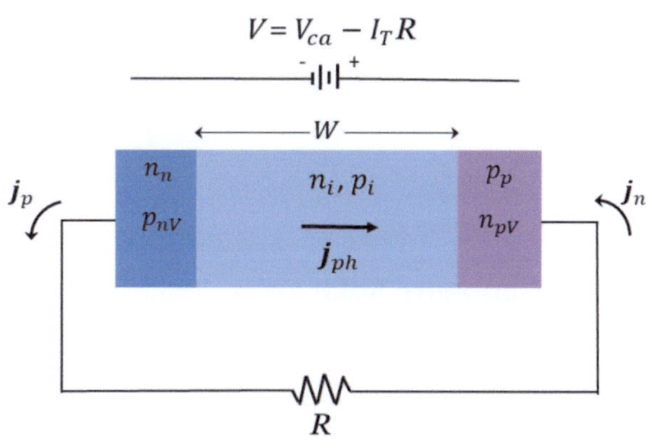

Fig. 9.8 Photon and dark currents through an illuminated cell connected to a device with effective resistance R. When current flows through a resistance R, the heights of the barriers in the cell partially decrease, as if a potential $V = V_{ca} - j_T AR$ were applied to the ends of the cell

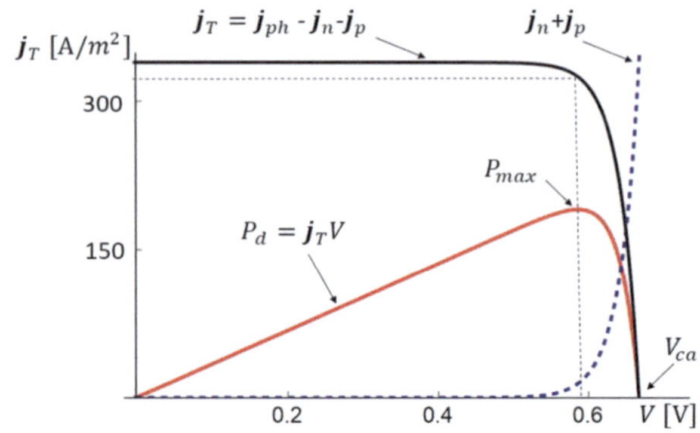

Fig. 9.9 The illuminated and dark currents and the dissipated power as functions of the applied potential. For potentials lower than the open circuit potential V_{ca}, the current is basically that generated by radiation. The dissipated power has its maximum for $V \sim 0.58$V

Note that for $V > 0$, these currents go in the opposite direction to the photocurrent j_{ph}, as shown in Fig. 9.8. The current $j_V = j_n + j_p$ is called the dark current to differentiate it from the illumination-induced photocurrent j_{ph}. The total current for a given value of R takes the form

$$j_T(V) = j_{ph} - J_0\big(e^{eV/k_BT} - 1\big). \tag{9.34}$$

In Fig. 9.9 we plot this current as a function of $V = V_{ca} - j_T AR$ and we plot also the dark current $j_V = j_n + j_p$, and the dissipated power per unit area, $P_d = j_T V$. Notice that the dissipated power has a maximum P_{max} when the applied potential is $V_m \sim 0.58$V, is close to the potential V_{ca} induced by solar radiation in the open circuit, and the current j_m is slightly lower than j_{ph}. These are some of the parameters that characterize commercial solar cells. It is also customary to define a factor composed of several parameters, the *fill factor*

$$FF = \frac{j_m V_m}{j_{ph} V_{ca}}. \tag{9.35}$$

The numerical value of this factor is generally between 0.8 and 0.9, depending on the semiconductor used to build the solar cell. Another important parameter in solar cells is the *conversion efficiency* η determined by the maximum dissipated power and the incident power $P_i(\lambda)$ (which is of the order of 1000 A/m^2 in the AM1.5 spectrum), specifically, defined by the ratio as

$$\eta = \frac{P_{max}}{P_i}. \tag{9.36}$$

9.7 Nanowire Solar Cells

The ability of some semiconductors with wurzite structure to establish an spontaneous polarization of charges is an advantageous property that has been used to enhance the absorption efficiency of quasi-1D (nanowire) photodetectors. Among the various semiconductors with this ability are the III-nitride semiconductors (GaN, AlN, InN) and ZnO. We will focus on the results obtained for nanowires containing axial GaN superlattices. This structure is similar to the blue laser superlattice studied in the previous chapter, but for a solar cell the superlattice must be an open superlattice. The purpose is to move the excited electrons outside the superlattice, before the electrons recombine.

For the analysis of the device performance, it is common to define the responsivity R_λ as the ratio of the photocurrent $I_{ph} = j_{ph} A_{opt}$ and the incident optical power $P_{opt} = \mathtt{I} A_{opt}$:

$$R_\lambda = \frac{I_{ph}}{P_{opt}} = \frac{j_{ph} A_{opt}}{\mathtt{I} A_{opt}} \tag{9.37}$$

where $\mathtt{I}$ is the illumination intensity, or photon flux density. We have defined the illumination intensity as $\mathtt{I} = \hbar\omega c\phi$, with ϕ the number of photons per unit volume. The photon induced current density can be written as

$$j_{ph} = \eta_g \mathcal{G} e c \phi \tag{9.38}$$

where η_g is the pair generation efficiency, and $\mathcal{G}$ the effective number of electrons detected per electron-hole pair generated. This factor is defined as the gain. The probability that an excited electron could be detected in nanowires with axial semiconductor heterostructures is given by the probability that electrons that are excited to the energy levels where they are allowed to stay, be transmitted from inside the SL to its surface. This probability depends also on the correlation between the tunneling time and the decay or mean lifetime. At the end of the last chapter we derived the transfer matrix and transmission coefficient for quantum particles through a system whose potential has a local Stark effect, similar to the system illustrated in Fig. 9.10. For this potential profile, the transmission coefficient was given as

$$T_n = \frac{1}{|\alpha_n|^2} = \frac{1}{|U_n - \delta\, U_{n-1}|^2} \tag{9.39}$$

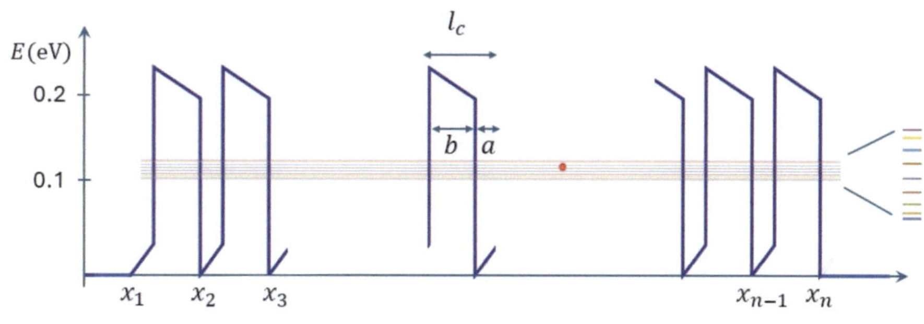

Fig. 9.10 The potential profile of a superlattice with local Stark fields. We show also the resonant energies, calculated by solving the (9.41)

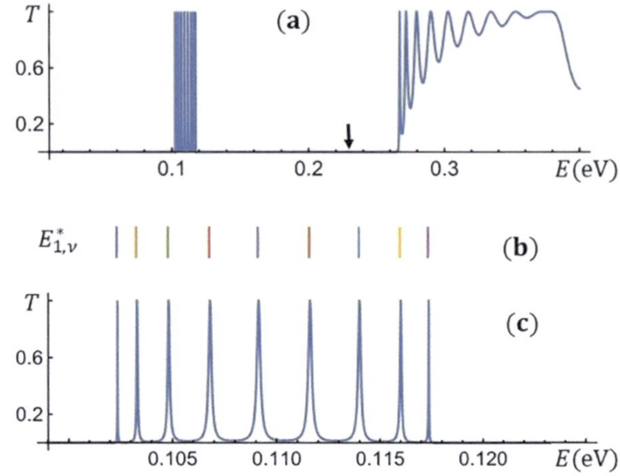

Fig. 9.11 Resonant transmission and resonant energies. In the upper panel we plot the resonant transmission coefficient for the superlattice with potential profile shown Fig. 9.10

where the number of unit cells is n. It was shown in Chap. 6 that the transmission coefficient can also be written as

$$T_n = \frac{1}{1 + |\beta|^2 \, U_{n-1}^2} \tag{9.40}$$

This equations makes evident that the resonances of the transmission coefficient shown in Fig. 9.11, copied also in Fig. 9.10, occur at energies in which the Chebyshev polynomial U_{n-1} is zero. In panel (a) of Fig. 9.11 we plot the transmission coefficient up to energy $E = 0.4$ eV, which is above the potential barrier $V_o + Fa \simeq 0.23$ eV, see the arrow. The resonant energies in the subband, below the potential barrier, are very close to the energy eigenvalues of the bounded superlattice $(In_xGa_{1-x}N/In_yGa_{1-y}N)^n In_xGa_{1-x}N$, in the active zone of the blue laser.[1] As explained with more detail in the literature of finite periodic systems,[2] the resonant energies $E^*_{\mu,\nu}$ are solutions of

$$\alpha_R = \cos \frac{\nu + n(\mu - 1)}{n}\pi, \tag{9.41}$$

more explicitly, solutions of

$$\frac{1}{2}(\theta_{l_c} + \chi_{l_c}) = \cos \frac{\nu + n(\mu - 1)}{n}\pi \tag{9.42}$$

with the functions θ_{l_c} and χ_{l_c} given in (8.60). The resonant energies $E_{1,\nu}$, with $\nu = 1, 2, \ldots, n\text{-}1$, predicted by the last equation, and the transmission coefficient

[1] See for example [1].

[2] See for example [2].

of the under-barrier subband are plotted in panels (*b*) and (*c*) of Fig. 9.11. Since the analytical results derived in this chapter are exact, we expect that the accuracies of the predicted resonances and the actual transmission-coefficient resonances will be good enough if the potential profile in the theoretical model is close to the potential profile of the actual superlattice. When the nanowire is exposed to solar light, electrons are excited to one of the n-1 resonant energy levels $E^*_{1,\nu}$, of the first subband. The gain defined above is then proportional to the probability of being transmitted throughout the superlattice and reach the left or right hand side layer. Since the probability to decay or recombine is also present, we can consider the time distribution derived in Chap. 7 (see (Sec. 7.5)). It was shown in [3] that the tunneling times τ_t of electrons in semiconductor structures is of the order of 1ps. If the tunneling time of an electron excited to the resonant energy level $E^*_{1,\nu}$ is $\tau_{t\nu}$, its mean lifetime $\tau_\nu = 1/2\gamma_\nu$, and the transmission probability $T_{n\nu}$ is defined as the transmission coefficient T_n in (9.40) evaluated for $E = E^*_{1,\nu}$, the gain $\mathcal{G}$ can be written as

$$\mathcal{G} = \sum_{\nu=1}^{n-1} T_{n\nu}\varrho(t,\tau_\nu)\Big|_{t=\tau_{t\nu}} = \sum_{\nu=1}^{n-1} T_{n\nu}2e^{-2\tau_{t\nu}\gamma_\nu}\big(-1+e^{\tau_{t\nu}\gamma_\nu}\big). \tag{9.43}$$

As can be seen in Fig. 9.11, the transmission coefficients at resonant energies are $T_{n\nu}(E^*_{1\nu}) = 1$. The tunneling times, even for large superlattices, due to Hartman effect, are two or three orders of magnitude smaller than the recombination times. Therefore,

$$\mathcal{G} \simeq 2(n-1). \tag{9.44}$$

Replacing this into (9.38) and (9.37), the responsivity of a superlattice with n unit cells is approximately given by

$$R_\lambda = \frac{j_{ph}A_{opt}}{\mathrm{I}A_{opt}} \simeq 2(n-1)\eta_g\frac{e\lambda}{hc}. \tag{9.45}$$

and the photon induced current density by

$$j_{ph} \simeq \eta_g 2(n-1)ec\phi. \tag{9.46}$$

These results show that, in ideal axial superlattices, the responsivity and current density, may increase by a factor proportional to the number of unit cells n. Nanowires with axial GaN heterostructures were fabricated by plasma assisted molecular beam epitaxy methods. In [4] and references therein, the reader can see an interesting review of wurzite nanowire photodetectors.

9.8 Solved Problems

1. Given the potential difference ΔU_o between the layers p^+ and n^+ of the solar cell, obtain the order of magnitude of w_n, if $e/2\epsilon\epsilon_o = 7.736\times10^{-10}$Vm and $N_D \sim 10^{23}m^{-3}$.

Solution

In (9.6) we have

$$\Delta U_o = \frac{e^2}{2\epsilon\epsilon_o}\left(N_D w_n^2 + n_a w_{pa}^2 + N_A w_p^2 + p_a w_{na}^2\right). \tag{9.47}$$

If $N_D w_n = n_a w_{pa}$ and $N_A w_p = p_a w_{na}$, and also $N_D w_n \sim N_A w_p$, the previous equation can be written as

$$\left(w_n + w_{pa} + w_p + w_{na}\right) w_n \simeq \Delta U_o \frac{2\epsilon\epsilon_o}{e^2 N_D}. \tag{9.48}$$

If we assume that the widths of the different sections are of the same order, we have

$$4w_n^2 \simeq \Delta U_o \frac{2\epsilon\epsilon_o}{e^2 N_D}. \tag{9.49}$$

So

$$w_n \simeq 0.57 \times 10^{-7} m = 0.057 \mu m. \tag{9.50}$$

2. Evaluate the rate of generation of pairs Γ of c-Si, with refractive index $n_s = 4.293 + 0.045i$ when the refractive index of air is $n_o = 0.9$ and when $n_o = 1.1$. In each case, evaluate the electron current j_e, the hole current j_h and the total current j_{ph}. Assume λ is in the range between 0.4μm and 0.92μm and the average power density is P_m=1000 kW/m^{2}μm.

Solution

If $n_o = 0.9$

$$Rs = \left|\frac{n_o - n_s}{n_o + n_s}\right|^2 = 0.42, \tag{9.51}$$

and

$$\Gamma = \frac{P_m}{2hc}(\lambda_2^2 - \lambda_1^2)(1 - R_s) = 9.9 \times 10^8 \frac{\text{photons}}{\mu\text{m}^2\text{s}}. \tag{9.52}$$

Since each photon excites an *e*-*h* pair, we have

$$j_e = e\Gamma = 158.6\,\text{A/m}^2 \qquad \text{and} \qquad j_h = ej_f = 158.6\,\text{A/m}^2 \tag{9.53}$$

and the total-current density is approximately j_{ph}=317 A/m^2.

If n_o =1.1

$$Rs = \left|\frac{n_o - n_s}{n_o + n_s}\right|^2 = 0.35, \tag{9.54}$$

and

$$\Gamma = \frac{P_m}{2hc}(\lambda_2^2 - \lambda_1^2)(1 - R_s) = 1.12 \times 10^9 \frac{\text{photons}}{\mu\text{m}^2\text{s}}. \tag{9.55}$$

Taking into account again that each photon excites an *e-h* pair, the electron and hole current densities are

$$j_e = e\Gamma = 179.7\,\text{A/m}^2 \qquad \text{and} \qquad j_h = ej_f = 179.7\,\text{A/m}^2 \tag{9.56}$$

and the total-current density is approximately j_{ph}=359 A/m^2.

3. If the solar cell is made of c-Si and the parameters are: L_e=600µm, L_h=50µm, $N_A = 10^{22}\text{m}^{-3}$, $N_D = 10^{23}\text{m}^{-3}$, $D_e = D_h = 2.6^{-3}\text{m}^2 s^{-1}$, and the temperature is 300K, obtain the saturation current J_0 and the circuit potential V_{ca} when the photonic current is j_{ph}=317 A/m^2.

Solution

In (9.25) and (9.27), the saturation current and the circuit potential are given by the expressions

$$J_0 = en_i^2\left(\frac{D_e}{N_D L_e} + \frac{D_h}{N_A L_h}\right) \qquad \text{with} \qquad V_{ca} = \frac{k_B T}{e}\left(\ln\frac{j_{ph}}{J_0} + 1\right). \tag{9.57}$$

If we remember that in Si n_i=1.5$10^{16}$m^{-3}, and we substitute the cell parameters, the saturation current becomes

$$J_0 = 1.89 \times 10^{-7}\text{A/m}^{-2}, \tag{9.58}$$

thus

$$V_{ca} = 0.57\text{V}. \tag{9.59}$$

9.9 Problems

1. Of what order of magnitude is the width w_n if $N_D = 10^{22}\text{m}^{-3}$, and of what order will the minimum length L of the cell be?
2. What is the relationship between the generation rate $G(x)$ of (9.17) and that of (9.20)? What important assumptions are made to get (9.20) from (9.16)?
3. Plot δ_d as a function of d_s for $\alpha = 1.0 \times 10^4\text{cm}^{-1}$.

4. Plot δ_d as a function of α for $d_s = 20\mu$m.
5. Obtain the pair generation rate Γ of c-Si, with refractive index $n_s = 4.293 + 0.045i$ when $n_o = 1.0$, and the electron and holes currents, j_e, j_h, and j_{ph}. Assume that λ is in the range between 0.45μm and 0.7μm and the average power density is P_m=1200 kW/m^{2}μm.
6. The refractive index of c-Si is generally written as $n_s = n_{sR} + in_{sI}$. Obtain the pair generation rate Γ as a function of n_{sR} if $n_{sI} = 0.045$. Assume that λ is in the range between 0.45μm and 0.7μm and that the average power density is P_m=1200 kW/m^{2}μm. What would be the optimal n_{sR} index?
7. If the refractive index of c-Si is written as $n_s = n_{sR} + in_{sI}$, obtain the pair generation rate Γ as a function of n_{sI} if n_{sR} has the optimal value of the previous problem. Assume that λ is in the range between 0.45μm and 0.7μm and that the average power density is P_m=1200 kW/m^{2}μm. What would be the optimal refractive index?
8. Obtain the graphs of Fig. 9.9.
9. Obtain the equivalent of the graphs in Fig. 9.9 for the cell of solved problem 2 with n_o =0.9, and determine the parameters P_d, V_m, FF and η.
10. Obtain the equivalent of the graphs in Fig. 9.9 for the cell of solved problem 2 with n_o =1.1, and determine the parameters P_d, V_m, FF and η.
11. Obtain the equivalent of the graphs in Fig. 9.9 for the cell of problem 7, and determine the parameters P_d, V_m, FF and η.

References

1. P. Pereyra, Ann. Phys. **397**, 159 (2018)
2. P. Pereyra, Ann. Phys. **378**, 264 (2017)
3. P. Pereyra, Phys. Rev. Lett. **84**, 1772 (2000)
4. M. Spies, E. Monroy, Semicond. Sci. Technol. **34**, 053002 (2019)

The Schottky, MS and MOS Structures 10

Abstract

In this chapter, we examine two essential semiconductor structures: the Schottky barrier and the metal-oxide-semiconductor (MOS) system. Our analysis builds on the fundamental semiconductor principles introduced in the earlier chapters of this book. We begin with the Schottky barrier, deriving the current across the metal-semiconductor interface and emphasizing its effectiveness as a current rectifier. The second part of the chapter presents a novel and rigorous analysis of the MOS, with particular focus on the inversion layer, a critical feature in field-effect devices. Traditionally MOS structure are treated as capacitors, and the spatial distribution of the electric potential in the semiconductor near the interface is either oversimplified or entirely neglected. This is largely due to the difficulty of solving the nonlinear Poisson equation in the inversion layer, an unresolved challenge since the work of Kingston and Neustadter in 1955 [1]. In contrast to this traditional approach. The present chapter introduces an alternative formulation based on an energy analysis of displacement charges and electric displacement fields, similar in spirit to Maxwell's reasoning in extending classical electromagnetism. This leads to a second-order nonlinear differential equation whose universal solution accurately captures the electric potential across the inversion layer [2,3]. It can be shown that this solution agrees with numerical solutions of the original Poisson equation, and in special cases where the charge distribution is directly integrable, it reduces exactly to the corresponding closed form result. This is analogous to solving mechanical problems using energy conservation rather than Newton's laws. To our knowledge, this is the first textbook to provide a complete analytical treatment of the inversion layer based on this approach. The universal solution is expressed in terms of impurity concentration, gate voltage, and a model-dependent parameter associated with the chosen charge profile. From this framework, we derive explicit expressions for the electric field, charge distribution, inversion layer width, ionized impurity concentration, and the Pao-Sah drift-diffusion current in both single-gate and double-gate junctionless MOSFETs.

P. Pereyra, *Contemporary Theory of Semiconductors and Devices*,
Graduate Texts in Physics, https://doi.org/10.1007/978-3-032-00296-9_10

10.1 Introduction

Other types of devices with properties similar to those of the *pn* junction are the metal-semiconductor(MS) structure, also known as the Schottky junction, and the MOS structure made up of layers of metal, oxide (insulator) and semiconductor. The insulating layer that grows between the semiconductor substrate and the metal film is generally silicon oxide (SiO_2). The metal layer is known as the gate. The semiconductor layer is usually doped with either n-type or p-type impurities.

To better understand the behavior of these systems, we will start by examining the characteristics of the band diagrams of these materials, drawn separately in Fig. 10.1. The distance (in energy) from the vacuum level to the Fermi levels in the metal and in the semiconductor is known as the work function, denoted Φ_m for the metal and Φ_s for the semiconductor. The work function is the energy necessary to remove an electron from the Fermi level to the vacuum. A key distinction between insulators and semiconductors lies in their energy gaps, E_g. In insulators the gap typically is of the order of 10 eV, while in semiconductors it is of the order of 1 eV. The distances from the conduction band edge to the vacuum level is characterized by the electron affinity, χ_i for the insulators and χ_s for the semiconductors.

In metals, the Fermi level, defined at $T = 0$ as the highest occupied energy level, is typically near the center of the valence band. This level can shift to the vacuum level E_o when an external voltage is applied: it moves closer under a negative potential, and further away under a positive potential. This response of metals is nearly instantaneous, unlike in semiconductors, where external potentials influence carrier concentrations more subtly. As previously discussed, applying an external field to a doped semiconductor changes the minority carrier population and can shift the Fermi level. However, once a semiconductor is doped, the doping level cannot be adjusted dynamically during operation, unlike external fields which can be tuned in real time. This ability to control the system via external potentials is central to the operation of MS and MOS devices, as well as field-effect transistors (FETs), which we will study in the next chapter. In isolated semiconductors and insulators, the work function Φ_s remains nearly unaffected by external potential. However, when

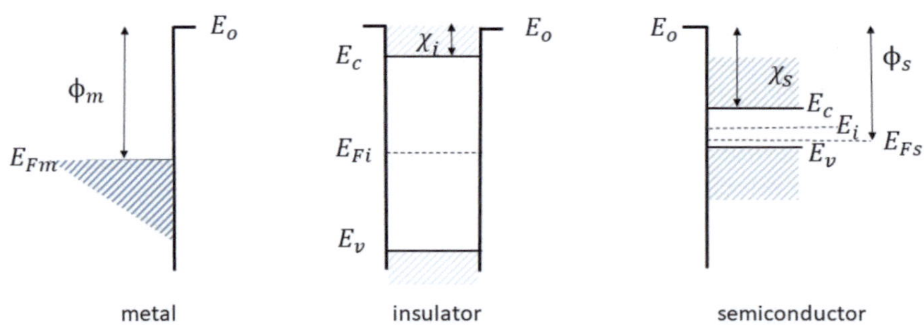

Fig. 10.1 Energy levels in a metal, an insulator and a semiconductor. E_o is the vacuum energy level in which electrons are free. Φ_m is the work function in the metal, Φ_s in the semiconductor. χ_i is the electron affinity in the insulator and χ_s is the electron affinity in the semiconductor

two materials are brought into contact, as in MS or MOS, new phenomena emerge, such as charge accumulation and band bending. We will begin by analyzing the metal-semiconductor junction.

10.2 Metal-Semiconductor Junction. The Schottky Junction

The metal-semiconductor junction is a fundamental and widely encountered structure, particularly in the electrical contacts of semiconductor devices. The behavior of such a junction, depends on the type of semiconductor (either n-type or p-type) and on the relative positions of the Fermi levels of the metal (E_{Fm}) and the semiconductor (E_{Fs}).

Figure 10.2, shows the case of an n-type semiconductor where $E_{Fs} > E_{Fm}$. In this configuration, electrons flow from the semiconductor into the metal due to the Fermi level difference. This movement creates a thin layer of negative charge on the metal side (a surface charge), and leaves behind a region of positively charged ionized donors in the semiconductor with concentration N_D. This charge redistribution leads to a band bending and a formation of a *space-charge* region (depletion region) in the semiconductor. Additionally, the concentration of minority carriers (holes) in the semiconductor increases slightly. This concentration is denoted as p_n^+. The system reaches thermal equilibrium when the electro-chemical potentials (Fermi levels) of both materials align. At equilibrium, the relationship between these potentials define the characteristics of the Schottky barrier and satisfy the relations

$$E_{Fm} = E_{Fs} + e\phi(0^+) \quad \text{then} \quad E_{Fm} = E_{Fs} - e\Delta\phi. \tag{10.1}$$

If $\Phi_m > \Phi_s$, the electric potential behaves as shown in Fig. 10.2c and the electron potential energies, dependent on the spatial coordinate z, causes the energy bands to bend as illustrated in Fig. 10.2b. Assuming a uniform positive charge density $\rho(z) = eN_D$ within the depletion region extending up to $z_B = w_S$, and that the electrical field (-$d\phi/dz$) vanishes at $z = w_S$, we can determine the electric potential by integrating Poisson's equation twice. If we further assume that both the potential and the electric field vanish at w_S, we obtain:

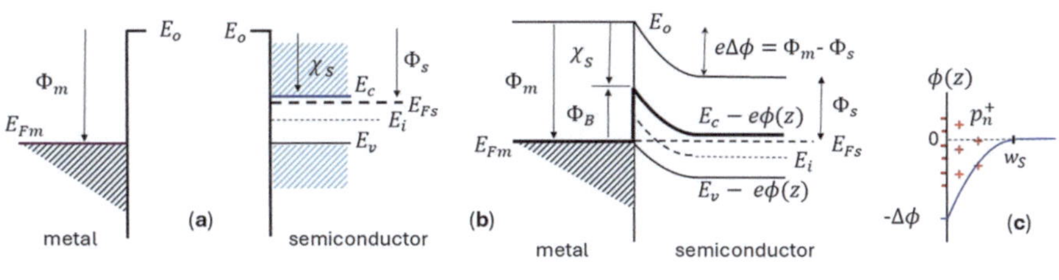

Fig. 10.2 Metal-semiconductor junction for the case where $E_{Fs} > E_{Fm}$ ($\Phi_m > \Phi_s$). Similar to the pn junction, charge redistribution occurs at the interface. A potential gradient forms near the junction, as illustrated in (**c**) and the Fermi levels align at equilibrium. A barrier, the Schottky barrier, is formed for the electrons in the metal. The height of the barrier is $\Phi_B = \Phi_m - \chi_s$

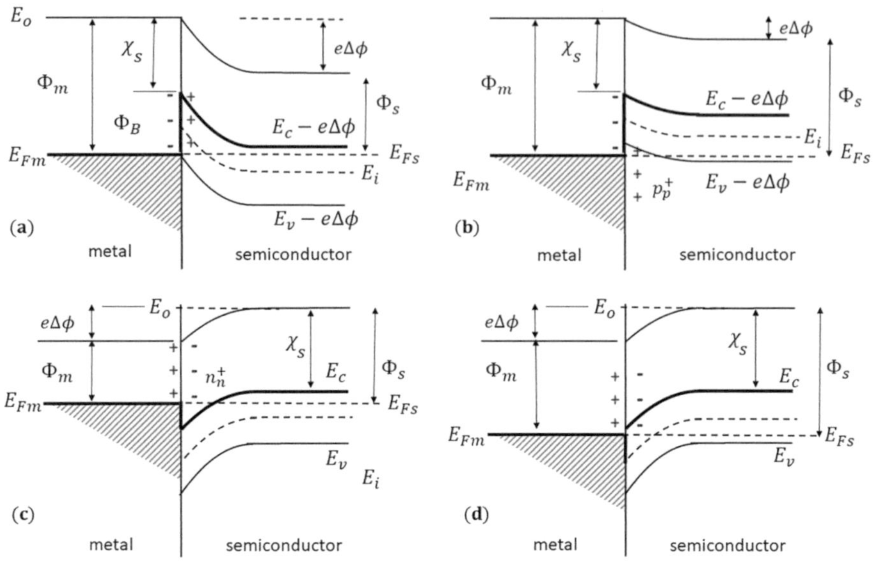

Fig. 10.3 Different possible scenarios depending on the type of semiconductor doping, n or p, and also depending on the order relationship between Φ_m and Φ_s

$$\phi(z) = -\frac{eN_D}{2\epsilon\epsilon_0}(w_S - z)^2 \quad \text{with} \quad w_S = \sqrt{\frac{2\epsilon\epsilon_0}{e^2 N_d}(\Phi_m - \Phi_s)}. \tag{10.2}$$

Figure 10.3 illustrates the various possible configurations of a metal-semiconductor junction, depending on the type of doping, n-type or p-type, and the relative magnitudes of Φ_m and Φ_s. When $\Phi_m > \Phi_s$ electrons are transferred from the semiconductor to the metal in both the n-type and p-type semiconductors. In an n-type semiconductor (case (a)), this leads to a depletion of the majority carriers, the formation of a potential barrier at the interface, and a reduction of the conductance. If, on the other hand, $\Phi_m < \Phi_s$, the displacement of the electrons flow from the metal into the semiconductor, with the consequence of a decrease in concentration if the semiconductor is type p as in (d). Also in this case, the majority concentration decreases (depletion of majority carriers) and the conductance decreases. In the other two cases, the majority concentrations increase, and the conductance increases.[1]

In Fig. 10.2, it can be seen that between the Fermi level of the metal and the conduction band of the semiconductor, a potential barrier is formed, the Schottky barrier, almost triangular whose height is given by

$$\Phi_B = \Phi_m - \chi_s. \tag{10.3}$$

[1] For further discussion on these effects in the Schottky junction, it is recommended to consult Shalimova's book, Physics of Semiconductors. Ed. MIR, 1980.

Experimental results show that the above expression is not exact. A way out has been sought by accepting, on the one hand, the possibility of a difference in potential energy $e\delta\phi_0$ such that

$$\Phi_B = \Phi_m - \chi_s + e\delta\phi_0, \tag{10.4}$$

which is justified by the defects and surface states at the metal-semiconductor interface. With a large concentration of surface states in the gap, the Fermi level of the semiconductor moves locally to the level of the surface states E_s. In this case the height of the barrier Φ_B can be written as

$$\Phi_B = \Phi_m - \Phi_s + E_g - E_s = e\Delta\phi + E_g - E_s \tag{10.5}$$

which is closer to the experimental results. Surface states were suggested a long time ago (by Tamm and Shockley), the exact calculation of surface levels is complicated, an important fraction of these results from the capture of electrons by surface atoms that do not find sufficient neighbors for electrochemical stability. Similar results are considered when the semiconductor of the Schottky junction is type p.

The transition region width w_S in the semiconductor is an important parameter in device operation and characterization. In terms of the Fermi energy E_s

$$w_S = \sqrt{\frac{2\epsilon\epsilon_0}{eN_D}\Delta\phi} = \sqrt{\frac{2\epsilon\epsilon_0}{eN_D}\big(\Phi_B - E_g - E_s\big)}. \tag{10.6}$$

When a potential difference V is applied, the width becomes

$$w_{SV} = \sqrt{\frac{2\epsilon\epsilon_0}{eN_D}\big(\Delta\phi - V\big)}. \tag{10.7}$$

It decreases in forward polarization and increases in reverse polarization.

10.3 The Schottky Junction as a Current Rectifier

The Schottky junction, like the pn junction, can be connected to a battery in forward and reverse bias. The left panel of Fig. 10.4 illustrates the junction under forward bias. If the semiconductor is type n and $\Phi_m > \Phi_s$, the height of the barrier that the majority electrons in the conduction band have to overcome is reduced from $e\Delta\phi$ to $e\Delta\phi - eV$, where V is the applied voltage. However, the Schottky barrier height for electrons in the metal remains unaffected by this external bias. In equilibrium (i.e., without an applied voltage), the thermal energy of electrons in semiconductor and metal, and to a lesser extent quantum tunneling, allows them to pass the barrier in either direction. When a forward bias is applied, the kinetic energy of the electrons in the semiconductor conduction band must satisfy the condition

$$\frac{\hbar^2 k^2}{2m^*} > e(\Delta\phi - V) = \Phi_B - (E_c - E_s) - eV. \tag{10.8}$$

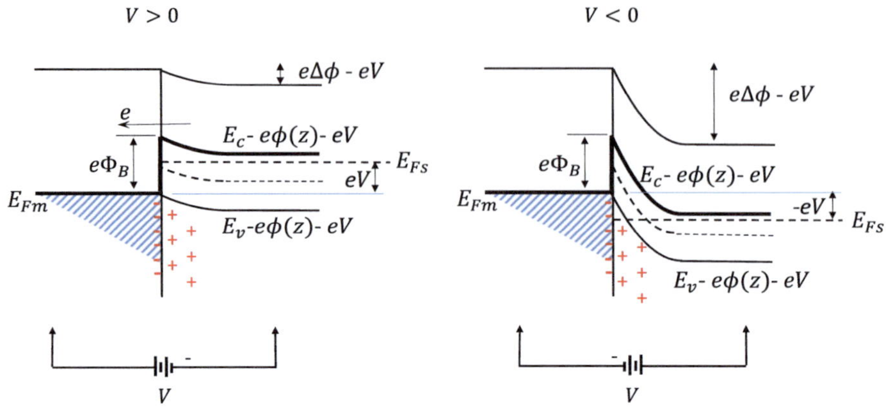

Fig. 10.4 The Schottky junction in forward bias (left) and reverse bias (right). In forward polarization the barrier for the CB electrons in the semiconductor decreases. The majority electrons define the current. In reverse polarization, the electrons in the metal have the Schottky barrier

making it more likely for them to overcome the barrier and flow into the metal. This probability increases with increasing forward bias ($V \geq 0$). The total energy of an electron that can surmount the barrier, measured from the valence band, is:

$$E = \frac{\hbar^2 k^2}{2m_e^*} + E_c = \frac{\hbar^2(k_x^2 + k_y^2 + k_z^2)}{2\,m^*} + E_c > \Phi_B + E_s - eV. \tag{10.9}$$

The probability that these energy states are occupied is determined by the Fermi-Dirac distribution:

$$f(E,T) = \frac{1}{e^{(E-E_s)/k_BT} + 1} \simeq e^{-(E-E_s)/k_BT} \tag{10.10}$$

$$= e^{-\left(\hbar^2\left(k_x^2+k_y^2+k_z^2\right)/2m_e^*+E_c-E_s\right)/k_BT}. \tag{10.11}$$

With this information, the electron current from the semiconductor to the metal, can be evaluated from the following expression;

$$j_{s\to m} = \frac{2}{(2\pi)^3}\int dk_x dk_y dk_z e j_z f(E,T). \tag{10.12}$$

Here $j_z = \hbar k_z/m_e^*$ is the quantum particle current density in a state described by the wave function $\varphi = e^{ik_z z}$, see Sect. 5.2. Substituting, we have

$$j_{s\to m} = \frac{2}{(2\pi)^3}\int dk_x dk_y dk_z e\frac{\hbar k_z}{m_e^*} e^{-\left(\hbar^2\left(k_x^2+k_y^2+k_z^2\right)/2m_e^*+E_c-E_s\right)/k_BT} \tag{10.13}$$

The integrals on k_x and k_y are easily evaluated and we obtain

$$\int_{-\infty}^{\infty}\int_{-\infty}^{\infty} dk_x dk_y e^{-\hbar^2\left(k_x^2+k_y^2\right)/2m_e^* k_B T} = \frac{2\pi m_e^* k_B T}{\hbar^2}. \tag{10.14}$$

The electron current takes the form

$$j_{s\to m} = \frac{2}{(2\pi)^3}\frac{2\pi m_e^* k_B T}{\hbar^2}\int_{\Phi_B-(E_c-E_s)-eV}^{\infty} dk_z \frac{e\hbar k_z}{m_e^*} e^{-\left(\hbar^2 k_z^2/2m_e^*+E_c-E_s\right)/k_B T}. \tag{10.15}$$

Making here the change of variable $\eta = \hbar^2 k_z^2/2m_e^* + E_g - E_s$ and performing the integration we have

$$j_{s\to m} = \frac{m_e^* e k_B^2 T^2}{2\pi^2\hbar^3} e^{-(\Phi_B - eV)/k_B T}. \tag{10.16}$$

If the applied potential is zero, then the electron current into the metal is

$$j_{s\to m}^0 = \frac{m_e^* e k_B^2 T^2}{2\pi^2\hbar^3} e^{-\Phi_B/k_B T}. \tag{10.17}$$

Since the total current is zero when no external potential is applied, the electron current from the metal to the semiconductor, which balances the thermoionic emission from the semiconductor, must be

$$j_{m\to s} = -\frac{m_e^* e k_B^2 T^2}{2\pi^2\hbar^3} e^{-\Phi_B/k_B T}. \tag{10.18}$$

This current remains unchanged when the electric potential is applied because the Schottky barrier height Φ_B seen by the electrons in the metal does not change. To find the total current across the metal-semiconductor (MS) junction under an applied potential V, we sum the current contributions in both directions. This gives the net current through the Schottky junction:

$$j_{sm} = \frac{m_e^* e k_B^2 T^2}{2\pi^2\hbar^3} e^{-\Phi_B/k_B T}\left(e^{eV/k_B T} - 1\right) = j_{sm}^0\left(e^{eV/k_B T} - 1\right). \tag{10.19}$$

Because $\Phi_B \gg k_B T$, the amplitude of the current $j_{s\to m}^0$, even at high temperatures, is very small. However, under forward bias ($V > 0$), the exponential term causes the current to increase rapidly. This exponential rise in current becomes significant when the applied voltage satisfies:

$$V \geq \frac{\Phi_B}{e} = \Delta\phi + \frac{E_g - E_s}{e}. \tag{10.20}$$

In materials like GaAs, the surface states can reduce the effective Schottky barrier height to approximately $2E_g/3e$. Under reverse bias ($V < 0$), the current remains practically constant, given approximately by

$$j_{sm} \approx -j^0_{sm}. \tag{10.21}$$

This asymmetry in current response makes the Schottky diode an effective rectifier of alternating current.

A key advantage of the Schottky junction is its nearly instantaneous response to forward bias, unlike the *pn* junction, where the recombination processes limit the minority carrier current. *The Schottky junction is a majority carrier device* and responds efficiently to small forward voltages.

10.4 The MOS Device

Metal-oxide-semiconductor (MOS) devices and field-effect transistors (FETs) are among the most fundamental components in modern semiconductor nanotechnology. In the remainder of this chapter we will examine the physical principles underlying MOS structures. In the next chapter, after introducing the conventional transistor, we will turn our attention to MOSFETs, field effect transistors that utilize the MOS structure to generate and control free charge carriers.

When metal-oxide-semiconductor devices were introduced, the experimental and theoretical challenges spurred extensive research work and efforts. Among the most critical areas of investigation were the analytical methods for calculating electric fields and carrier currents within the device. Particular attention was directed toward determining the electric potential, field, current and charge distribution in the semiconducting layer, specifically within the inversion layer that forms under certain biasing conditions. As we will see, the nonlinear nature of the Poisson equation posed a major obstacle to obtaining a complete analytical solution. In this chapter, we will present a method for solving this problem.

Given the relevant role that MOS devices have played in the growing electronics industry, the characterization of operational parameters, empirical formulas and numerical simulations were useful and generally much more effective than the complete understanding of basic physics underlying the behavior of the system, which under certain conditions emerged and defined new properties whose understanding was unfortunately hindered by the inherent mathematical complexity of the governing equations.

When a metal, insulator and semiconductor are brought into contact to form a MOS structure, the system rapidly seeks electrostatic equilibrium. In this process, the Fermi levels in the metal and the semiconductor tend to align. To achieve this alignment, charge redistribution occurs, electrons or holes move toward the semiconductor surface, leading to the formation of an extremely thin charged layer at the metal-oxide interface. In this process a field is established across the structure, as illustrated in the left panel of Fig. 10.5. This field causes bending of the energy bands within the semiconductor. In the insulating oxide layer, the energy bands tilt linearly, reflecting the presence of a uniform electric field, while in the metal, the Fermi level remains flat, indicating the absence of an internal field (see Fig. 10.5).

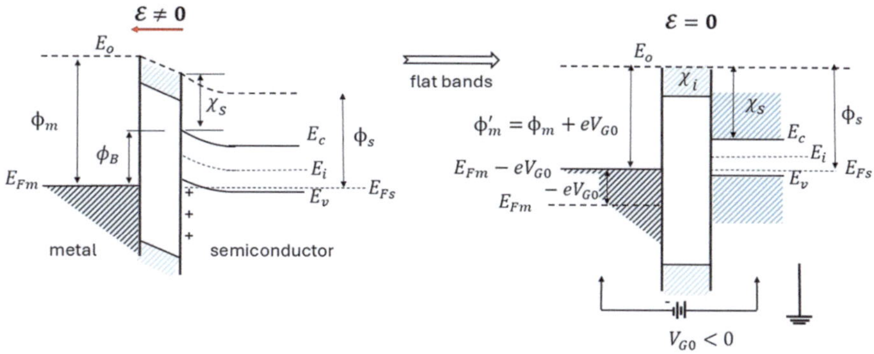

Fig. 10.5 Energy levels when the metal, the insulator and the semiconductor are brought into contact. The semiconductor bands curve in an attempt to balance the Fermi levels. In the right panel, a gate potential V_{G0} is applied, positive in this case, which lowers the potential energy on the left side until the bands align. When this happens the electric fields cancel out

10.5 Gate Potential, Flat Band Structure

To facilitate the analysis and operation of this device, it is customary to ground the semiconductor and establish a potential difference with the metal that is called gate potential or 'gate'. If this gate potential is equal to

$$V_G \simeq (\Phi_m - \Phi_s)/e \equiv V_{G0}, \tag{10.22}$$

the horizontality of the bands, return, as shown on the right panel of Fig. 10.5. Simultaneously, the Fermi levels shift and their difference will be: $E_F - E_{Fm} = eV_{G0}$. This band configuration is called *flat band or ideal MOS*. In the ideal MOS the electric field in the system is zero.

Starting from the ideal MOS, and modifying the gate potential V_G, a very important phenomenon is achieved in a controlled manner, the accumulation of charge at the semiconductor-oxide (S-O) interface. As indicated in Fig. 10.6a and b, the application of a gate potential $V_G > 0$ lowers the Fermi level in the metal and simultaneously electrons accumulate at the interface S-O, while if a potential $V_G < 0$ is applied, there is an accumulation of positive charge on the S-O interface.

To establish the notation and some of the parameters used to characterize the system, and to avoid confusion, we will consider in what follows that the semiconductor is type p. If the semiconductor is type p, the majority concentration is p_p and the minority concentration is n_p. If a potential V_G is applied, from the flat band configuration, in addition to the indicated accumulations, two other interesting phenomena can occur: charge suppression when $V_G > 0$ and the charge population inversion when $V_G \gg 0$. Figure 10.7 indicates the behavior of the concentrations n_p and p_p in three different cases. When a potential $V_G < 0$ is applied, an accumulation of positive charge is induced at the O-S interface with variations in concentrations as qualitatively indicated in the left panel. In reverse bias, $V_G > 0$, the concentration

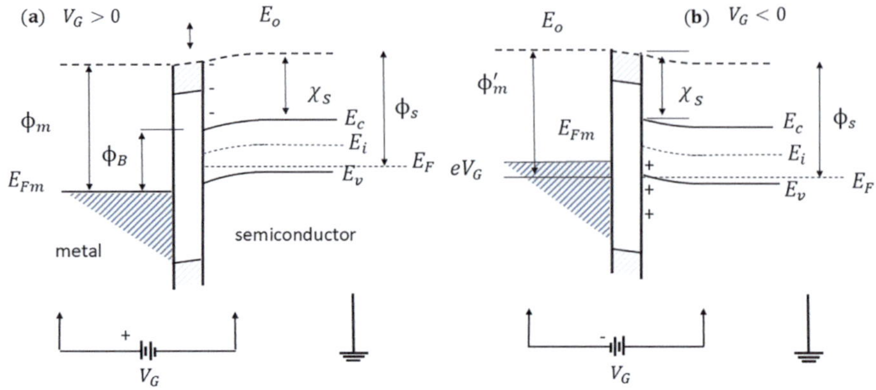

Fig. 10.6 Accumulation of electrons when $V_G > 0$ and of holes if $V_G < 0$

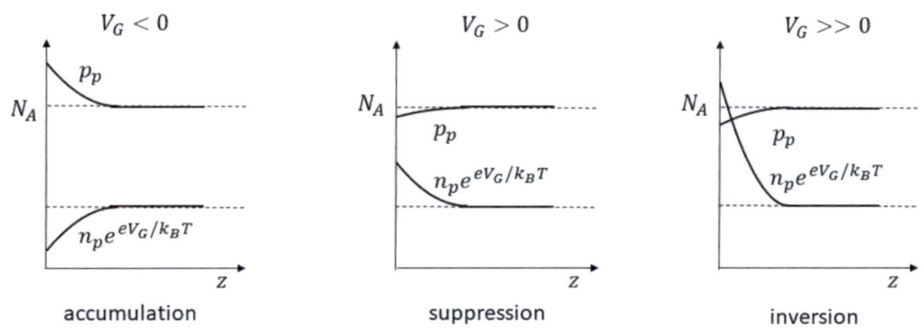

Fig. 10.7 The accumulation, suppression and inversion phenomena in a $MOSp^+$ when $V_G < 0$, $V_G > 0$ and $V_G \gg 0$, respectively

p_p in the vicinity of the O-S interface decreases (holes are repelled and impurities that receive electrons are ionized), and the minority concentration $n_p e^{eV_G/k_BT}$ grows; the net charge $-en_{pV} + ep_p \simeq 0$, as seen in the center panel, and we have the phenomenon of suppression or depletion. If $V_G \gg 0$ and $n_{pV} > p_p$ we reach the population inversion regime.

Suppose a positive gate potential V_G is applied as shown in Fig. 10.8. The positive charge on the metallic side creates an electric field $\mathscr{E}(z)$ that penetrates the semiconductor side, attracts electrons and repels holes and, in the end, establishes an electric potential $\phi(z)$ such that the valence and conduction band edges curves are described by the functions $E_v - e\phi(z)$ and $E_c - e\phi(z)$, respectively. When the band-edge bending is such that the lower part of the conduction band falls below the Fermi level, the population inversion process begins, fueled by the large number of electrons in the ionized impurities and those that pass directly from the upper part of the valence band by tunneling. The gate potential with which the beginning of the population inversion process starts is the threshold potential V_{Gt}. If we look at Fig. 10.8, we can conclude that the threshold potential satisfies the relation

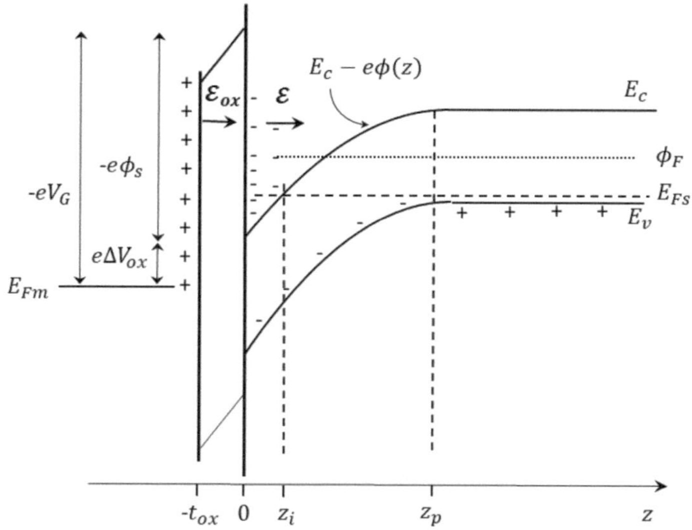

Fig. 10.8 The band structure of a MOS with a p type semiconductor layer, when a positive gate potential V_G greater than the threshold potential is applied. Several parameters are shown that are used to characterize the physics of this system. $z = 0$ is the oxide-semiconductor interface

$$V_{Gt} = (E_g - E_{Fs})/e + \mathcal{E}_{ox}^t t_{ox}. \tag{10.23}$$

E_{Fs} is the Fermi energy of the semiconductor, $\mathcal{E}_{ox}^t$ is the electric field in the oxide layer at threshold, and t_{ox} is the width of the oxide layer. Also shown in Fig. 10.8 are the surface potential $\phi_s \equiv V_G - \mathcal{E}_{ox} t_{ox}$, the width of the inversion layer z_i and a transition or depletion layer, which extends from z_i to z_p. We assume that z_p is less than or equal to the width of the semiconductor layer W_z.

When $V_G > V_{Gt}$, the system acquires not only an additional layer, a layer whose properties make it the very heart of MOS and MOSFETs.

As in the pn junction and the Schottky junction, an important objective is to obtain the electrical potential throughout the entire system. The external potential modifies the distribution of the majority and minority carriers. In the pn junction and in the Schottky junction, the solutions to the Poisson equation could be determined without difficulty because the charge density is basically constant and depends on the concentration of the ionized impurities. In the MOS the charge density in the vicinity of the oxide-semiconductor interface is highly sensitive to the applied gate potential, and charges, due to the oxide layer, have to accumulate at the interface.

We begin in the next section by recalling some obvious and already introduced quantities related to the oxide and depletion layers, and we will discuss the physics related to the inversion layer separately later.

As will be seen in Sect. 10.7, the charge density in the inversion layer depends on the local potential, which is a solution of Poisson's equation, and the local potential depends on the charge density. This interdependence makes the MOS potential subtle and mathematically complicated problem. This is one of the reasons why the problem of understanding the physics of MOS was transformed over time, due to the enormous importance that this system has in current technology, into a problem where describing the phenomenology and obtaining simple empirical formulas to characterize, control and improve the main working properties of the MOS is more important. Intermittently attempts were made to solve the inversion-layer Poisson equation. We will refer to this issue a bit later.

10.6 The Fiel in the Oxide and Depletion Layers of the MOS

We will present here results that are well established for the oxide layer and for the suppression layer in the MOS. The potential inside the oxide layer is given by the function

$$\phi_{ox}(z) = V_G - (z + t_{ox})\mathscr{E}_{ox} \quad - t_{ox} < z < 0, \tag{10.24}$$

and the capacitance per unit area by

$$C_{ox} = \frac{\epsilon_{ox}}{t_{ox}}, \tag{10.25}$$

where ϵ_{ox} is the electrical permittivity in the oxide layer and t_{ox} is the thickness of the oxide layer.

The boundary conditions at the oxide-semiconductor interface, i.e., at $z = 0$, and at the edge of the suppression layer, that we denote as z_p, are

$$\phi_{ox}(0) = V_G - \mathscr{E}_{ox} t_{ox}, \tag{10.26}$$

and

$$\phi_d(z_p) = 0, \qquad d\phi_d(z_p)/dz = 0, \tag{10.27}$$

respectively. Under these conditions, the electric potential for a constant charge concentration N_d is given by

$$\phi_d(z) = \frac{eN_d}{2\epsilon_s}(z_p - z)^2 + \phi(z_p), \qquad z_i < z < z_p, \tag{10.28}$$

and the electric field, as a function of the electric potential, can be written as

$$\mathscr{E}_d(z, \phi_d(z)) = \sqrt{\frac{2eN_d\phi_d(z)}{\epsilon_s}}, \qquad z_i < z < z_p. \tag{10.29}$$

Here and in what follows, we write ϵ_s instead of $\epsilon_r\epsilon_0$, and for the charge concentration in the depletion layer we will use N_d. We will see in later sections that

$N_d \leq N_A$. The precise value of this concentration of ionized impurity atoms will be determined by continuity requirements. A closed formula will be obtained.

The population inversion process begins, as mentioned above, when the bending of the conduction band reaches the Fermi level. Once this inversion regime is established and the gate potential is further increased, a highly charged inversion layer emerges. The thickness of this layer, or width of the inversion layer z_i, depends on the gate potential and on the concentration of impurities. To obtain this width, we have to first solve the Poisson equation. In terms of the parameters defined above, the condition for determining the width of the inversion layer is

$$V_{Gt} = \mathscr{E}^t_{ox} t_{ox} + \phi(z_i), \tag{10.30}$$

where

$$\phi(z_i) = \frac{E_c - E_{Fs}}{e} \equiv V_{Gu}. \tag{10.31}$$

Notice that to distinguish from the threshold potential V_{Gt}, for the gate potential, we denote the threshold of the surface potential as V_{Gu}, with u for *umbral* (from the Latin word *umbra*∼ shadow, and *liminaris*∼ limit).

Because the importance that the inversion layer has in the operation of MOS and MOSFETs, we will discuss in the next sections the physics of the inversion layer.

10.7 Charge Densities and Nonlinearity Problem in MOS

The inversion population process in the semiconductor layer of a MOS is the most important phenomenon in the operation of this device as part of the field effect transistor, the MOSFET. A device that we shall study in the next chapter. In this section we will present the most important models proposed for the charge density distribution in the semiconductor layer of a biased MOS structure. We will meet the nonlinear problem when trying to solve the Poisson equations, and we will briefly refer to the approximate and the alternative solutions introduced to characterize MOS structures. Due to the importance that the inversion layer has, we shall devote the next sections to understand this system. We will explain the charge-field displacement approach that made possible to overcome the nonlinearity problem and to determine, among other properties, the potential, field and charge distributions in the inversion layer. We will end up with a brief analysis of the metallic capacitor limit of the MOS.

The more accurate charge-density models introduced in the 1950s were the W. Shockley model and the Kingston-Neustadter (KN) model. Although the Poisson equations for these models could not be completely solved, the physical quantities obtained by integrating once the nonlinear Poisson equation, together with the charge and MOS capacitance, became part of the standard theory to describe and characterize MOSFETs. In the 1960s, Hauser and Littlejohn (HL) studied a simplified model and solved the approximate nonlinear Poisson equation. We will briefly refer to these models now.

10.7.1 The Shockley Model

To determine the electric potential in the semiconductor layer with donor and acceptor impurities, one has to introduce the charge density that takes into account the distribution of these charges in the presence of the gate potential V_G, and solve the Poisson equation. In the Shockley approach, the Poisson equation and charge density are

$$\frac{d^2\phi}{dz^2} = -\frac{q(p(z) - n(z) + \delta N)}{\epsilon_s}, \tag{10.32}$$

where $q = e$, $\delta N = N_D - N_A$ is the effective doping charge concentration. When the donor and acceptor impurities are fully ionized, the charge of the ionized impurities is $e(N_D - N_A)$. The free electron and hole distributions, are $n(z)$ and $p(z)$, respectively. These distributions are given by

$$n(z) = n_i e^{(e\phi(z)-\mu_n)/k_B T}, \tag{10.33}$$

and

$$p(z) = p_i e^{(\mu_p - e\phi(z))/k_B T}, \tag{10.34}$$

with μ_n and μ_p the quasi-Fermi energy levels and $e\phi(z)$ the electrostatic potential energy. Given these charge distributions, it is clear that the Poisson equation is a non linear differential equation. W. Shockley, by assuming a charge-neutrality condition in the bulk, introduced the charge density

$$\delta N = n_b - p_b = n_i e^{q\phi_B/k_B T} - p_i e^{-q\phi_B/k_B T}, \tag{10.35}$$

where ϕ_B is the potential in the bulk. Assuming

$$n(z) - p(z) = n_i e^{q\phi(z)/k_B T} - p_i e^{-q\phi(z)/k_B T}, \tag{10.36}$$

and using the identity

$$\frac{d}{dz}\left(\frac{d\phi}{dz}\right)^2 = 2\frac{d\phi}{dz}\frac{d^2\phi}{dz^2}, \tag{10.37}$$

transformed the Poisson equation into

$$\frac{d\phi}{dz} d\left(\frac{d\phi}{dz}\right) = -d\phi\frac{2qn_i}{\epsilon_s}\left(\sinh\frac{q\phi_B}{k_B T} - \sinh\frac{q\phi(z)}{k_B T}\right). \tag{10.38}$$

A first integration of this equation can be easily carried out, with $\phi'(z)$ and ϕ as integration variables. For the integration limits at z and $z = \infty$, with $\phi(\infty) = \phi'(\infty) = 0$, the first integration gives

$$\frac{1}{2}\left(\frac{d\phi}{dz}\right)^2 = \frac{2qn_i}{\epsilon_s}\left(\phi(z)\sinh\frac{q\phi_B}{k_B T} - \frac{q}{k_B T}\cosh\frac{q\phi(z)}{k_B T}\right)\Bigg|_0^{\phi(z)}, \tag{10.39}$$

from which we have the electric field at z,

$$\mathscr{E}_S(z) = -\frac{d\phi}{dz} = \pm\left[\frac{4qn_i}{\epsilon_s}\left(\phi(z)\sinh\frac{q\phi_B}{k_BT} - \frac{q}{k_BT}\cosh\frac{q\phi(z)}{k_BT}\right) + \frac{q}{k_BT}\right]^{1/2}$$
$$= \pm\frac{k_BT}{qL_D}\sqrt{F(\phi_B,\phi(z))}. \qquad (10.40)$$

where $F(\phi_B,\phi(z))$ is a nonlinear function of $\phi(z)$. $L_D = \sqrt{\epsilon_s k_B T/2q^2 n_i}$ is the *Debye length*, a distance normalization constant defined by Shockley [4,5]. This first order nonlinear differential equation could not be further integrated, analytically. Thus the potential distribution $\phi(z)$ and several quantities related to this quantity, could not be obtained in this approximation. However, evaluating at $z = 0$ one has the electric field at the oxide-semiconductor interface

$$\mathscr{E}_{Ss} = -\frac{d\phi}{dz}\bigg|_{\phi=\phi_s} = \frac{k_BT}{qL_D}\sqrt{F(\phi_B,\phi_s)}, \qquad (10.41)$$

with ϕ_s the surface potential. Given the electric field at the surface of the oxide layer and the capacitance, the total charge Q_T in the semiconductor layer, could be obtained. Thus the analysis in terms of MOS capacitance become relevant to characterize MOS structures. Notice that choosing this limits, implies that the charge density in (10.32) is defined from $z = 0$ to $z = \infty$.

10.7.2 The Kingston-Neustadter Model

In 1955 a new attempt to solve the non-linear Poisson equation, in a more accurate approach, was the work of Kingston and Neustadter [1], where the main assumption is that the semiconductor is type p, and the impurities are fully ionized. In this case, the Shockley charge distribution becomes [6]

$$\rho(z) = q\big(p(z) - n(z) - N_A\big), \qquad (10.42)$$

with N_A the impurities concentration. In the presence of a gate potential V_G and an electric potential $\phi(z)$, the charge concentrations $p(z)$ and $n(z)$, shown in (4.19) and (4.20), are

$$n(z) = n_p e^{e\phi(z)/k_BT}, \qquad p(z) = p_p e^{-e\phi(z)/k_BT}. \qquad (10.43)$$

with n_p and p_p the minority and majority carrier concentrations. It can be shown[2] that for $p_p \simeq N_A, n_p \simeq n_i^2/N_A$ and the Fermi potential Φ_F from (2.51), the Shockley density in (10.42), becomes

$$\rho(z) = eN_A\left[e^{-\phi(z)/\phi_T} - 1 - e^{-2\phi_F/\phi_T}\left(e^{\phi(z)/\phi_T} - 1\right)\right], \qquad (10.44)$$

[2] See for example Y. Tsividis, Operation and modeling of the MOS transistor, (Oxford University Press, UK, 1987).

where $\phi_T = k_B T/e$ is the thermal energy. This density can also be written in the form

$$\rho(z) = 2eN_A e^{-\phi_F/\phi_T}\left[\sinh\frac{\phi_F - \phi(z)}{\phi_T} - \sinh\frac{\phi_F}{\phi_T}\right]. \tag{10.45}$$

Using the identity (10.37), one can perform the first integration of the Poisson equation over the variables $\phi'(z)$ on the left and ϕ on the right. Choosing the limits as in Shockley's model, i.e. at z and $z = \infty$, with $\phi(\infty) = 0$ and $\phi'(\infty) = 0$, we have[3]

$$\begin{aligned}\frac{1}{2}\left(\frac{d\phi(z)}{dz}\right)^2 &= \frac{eN_A}{\epsilon_s}\left(\phi_T e^{-\phi(z)/\phi_T} + \phi(z) - \phi_T + e^{-2\phi_F/\phi_T}\left(\phi_T e^{\phi(z)/\phi_T} - \phi(z) - \phi_T\right)\right)\\ &= \frac{2eN_A}{\epsilon_s} e^{-\phi_F/\phi_T}\left(\phi_T \cosh\frac{\phi(z) - \phi_F}{\phi_T} + \phi(z)\sinh\frac{\phi_F}{\phi_T} - \phi_T\cosh\frac{\phi_F}{\phi_T}\right)\\ &\equiv G(z, \phi(z)).\end{aligned} \tag{10.46}$$

Again, from this result one can write the electric field at a depth z inside the semiconductor as a function of the still unknown electric potential $\phi(z)$. This field is

$$\mathscr{E}(z) = -\frac{d\phi}{dz} = \pm\left[\frac{4eN_A}{\epsilon_s} e^{-\phi_F/\phi_T}\left(\phi_T\cosh\frac{\phi(z) - \phi_F}{\phi_T} + \phi(z)\sinh\frac{\phi_F}{\phi_T} - \phi_T\cosh\frac{\phi_F}{\phi_T}\right)\right]^{1/2}. \tag{10.47}$$

As in the Shockley model, this nonlinear differential equation could not be integrated analytically, and again the MOS capacitance and surface fields were considered to characterize MOSs and MOSFETs. In the analysis of the next sections we will use, for specific calculations, the charge density in (10.42).

Two important quantities that appear in this formalism and frequently later are the surface electric potential

$$\phi_s = \phi(0) \tag{10.48}$$

and the surface electric field

$$\mathscr{E}_s = \left[\frac{4eN_A}{\epsilon_s} e^{-\phi_F/\phi_T}\left(\phi_T\cosh\frac{\phi_s - \phi_F}{phi_T} + \phi_s\sinh\frac{\phi_F}{\phi_T} - \phi_T\cosh\frac{\phi_F}{\phi_T}\right)\right]^{1/2}. \tag{10.49}$$

[3] Although a semiconductor of infinite width is not real, several authors follow this strategy to obtain the electric field at z. Others like Pao and Sah considered, instead, the limits at $z = 0$ and z. In any case, the results are the same. Notice also that assuming these limits, one is assuming that the charge density $\rho(z)$ in (10.44) is valid from $z = 0$ to $z = \infty$.

The relationships between these quantities and the gate potential V_G, defined by the continuity conditions at the oxide-semiconductor interface ($z = 0$) are determined by the following equations

$$\phi_s = V_G - \mathscr{E}_{ox} t_{ox}, \qquad \text{and} \qquad \epsilon_{ox}\mathscr{E}_{ox} = \epsilon_s \mathscr{E}_s. \tag{10.50}$$

The first of these equations relates the gate potential to the surface potential ϕ_s, and to the potential drop through the oxide layer $\Delta V_{ox} = \mathscr{E}_{ox} t_{ox}$. Note that the electric field in the oxide layer at threshold $\mathscr{E}^t_{ox}$, mentioned above, is $\mathscr{E}^t_{ox} = (\epsilon_s/\epsilon_{ox})\mathscr{E}^t_s$, with $\mathscr{E}^t_s$ given by

$$\mathscr{E}^t_s = \left[\frac{4eN_A}{\epsilon_s} e^{-\phi_F/\phi_T}\left(\phi_T \cosh\frac{V_{Gu} - \phi_F}{\phi_T} + V_{Gu}\sinh\frac{\phi_F}{\phi_T} - \phi_T\cosh\frac{\phi_F}{\phi_T}\right)\right]^{1/2}. \tag{10.51}$$

In Fig. 10.9a and b we show specific examples of these relationships. They are shown for different values of the impurity concentrations. When the gate potential V_G is less than V_{Gt}, the relation between V_G and ϕ_s is almost linear, but when V_G grows beyond the threshold V_{Gt}, it becomes nonlinear. This is related to a saturation phenomena that will consider below, and manifests also in the nonlinear growth of the potential drop ΔV_{ox}, seen in Fig. 10.9b.

In Fig. 10.10 we plot the surface electric field $\mathscr{E}_s$ as a function of the surface potential ϕ_s, in the inversion population regime. Most of the results obtained below will be written as functions of ϕ_s and $\mathscr{E}_s$. However, knowing the relationship $V_G = \phi_s + \kappa t_{ox}\mathscr{E}_s$, where $\kappa = \epsilon_s/\epsilon_{ox}$, it is also possible to express them in terms of V_G.

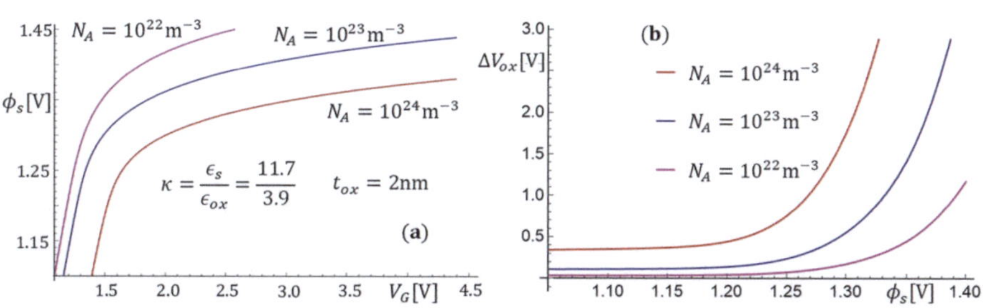

Fig. 10.9 The surface potential as a function of the gate potential (**a**) and the potential drop across the oxide layer (**b**), for different concentrations of impurities. The oxide layer parameters are indicated in the left panel

10.7.3 The Hauser-Littlejohn Model

In 1968, Hauser and Littlejohn (HL) considered a simplified and solvable model for charge distribution in the inversion layer of the metal-oxide-semiconductor. Their proposal was to assume that only electrons or only holes are present [7]. In this model the Poisson equation can be solved, and the resulting analytical solution, despite its approximate nature, has been widely used.

In the HL model, the charge density when the semiconductor is type p is

$$\rho_H(z) = -eN_0 e^{e\phi_H/k_BT}. \tag{10.52}$$

Hauser and Littlejohn assume also that at the boundary between the inversion and depletion layer, defined here as the point $z = z_b$, the electric potential $\phi_H(z_b)$ vanishes. From the continuity of the electric field at this point, one also has that $\phi'_H(z_b) = 2qN_0VGu/\epsilon_s \equiv -\mathscr{E}_b$. Using the identity (10.37), the Poisson equation for the Hauser-Littlejohn model [7], written as

$$\int_{\mathscr{E}_b}^{\phi'_H(z)} \frac{d\phi_H(z)}{dz} d\left(\frac{d\phi_H(z)}{dz}\right) = \frac{eN_0}{\epsilon_s}\int_0^{\phi_H(z)} e^{\phi_H/\phi_T}\, d\phi_H, \tag{10.53}$$

gives

$$\frac{1}{2}\left(\frac{d\phi_H(z)}{dz}\right)^2 = \frac{eN_0}{\epsilon_s}\phi_T\left(e^{\phi_H(z)/\phi_T} - 1\right) + \frac{1}{2}\mathscr{E}_b^2. \tag{10.54}$$

A second integration of this equation is possible. It is indeed easy to verify that the electric potential becomes

$$\phi_H(z) = \phi_T \ln\left[-\frac{b}{a}\operatorname{sech}^2\left(\frac{(z-z_b)\sqrt{b}}{2\phi_T} + \tanh^{-1}\sqrt{\frac{a}{b}+1}\right)\right], \tag{10.55}$$

with

$$a = \frac{2eN_0\phi_T}{\epsilon_s}, \quad \text{and} \quad b = \mathscr{E}_b^2 - a. \tag{10.56}$$

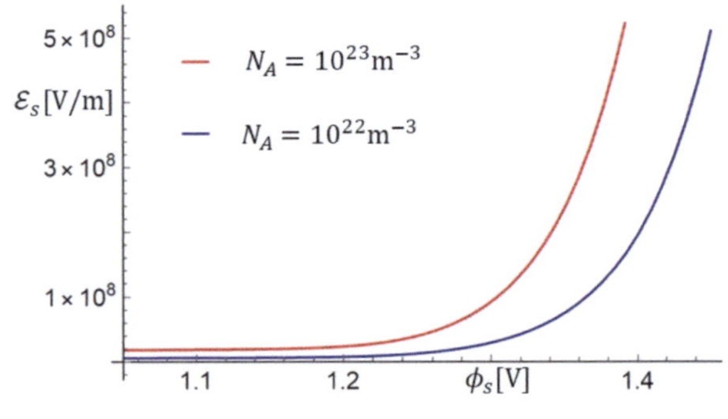

Fig. 10.10 The surface electric field $\mathscr{E}_s$ as a function of the surface potential, for two different values of the impurity concentration

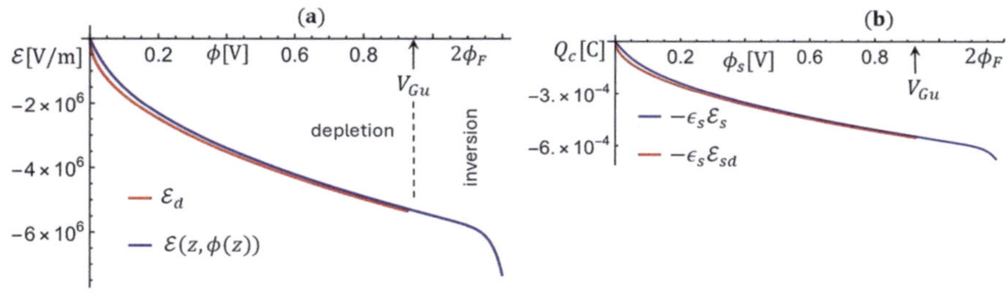

Fig. 10.11 The electric fields $\mathscr{E}(z, \phi(z))$ and $\mathscr{E}_d = -\sqrt{2eN_A\phi(z)/\epsilon_s}$ as functions of $\phi(z)$. While $\mathscr{E}$ vanishes at $z = \infty$, $\mathscr{E}_d$ vanishes at $z = z_p$. In (**b**) we plot the product of the surface electric fields with the electric permittivity. The behavior looks like the graph shown in Appendix F of [6]

We will refer to this solution later, only to show that when the analytical method, introduced in the next section, is applied to this solvable model, we obtain exactly the same solution as from direct integration.

10.8 Charge-Field at Threshold and Inversion-Population Process

We will introduce a different approach based on the phenomenology of the inversion population process. In the MOS and for the formalism that we will introduce, we need to clearly distinguish the inversion region from the rest of the semiconductor. Generally these layers correspond to the channel and the depletion layer. The existence and width of these layers depends on the magnitude of the gate potential V_G. If $V_G < V_{Gt}$, there is no inversion layer, the charge density is that of the ionized and uniformly distributed impurity atoms, but if $V_G > V_{Gt}$, there are two physically different layers. The inversion layer with the Shockley charge density introduced in (10.44), and the depletion or suppression layer of ionized impurity atoms uniformly distributed. The physical properties of the latter were considered in Sect. 10.6, and written for the concentration N_d that will be determined for the specific models by the continuity conditions.

To justify this two-layer model, we plot in Fig. 10.11a, the field $\mathscr{E}(z)$ of (10.47) as a function of the electric potential ϕ, blue curve, and the field $\mathscr{E}_d$ of (10.29), also as a function of the potential $\phi = \phi_d$, red curve. It is interesting to notice that the behavior of $\mathscr{E}_d$ is similar to that of $\mathscr{E}(z)$. Both fields go to zero when ϕ goes to zero, however, it should be noticed that while in $\mathscr{E}(z)$ the potential $\phi = 0$ corresponds to $z = \infty$, in $\mathscr{E}_d$ the zero of the potential ($\phi = 0$) occurs at $z = z_p$. The qualitative and quantitative agreement is encouraging and justifies the constant density model for the depletion layer. When the electric potential is greater than $2\phi_F$, and also greater than the threshold potential V_{Gu}, the function $\mathscr{E}(z)$ grows rapidly. This is the population inversion regime. For all graphs considered here and in the following, we will assume that the intrinsic Fermi potential is $\phi_F = 0.56$ V. In Fig. 10.11b we also plot the products $\epsilon_s\mathscr{E}_s$ and $\epsilon_s\mathscr{E}_{sd}$, of the surface electric fields with the electric permittivity, as functions of the surface potential ϕ_s. The behavior of these functions resembles the graph presented in Appendix F of [6] for the charge $Q_C \simeq \epsilon_s \mathscr{E}_s$ on the semiconductor.

10.8.1 Charge and Field Displacement at Threshold

We shall now introduce a physical argument that will allow us to solve the Poisson equation in the inversion layer. The new method is based on the analogy between the inversion population process and the charging and discharging of a capacitor, characterized by the appearance of transient currents [2]. It was in this context that the argument of continuity of currents led Maxwell to admit that as the capacitor charges, a temporal variation of the displacement field $\mathscr{D}$ also occurs. When a positive gate potential is applied, and the surface potential ϕ_s becomes greater than the threshold potential $V_{Gu} \sim 2\phi_F$, two important processes occur in the semiconductor: a transient current of free charges (electrons attracted to the interface and holes repelled from it), and accompanying this process, a temporal variation of the displacement vector $\mathscr{D}$. If $\mathscr{N}_i(z, T)$ is the concentration of charge carriers that move or tunnel in this process, and their energy, according to the Fermi distribution, is in the range $k_B T$ of the Fermi level, the energy density of these carriers will be of the order of $k_B T \mathscr{N}_i(z, \phi, T)$. Assuming that the semiconductor is linear and the charge concentration is highly localized, one can introduce the *ansatz* that the rate of temporal variation of this energy is equal to the rate of temporal variation of the energy density $u_E = \mathscr{E} \cdot \mathscr{D}$ of the electric field. More precisely, that

$$k_B T \frac{\delta \mathscr{N}_i(z, \phi, T)}{\delta t} = \mathscr{E} \cdot \frac{\delta \mathscr{D}}{\delta t} = \frac{1}{2} \frac{\delta u_E}{\delta t}. \tag{10.57}$$

Multiplying and dividing by the electric charge e, and using the notation $\phi_T = k_B T/e$, and $\rho_i(\phi, T) = -e\mathscr{N}_i(z, \phi, T)$, for the charge carrier density in the inversion layer, (10.57) can be written as

$$-\phi_T \int_{\rho_i(2\phi_F, T)}^{\rho_i(\phi, T)} d\rho_i(\phi, T) = \frac{1}{2} \int_{u_E(2\phi_F, T)}^{u_E(\phi, T)} du_E. \tag{10.58}$$

The limits are chosen for a transient process where the energy densities of charges and field start to vary at threshold, taken as the reference state, a state in which the electric potential $\phi(z)$ is of the order of $2\phi_F$.

If we also assume that $\mathscr{E} = -d\phi(z)/dz\hat{\mathbf{z}}$, we have

$$\frac{\epsilon_s}{2}\left(\frac{d\phi(z)}{dz}\right)^2 - \frac{\epsilon_s}{2}\left(\frac{d\phi(z)}{dz}\right)^2\Big|_{2\phi_F} = -\phi_T\big(\rho_i(\phi(z), T) - \rho_i(2\phi_F, T)\big), \tag{10.59}$$

which we write as

$$\frac{1}{2}\left(\frac{d\phi(z)}{dz}\right)^2 = -\phi_T \frac{\rho_i(\phi(z), T)}{\epsilon_s} + \frac{1}{2}\left(\frac{d\phi(z)}{dz}\right)^2\Big|_{2\phi_F} + \phi_T \frac{\rho_i(2\phi_F, T)}{\epsilon_s}. \tag{10.60}$$

This equation is similar to those obtained after the first integration of the Poisson equations in the Shockley, Kingston-Neustadter and Hauser-Littlejohn models. The question that arises is whether the functions on the right side of this equation and on the right side of (10.46), coincide when the charge density $\rho_i(z, \phi, T)$ is equal to the charge density $\rho(z, \phi, T)$ in (10.44). The answer is yes. In Fig. 10.12 we plot both functions and show that they certainly coincide. In this figure, the function $G(z, \phi(z))$ defined in (10.46), is drawn with a blue dashed line, and the function on the right side of (10.60) is drawn with a solid red curve. Both functions are represented in units of $2eN_A/\epsilon_s$. Thus, the agreement in the figure is independent of the impurity concentration N_A. This and other checking properties, that will be mentioned below, show that defining the parameter

$$\xi^2 = \frac{1}{2}\left(\frac{d\phi(z)}{dz}\right)^2\bigg|_{2\phi_F} + \phi_T \frac{\rho_i(2\phi_F, T)}{\epsilon_s}, \tag{10.61}$$

which depends on the charge density model, and taking into account the Poisson equation, we end up with the nonlinear differential [2] equation

$$\phi_T \frac{d^2\phi(z)}{dz^2} = \frac{1}{2}\left(\frac{d\phi(z)}{dz}\right)^2 - \frac{1}{2}\xi^2. \tag{10.62}$$

This equation replaces the original Poisson equation, whose specific properties are contained in the parameter ξ. We shall refer to this equation as the modified Poisson equation for the inversion layer of the MOS. In principle, all Poisson equations of MOS systems can be represented in the form of the modified Poisson equation (10.62), with the parameter ξ determined by the corresponding charge density. We have yet to show that the solution of this equation solves also the original Poisson equation. We will come back to this issue after solving (10.62), and applying the general solution for specific charge density models. With ξ as a parameter, we will be solving a family of Poisson equations.

Note that when defining the energy density of attracted and repelled charges $u_\rho = 2\phi_T \rho_i(\phi)$, (10.58) can also be written as

$$u_E(\phi, T) + u_\rho(\phi, T) = u_E(2\phi_F, T) + u_\rho(2\phi_F, T) \tag{10.63}$$

and

$$\xi^2 = \frac{2}{\epsilon_s}\Big[u_E(2\phi_F, T) + u_\rho(2\phi_F, T)\Big] = \mathscr{E}_F^2 + 2\frac{\phi_T \rho(2\phi_F)}{\epsilon_s}, \tag{10.64}$$

where $\mathscr{E}_F$ is the electric field at threshold. The parameter ξ plays an important role in the physics of MOS devices. In this case, and in the solvable model considered in the Appendix H, the parameter ξ is $\propto \phi_T/L_D$; has units of electric field; is independent of z and carries information about the charge density, the band structure, the electrical permittivity of the semiconductors and the temperature. When $\phi_s \geq 2\phi_F = E_g$, the

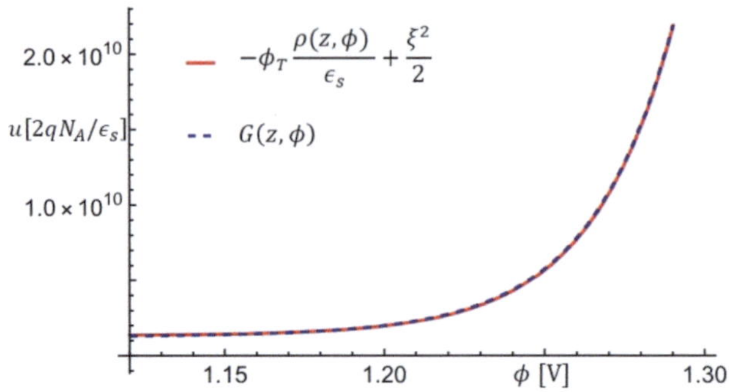

Fig. 10.12 The functions $G(z, \phi(z))$ and -$\phi_T \rho/\epsilon_s + \xi^2/2$, on the right sides of (10.46) and (10.54). Both functions are represented in units of $2eN_A/\epsilon_s$. The graphs show perfect agreement

energy of electrons in the impurity atoms and in the valence band is above the lower edge of the conduction band, and they may move toward the oxide-semiconductor interface, and even undergo quantum tunneling through the Esaki barrier, contributing to the population inversion process.

10.8.2 Solving the Modified Poisson Equation

In the last section we have shown that the Poisson equation can be written as the nonlinear differential equation

$$\frac{d^2\phi(z)}{dz^2} = \frac{1}{2\phi_T}\left(\frac{d\phi(z)}{dz}\right)^2 - \frac{\xi^2}{2\phi_T}. \tag{10.65}$$

This equation is equivalent to the following pair of equations

$$\frac{d\phi(z)}{dz} = f(z) \tag{10.66}$$

and

$$\frac{df(z)}{dz} = \frac{f(z)^2 - \xi^2}{2\phi_T}. \tag{10.67}$$

From the last equation, we have

$$\int_{f_0}^{f(z)} \frac{df}{f^2 - \xi^2} = \int_{z_0}^{z} \frac{dz}{2\phi_T}. \tag{10.68}$$

Here we write f and f_0 for $f(z)$ and $f(0)$, respectively. Note that $f(z)$ corresponds to $-\mathscr{E}(z)$ and, in the lower limit, f_0 corresponds to the electric field at z_0. We will choose this point at the oxide-semiconductor interface. The integrations are simple and we obtain

$$\frac{1}{\xi}\tanh^{-1}\frac{\xi(f-f_0)}{\xi^2-ff_0}=\frac{z-z_0}{2\phi_T}. \tag{10.69}$$

If we solve for $f(z)$ in the previous equation, we have

$$f(z)=\xi\frac{f_0-\xi\tanh\xi(z-z_0)/2\phi_T}{\xi-f_0\tanh\xi(z-z_0)/2\phi_T}. \tag{10.70}$$

Replacing this function into (10.66), we have the first order differential equation

$$\frac{d\phi(z)}{dz}=\xi\frac{f_0-\xi\tanh\xi(z-z_0)/2\phi_T}{\xi-f_0\tanh\xi(z-z_0)/2\phi_T}, \tag{10.71}$$

which can easily be integrated and gives us the elusive electric potential in the inversion layer that can be written as

$$\phi(z)=\phi_0-2\phi_T\ln\left(\cosh\frac{\xi(z-z_0)}{2\phi_T}-\frac{f_0}{\xi}\sinh\frac{\xi(z-z_0)}{2\phi_T}\right), \tag{10.72}$$

which for $\phi_0=\phi_s$, $f_0=-\mathscr{E}_s$ and $z_0=0$, becomes [2]

$$\phi(z)=\phi_s-2\phi_T\ln\left(\cosh\frac{\xi z}{2\phi_T}+\frac{\mathscr{E}_s}{\xi}\sinh\frac{\xi z}{2\phi_T}\right). \tag{10.73}$$

This is an important result that can be used to calculate other quantities, among them, the electric field

$$\mathscr{E}(z)=\xi\frac{\mathscr{E}_s+\xi\tanh\dfrac{\xi z}{2\phi_T}}{\xi+\mathscr{E}_s\tanh\dfrac{\xi z}{2\phi_T}}. \tag{10.74}$$

Before using these quantities, it is worth to check the accuracy of these results. A first checking is to compare the analytical solution $\phi(z)$ with the numerical solution of the original Poisson equation $\phi_n(z)$. For this and the following comparisons we will consider the Kingston-Neustadter model. We will then compare the first derivative $d\phi(z)/dz$ with the electric field $\mathscr{E}(z)=\sqrt{2G(z,\phi(z)}$ of (10.47), obtained by direct integration of the original Poisson equation, and finally we will compare the two sides of the original Poisson equation, i.e. the second derivative $d^2\phi(z)/dz^2$ and the charge density $\rho(\phi(z))/\epsilon_s$.

In Fig. 10.13 we plot the numerical solutions (continuous curves) and the analytical solution (dashed blue curves) for two values of the surface potential ϕ_s. In the left panel, for $\phi_s=1.3$ V, and in the right panel, for $\phi_s=1.4$ V. In each case

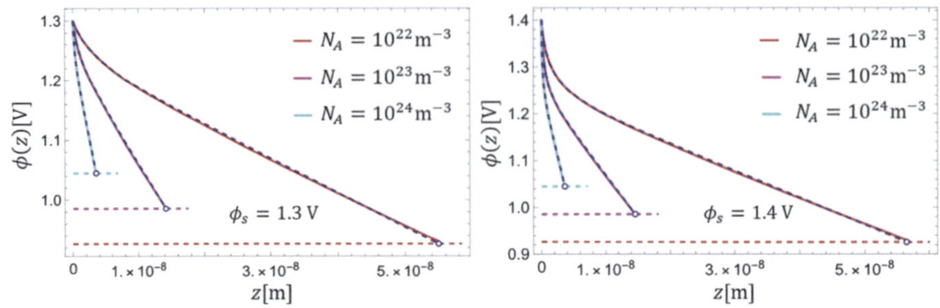

Fig. 10.13 Numerical and analytical solutions of the Poisson equation, continuous and dashed curves, respectively. For these graphs, the charge density is of the KN model, and the impurity concentrations and gate (and surface) potentials indicated in the graphs. The agreement is excellent, above 99.7%. In the left for $\phi_s = 1.3\,\mathrm{V}$ and in the right for $\phi_s = 1.4\,\mathrm{V}$. The author gratefully acknowledges Jürgen Reiter for providing this figure

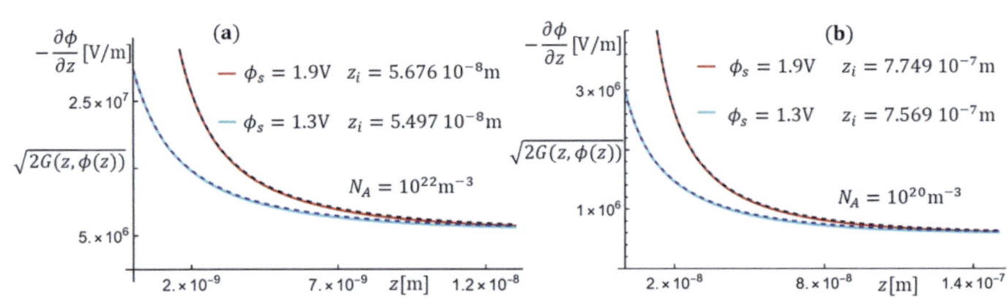

Fig. 10.14 The electric fields $\sqrt{2G(z, \phi(z))}$ (solid curves) and $-\partial\phi(z)/\partial z$ (dashed lines) as functions of z, for different values of N_A and the surface potential ϕ_s. The field $\sqrt{2G(z, \phi(z))}$ obtained by integration of the Poisson equation agrees with the electric field $-\partial\phi(z)/\partial z$ obtained in the present work from the first derivative of the solution in (10.73)

we considered three impurity concentrations: $N_A = 10^{22}\mathrm{m}^{-3}$, $N_A = 10^{23}\mathrm{m}^{-3}$ and $N_A = 10^{24}\mathrm{m}^{-3}$. The agreement is of the order of 99.7% or better.[4] The horizontal dotted lines indicate the threshold potential and the small circles, at the intersection of threshold and inversion layer potential, indicate the inversion layer widths, calculated with the inversion layer width formula determined below.

In Fig. 10.14 the first derivative of $\phi(z)$ and the function $\sqrt{2G(z, \phi(z))}$, defined in (10.46), are plotted as functions of z, for two different values of the impurities concentration and for the surface potentials ϕ_s indicated in the graphs. The agreement is very good. We also indicate the widths of the corresponding inversion layers obtained with the formula derived below.

In Fig. 10.15a and b, we plot the left and right sides of the Poisson equation as functions of z. The second derivative is represented with solid curves and the

[4] The numerical solutions in this graphs were obtained by Jürgen Reiter.

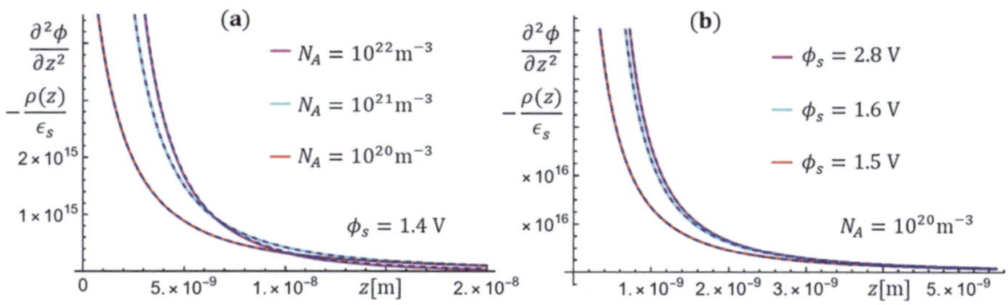

Fig. 10.15 The functions on the left and right sides of the Poisson equation. These functions are evaluated as functions of z, using the electric potential $\phi(z)$ in (10.73), and for different values of N_A and ϕ_s. The solid curves are the second derivatives of $\phi(z)$, and the dashed curves are the charge densities in the KN model (divided by ϵ_s) when the electric potential is the solution $\phi(z)$ in (10.73). Note that the greater the concentration of impurities, the greater and more localized is the charge density

charge density (divided by ϵ_s) with dashed curves. In Fig. 10.15a the graphs are for different values of the impurity concentration N_A, while in (*b*), the graphs are for different values of the surface potential ϕ_s. Since $\phi_s = V_G - \kappa t_{ox}\mathscr{E}_s$, this is also a check for different values of the gate potential. The agreement is also very good and supports the alternative procedure presented here. In addition to these tests, we show, in Appendix H that by applying this method to the solvable Hauser-Littlejohn (HL) model, where the charge density is $qN_0e^{q\phi/k_BT}$, or to the Taur model, with n_i instead of N_0, the result is identical to the function obtained from direct integration of the original HL Poisson equation.

With the result obtained in this section, the set of electric potentials that are solutions in each of the metal-oxide-semiconductor layers is [2]

$$\phi(z) = \begin{cases} V_G - (z + t_{ox})\mathscr{E}_{ox} & -t_{ox} < z < 0 \\ \phi_s - 2\phi_T \ln\left(\cosh\dfrac{\xi z}{2\phi_T} + \dfrac{\mathscr{E}_s}{\xi}\sinh\dfrac{\xi z}{2\phi_T}\right) & 0 < z < z_i \\ \dfrac{eN_d}{2\epsilon_s}(z_i + w_p - z)^2 & z_i < z < z_p. \end{cases} \tag{10.75}$$

In the next section we will determine new formulas.

10.9 Inversion Layer Width and the Potential Energy

Given the electric potentials and electric fields in the inversion and depletion layers, we have still to impose the continuity conditions of these functions. The continuity of the electric potential at $z = z_i$ implies the relationship

$$\phi_s - 2\phi_T \ln\Big(\cosh(\xi z_i/2\phi_T) + \frac{\mathscr{E}_s}{\xi}\sinh(\xi z_i/2\phi_T)\Big) = V_{Gu}, \tag{10.76}$$

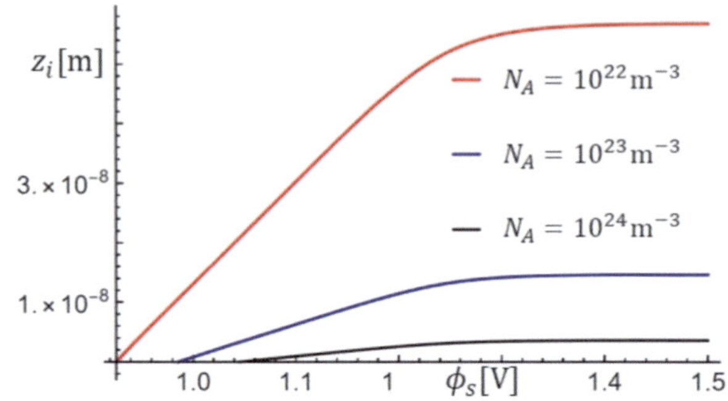

Fig. 10.16 The width of the inversion layer as a function of the surface potential ϕ_s and for different concentrations of impurities

This equation allows us to obtain an important quantity that was also elusive, the width of the inversion layer[2]

$$z_i = \frac{2\phi_T}{\xi}\cosh^{-1}\left(\frac{-e^{\theta/2\phi_T}\xi^2 + \sqrt{\mathcal{E}_s^4 - \mathcal{E}_s^2\xi^2(1 - e^{\theta/\phi_T})}}{\mathcal{E}_s^2 - \xi^2}\right). \tag{10.77}$$

Here $\theta = \phi_s - V_{Gu}$. In Fig. 10.16 we plot the width of the inversion layer z_i as a function of the surface potential ϕ_s, for different values of the impurity concentration N_A.

At threshold, i.e. at $\phi_s = V_{Gu}$, the inversion layer begins to grow. When the values of the difference $\phi_s - V_{Gu}$ are small, which also means small values of $V_G - V_{Gu}$, the width of the inversion layer increases linearly with ϕ_s. For $\phi_s \sim 1.2$ V, the inversion layer widths, almost independently of the impurity concentration, reach their maximum values. For $N_A \simeq 10^{24}\mathrm{m}^{-3}$, the maximum width of the inversion layer is about 4 nm, while for $N_A \simeq 10^{22}\mathrm{m}^{-3}$ the maximum is of the order of 60 nm. The width of the inversion layer z_i decreases by one order of magnitude when N_A increases by two orders of magnitude. In Fig. 10.15 we also see this localization effect. The charge density increases due to the increase in impurity concentration and due to the fact that, simultaneously, the charge is localized near the oxide-semiconductor interface. We will come back to these limiting behaviors in the next section.

The continuity of the electric field at z_i, assuming that $\phi_s \geq V_{Gu}$, is expressed by the equation

$$-\xi\frac{f_0\cosh\left(\xi z_i/2\phi_T\right) - \xi\sinh\left(\xi z_i/2\phi_T\right)}{\xi\cosh\left(\xi z_i/2\phi_T\right) - f_0\sinh\left(\xi z_i/2\phi_T\right)} = \frac{eN_d}{\epsilon_s}w_p, \tag{10.78}$$

which allows us to obtain another important result, the charge density N_d in the depletion layer. After a little algebra, the last equation gives us the concentration of ionized atoms

$$N_{dl} = \frac{\epsilon_s}{2eV_{Gu}}\left(\xi^2 + e^{-(\phi_s - V_{Gu})/\phi_T}\left(\mathcal{E}_s^2 - \xi^2\right)\right). \tag{10.79}$$

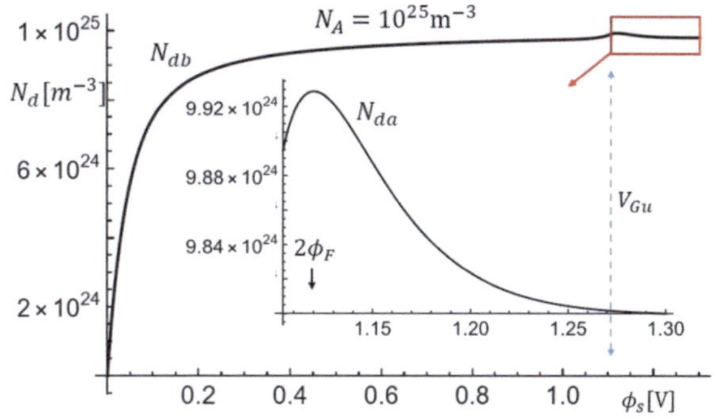

Fig. 10.17 Charge concentrations N_{ds} and N_{dl} in the depletion layer as functions of ϕ_s, for $\phi_s < V_{Gu}$ and $\phi_s > V_{Gu}$, respectively, and for a silicon semiconductor of type p with impurity concentration $N_A = 10^{25}\mathrm{m}^{-3}$, threshold potential $V_{Gu} = 1.1045\,\mathrm{eV}$ and gap energy $E_g = 2\phi_F = 1.12\,\mathrm{eV}$. The inset shows N_{dl} with a maximum at $\phi_s = 2\Phi_F$, that is at the edge of the valence band

The subscript d was changed here to dl to distinguish this concentration in the population inversion regime, where ϕ_s is "larger" than V_{Gu}, of the concentration N_{ds} expected for the depletion layer when ϕ_s is "smaller" than V_{Gu}, and the inversion layer does not exist yet. In this case, when the inversion layer is not yet present, the suppression layer extends to the oxide-semiconductor interface, so the continuity conditions at $z = 0$ require that

$$N_{ds} = \frac{\epsilon_s \mathcal{E}_s^2}{2e\phi_s}. \tag{10.80}$$

This function is compatible with well-known numerical calculations for the charge in a capacitor, when the MOS is seen as a capacitor. In Fig. 10.17 we represent the concentrations N_{dl} and N_{ds}, as functions of the surface potential ϕ_s, for silicon with impurity concentration $N_A = 10^{24}\ \mathrm{m}^{-3}$, intrinsic Fermi level $\phi_F = 0.56\,\mathrm{eV}$, and threshold potential $V_{Gu} = 1.045\,\mathrm{eV}$. As shown in the inset, the ionization of the impurity atoms reaches almost 100% but as soon as the inversion layer emerges, a fraction of the electrons move towards the inversion layer (Fig. 10.18).

With the analytical expressions for the electric potential in the inversion layer of a biased MOS, we have also the charge and field distributions in the inversion layer, the width of the inversion layer, and how the charge concentration in the depletion layer varies with the surface and gate potentials. We can also see how the potential energy behaves. We will plot the potential energy $-e\phi(z)$ as a function of z for different values of the impurity concentration N_A and for different values of the surface potential (or gate potential).

In Fig. 10.18, we present the potential energy as a function of z for different values of the surface potential and for a fixed value of the impurity concentration. The growth of the potential energy as the surface potential increases is accompanied by a growth of the inversion layer width. In this figure, the open circles at the intersection of the conduction band edge and the Fermi level E_{Fs} mark the boundary between the inversion and depletion layers. Both the inversion layer widths z_i and the potential energy reach a maximum as the surface potential (or gate) increases. The inversion layer widths for $N_A = 10^{24}\text{m}^{-3}$ and ϕ_s =1.15 V, 1.25 V and 1.4 V are, respectively, 1.40 nm, 2.74 nm and 3.18 nm. The values of the gate potential V_G are 1.52 V, 2.00 V and 13.13 V, respectively. This behavior of the width of the inversion layer reaches a limit, once the gate potential grows beyond a certain value (which depends on N_A). When this occurs there is a greater density of charges near the interface and also an increase in the surface field [2,3].

In Fig. 10.19, we plot the potential energy again as a function of z but now for different values of the impurity concentration and a fixed value of the surface potential $\phi_s = 1.4$ V. It is well known from standard semiconductor theory that the Fermi energy E_{Fs} of the type p semiconductor is given by $\simeq k_B T \ln N_v/N_A$, with N_v being the effective intrinsic concentration in the valence band. For ϕ_s =1.4 V, and $N_A = 10^{22}$, 10^{23} and 10^{24}m^{-3}, the Fermi energy levels are at 0.17 eV, 0.11 eV and 0.05 eV, above the edge of the valence band, and the widths of the inversion layer are 52.4 nm, 13.3 nm and 3.18 nm, respectively. Different impurity concentrations and different Fermi energies E_{Fs}, imply different potential thresholds V_{Gu} and different widths z_i of the inversion layer. As mentioned before, the open circles at the intersection of the band edge with E_{Fs} determine the point where the inversion layers begin. It is clear from these figures that the width of the inversion layer decreases as the impurity concentration increases. This is consistent with a well-known property for impurity concentration and field penetration in semiconductors.

In Fig. 10.20, we have the potential energy for three examples where both the impurity concentrations and the surface potentials are different. We saw in Fig. 10.19 that increasing the impurity concentration N_A decreases the width z_i and we also saw in Fig. 10.18 that increasing the surface potential ϕ_s increases also the width z_i.

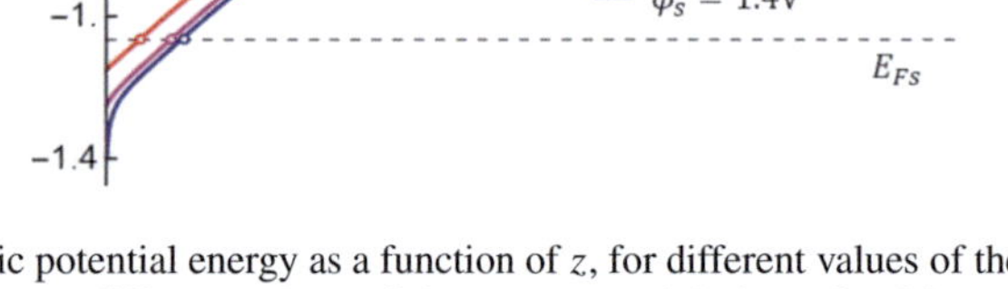

Fig. 10.18 Electric potential energy as a function of z, for different values of the surface potential ϕ_s, which also means different values of the gate potential, for a fixed impurity concentration $N_A = 10^{24}/\text{cm}^3$ and fixed threshold potential V_{Gu}. The open circles, at the intersection of the conduction band edge with the Fermi level E_{Fs}, mark the boundary between the inversion and depletion layers. Increasing the surface potential ϕ_s increases the width z_i of the inversion layer. This width tends to a limit value. For the surface potentials, for which the graphs are drawn, the inversion layer widths z_i are 1.40 nm, 2.74 nm and 3.18 nm, and the gate potentials are 1.52 V, 2.00 V and 13.13 V, respectively

In this figure we increase the impurity by an order of magnitude and decrease the surface potential ϕ_s from 1.5 V (V_G ~9.52 V) to $\phi_s = 1.3$ V (V_G ~3 V). Although their effects are opposite, for the quantities chosen here there are still clear differences between the widths of the inversion layers, ranging from ~55 nm to ~3 nm. The complex relationship between channel width, gate potential and impurity concentration observed in these figures and in Fig. 10.16 has effects on the charge density and drift currents that we analyze in the next chapter; effects that can be better understood from these results.

The solution obtained here for one MOS structure can be used for more complex devices containing more than one MOS structure. It is appropriate to emphasize, on the one hand, that the free electrons at the oxide-semiconductor interface are confined electrons and therefore their energies in the z direction are quantized; on the other hand, a consequence of the behavior z_i and that of the potential energy when $\phi_s >$ 1.24 V (in Si), is that the charge density also begins to concentrate near the surface. We will briefly illustrate the last case and in the final section we will comment on the quantization of the energy in the linear approximation of the confinement potential. A better approach is, for example, the parabolic approach which is discussed elsewhere because it is beyond the purpose of this book.

10.10 Two-Dimensional Charge Density and MOS Capacitance Limit

An important quantity whose distribution was assumed to be two-dimensional is the charge distribution $\rho(z)$. By determining the electric potential more accurately we have a better representation of the charge density in the semiconductor layer. In

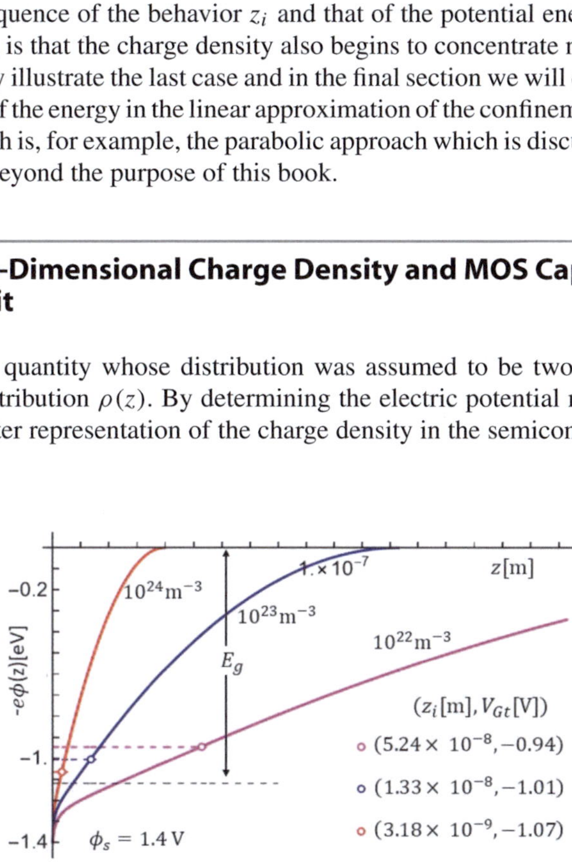

Fig. 10.19 The potential energy as a function of z for different values of the impurity concentration. In these graphs, the surface potential is kept constant and equal to ϕ_s =1.4 V. Since different concentrations of impurities imply different Fermi energies E_{Fs}, and different threshold potentials V_{Gu}, we also have different widths of the inversion layers z_i (see open circles). For ϕ_s =1.4 V, and N_A =10^{22}, 10^{23} and 10^{24}m^{-3}, the Fermi energy levels are at 0.17 eV, 0.11 eV and 0.05 eV, above the edge of the valence band, respectively

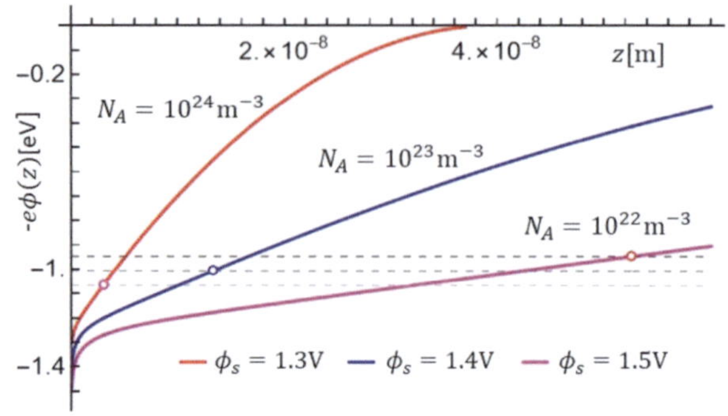

Fig. 10.20 The potential energy as a function of z for different values of the impurity concentrations and different values of the surface potential. Here, an order of magnitude difference in the impurity concentration and surface potentials that differ by only a tenth of a volt give rise to large differences in the inversion layer widths

Fig. 10.15a the Shockley density was plotted as a function of z for different impurity concentrations, and in Fig. 10.15b for different surface potentials (or gate potentials). It is clear from these figures that, approaching the oxide-semiconductor interface ($z = 0$), the density becomes greater. To visualize how the charge distribution approaches the widely assumed two-dimensional electron gas limit, we plot in Fig. 10.21 the charge density $\rho(z)/\epsilon_s$ as a function of z for $z < z_i$, and for three different values of the surface potential ϕ_s =1.12, 1.25 and 1.35 V. The corresponding values of V_G are also given. The figure shows that the charge density $\rho(z)/\epsilon_s$, which is almost constant for $\phi_s = 1.12$ V, when the conduction-band bending reaches the valence band edge, changes rapidly to a distribution that tends to the 2DEG limit. For ϕ_s =1.35 V the density is several orders of magnitude larger and, at the same time, becomes highly

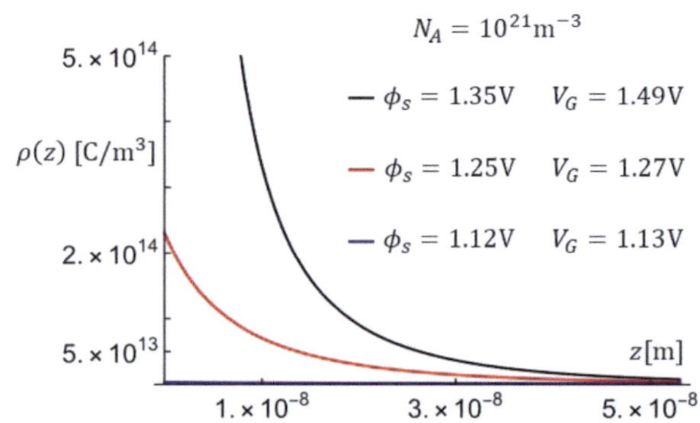

Fig. 10.21 The charge concentration $\rho(z)$ as a function of z for three different values of the surface potential. The corresponding gate potential is also indicated. As the gate potential increases, the charge density varies from a nearly constant distribution to a two-dimensional distribution known as 2DEG

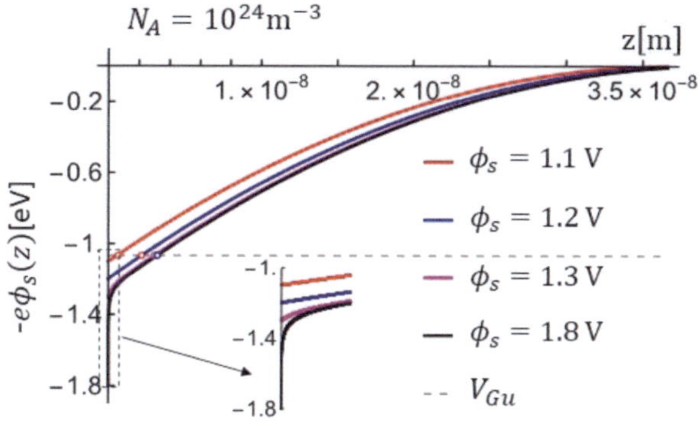

Fig. 10.22 Electric potential energy as a function of z, for different values of surface potential ϕ_s, impurity concentration $N_A = 10^{24}/\mathrm{cm}^3$ and threshold potential V_{Gu}. The open circles indicate the inversion-layer width z_i. For the surface potentials, whose graphs are plotted here, the corresponding gate potentials are 1.45 V, 1.64 V, 3.05 V and ~25700 V, respectively. Reproduced with permission of AIP

localized. This is precisely the range of potentials in which the width of the inversion layer stops growing. To understand better the features of this transition, let us see now what happens with the electric potential distribution. In Fig. 10.22, the potential energy $-e\phi(z)$ is plotted as a function of z for $N_A = 10^{24}\mathrm{m}^{-3}$ and for different values of the surface (or gate) potential. We see in this graph that changing the gate potential from ~3 V to ~26000 V, has no effect on the shape of the potential energy. This means that even though the accumulation of quasi-free electrons at the semiconductor surface reaches high values, the electric field in the inversion layer remains at its maximum value. This is an important property that will help us to understand, why and how is that the accumulation of charge at the oxide-semiconductor interface, and the shift of the MOS response, occur.

As mentioned in the abstract and in the introduction, the transition from a growing inversion-layer width to a rapid increase of the charge accumulation occurs as soon as the quantum tunneling phenomenon through the type of Esaki barrier

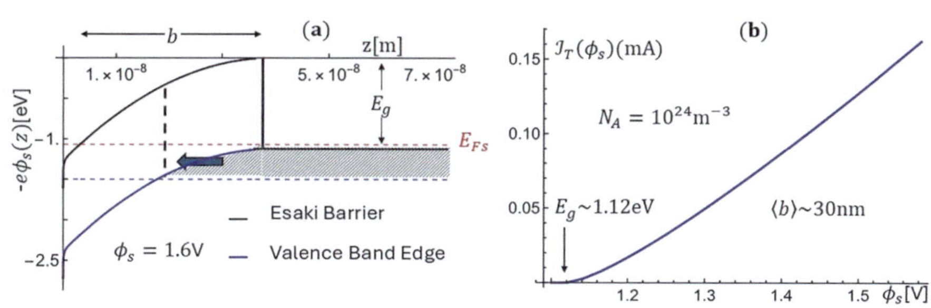

Fig. 10.23 Esaki barriers and tunneling current. In (**a**) electrons of the valence band capable of tunneling through the Esaki barriers whose widths diminish as the gate potential grows. In (**b**) the tunneling current calculated assuming a triangular shape for the Esaki barrier and an average width of 30 nm. Reproduced with permission of AIP

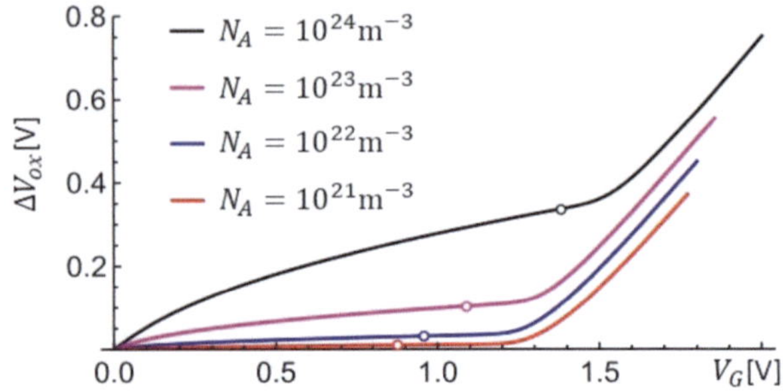

Fig. 10.24 The potential drop in the oxide layer as a function of the gate potential. It shows the transition to the capacitor regime of the MOS

becomes possible for the valence band electrons. The sudden appearance of the tunneling probability bursts the flux of electrons towards the, almost invisible, thin well between the conduction band-edge and the oxide-semiconductor interface, see the inset in Fig. 10.22. To check quantitatively this mechanism of feeding the oxide-semiconductor interface with electrons from the valence band, we made a simple calculation of the tunneling current for the Esaki barriers that the electrons have to pass through. Although the exact barrier shape could be taken into account by using, for example, the WKB approach,[8] for the purpose here it is good enough to assume a triangular Esaki barrier with an average barrier width $\langle b \rangle = 30\,\text{nm}$, that corresponds to the impurity concentration considered in Fig. 10.23a. It is evident from this figure that the larger V_G is, the smaller the width of the Esaki barrier is. Thus our calculation underestimates the real tunnelling current evaluated from

$$\mathscr{I}_T = \frac{e}{\pi \hbar} \int_{-e\phi_s}^{E_v} f(E,T)\mathscr{T}(E)dE \tag{10.81}$$

where the 1D systems' density of levels was assumed. To avoid confusion with the temperature, the transmission coefficient is written as $\mathscr{T}$. The tunneling current in Fig. 10.23b was obtained after calculating the transmission coefficient as a function of the barrier height, an occupation probability equal to 1 and E_v=-1.12 eV. This result shows that the tunneling process can effectively supersede the inversion process that was determined by the minority charge concentration. Since the effective Esaki barrier width is thinner at larger V_G and the electric field is kept almost constant, the tunneling mechanism facilitates the charge accumulation without further growing of the inversion-layer width and the electric field in the semiconductor side.

To conclude this analysis, we plot in Fig. 10.24 the potential drop as a function of V_G for different impurity concentrations. The slow growing of ΔV_{ox}, for V_G smaller than V_{Gt}, transforms a few tenths of volt above into the linear behavior of ΔV_{ox} as a function of V_G, a behavior that characterizes the inversion process determined by the inversion tunneling in the capacitor limit of the MOS.

10.11 Quantization in the Inversion Layer

In Figs. 10.19 and 10.22 it is seen that, in the inversion layer and fixed impurity concentration, the depth and the width z_i, grow with V_G. In Fig. 10.19 it is seen that as the concentration increases, z_i decreases and the slope increases. This behavior is related to that of z_i. As we said before, these potentials are of nanoscopic dimensions and quantize the energies in the z direction. An exact calculation of these levels and their effect on the rest of the physical properties of the MOS is possible but is relatively complex. For quick calculation purposes, a linear approximation of the potential can be considered and also for z_i. Indeed, if $N_A = N_c 10^{\lambda}$ (for Si $N_c \sim 2.7810^{19}$ when $T = 300$K), an expression that allows obtaining the values of z_i with good approximation in the linear region (see Fig. 10.25), is

$$z_{ia} = \frac{\phi_s - V_{Gu}}{\xi}. \tag{10.82}$$

The potential energy in the inversion layer, measured in eVs, can be modeled in a linear approximation as

$$V(z) = -\phi_s + \frac{\phi_s - V_{Gu}}{z_i} z. \tag{10.83}$$

The Schrödinger equation for this potential is

$$\frac{\partial^2 \varphi(z)}{\partial z^2} + \frac{2m_e^*}{\hbar^2} \left(E_z + \phi_s - \frac{\phi_s - V_{Gu}}{z_i} z \right) \varphi(z) = 0. \tag{10.84}$$

E_z is the energy for the motion along the z axis. This equation can be solved by using the WKB method or Airy functions. We will use the WKB approximation.[5]

If we denote as region I the oxide region, II the inversion layer up to the return point for the energy of the electrons, and as region III beyond the return point, the solutions that satisfy the conditions of continuity at the return points $z_r = z_i(E_z + \phi_s)/(\phi_s - V_{Gu})$ and the boundary at $z = 0$, are

$$\varphi_{\rm I}(z)|_{z=0} = 0 \quad \text{and} \quad \varphi_{\rm II}(z)|_{z=z_r} = \varphi_{\rm III}(z)|_{z=z_r}. \tag{10.85}$$

In the WKB approximation the solutions in regions II and III are

$$\varphi_{\rm II}(z) = A \frac{e^{i\xi_2(0)}}{\sqrt{k(z)}} \sin\left(\xi_2(z) - \xi_2(0)\right) \tag{10.86}$$

[5] For the notation you can consult Chap. 6 of *Fundamentals of Quantum Physics*, P. Pereyra. For solutions based on the Airy functions you can consult sections §47 and §48 of *Quantum Mechanics*, L.D. Landau and E.M. Lifshitz.

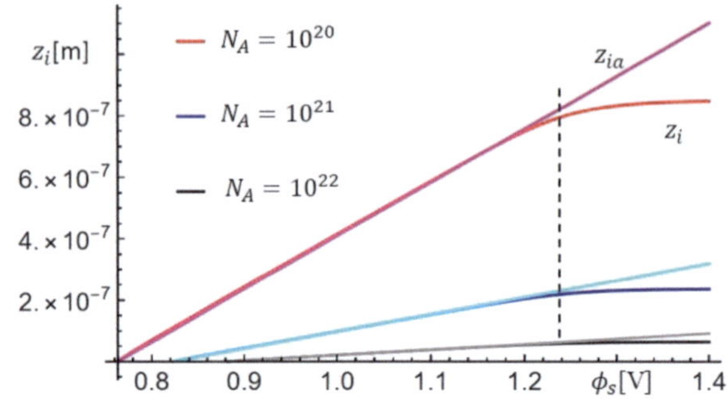

Fig. 10.25 Exact width and approximate width of the inversion layer. The linear behavior with a slope that depends on the concentration of impurities fits very well the exact result for surface potentials $\phi_s \leq 1.24\,\text{V}$

where

$$\xi_2(z) = \int_z^{z_r} k(z)dz, \qquad k(z) = \sqrt{\frac{2m_e^*}{\hbar^2}\big(E_z - V(z)\big)} \tag{10.87}$$

and

$$\varphi_{\text{III}}(z) = A\frac{e^{i\xi_2(0)}}{2\sqrt{q(z)}}e^{-\zeta_2(z)+\zeta_2(0)}\big(\sin(\Delta\zeta_{r0}) - \cos(\Delta\zeta_{r0})\big), \tag{10.88}$$

where

$$\zeta_2(z) = \int_{z_r}^{z} q(z)dz, \qquad q(z) = \sqrt{\frac{2m_e^*}{\hbar^2}\big(V(z) - E_z\big)} \quad \text{and} \quad \Delta\zeta_{r0} = \zeta_2(z_r) - \zeta_2(0). \tag{10.89}$$

The eigenvalue equation results from the boundary and continuity conditions.

$$\cot\left(-\frac{2\sqrt{2m_e^*}}{3\hbar}\frac{(E_z + \phi_s)^{3/2}}{\phi_s - V_{Gu}}z_i\right) = -1, \tag{10.90}$$

whose solutions are the eigenvalues E_n that are easily obtained from

$$\big(E_n + \phi_s\big)^{3/2} = \left(n + \frac{1}{4}\right)\frac{3\pi\hbar}{\sqrt{2m_e^*}}\big(\phi_s - V_{Gu}\big)\frac{1}{z_i}. \tag{10.91}$$

Here we dropped the subindex z of the energy E_z, to simplify the notation. Figure 10.26 shows the particle densities $|\varphi_n(z)|^2$ of the first six energy levels in a Si MOS, with $V_{Gu} \simeq$-0.94 eV for the concentration $N_A = 10^{22}\text{m}^{-3}$ in the left panel, and $V_{Gu} \simeq$-0.88 eV for the concentration $N_A = 10^{21}\text{m}^{-3}$ in the right panel. The eigenvalues and eigenfunctions shown in this figure are for the surface potential $\phi_s = 1.24\,\text{eV}$. In the left panel of higher concentration, the width of the inversion layer is $z_i = 52.1\,\text{nm}$, and on the right panel the inversion layer is wider, with $z_i = 197\,\text{nm}$.

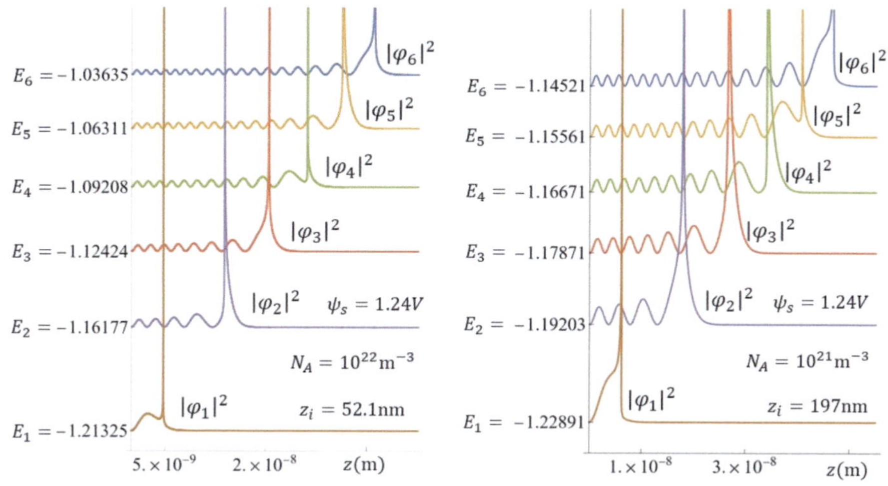

Fig. 10.26 Particle densities at the energy levels of the inversion layer. The left panel is for $N_A = 10^{22}\text{m}^{-3}$ and in the right panel is for $N_A = 10^{21}\text{m}^{-3}$. Only the first six energy levels and the corresponding particle densities $|\varphi_n(z)|^2$ are shown. The solutions are wavelike in the confinement well and decay exponentially in the gap. In the left panel the layer is narrower, the energy levels are more widely spaced than in the right panel

In the narrow layer the energy eigenvalues separation is larger than in the wider layer. These energy levels are highly degenerate. At each level there is a two-dimensional density of levels that depends on the transverse dimensions, similar to those of the quantum well in Chap. 1.

MOS devices, like the one we study here, are part of the electric field transistors called MOSFETs, in which the field of the inversion layer is used to regulate the concentration of charge carriers in the current channel. In the next section we will consider this device.

10.12 Solved Problem

Suppose you have a tungsten-silicon Schottky diode, W-nSi, with a doping of 10^{16}cm^{-3} and a cross-sectional area of 10^{-3}cm^2. The affinity of nSi is approximately the same as that of Si (see Table 10.1) and $e\delta\phi = 0.13\,\text{eV}$. Calculate:

1. The diode current at T = $300K$ and with 0.3 V forward bias.
2. Consider a joint p^+n of Si with the same cross-sectional area A, concentrations $N_A = 10^{19}\text{cm}^{-3}$ and $N_D = 10^{16}\text{cm}^{-3}$, $\tau_p = \tau_n = 10^{-6}s$ and $D_p = 10.5\,\text{cm}^2\,s^{-1}$. For what forward-bias potential does the junction pn allows a current equal to that allowed by the Schottky junction?

Solution We saw that at the Schottky junction

$$j_{sm} = j_{sm}^0 \left(e^{eV/k_BT} - 1\right)$$

where

$$j_{sm}^0 = \frac{m_e^* e k_B^2}{2\pi^2\hbar^3} T^2 e^{-\Phi_B/k_BT}. \tag{10.92}$$

Here

$$\frac{m_e^* e k_B^2}{2\pi^2\hbar^3} = R = 120\left(\frac{m_e^*}{m_0}\right)\left[\text{Acm}^{-2}\text{K}^{-2}\right]. \tag{10.93}$$

If $m_e^* = 0.98m_0$, Φ_B =0.67 eV (for tungsten), then

$$j_{sm}^0 = 6.81 \times 10^{-5}\text{Acm}^{-2}, \tag{10.94}$$

which substituted into the current density leaves us with

$$j_{sm} = 6.81 \times 10^{-5}\left[e^{eV/k_BT} - 1\right] \frac{\text{C}}{\text{s}\,\text{cm}^2}. \tag{10.95}$$

When $V = 0.3$ V, and the surface cross section is A cm^2, we have an approximate current of

$$I_{sm} = j_{sm}A \approx A\, 6.639 \times 10^{-10}\frac{\text{C}}{\text{s}}. \tag{10.96}$$

For the diode p^+n we need to know the Fermi levels on each side and coefficients and diffusion lengths. If

$$\mu_n = E_g - k_BT\ln(N_c/N_D) \simeq 0.913\text{eV}, \quad \mu_p = k_BT\ln(N_v/N_A) \simeq 0.0265\text{eV}, \tag{10.97}$$

$$p_n = N_v e^{-\mu_n/k_BT} = 5.319 \times 10^3\text{cm}^{-3} \quad \text{and} \quad n_p = N_c e^{(\mu_p - 1.12)/k_BT} = 1.5 \times 10\,\text{cm}^{-3}. \tag{10.98}$$

Table 10.1 Work functions and affinities

Elements	Work function [eV]	Semiconductor	Affinity [eV]
Cupper	4.7	Germanium	4.13
Gold	5.1	Silicon	4.01
Nickel	5.15	GaAs	4.07
Platinum	6.35	AlAs	3.5
Silver	4.73	GaN	4.1
Tungsten	4.55	CdSe	3.93
Aluminum	4.08	pCdTe	4.28
Cadmium	4.07	nCdS	4.5

* Many parameters in these tables were obtained from Ioffe's Institute page, https://www.ioffe.ru/SVA/NSM/Semicond/

If the current density at the *pn* junction is

$$j_0 = e\left(\frac{D_p p_n}{L_p} + \frac{D_e n_p}{L_e}\right) \tag{10.99}$$

since $n_p \ll p_n$ the current is approximately

$$j_0 \approx e\frac{D_p p_n}{L_p}. \tag{10.100}$$

With $D_p = 10.5\,\text{cm}^2/\text{s}$, $\tau_p = 10^{-6}\text{s}$ $L_p = 3.24 \times 10^{-3}\,\text{cm}$, we have

$$j_0 \approx 1.602 \times 10^{-19}\frac{10.5\times}{3.24 \times 10^{-3}}5.319 \times 10^3\frac{\text{C}}{\text{s cm}^2} \tag{10.101}$$

$$j_0 = 27.61 \times 10^{-13}\,\frac{\text{C}}{\text{s cm}^2}. \tag{10.102}$$

For V =0.3 V and an area A in cm^2 the current in the diode *pn* is

$$I = j_0 A \approx j_0 A e^{eV/k_B T} \approx A\,2.69 \times 10^{-17}\frac{\text{C}}{\text{s}}. \tag{10.103}$$

For the current to be the same as in the Schottky junction, a potential $V \simeq$0.74 V is needed. This exhibits the difference between Schottky junctions and p^+n junctions. It was shown here that to establish a current density of ~6.5 mA/cm^2, the Schottky diode requires a potential of V =0.3 V while the junction p^+n requires a potential V =0.7 V.

10.13 Problems

1. Consider the Schottky diode $Pt - nCdS$ (see Table 10.1) with $e\delta\phi = 0.08$e V and doping $n = 10^{17}\text{cm}^{-3}$. Calculate:

   (a) The diode current at T = 300K and with 0.3 V forward bias.
   (b) Consider a joint p^+n of Si with the same cross-sectional area A, concentrations $N_A = 10^{18}\text{cm}^{-3}$ and $N_D = 10^{17}\text{cm}^{-3}$, $\tau_p = \tau_n = 10^{-6}s$, $D_p = 10.5\,\text{cm}^2\,s^{-1}$ and $D_n = 30.1\,\text{cm}^2\,s^{-1}$. For what forward-bias potential does the *pn* junction allows a current equal to that allowed by the Schottky junction?

2. Suppose you have a MOS device and that the semiconductor layer is Si with the parameters $\Phi_F = 0.56$eV, ϵ_s =11.7, $t_{ox} = 3$ nm and $N_A = 10^{21}\,\text{m}^{-3}$. Consider the relations (10.51) and (10.50) and obtain the surface potentials when V_G =1.1, 1.2 and 1.3 V.

3. Plot the width of the inversion layer for the three values of the potential surface in Problem 2.
4. Plot the potential energy for the system and the surface potentials indicated in Problem 2.
5. Plot the charge concentration in the system and the surface potentials indicated in Problem 2.
6. Compare the widths of the inversion layer in Problem 3 with the behavior of the approximate width in (10.82).

References

1. R.H. Kingston, S.F. Neustadter, J. Appl. Phys. **26**, 718–720 (1955)
2. P. Pereyra, Semicond. Sci. Technol. **39**, 095006 (2024)
3. P. Pereyra, IntechOpen 1001645, 79 (2024); *Chapter in MOSFETS Developments and Trends*, ed. by Y. Tu, R. Abd-Alhameed, A. Rayit (IntecOpen, London, 2024)
4. W. Shockley, Bell Syst. Tech. J. **28**, 435 (1949)
5. W. Shockley, *Electrons and Holes in Semiconductors* (D. Van Nostrand, New York, 1950). Chaps. 9 and 10
6. Y. Tsividis, *Operation and Modelling of MOS Transistor*. The Oxford Series in Electrical and Computer Engineering (Oxford, UK, 2003)
7. J.R. Hauser, M.A. Littlejohn, Solid State Electron **11**, 667 (1968)
8. P. Pereyra, D. Weiss, Phys. B **455**, 96 (2014)

Bipolar Junction Transistor and MOSFET

11

Abstract

In this chapter, we examine two cornerstone semiconductor devices: the bipolar junction transistor (BJT) and the field-effect transistor (FET), with particular emphasis on the MOSFET. Building on the foundational semiconductor principles introduced in the earlier chapters, we begin with a discussion of the BJT, focusing on its operation and current control mechanisms. For the MOSFET, while we revisit conventional approximate models that have historically provided a qualitative understanding of current saturation as a function of the source-drain voltage, we go beyond them by incorporating the analytical methods introduced in Chap. 10. These methods allow us to compute the drift-diffusion current explicitly for both single- and double-gate MOSFETs. We present plots of the current as a function of both source-drain and gate voltages across different impurity concentrations, demonstrating how operational threshold voltages can be accurately predicted. This analytical approach provides a more rigorous and insightful characterization of MOSFET behavior.

11.1 Introduction

A bipolar junction transistor (BJT) consists of two semiconductor layers of the same type, either n-type or p-type, separated by a layer of the opposite type, forming an *npn* or *pnp* structure, respectively. To operate, a BJT requires two power supplies with opposite biasing: one for the base-emitter junction (forward bias) and another for the base-collector junction (reverse biased). A MOSFET (or more generally a MISFET in which the oxide layer is replaced by I for insulator) is also based on an *npn* or *pnp* structure, however, it operates with a single power supply and utilizes the MOS (or MIS) structure to modulate charge carrier flow between the source and the drain.

We begin by studying the bipolar transistor. In this discussion we consider the *npn* transistor in the configuration shown in Fig. 11.1. To understand how the transistor functions as a power amplifier, we first analyze the current components in each

P. Pereyra, *Contemporary Theory of Semiconductors and Devices*,
Graduate Texts in Physics, https://doi.org/10.1007/978-3-032-00296-9_11

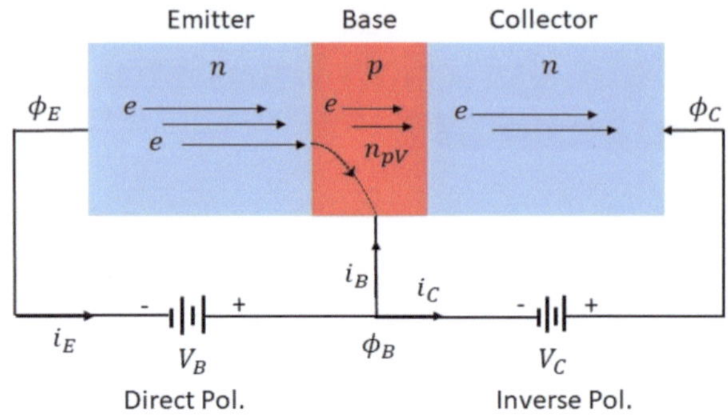

Fig. 11.1 Bipolar transistor *npn*. In operation the potentials satisfy the relation $\phi_C > \phi_B > \phi_E$

branch of the device and then explain how the collector current i_C can effectively be controlled and amplified through modulation of the base current.

11.2 The Currents in the Emitter, Base and Collector

In an *npn* device like the one shown in Fig. 11.2, connected to sources V_B and V_C, with $V_C > V_B$, the carrier currents that flow throughout the device are known as the emitter, base and collector currents. From Kirchhoff's current law, the emitter currents i_E, base current i_B and the current i_C circulating through the collector, satisfy the relation

$$i_E = i_B + i_C. \tag{11.1}$$

In the *np* junction, the potential V_B is forward biased, while the potential V_C in the *pn* junction, known also as the blocking voltage, is reverse biased. We will denote by z_p and $-z_n$, the ends of the transition region in the *np* junction. The current density j_E in this junction is composed by the current of holes $j_p(-z_n)$ and the current of electrons $j_n(z_p)$ that are injected into the layer *p* where they form part of the minority carriers. Holes arrive at layer *n* and move towards the left of $-z_n$ as minority holes. The electrons that arrive at layer *p* divide into three parts, one part contributes to the base current i_B, another to the recombination processes, and the other part reaches the transition zone of the *pn* junction, where the electric field $\mathcal{E}'$ drives them towards the collector and become part of the current i_C. Taking these contributions into account, the current density j_E can be written as

$$j_E = j_n(z_p) + j_p(-z_n) + e \int_{-z_n}^{z_p} dz R_{gr}(z). \tag{11.2}$$

Here $j_n(z_p)$ is the electron current minority injected into the base, $j_p(-z_n)$ is the current of minority holes injected into the emitter, and the integral is the contribution

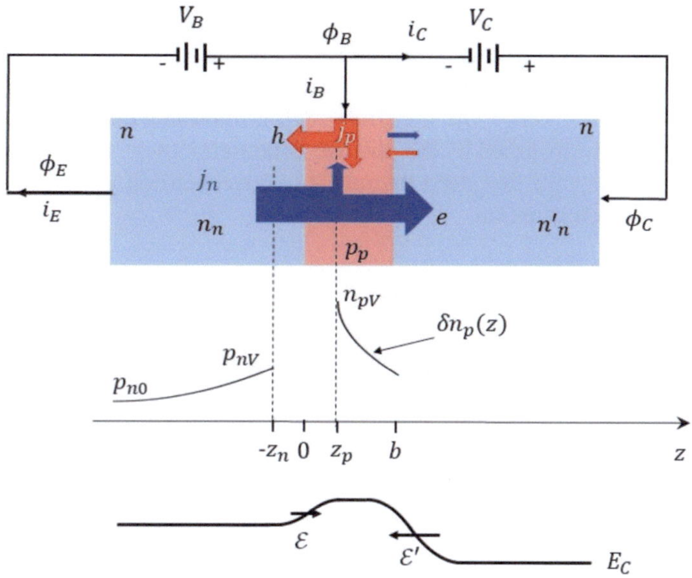

Fig. 11.2 Electron currents in the npn transistor. An important ingredient is the minority concentration that reaches the base and the transition zone of the pn junction. There, the field $\mathcal{E}'$ drives the carriers towards the layer n on the right, which works as a collector. These electrons contribute to the current i_C. Another part recombines and many return through the branch of the potential V_B that is forward biased

to the current related to the carriers that are created and annihilated in the generation-recombination processes that occur in the transition region, which can be neglected when $L_e \gg b$. If this is the case, we have

$$i_E = j_n(z_p)A + j_p(-z_n)A. \tag{11.3}$$

Here, it is convenient to note that the current $j_n(z_p)$ is a diffusion current that depends on the gradient of the excess of minority electrons δn_p whose population grows with the potential V_B in forward polarization. We will determine the concentration $j_n(z)$ from the diffusion equation (4.88), introduced in Chap. 4, when the thickness b of the layer p is much smaller than L_e, i.e., from the equation

$$\frac{1}{eD_e}\mathbf{j}_n(z) = \nabla n_p(z). \tag{11.4}$$

For the minority holes concentration, the situation changes a little if the thickness of the layer n is $\gg L_h$. In this case, we can use (4.56), with the changes that correspond to the configuration in Fig. 11.2 (here we have changed V to V_B, and $> w_n$ to $< -z_n$), thus,

$$\delta p_n(z) = p_n\big(e^{eV_B/k_BT} - 1\big)e^{(z+z_n)/L_h} \qquad \text{for} \qquad z < -z_n < 0. \tag{11.5}$$

Before obtaining the currents j_p and j_n we need to determine the concentration $\delta n_p(z)$. To this purpose we must take into account the finiteness of layer p width. Due to this fact, only a part of the injected electrons recombine in the p region. The rest diffuse, passing through the transition region into the collector, driven by the field, and are finally absorbed by the positive potential.

To obtain $\delta n_p(z)$, we first need the minority concentartion n_p as function of z, solving the diffusion equation

$$\frac{\partial^2 n_p(z)}{\partial z^2} - \frac{n_p(z)}{D_e\tau_e} = 0, \qquad 0 < x < b, \tag{11.6}$$

when the width b of the base is $\ll L_e$. If a solution of the form

$$n_p = e^{rx}, \tag{11.7}$$

is proposed, the diffusion equation becomes the algebraic equation

$$r^2 - \frac{1}{D_e\tau_e} = 0, \tag{11.8}$$

whose roots are

$$r = \pm\frac{1}{\sqrt{D_e\tau_e}} = \pm\frac{1}{L_e}. \tag{11.9}$$

Therefore, the most general solution for the minority electron concentration is

$$n_p(z) = C_1 e^{z/L_e} + C_2 e^{-z/L_e}. \tag{11.10}$$

We still have to obtain the coefficients C_1 and C_2. Evaluating the concentration $n_p(z)$ at the edge of the p layer, we have

$$n_p(b) = n_{pb} = C_1 e^{b/L_e} + C_2 e^{-b/L_e}, \tag{11.11}$$

which we use to solve for one of the coefficients as

$$C_1 = n_{pb} e^{-b/L_e} - C_2 e^{-2b/L_e}. \tag{11.12}$$

If we substitute this coefficient into (11.10), we have

$$\begin{aligned} n_p(z) &= n_{pb} e^{(z-b)/L_e} - C_2 e^{-b/L_e}\left(e^{(z-b)/L_e} - e^{(b-z)/L_e}\right) \\ &= n_{pb} e^{(z-b)/L_e} - 2C_2 e^{-b/L_e} \sinh\left[(z-b)/L_e\right]. \end{aligned} \tag{11.13}$$

To obtain C_2 we evaluate $n_p(z)$ at z_p, i.e. at the edge of the transition zone of the np junction. The concentration at z_p is then

$$n_p(z_p) = n_{pb} e^{(z_p-b)/L_e} - 2C_2 e^{-b/L_e} \sinh\left[(z_p - b)/L_e\right], \tag{11.14}$$

from which results the coefficient

$$2C_2 e^{-b/L_e} = -\frac{n_p(z_p) - n_{pb}e^{(z_p-b)/L_e}}{\sinh\left[(z_p - b)/L_e\right]}. \tag{11.15}$$

Therefore,

$$n_p(z) = n_{pb}e^{(z-b)/L_e} + \left(n_p(z_p) - n_{pb}e^{(z_p-b)/L_e}\right)\frac{\sinh\left[(b-z)/L_e\right]}{\sinh\left[(b-z_p)/L_e\right]}. \tag{11.16}$$

We can now obtain the excess of minority electrons, however it is convenient to simplify the previous expression. Since $z_p \sim b$, we can take $e^{(z_p-b)/L_e} \simeq 1$. If in addition $n_{pb} \cong n_p$ and $n_p(z_p) = n_p e^{eV_B/k_BT}$, the excess minority electrons, defined as

$$\delta n_p(x) = n_p(z) - n_{pb} \tag{11.17}$$

becomes

$$\delta n_p(z) = n_p\left(e^{eV_B/k_BT} - 1\right)\frac{\sinh\left[(b-z)/L_e\right]}{\sinh\left[(b-z_p)/L_e\right]} \qquad \text{for} \qquad z > z_p > 0. \tag{11.18}$$

With this result and (11.5) we can easily evaluate the diffusion currents. For j_p we have

$$j_p(z) = -eD_h\nabla\delta p_n(z) = \frac{eD_h p_n}{L_h}\left(e^{eV_B/k_B} - 1\right)e^{(z+z_n)/L_h}, \qquad z < -z_n < 0. \tag{11.19}$$

For j_n we have

$$j_n(z) = -eD_e\nabla\delta n_p(z) = \frac{eD_e n_p}{L_e}\left(e^{eV_B/k_B} - 1\right)\frac{\cosh\left[(b-z)/L_e\right]}{\sinh\left[(b-z_p)/L_e\right]}, \quad z > z_p > 0. \tag{11.20}$$

To obtain the current that effectively flows through the device, which corresponds to the emitted current, we have to discount the currents that recombine between $-z_n$ and z_p. To this purpose, we evaluate the currents j_n and j_p at the exits of this region. The sum of these currents, must equal the emitter current, i.e.,

$$i_E = j_n(z_p)A + j_p(-z_n)A. \tag{11.21}$$

Substituting the explicit functions obtained above, we obtain

$$i_E = \left[\frac{en_pD_eA}{L_e}\coth[(b-z_p)/L_e] + \frac{ep_nD_hA}{L_h}\right]\left(e^{eV_B/k_BT} - 1\right). \tag{11.22}$$

Before we calculate the current i_B, through the base, we will calculate the current i_C through the collector. This current is equal to the current that crosses the surface at $z = b$, i.e., to $j_n(z = b)A$. Thus

$$i_C = \frac{en_pD_eA}{L_e}\left(e^{eV_B/k_BT} - 1\right)\frac{1}{\sinh[(b - z_p)/L_e]}. \tag{11.23}$$

The current through the collector depends also on V_B, therefore, it can be controlled with the potential V_B. This is an important property of the device. Given i_E and i_C and the Kirchhoff law, the base current becomes

$$i_B = \left[\frac{en_pD_eA}{L_e}\frac{\cosh[(b - z_p)/L_e] - 1}{\sinh[(b - z_p)/L_e]} + \frac{ep_nD_hA}{L_h}\right]\left(e^{eV_B/k_BT} - 1\right). \tag{11.24}$$

This current consists of a part of the hole current that reaches $-z_n$, that is, $j_p(-z_n)A$, plus the contribution of the minority electrons that do not reach b and do not recombine between $-z_n$ and z_p.

With the currents $j_n(z)$ and $j_p(z)$ we can also determine the currents that do recombine in the transition zone. In fact, the current density of minority electrons that recombine between $-z_n$ and z_p is given by the difference

$$j_{nr} = j_n(-z_n) - j_n(z_p), \tag{11.25}$$

which, explicitly, is given by

$$j_{nr} = \frac{en_pD_e}{L_e}\left(e^{eV_B/k_BT} - 1\right)\frac{\cosh[(b + z_n)/L_e] - \cosh[(b - z_p)/L_e]}{\sinh[(b - z_p)/L_e]}. \tag{11.26}$$

When the width of the transition region in the np contact is less that the electron mean free path, i.e., when $z_n + z_p << L_e$, an approximate expression for this current is

$$j_{nr} = \frac{en_pD_e}{L_e}\left(e^{eV_B/k_BT} - 1\right)\frac{z_n + z_p}{L_e}. \tag{11.27}$$

In the same way, the current density of holes that recombine between z_p and $-z_n$ is given by the difference

$$j_{pr} = j_p(z_p) - j_p(-z_n), \tag{11.28}$$

from which we have

$$j_{pr} = \frac{ep_nD_h}{L_h}\left(e^{eV_B/k_BT} - 1\right)\left[e^{(z_p+z_n)/L_h} - 1\right], \tag{11.29}$$

This expression can also be approximated as

$$j_{pr} = \frac{ep_nD_h}{L_h}\left(e^{eV_B/k_BT} - 1\right)\frac{z_n + z_p}{L_h}. \tag{11.30}$$

Therefore, the total current that recombines is

$$j_r = j_{nr} + j_{pr} = e\left(\frac{p_n D_h}{L_h^2} + \frac{n_p D_e}{L_e^2}\right)\left(e^{eV_B/k_BT} - 1\right)(z_n + z_p). \tag{11.31}$$

11.2.1 Some Useful Coefficients

From the currents flowing in the transistor, currents that were obtained in the previous section, it is common to define coefficients used to characterize the transistors. Some of them are the following:[1]

1. The **current transfer ratio** (or transfer coefficient) defined as

$$\alpha = \frac{i_C}{i_E}. \tag{11.32}$$

For highly doped transistors n^+pn^+, we can obtain simpler approximate results. In these devices $p_n \ll n_p$. If that is the case

$$i_E \to i_E = \frac{en_p D_e A}{L_e} \coth[(b - z_p)/L_e]\left(e^{eV_B/k_BT} - 1\right), \tag{11.33}$$

while for the currents i_C and i_B we have

$$i_C = \frac{en_p D_e A}{L_e}\left(e^{eV_B/k_BT} - 1\right)\frac{1}{\sinh[(b - z_p)/L_e]}, \tag{11.34}$$

and

$$i_B = \frac{en_p D_e A}{L_e}\left(e^{eV_B/k_BT} - 1\right)\frac{\cosh[(b - z_p)/L_e] - 1}{\sinh[(b - z_p)/L_e]}. \tag{11.35}$$

Thus, for n^+pn^+ devices the transfer ratio is given by

$$\alpha \approx \frac{1}{\cosh[(b - z_p)/L_e]}. \tag{11.36}$$

If additionally $(b - z_p) \ll L_e$ the transfer ratio is approximately given by

$$\alpha \approx \frac{2L_e^2}{2L_e^2 + (b - z_p)^2} \sim 1. \tag{11.37}$$

[1] The use of α and β in this field has a long tradition and should not be confused with the matrix elements of the transfer matrix.

2. The **amplification factor** of the current (from base to collector) is defined as

$$\beta = \frac{i_C}{i_B}, \tag{11.38}$$

This factor, in terms of the transfer ratio, can be written, in the form

$$\beta = \frac{\alpha}{1-\alpha} \tag{11.39}$$

which shows that this factor can become very large if $\alpha \to 1$. Note that the transfer ratio and the amplification factor can also be defined as

$$\left.\frac{\partial i_C/\partial V_B}{\partial i_E/\partial V_B}\right|_{V_C=cte} = \frac{i_C}{i_E} = \alpha, \tag{11.40}$$

and as

$$\beta = \frac{i_C}{i_B} = \left.\frac{\partial i_C/\partial V_B}{\partial i_B/\partial V_B}\right|_{V_C=cte} = \frac{\alpha}{1-\alpha}. \tag{11.41}$$

Writing α and β as ratios of partial derivatives is useful, as this form allows to analyze the sensitivity of the device parameters, which is the purpose of the next section.

11.3 Power Amplification Factor

In order to obtain a useful power device, an element that performs work on the collector circuit must be included. For example, a resistor R_L, as shown in Fig. 11.3. The potential drop $i_C R_L$ is equal to V_C. So, $P_L = i_C V_C$ is the dissipated power in the work element. Our purpose now is to see how large is the change dP_L when the base potential V_B varies a dV_B, while V_C is kept constant. In this case, the change dP_L in the power device is given by

$$dP_L = d(V_C i_C)|_{V_C=cte} = V_C \left.\left(\frac{\partial i_C}{\partial V_B}\right)\right|_{V_C} dV_B, \tag{11.42}$$

while the variation in power $P_B = i_B V_B$, due to the same change dV_B, is given by

$$dP_B = d(V_B i_B)|_{V_C=cte} = \left[i_B + V_B \left.\left(\frac{\partial i_B}{\partial V_B}\right)\right|_{V_C}\right] dV_B. \tag{11.43}$$

If, in the last expression, $\frac{\partial i_B}{\partial V_B} = i_B \frac{e}{k_B T}$, the relation between the second term and the first is equivalent to $\frac{eV_B}{k_B T} \gg 1$, so we can write the power change in the base as

$$dP_B \approx V_B \left(\frac{\partial i_B}{\partial V_B} \right) \Bigg|_{V_C} dV_B. \tag{11.44}$$

With the ratio of the dissipated power dP_L and the power change in the base, one can determine the power amplification, which can be written as:

$$\frac{dP_L}{dP_B} = \beta \frac{V_C}{V_B} = \frac{\alpha}{1-\alpha} \frac{V_C}{V_B}. \tag{11.45}$$

This shows that the power amplification factor is proportional to the current amplification factor. It should be emphasized that this is not a system for power generation, but of two circuits in which the output power is tuned by the base circuit. This is in fact the role that the transistor plays in electronic circuits.

Summary of useful results for Bipolar Transistors

In summary, the expressions of the currents through the emitter, the base and the collector, for a narrow base, are:

$$i_E = A \left[en_p \frac{D_e}{L_e} \coth\left(\frac{b - z_p}{L_e} \right) + ep_n \frac{D_h}{L_h} \right] \left(e^{\frac{eV_B}{k_B T}} - 1 \right);$$

$$i_B = A \left[en_p \frac{D_e}{L_e} \frac{\cosh\left(\frac{b-z_p}{L_e} \right) - 1}{\sinh\left(\frac{b-z_p}{L_e} \right)} + ep_n \frac{D_h}{L_h} \right] \left(e^{\frac{eV_B}{k_B T}} - 1 \right);$$

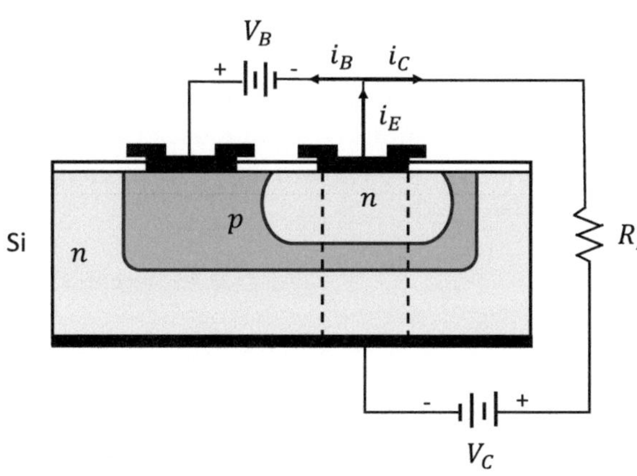

Fig. 11.3 If in the circuit of current i_C, there is a device such that the effective resistance in the circuit is R_L, the potential drop is $V_C = -i_C R_L$, and the power $P_L = i_C^2 R_L = i_C V_C$

$$i_C = en_p \frac{D_e}{L_e} \frac{A}{\sinh\left(\frac{b-z_p}{L_e}\right)} \left(e^{\frac{eV_B}{k_BT}} - 1\right). \tag{11.46}$$

The coefficients that characterize the transistors are:

$$\alpha = \frac{i_c}{i_E} \simeq \frac{2L_e^2}{2L_e^2 + (b - z_p)^2}; \tag{11.47}$$

$$\beta = \frac{\alpha}{1 - \alpha} \simeq 2\frac{L_e^2}{(b - z_p)^2}. \tag{11.48}$$

11.4 MOSFETs

Another system widely used in the current electronic devices is the MOSFET. This acronym results from MOS, metal-oxide-semiconductor, which we introduced in the previous chapter, and FET for *field-effect transistor*. In the MOSFET, the function of the potential V_B in the bipolar transistor is played by the gate potential V_G. This potential regulates the population of the attracted charges to the inversion layer, charges that will move in the MOSFET channel. Before referring to the MOSFET configuration, we will briefly review some properties of the MOS, especially those related to the inversion layer.

In the MOSFET of Fig. 11.4, the MOS structure is in the center of the device. The inversion layer of the MOS is precisely the channel of the MOSFET, through which most of the current moves from the source to the drain. By modifying the field produced by the gate potential V_G, and hence the channel thickness z_i, one can control the current I_D in the MOSFET. It was mentioned before that to simplify the analysis, it is convenient to consider the flat-band configuration as a reference. In this configuration the gate potential is written as

$$V_G = \phi_s + \Delta V_{ox}. \tag{11.49}$$

where ΔV_{ox} is the potential drop in the insulating oxide film, and ϕ_s is the surface potential. These quantities were introduced and defined in the last chapter. When the potential V_{G0}, required to reach the flat-band configuration is taken into account, the gate potential is written as

$$V_G = \phi_s + V_{G0} + \Delta V_{ox}. \tag{11.50}$$

We will, in general, refer to the gate potential V_G in (11.49), measured from the flat-band configuration. When we assume the existence of the MOSFET channel, we are also assuming that the surface potential ϕ_s is greater than the threshold potential V_{Gu}. When the gate potential V_G is greater than $V_{Gu} + \Delta V_{ox}$, the inversion layer

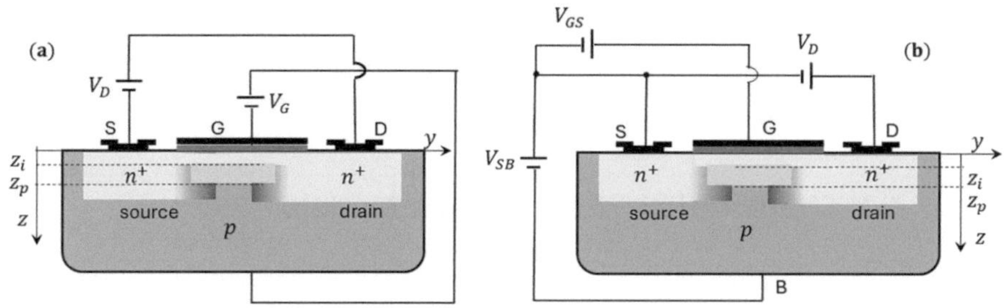

Fig. 11.4 An *npn* structure with a MOS device in the *p* layer in which the inversion layer with a quasi-surface charge distribution is created. In (**a**) the simplest configuration with independent gate potentials V_G and a "drift-diffusion" potential V_D that moves the charges through the channel constituted by the depleted layer and, mainly, by the inversion layer, from source to drain. In (**b**) a potential is added between the source and the base (or "bulk"). These potentials regulate the current through the channel in the same way as in the bipolar transistor

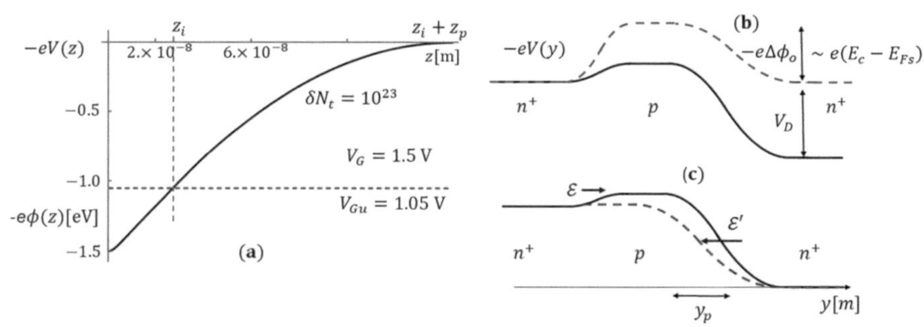

Fig. 11.5 Potentials in the *z* and *y* directions of a MOSFET. In (**a**) the potential as a function of *z*. In (**b**) the potential energy profile before (blue curve) and when the potential V_D is applied (black curve). In (**c**) the profile when a potential V_{SB} is also applied, as indicated in panel (*b*) of Fig. 11.4. This allows us to get a profile like the one shown with the red curve

widens. In Chap. 10 we found the width z_i as function of ϕ_s, and using the relation (11.49), one can express it as a function of V_G. The inversion layer width, also depends on the impurities concentration. But one can use the approximate form that was introduced in (10.82) of the previous chapter.

In a MOSFET, in addition to the potentials in the MOS, a potential V_D is applied between the source and the drain. If this potential has the polarization indicated in Fig. 11.4, the potential energy profile along the *y* axis will be as shown in the lower part of Fig. 11.2. For easy reference to the potential energies in the MOSFET, we copy in Fig. 11.5 the potential energies of the MOS, in the *z* direction, and those of the n^+pn^+ structure in the *y* direction. In panel (*b*) we see the profile before and when the potential V_D is applied (blue an black curves, respectively). In panel (*c*) the profile is modified (red curve) when a potential V_{SB} is applied, as shown in panel (*b*) of Fig. 11.4, between the source and the base (or "bulk"). Taking into account these potentials, we will determine in a simplified approach the behavior of the current

I_D as a function of the potential V_D applied between source and drain. An approach that, even though it is approximate, gives not only a good qualitative but also a good quantitative description.

11.5 I-V Characteristic. Approximate Models

The relationship between current and potentials defines what is called the I-V characteristic of a device. In the last chapter we have seen that a positive gate potential V_G attracts electrons towards the oxide-semiconductor interface. We saw also that when V_G is greater than the threshold potential, a population inversion process starts, and that increasing further the gate potential, the MOS tends to behave as capacitor with a two dimensional charge distribution at the oxide-semiconductor interface. If in addition a potential $V_D(y)$ is also established between the source and the drain, the electric charges in the MOSFET channel move along the y axis as drift and diffusion currents. In the calculations that presented here, there is no need of knowing the explicit function $V_D(y)$). The general assumption that $V_D(y)$ varies smoothly with y, from 0 at $y = 0$ to V_D at $y = L$, is sufficient.[2] If the charge density in the channel is $\rho_c(z, y)$, the current dI_D through a film of width dz is given by

$$dI_D = \mu_e \rho_c(z, y) W_x dz \frac{dV_D(y)}{dy} + \mu_e W_x dz \phi_T \frac{d\rho_c(z, y)}{dy}, \tag{11.51}$$

here $\phi_T = k_B T/e$, W_x is the MOSFET-channel width in the transverse direction x, and μ_e is the electrons' mobility in the channel. This mobility is a physical quantity that can change with y. At the surface it is considerably smaller than inside the semiconductor. Most of the charge moving in the channel, is charge of the inversion layer (with thickness z_i). A smaller amount is in the suppressed or depleted layer.

In the next sections we derive the currents: first, the current generally obtained in the two-dimensional approximation (charge-sheet model) and then two expressions for the current using the more accurate results obtained in Chap. 10 for the inversion layer potential. In all cases we use the procedure of adding to the local potential, determined by the gate, the source-drain potential $V_D(y)$. In the more accurate cases, we obtain the currents using, in one case, the standard capacitor charge-density and Gauss-law procedure, and in the second, starting from the Pao-Sah method.

11.5.1 Currents in the Two-Dimensional Approximation

It is common to assume that the thickness of the inversion layer tends to zero, with a two-dimensional density σ_I. To obtain this density some of the well-known relations for the capacitance, and Gauss law, are used.

[2] This is known as the gradual approximation of W. Shockley. See W. Shockley, A unipolar "Field-Effect" Transistor, Proceedings of the IRE, 1365 (1952).

By Gauss's Law, the charge contained in a volume limited by two planes (at z_1 and z_2) and a transverse surface A, can be written as

$$\frac{Q_{12}}{\epsilon_s} = \big(\mathscr{E}(z_2) - \mathscr{E}(z_1)\big)A. \qquad (11.52)$$

If $z_2 = \infty$ and $z_1 = 0$, the charge Q_{12} is the total charge Q_T in the semiconductor.[3] This negative charge is, by the neutrality condition, equal in magnitude to charge $Q_G = \sigma_G A$ at the gate-oxide interface, this means that $Q_T = -\sigma_G A$. Furthermore, if the charge in the depletion layer, between the inversion layer limit at z_i and ∞, is Q_d, the total charge can be written as

$$Q_T = \sigma_I A + Q_d = -\sigma_G A, \qquad (11.53)$$

Hence

$$\sigma_I = -\sigma_G - \frac{Q_d}{A}. \qquad (11.54)$$

To obtain this charge density we need σ_G and Q_d. For σ_G we use the relation with the capacitance C_{ox} and for Q_d Gauss's Law, that is:

$$\sigma_I = -C_{ox}\Delta V_{ox} - \epsilon_s\Big[\mathscr{E}(\infty) - \mathscr{E}(z_i)\Big] \qquad (11.55)$$

We mentioned before that when potentials are measured from the flat band configuration, the gate potential is $V_G = \phi_s + \Delta V_{ox}$. In Sect. 10.6, the electric field in the depleted layer of the MOS was obtained. For a type p semiconductor, the electric field at $z = z_i$, with $\phi(z = z_i) = V_{Gu}$ and $N_d \sim N_A$, is

$$\mathscr{E}(z_i) = \sqrt{\frac{2eN_A V_{Gu}}{\epsilon_s}} \qquad (11.56)$$

Replacing these quantities in (11.55), and taking into account that $\mathscr{E}(\infty) = 0$, we have

$$\sigma_I = -C_{ox}\big(V_G - \phi_s\big) + \sqrt{2e\epsilon_s N_A V_{Gu}}. \qquad (11.57)$$

To obtain this density, we did not take into account the potential $V_D(y)$ from source to drain. To include this potential into σ_I in a simple way, the common procedure is to add $V_D(y)$ to ϕ_s and V_{Gu}. With this change, the density of charges that move in the channel becomes

$$\sigma_I(y) = -C_{ox}\big(V_G - \phi_s - V_D(y)\big) + \sqrt{2e\epsilon_s N_A(V_{Gu} + V_D(y))}. \qquad (11.58)$$

---

[3] Assuming, of course, that the repelled holes move away of the semiconductor through the opposite contact towards the gate-oxide interface.

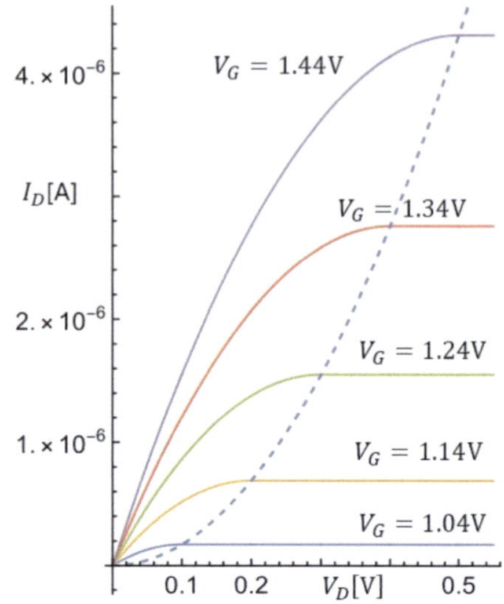

Fig. 11.6 The current I_D, from source to drain, as a function of the potential V_D for different gate potential values. For these graphs we considered $N_A = 10^{22}\mathrm{m}^{-3}$ and $\mu_e = 500\,\mathrm{cm}^2/\mathrm{Vs}$. This figure shows the almost linear growth of current when V_D is relatively small. In each case, the current reaches a maximum value known as the "pinch off" or saturation limit. Formula (11.81) describes the current up to the saturation point, indicated by the dashed curve

If we introduce this charge density into the drift-diffusion current, and neglect the diffusion current, we have

$$I_D dy = -\mu W_x \left[-C_{ox}\big(V_G - \phi_s - V_D(y)\big) + \sqrt{2e\epsilon_s N_A (V_{Gu} + V_D(y))}\right] dV_D(y) \tag{11.59}$$

Integrating over y, from $y = 0$ to $y = L$ and $V_D(y)$ from 0 to V_D, we obtain

$$I_D = \frac{\mu W_x C_{ox}}{L}\left[\big(V_G - \phi_s\big)V_D - \frac{V_D^2}{2} - \frac{2\sqrt{2e\epsilon_s N_A}}{3C_{ox}}\Big(\big(V_{Gu} + V_D\big)^{3/2} - V_{Gu}^{3/2}\Big)\right] \tag{11.60}$$

It is common to replace the surface potential ϕ_s by $2\phi_F$, which is of the order of the threshold potential V_{Gu}. In Fig. 11.6 the current I_D is plotted, as a function of the applied potential V_D for different fixed values of the gate potential V_G. This figure shows the almost linear growth of current when V_D is relatively small. In each curve, the current reaches a maximum value known as the "pinch off" or current saturation limit. The dashed curve indicates that limit. Equation (11.60) corresponds to an inverted parabola and describes the current up to the saturation point. It is known that beyond the saturation point, the current is approximately constant, an assumption used in the plots of Fig. 11.6. In the next sections, using more accurate derivations, we will see that in the saturation region the current is indeed almost constant. The saturation potential is

$$V_{Dsat} = V_G - V_{Gu} + V_s\left(1 - \sqrt{1 + 2\frac{V_{Gu}}{V_s}}\right) \quad \text{with} \quad V_s = \frac{e\epsilon_s N_A}{C_x^2}. \tag{11.61}$$

11.5.1.1 Linear Approximation, Turn-on Potential

In semiconductor device literature there is a tendency toward simple formulas that approximately describe the device behavior. For this reason it is common to separate the linear growth zone from the saturation zone. In the linear region an approximation is made to first order in V_D, if

$$V_D \ll V_G - V_{Gu} \tag{11.62}$$

the term $V^2/2$ is neglected, and in the second term of the (11.60) a Taylor expansion is performed such that a contribution of order V_D is preserved. In this linear approximation the current is

$$I_D = \frac{\mu_n W x C_{ox}}{L}\left(V_G - V_{Gu} - \frac{\sqrt{2e\epsilon_s N_A}}{C_{ox}}\sqrt{V_{Gu}}\right) V_D. \tag{11.63}$$

It is common to define the *turn-on potential* (or threshold voltage of the MOSFET) as

$$V_T = V_{Gu} + \frac{\sqrt{2e\epsilon_s \delta N_A}}{C_{ox}}\sqrt{V_{Gu}}, \tag{11.64}$$

with which the current remains as

$$I_D = \frac{\mu_n W_x C_{ox}}{L}\left(V_G - V_T\right) V_D. \tag{11.65}$$

These simplified formulas are widely used. With these formulas, and the I-V curves measured experimentally, some parameters such as the turn-on potential V_T are obtained. In the last section of this chapter we will derive two new formulas for the source-drain current based on the analytical solutions found in the last chapter for the electric potential in the inversion layer.

11.5.1.2 MOSFET Conductance and Transconductance

Two parameters that are used in the characterization and behavior of a MOSFET are conductance and transconductance. The conductance is defined as

$$g_D = \left.\frac{\partial I_D}{\partial V_D}\right|_{V_G=const.} \tag{11.66}$$

If the linear expression of the ohmic zone is considered, the conductance has the simple form

$$g_D = \frac{\mu_n W x C_{ox}}{L}\left(V_G - V_T\right) \tag{11.67}$$

Transconductance is defined as

$$g_m = \left.\frac{\partial I_D}{\partial V_G}\right|_{V_D=const}. \tag{11.68}$$

This function gives an idea of the device sensitivity. Explicitly it is given by

$$g_m = \frac{C_{ox} W_x \mu_n}{L} V_D. \tag{11.69}$$

11.5.2 Drift Currents Using MOS Inversion Layer Charge

We shall now derive a different formula for the source-drain current I_D. We will use the analytical solutions obtained in Chap. 10 for the electric potential in the inversion layer, and for the Shockley charge density in the Kingston-Neustadter model. To include the source-drain potential in the charge density σ_I, we will use the same method used in the last section, i.e. by direct addition of $V_D(y)$ to the electric potentials in σ_I. In the next section we obtain a more accurate formula, using the Pao-Sah method.

In this section we abandon the two-dimensional charge assumption and consider, more consistently, the Shockley density and the analytical solution of the Poisson equation, with some simplifications. We will see that the current obtained using the Pao-Sah method also exhibits saturation and describes the drift and depletion currents in the entire range of the drain voltage.[4] The formulas obtained in this section *likewise* describe the saturation of the source-drain current.

In (10.45), the Shockley's charge density in the inversion layer was written as

$$\rho(z) = -2eN_A e^{-\phi_F/\phi_T} \left[\sinh \frac{\phi_F - \phi(z)}{\phi_T} - \sinh \frac{\phi_F}{\phi_T} \right], \tag{11.70}$$

and the solution of the Poisson equation in the inversion layer was given as

$$\phi(z) = \phi_s - 2\phi_T \ln \left(\cosh \frac{\xi z}{2\phi_T} + \frac{\mathscr{E}_s}{\xi} \sinh \frac{\xi z}{2\phi_T} \right). \tag{11.71}$$

Substituting the potential $\phi(z)$ into the density $\rho(z)$, the charge density in the MOSFET channel, i.e., the charge per unit area in the inversion layer can be obtained from

$$\sigma_i = \int_o^{z_i} \rho(z) dz. \tag{11.72}$$

If we take into account only the leading order terms (assuming $\phi_s > \phi_F$), and call σ_I to the leading term of the density σ_i, we have

$$\sigma_I \simeq -2eN_A e^{-\phi_F/\phi_T} \left[\frac{\phi_T e^{(\phi_s - \phi_F)/\phi_T}}{\mathscr{E}_s + \xi \coth(\xi z_i / 2\phi_T)} + z_i \sinh \frac{\phi_F}{\phi_T} \right] \tag{11.73}$$

[4] See H.C. Pao and C.T. Sah, *Solid-State Electronics*, **9**, 927 (1966).

These quantities depend on the surface electric field $\mathscr{E}_s$, the field parameter ξ and the width z_i of the inversion layer. In Chap. 10, the electric field was obtained, in general, as

$$\mathscr{E}_s = \frac{\xi e^{(\phi_s - V_{Gu})/2\phi_T}}{\sinh(\xi z_i/2\phi_T)} - \xi \coth(\xi z_i/2\phi_T) \tag{11.74}$$

Replacing this field into σ_I, the charge density per unit area in the inversion layer becomes

$$\sigma_I \simeq -4eN_A e^{-\phi_F/\phi_T} \left[\frac{\phi_T e^{(\phi_s - 2\phi_F + V_{Gu})/2\phi_T} \sinh(\xi z_i/2\phi_T)}{\xi} + z_i \sinh \frac{\phi_F}{\phi_T} \right]. \tag{11.75}$$

To be specific, we will consider the ξ parameter and the inversion layer width z_i of the Kingston-Neustadter model, which were obtained exactly in (10.64) and (10.77). To simplify the calculations, we will consider the approximate expression for the inversion layer width

$$z_{ia} = \frac{\phi_s - V_{Gu}}{\xi}. \tag{11.76}$$

This formula describes correctly the linear part of the inversion layer width. For easy reference we will write again the field parameter ξ in the Kingston-Neustadter model. It was given in (10.61) as

$$\xi = e^{-\phi_F/2\phi_T} \left(\frac{8eN_A}{\epsilon_s} (\phi_F - \phi_T) \sinh \frac{\phi_F}{\phi_T} \right)^{1/2}. \tag{11.77}$$

Before evaluating the current in the inversion layer, we must introduce the source-drain potential $V_D(y)$ into σ_I. We have seen in the first part of this text that the applied potentials modify the Fermi level. Likewise, the carrier concentration in the presence of the drain-source potential $V_D(y)$ changes from[5]

$$n(z) \propto n_o e^{\phi(z)/\phi_T} \quad \text{to} \quad n(y,z) \propto n_o e^{\left(\phi(z) - V_D(y)\right)/\phi_T}. \tag{11.78}$$

This is a change compatible with the transformation found in Sect. 4.5, when a polarizing potential is applied. Thus, a simple procedure to include the source-drain potential is to replace ϕ_s and V_{Gu} by $\phi_s - V_D(y)$ and $V_{Gu} - V_D(y)$. Since the surface field and the inversion layer are functions of the difference $\phi_s - V_{Gu}$, they

---

[5] When no source-drain potential is applied, the vanishing of the drift-diffusion current in the z direction, expressed as

$$j_z = -e\mu_n n(z)\mathscr{E}_z - eD_n \frac{dn(z)}{dz} = eD_n n(z) \frac{d}{dz} \left(\frac{e\phi(z)}{k_B T} - \ln n(z) \right) = 0$$

implies that $n(z) \propto n_o e^{\phi(z)/\phi_T}$. See Problem 10.

do not change when the source-drain is applied. Only the first term of σ_I changes. As expected, the surface field, perpendicular to the oxide-semiconductor interface and the inversion-layer width are not affected by the source-drain potential. This does not imply absence of saturation. As discussed in the last section of Chap. 10, saturation occurs in several quantities that characterize the inversion layer, including the inversion layer current, which tend to limiting values.

After replacing ϕ_s and V_{Gu} by $\phi_s - V_D(y)$ and $V_{Gu} - V_D(y)$, respectively, the inversion layer drift current, evaluated from

$$I_{Di} = \frac{\mu W_x}{L} \int_0^{V_D} \sigma_I dV_D, \tag{11.79}$$

becomes

$$\begin{aligned} I_{Di} = \frac{\mu W_x e N_A}{L\xi} \Big[& \phi_T^2 \big(1 - e^{-V_D/\phi_T}\big)\big(e^{\phi_s/\phi_T} - e^{V_{Gu}/\phi_T}\big) \\ & + 2V_D\big(\phi_s - V_{Gu}\big) e^{-\phi_F/\phi_T} \sinh\frac{\phi_F}{\phi_T}\Big]. \end{aligned} \tag{11.80}$$

This is just part of the current in the MOSFET channel. The current moving in the depleted layer is the other part, and it is the same as in the previous section, i.e.

$$I_{Dd} = \frac{\mu W_x}{L}\left[-\frac{2\sqrt{2e\epsilon_s N_A}}{3}\Big(\big(V_{Gu} + V_D\big)^{3/2} - V_{Gu}^{3/2}\Big)\right] \tag{11.81}$$

In Fig. 11.7 the current I_{Di} in (11.80) is plotted in panel (*a*) as a function of the potential V_D, and in panel (*b*) as a function of ϕ_s. In both graphs the qualitative behavior of the current is correctly described.

11.6 Drift-Diffusion Current in a MOSFET Channel. The Pao-Sah Method

We will derive here the source-drain current based on the Pao-Sah method applied to the so-called junction-less MOSFET, where no extra doping is added to the source or drain contacts. Here we consider the single-gate JMOSFET, and in the Appendix I we will consider the current I_D in the double-gate JMOSFET.

When the biased MOS system is part of single or multiple-gate MOSFETs, the charge and potential distributions obtained for the inversion layer in Chap. 10 play an important role in the nature and magnitude of the drift currents moving in the MOSFET channel(s), parallel to the oxide-semiconductor interface(s). We will see here that the electron currents of the inversion layer play the most important role. A well-established formula for the calculation of the source-drain drift-diffusion current is Pao-Sah's formula [1]

$$I_D = \frac{eD_n W_x}{L} \int_0^{z_p} n(y,z)dz \int_0^L dy \frac{d\eta(y)}{dy}. \tag{11.82}$$

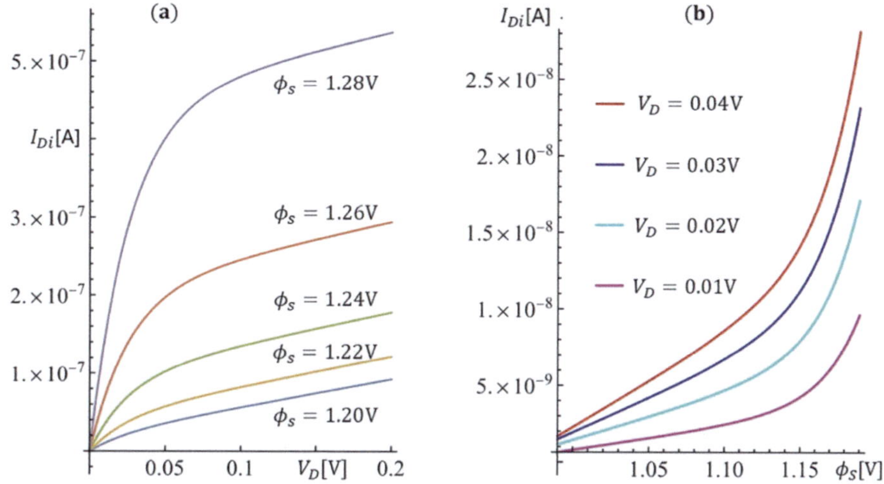

Fig. 11.7 The current I_{Di}, from source to drain, as a function of the potential V_D in panel (a) and as a function of the gate potential in panel (b). For these graphs $N_A = 10^{22}\text{m}^{-3}$ and $\mu_e = 500\,\text{cm}^2/\text{Vs}$. This figure shows that the formula (11.80) describes the linear and saturation behavior of the current

Here L is the channel length along the y direction, W_x is the channel width, D_n the diffusion coefficient, and $z_p = z_i + w_p$ the inversion plus depletion layers widths. When the drift and diffusion currents are taken into account, the distribution $n(y, z)$ becomes

$$n(y, z) = n_A e^{(\phi(z) - \eta(y) - \phi_{Fs})/\phi_T}, \tag{11.83}$$

where $\phi(z)$ is the electric potential due to the gate potential, summarized in (10.75), $\eta(y)$ is the quasi-Fermi level modified by the drain-source potential U_D along the channel, and $\phi_{Fs} = (E_c - E_{Fs})/e$ the equilibrium Fermi potential in the bulk, which we will assume to be measured from the conduction band edge. Usually dz is replaced by $-d\phi/\mathscr{E}(z, \phi)$, where $\mathscr{E}(z, \phi) = \sqrt{2G(z, \phi)}$. We do not need to make this change because we do have the electric potential as an explicit function of z. We keep the integration over z, but we will change the integration on y by the integration on η. We also split the integration in z in two parts, one for the inversion region and one for the depletion layer. In this representation,

$$I_D = \frac{eD_n w_x n_i}{L}\left[\int_0^{z_i} e^{\phi_i(z)/\phi_T} dz + \int_{z_i}^{z_i + w_p} e^{\phi_d(z)/\phi_T} dz\right]\int_0^{U_D} e^{(-\eta(y) - \phi_{Fs})/\phi_T} d\eta, \tag{11.84}$$

with the electric potentials as defined in (10.75), denoted here as ϕ_i and ϕ_d for the corresponding regions. The integrations are direct and we obtain

$$
\begin{aligned}
I_D = \frac{eD_n w_x n_i}{L}\Bigg[& e^{(\phi_s-\phi_{Fs})/\phi_T}\left.\frac{2\phi_T \sinh\frac{\xi z}{2\phi_T}}{\xi\left(\cosh\frac{\xi z}{2\phi_T}+\frac{\mathscr{E}_s}{\xi}\sinh\frac{\xi z}{2\phi_T}\right)}\right|_0^{z_i} \\
& + \frac{\sqrt{\pi\epsilon_s\phi_T}}{\sqrt{eN_d}}e^{-\phi_F/2\phi_T}\operatorname{Erf}\left[\frac{\sqrt{eN_d}w_p}{2\sqrt{\epsilon_s\phi_T}}\right]\Bigg] \times \left(1-e^{-U_D/\phi_T}\right),
\end{aligned} \tag{11.85}
$$

which reduces to

$$
\begin{aligned}
I_D = \frac{eD_n w_x n_i}{L}\Bigg[& \frac{2\phi_T}{\xi}e^{(\phi_s-\phi_{Fs})/\phi_T}e^{(V_{Gu}-\phi_s)/2\phi_T}\sinh\frac{\xi z_i}{2\phi_T} \\
& + \frac{\sqrt{\pi\epsilon_s\phi_T}}{\sqrt{eN_d}}e^{-\phi_F/2\phi_T}\operatorname{Erf}\left[\frac{\sqrt{eN_d}w_p}{2\sqrt{\epsilon_s\phi_T}}\right]\Bigg] \times \left(1-e^{-U_D/\phi_T}\right).
\end{aligned} \tag{11.86}
$$

Since $V_{Gu} = \phi_{Fs}$, this becomes

$$
\begin{aligned}
I_D = \frac{eD_n w_x n_i}{L}\Bigg[& \frac{2\phi_T}{\xi}e^{(\phi_s-V_{Gu})/2\phi_T}\sinh\frac{\xi z_i}{2\phi_T} \\
& + \frac{\sqrt{\pi\epsilon_s\phi_T}}{\sqrt{eN_d}}e^{-\phi_F/2\phi_T}\operatorname{Erf}\left[\frac{\sqrt{eN_d}w_p}{2\sqrt{\epsilon_s\phi_T}}\right]\Bigg]\left(1-e^{-U_D/\phi_T}\right) \equiv I_{Di}+I_{Dd}.
\end{aligned} \tag{11.87}
$$

As mentioned before, the concentration N_d in this equation should be replaced by N_{ds}, when $\phi_s < V_{Gu}$, and by N_{dl} when $\phi_s > V_{Gu}$. Let us now present some graphs to visualize the behavior of the currents when both the gate and the source-drain potentials are applied. For all graphs shown in the following, we assume that $W_x/L = 2.5$, $L = 0.4\mu$m, and the same mobility $\mu = 400\,\text{cm}^2/\text{Vs}$ in the depletion and inversion layers. These quantities can of course be changed. The impurity concentrations N_A will be indicated in the graphs.

We will plot the currents for different physical realizations or configurations. We start with the currents in Figs. 11.8 and 11.9a and b, plotted as functions of the surface potential ϕ_s, gate potential V_G and of the source-drain potential U_D. These graphs give a broader view of the behavior of the depletion and inversion currents. The depletion current I_{Dd} in Fig. 11.8a is several orders of magnitude smaller than the inversion layer current. As shown in Fig. 11.8 and in Fig. 11.9, for specific values of the source-drain voltage U_D, the behavior of I_{Dd} and of I_{Di}, as functions of the gate potential V_G (or the surface potential ϕ_s), are different. The depletion current I_{Dd} grows steadily up until the gate potential V_G becomes equal to the threshold potential V_{Gt} (or ϕ_s equals $\phi_{st} = V_{Gu}$). Recall that $V_{Gt} = V_{Gu} + \kappa t_{ox}\mathscr{E}_s$. Close to this point, the depletion current diminishes a bit, following the behavior of the ionized impurities concentration N_d shown in Fig. 10.17, and beyond the thresholds, the current I_{Dd} remains constant, reflecting the well known saturation effect. At variance with this behavior, the inversion layer current starts growing, as a function of V_G and of ϕ_s, at

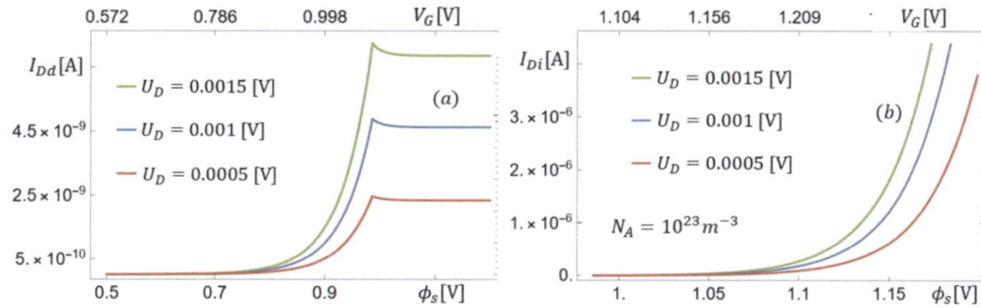

Fig. 11.8 Depletion and inversion layer currents as functions of the surface potential. The depletion current I_{Dd} in (**a**) for three values of the source-drain potential U_D, experiences a sharp change in its steady growth when the gate potential reaches the value of threshold potential. Above this threshold the depletion current becomes independent of the gate potential. The current I_{Di} in the inversion layer shown in (**b**) for the same values of U_D grows monotonously as a function of the gate and of the source-drain potentials. The inversion layer current is several orders of magnitude larger than the depletion layer current

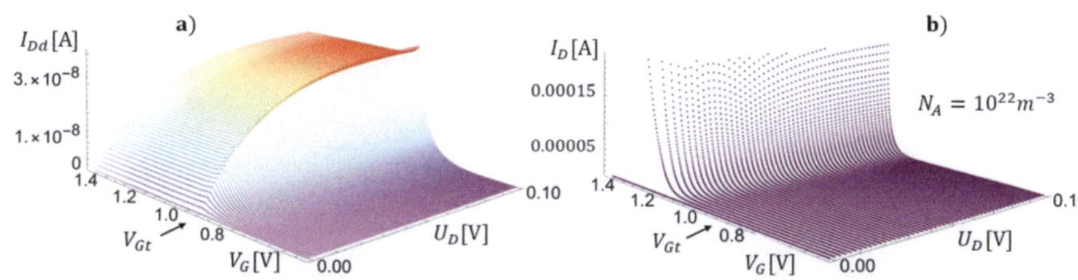

Fig. 11.9 Depletion and total drift current as functions of the gate and source-drain potential. For these plots we assume the same mobility $\mu = 400\,\mathrm{cm}^2/\mathrm{Vs}$ in the inversion and depletion layers. The depletion current I_{Dd} in (**a**) is several orders of magnitude smaller than the inversion current I_{Di}. For this reason the saturation current effect seen in (**a**), at the threshold potential $V_{Gt} = V_{Gu} + \kappa t_{ox}\mathscr{E}_s$, can not be seen in the total current $I_D = I_{Dd} + I_{Di}$

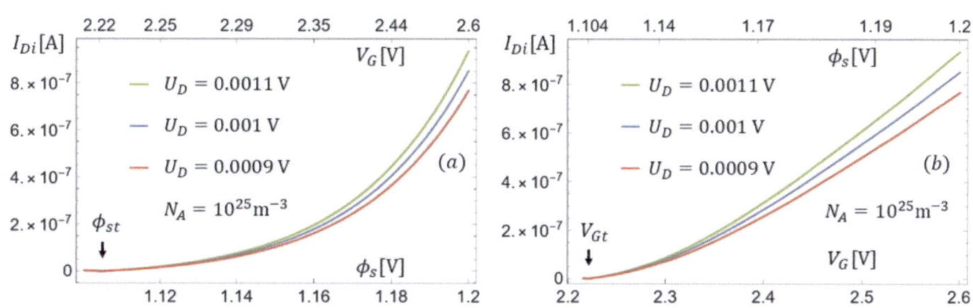

Fig. 11.10 The inversion-layer current I_{Di} as functions of the surface potential ϕ_s and the gate potential V_G. In (**a**) the current is plotted as a function of equidistant values of ϕ_s and in (**b**) as a function of equidistant values of V_G. In the last case it is evident that at larger values of V_G the current becomes a linear function of V_G

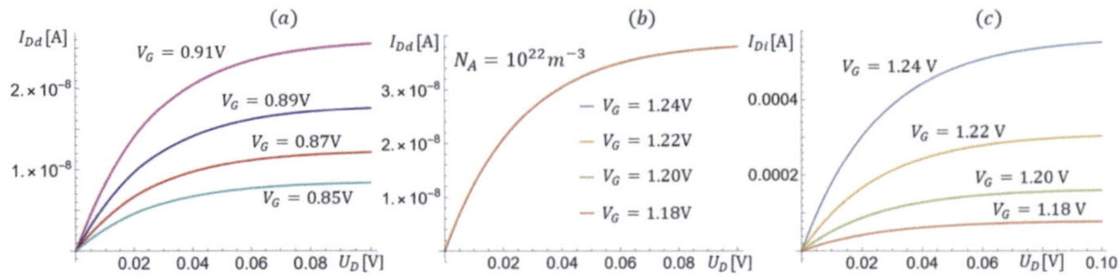

Fig. 11.11 Depletion and inversion currents as functions of the source-drain potential and in different gate potential domains. In (**a**) the gate potential is below the threshold, and in (**b**) the gate potential is above the threshold. In this domain I_{Dd} grows with U_D but is independent of V_G. In (**c**) the current I_{Di} also grows monotonously with U_D and shows also that it tends to a saturation at higher values of the source-drain potential

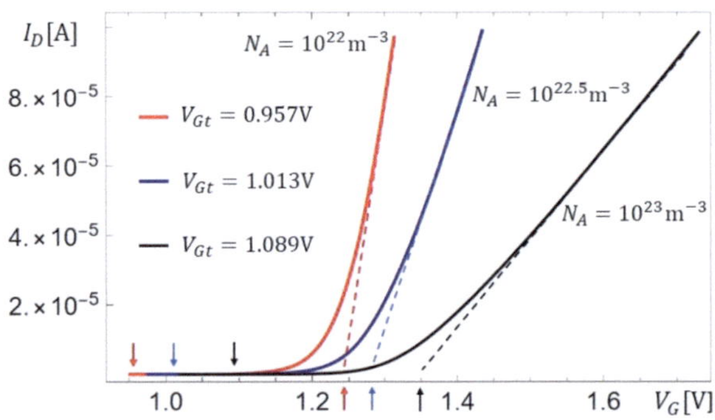

Fig. 11.12 The total source-drain current $I_D = I_{Dd} + I_{Di}$ in silicon MOSFETs as a function of the gate potential, for three different values of the impurity concentration. Here we assume that the mobility, in the depletion and inversion layers, is $\mu = 400\,\text{cm}^2/\text{Vs}$, U_D =0.001 V, and channel length $L = 0.4\mu$m. The gate potential thresholds V_{Gt} are given on the graph. These thresholds and the so-called operational threshold potentials, defined generally at the intersection of the abscissa with the extension of the linear current (see dashed lines), are indicated with arrows

the threshold potentials. It is worth noticing here that due to the non-linear relation between the gate and the surface potential ($V_G = \phi_s + \kappa t_{ox}\mathscr{E}_s$) and the non-linear relation between the surface electric field $\mathscr{E}_s$ and ϕ_s, the growth of the inversion current as a function of ϕ_s and the growth as a function of V_G are different. This difference can be seen in Fig. 11.10a and b, where the inversion current is plotted, in one case as a function of ϕ_s, i.e. as a function of equidistant values of ϕ_s, and in the other case, as a function of V_G. In Fig. 11.10a the current I_{Di} grows exponentially as a function of ϕ_s while in Fig. 11.10b it grows almost linearly as a function of V_G. By extrapolating the linear regime of I_{Di} one can derive the useful operational threshold potential.

As shown in Figs. 11.9a and b and 11.11a, b and c, both currents grow with U_D. Larger currents when the gate potential is larger, except for the depletion current above the threshold potential where it becomes insensitive to the gate potential.

We end up this section calculating the total drift currents I_D for different impurity concentrations and as functions of the gate potential V_G. As was shown before and can be seen in Fig. 11.12, the threshold potential for higher impurity concentration is also larger. In Fig. 11.12 we give the numerical values of the threshold potentials V_{Gt} in each case, and we indicate with arrows these potentials and the operational threshold potentials, generally determined at the intersection of the extended linear current with the abscissa. The operational threshold potentials are larger than the gate potential thresholds. In the particular case considered here, the difference is ~ 0.3 V. Given the appropriate parameters, these quantities can also be predicted.

It is important to notice that the charge and potential distributions considered here in the MOSFET's channel along the direction of the y axes, follow the widely used Pao-Sah model, and should be modified if actual devices' characteristics require. This may be independent of the potential function $\phi(z)$ obtained in chap. 10. It is also worth noting that the procedure introduced there for transforming Poisson's equation into a second-order nonlinear differential equation can also be applied to other types of charge distributions. In that case, the parameter ξ will change. The parameter ξ is an important parameter that carries subtle characteristics of the charge distribution model.

The number of different applications and calculations that can be performed using the analytical formulas obtained here is large. Among them for multi-gate devices.

11.7 Problems

1. Consider a silicon npn bipolar transistor. To have a reasonable base resistance, the base is doped with a concentration $N_A = 10^{16} cm^{-3}$. Calculate the concentration of type n impurities necessary in the emitter for an efficiency $\gamma = 0.995$. Suppose $b - z_p \sim L_e$ and $D_e \simeq D_h$
2. A pnp silicon transistor, at 300K, has an area A = 10^{-3} cm^2, b = 1μm and diffusion lengths $L_e = L_h = 10\mu$ m, $N_{Ae} = 2\times10^{18}$ cm^{-3}, $N_D = 10^{17}$, $N_{Ac} = 10^{16}$ cm^{-3}. Calculate I_c when $U_{EB} = 0.6V$. Suppose that in the base $D_e = 6$ cm^{2}s^{-1}
3. Consider the transistor in Problem 2. Calculate the gain and the transconductance defined as

$$g_m = \frac{\partial I_c}{\partial U_B} = \frac{e\beta I_B}{k_B T} \tag{11.88}$$

4. In an n type GaAs crystal at $T = 300$K the electron concentration varies as

$$n(z) = n_o e^{-z/L} \qquad z > 0. \tag{11.89}$$

with $n_o = 10^{16}$ cm^{-3} and $L = 1\mu$m. Calculate the diffusion current density at $z = 0$ if the diffusion constant is 220 cm^2/s.
5. If E_{Fi} is the intrinsic Fermi level of the semiconductor in the MOS and E_A is the Fermi level when the conductor is doped with acceptors, it is common to define

the energy Φ_F for the difference $E_{Fi} - E_A$. Show that the threshold potential to start the population inversion can be written as $eV_{Gu} = 2\Phi_F$.

6. Assume that the inversion occurs when $eV = 2\Phi_F$. ¿What is the width of the transition zone at $T = 300$K of a MOS where the semiconductor is type p silicon, with $N_A = 10^{16}\,\text{cm}^{-3}$ and $n_i = 1.5\,10^{10}\,\text{cm}^{-3}$.
7. Consider a transistor with the following properties at $T = 300$K, $N_D = 5\,10^{17}\,\text{cm}^{-3}$, $N_A = 10^{17}\,\text{cm}^{-3}$, $b = 10\mu$m and the diffusion coefficients $D_e = D_h = 20\,\text{cm}^2/\text{s}$. Calculate the current gain in the following cases: (a) when the lifetime of the minority electron and hole carriers is 10^{-6}s; (b) the lifetime is 10^{-8}s.
8. Be a transistor npn at $T = 300$K with $N_{DE} = 8\,10^{19}\,\text{cm}^{-3}$, $N_{AB} = 10^{17}\,\text{cm}^{-3}$, $N_{DC} = 10^{16}\,\text{cm}^{-3}$. The transistor operates in forward bias with $V_B = 0.5\,\text{eV}$, $V_C = 2\,\text{eV}$. Plot the minority carrier distributions at the emitter, base and collector ($p_{nE}(z), n_{pB}(z)$ and $p_{nC}(z)$). Assume the following parameters. $b = 10\mu$m, $D_{eB} = 30\,\text{cm}^2/\text{s}$, $\tau_B = 10^{-7}\text{s} = \tau_E = \tau_C$, $D_{hE} = D_{hC} = 10\,\text{cm}^2/\text{s}$.
9. Consider an npn transistor with $D_{eB} = 20\,\text{cm}^2/\text{s}$, $D_{eE} = 10\,\text{cm}^2/\text{s}$, $\tau_B = \tau_E = 10^{-7}s =$, $N_{DE} = 5\,10^{18}\,\text{cm}^{-3}$, $N_{AB} = 5\,10^{16}\,\text{cm}^{-3}$, $N_{DC} = 5\,10^{5}\,\text{cm}^{-3}$, $b = 1\mu$m, $n_i^2 = 2.25\,10^{20}$, $A = 10^{-2}\,\text{cm}^2$:

   (a) Calculate the forward bias current I_c with $V_B = 0.5$ V.
   (b) Determine the current when V_B is increased by 20%.

10. Show that the drift diffusion current along the z axis, perpendicular to the oxide-semiconductor interface, assuming that $j_p \ll j_n$, can be written as

$$j_{nz} = -e\mu_n n(z)\mathscr{E}_z - eD_n\frac{dn(z)}{dz} = eD_n n(z)\frac{d}{dz}\left(\frac{e\phi(z)}{k_BT} - \ln n(z)\right). \quad (11.90)$$

Show also that the vanishing of this current implies that $n(z) \propto n_o e^{\phi(z)/\phi_T}$.

Reference

1. H.C. Pao, C.T. Sah, Solid State Electron **9**, 927 (1966)

Multichannel Transport and Conductance Quantization

12

Abstract

In this chapter, we explore electronic transport in a series of relatively simple yet illustrative quantum systems. We begin with one-dimensional (1D) double-barrier structures and extend the analysis to polarized three-dimensional (3D) double barriers, where multiple propagating modes emerge. In these multichannel systems, we discuss channel mixing, Landauer conductance quantization, and a previously unexplained plateau that appears adjacent to the first resonance in I-V characteristics. We show that this feature naturally arises within the multichannel formalism under longitudinal polarization. We also briefly address transverse field-induced channel mixing and examine ballistic transport in resonant tunneling transistors and superlattices. A further example of channel coupling is presented through multichannel ballistic transmission in a 3D periodic array of δ scatterers. The chapter concludes with a theoretical review of the integer quantum Hall effect. Revisiting Büttiker's edge state model and quantum solutions, we introduce an alternative interpretation and a new formula for the Hall resistance that yields accurate predictions for the widths of the resistance plateaus, and a new theoretical perspective.

12.1 The Double Barrier

The double barrier potential that was originally introduced as an illustrative textbook example, [1,2] had an early experimental realization in Mead's triode tunnel-emission device, [3] and became afterwards the semiconductor device with the highest operational frequency. First evidences of resonant tunneling in double-barrier semiconductor structures were published by Chang et al. [4,5] in 1974, and some years later Sollner et al. [6,7] reported the first double-barrier resonant-tunneling (DBRT) oscillators with operation frequencies of the order of THz, well above the GHz frequencies estimated by Ricco et al. [8] and Luryi [9]. Since then, an overwhelming number of theoretical, experimental, and applied articles have been pub-

P. Pereyra, *Contemporary Theory of Semiconductors and Devices*,
Graduate Texts in Physics, https://doi.org/10.1007/978-3-032-00296-9_12

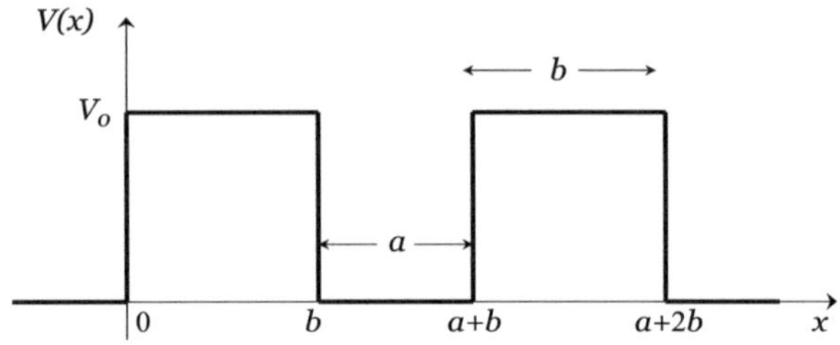

Fig. 12.1 Parameters of the double rectangular barrier

lished trying to unravel the basic properties of this system like the transmission and relaxation times involved in the tunneling process and the enhancing of the tunneling probability mechanism [10–23]. Issues like the emitter and quantum well dimensions, [24–26] the parameters' effects in the device performance and peak to valley ratio, [16,27–30], the shot noise, [31–34] and many other electric and optical properties have been extensively studied.

The one-dimensional double barrier, like the one seen in Fig. 12.1, is a simple system, for which the basic features are easily determined and the possible applications in optoelectronics and electronic transport envisioned. The main purpose is to solve the Schrödinger equation and to calculate the double barrier (DB) resonant transmission and conductance. We will be able to determine, in general, the reflected and transmitted coefficients as functions of the incoming particles' energy. In this system, the coefficient has a resonant behavior, a feature that has been exploited in the design and application of electronic devices. We again use the transfer matrix method to obtain information about the resonant energy levels.

In Chap. 5 we introduced the state vectors as an alternative representation of the wave function. We have seen also that the state vectors at external points on the left and right of a rectangular potential barrier are related by the transfer matrix M_b of (5.122) which, in (5.123), we denote in the form

$$M_b = \begin{pmatrix} \alpha_b & \beta_b \\ \beta_b^* & \alpha_b^* \end{pmatrix}. \tag{12.1}$$

This means that

$$\phi(b) = \begin{pmatrix} a_3 e^{ikb} \\ b_3 e^{-ikb} \end{pmatrix} = \begin{pmatrix} \alpha_b & \beta_b \\ \beta_b^* & \alpha_b^* \end{pmatrix} \begin{pmatrix} a_1 \\ b_1 \end{pmatrix} = \begin{pmatrix} \alpha_b & \beta_b \\ \beta_b^* & \alpha_b^* \end{pmatrix} \phi(0), \tag{12.2}$$

Similarly, for the second barrier we have

$$\phi(a+2b) = \begin{pmatrix} a_5 e^{ik(a+2b)} \\ b_5 e^{-ik(a+2b)} \end{pmatrix} = \begin{pmatrix} \alpha_b & \beta_b \\ \beta_b^* & \alpha_b^* \end{pmatrix} \begin{pmatrix} a_3 e^{ik(a+b)} \\ b_3 e^{-ik(a+b)} \end{pmatrix};$$

or just

$$\phi(a+2b) = \begin{pmatrix} \alpha_b & \beta_b \\ \beta_b^* & \alpha_b^* \end{pmatrix} \phi(a+b). \tag{12.3}$$

The explicit form of functions α_b and β_b was given in (5.122) and in the expression preceding (5.144), when E is less than and greater than V_o, respectively. To obtain the transfer matrix of the double barrier we still need to relate $\phi(a+b)$ with $\phi(b)$. Since both vectors are defined at points in the same region of constant potential, we have

$$\phi(a+b)=\begin{pmatrix} a_3e^{ik(a+b)} \\ b_3e^{-ik(a+b)} \end{pmatrix}=\begin{pmatrix} e^{ika} & 0 \\ 0 & e^{-ika} \end{pmatrix}\begin{pmatrix} a_3e^{ikb} \\ b_3e^{-ikb} \end{pmatrix}=\begin{pmatrix} e^{ika} & 0 \\ 0 & e^{-ika} \end{pmatrix}\phi(b). \tag{12.4}$$

Therefore

$$\begin{pmatrix} a_5e^{ik(2b+a)} \\ b_5e^{-ik(2b+a)} \end{pmatrix} = \begin{pmatrix} \alpha_b & \beta_b \\ \beta_b^* & \alpha_b^* \end{pmatrix}\begin{pmatrix} e^{ika} & 0 \\ 0 & e^{-ika} \end{pmatrix}\begin{pmatrix} \alpha_b & \beta_b \\ \beta_b^* & \alpha_b^* \end{pmatrix}\begin{pmatrix} a_1 \\ b_1 \end{pmatrix}. \tag{12.5}$$

By multiplying the transfer matrices, we obtain another transfer matrix; the matrix of the double barrier. It can be verified that this matrix will have the same symmetries as the transfer matrix of a single barrier, that is, a matrix with the following structure

$$M_D = \begin{pmatrix} \alpha_D & \beta_D \\ \beta_D^* & \alpha_D^* \end{pmatrix}. \tag{12.6}$$

Here

$$\alpha_D = \alpha_b^2 e^{ika} + |\beta_b|^2 e^{-ika} \qquad \text{and} \qquad \beta_D = \alpha_b\beta_b e^{ika} + \alpha_b^*\beta_b e^{-ika}. \tag{12.7}$$

We have seen earlier that the transmission and reflection scattering amplitudes are related with the elements of the transfer matrix. Using the relations in equation (5.107), we have the transmission and reflection amplitudes, t_D and r_D, of the double barrier given by

$$t_D = \frac{1}{\alpha_D^*} \qquad \text{and} \qquad r_D = -\frac{\beta_D^*}{\alpha_D^*}. \tag{12.8}$$

Hence, the transmission coefficient of the double barrier is given by the expression

$$T_D = \frac{1}{|\alpha_b|^4 + |\beta_b|^4 + 2\,|\beta_b|^2\,\Re\mathrm{e}\big[\alpha_b^2 e^{i2ka}\big]} \equiv T_2. \tag{12.9}$$

In Fig. 12.2 we plot this coefficient (red curve) alongside with the coefficient of a single barrier (black curve). The coefficient $T_D \equiv T_2$, takes values smaller than the single barrier coefficient T_1, except at and near the resonances. At the resonant energies, E_μ^*, the transmission coefficient reaches the maximum value 1. To understand

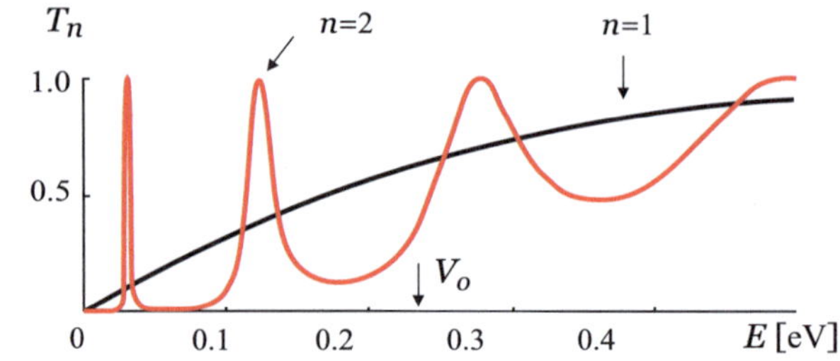

Fig. 12.2 The transmission coefficient of the double barrier (red curve) and that of a single barrier (black curve) when the parameters of the rectangular barrier are $a = 2$ nm, $b = 10$ nm and $V_o = 0.23$ eV

the origin of this resonant behavior, as a function of the energy, it is convenient to rewrite the transmission coefficient in the form

$$T_D = \frac{1}{1 + 2\,|\beta_b|^2 \left(|\alpha_b|^2 + \Re\text{e}\left[\,\alpha_b^2 e^{i2ka}\right]\right)}, \tag{12.10}$$

where $\Re$e stands for the real part. It is evident here that the transmission coefficient will oscillate with values less than or equal to 1. It will be equal to 1 when

$$|\alpha_b|^2 + \Re\text{e}\left[\alpha_b^2 e^{i2ka}\right] = 0. \tag{12.11}$$

The roots of this equation, are the resonant energies E_μ^* with $\mu = 1, 2, 3, ...$ The resonant behavior is a consequence of the coherent superposition of the right and left moving wave functions. When the energy of the incident particles is equal to a resonant energy, the transmission is complete, T_D is equal to 1. It is worth mentioning that if we make the barriers width large or ∞, we have a quantum well limited by the two potential barriers. The resonant energies E_μ^* are *close to* the energy eigenvalues E_μ of the potential well between the barriers, that is, $E_\mu^* \simeq E_\mu$. Unlike the true eigenfunctions, which correspond to stable states, the resonant functions correspond to meta-stable states. It is interesting to note that the resonant behavior occurs even for energies greater than the height of the barriers, which in our example is $V_o = 0.23$ eV, indicated by an arrow on the Fig. 12.2. The Schrödinger equation admits solutions for any value of the incident energy, and the probability that the particle gets transmitted is different from zero regardless of the incident energy.

12.2 Biased Multichannel DBs and Negative Resistance

Searching for higher frequency oscillators with larger peak-to-valley ratios, new devices like the hot electron injectors and the lateral confined structures have been produced [25,38–40]. The advances in microfabrication technology made feasible the addition of lateral constrictions on two-dimensional (2D) systems seeking for quasi one-dimensional (1D) wires with electronic transport properties determined by a finite number of propagating modes, and size scales reaching the electron's de Broglie wave length. Reed et al. [25] observed a discrete spectrum and a series of peaks in the current voltage characteristics of InGaAs double-barrier structure with lateral dimension of 100 nm, which superimpose on the negative-differential-resistance peak as the temperature increases. Double barrier structures, grown in columns with diameters of 100–200 nm, were fabricated by the so-called bottom-up technique. Different groups, using different methods, have observed peaks in the current voltage characteristics [41–48].

An alternative interpretation of the negative differential resistance (NDR) based on the momentum conservation picture was introduced by Luryi [9]. This picture was then extended by Liu et al. [21] and by Bagwell et al. [24] to explain the emission from 3D, 2D and 1D sources into quantum wells, quantum wires and quantum dots. The question of whether the quantum-states coherence or the sequential (two-step) tunneling mechanism is behind the resonant tunneling, has been of great concern [4,49–51]. A large number of structures were designed to study this problem; transmission probabilities of symmetric and asymmetric single and double barrier structures were measured, and resonant lifetimes estimated in order to elucidate the quantum-states coherence role [8,52–54].

Recently the resonant states, transmission coefficients and tunneling times of superlattices and DB structures, with arbitrary potential profile, and arbitrary number of propagating modes, were analytically studied and tunneling times ranging from ps to a few fs, implying frequencies of the order of THz, were obtained [55]. These results made clear the relevance of the quantum-states coherence to the resonant characteristics of transmission probabilities in open and bounded systems, as well as to the transmission times [56].

The biased double barrier is a simple system and, as mentioned above, the resonant tunneling and negative resistance behavior have been amply and extensively studied. The DB was for some decades the principal research field of semiconductor tunneling heterostructures, and little or nothing had to be added after 50 years of research and applications. However, it turns out that a shoulder superimposed in the negative resistance domain, appeared in numerous experimental reports, [7,10,14,16,30,36,37,49,57–61] and remained as an unsolved problem with questionable partial explanations. That feature, that can be seen in the inset of Fig. 12.3 and in the I-V characteristics shown in Fig. 12.4, had different explanations. In [7] the shoulder is considered a consequence of a self-detection; in [37] it is attributed to serial resistance, and, in [35], to oscillations due to the bias field. The accounting of the shoulder in the conductance and I-V characteristics of biased double barriers is a good example to test the theoretical approach for the calculation of transport properties in DB structures.

We will refer in the next subsections to the biased double barrier structure in a multichannel approach in the presence of longitudinal and transverse fields. We will first introduce the most general approach where longitudinal and transverse channel mixing bias are present. We will then be more specific and consider the simplest case of longitudinal bias only, which, as will be seen below, is enough to account for the shoulder experimentally observed in the negative resistance domain.

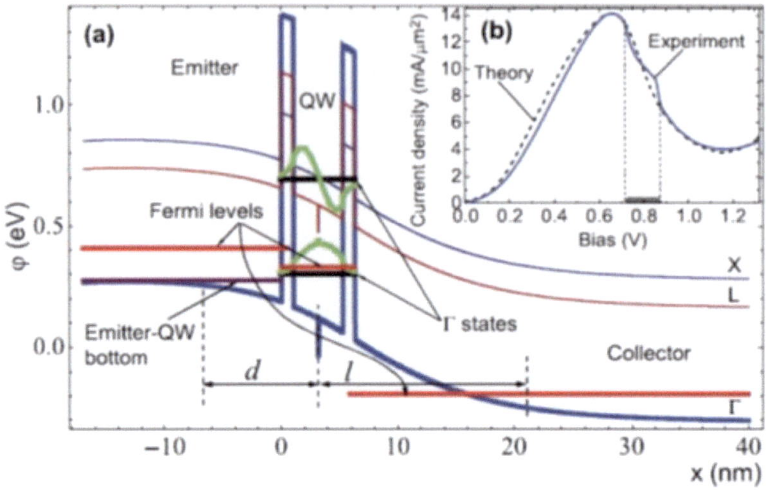

Fig. 12.3 Experimental set up of a biased double barrier. We show here a sketch of a biased double barrier potential with an I–V characteristic in the inset, reported by Feiginov et al. (2011). Reproduced from [35], with the permission of AIP Publishing

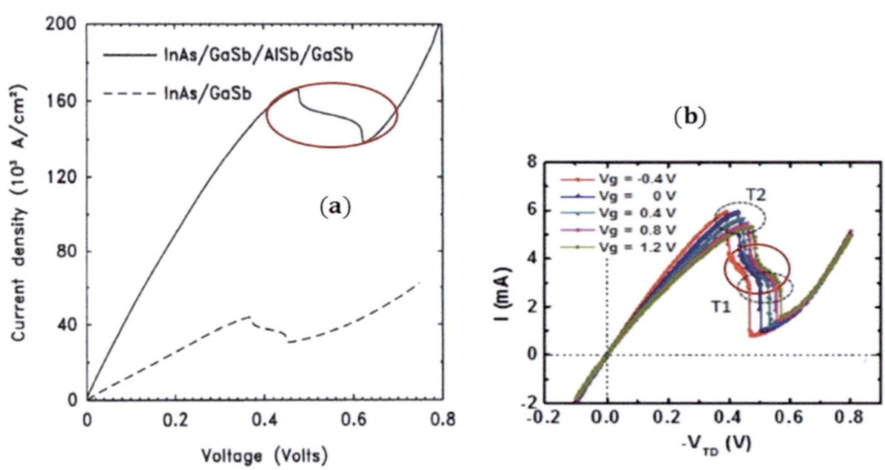

Fig. 12.4 Experimental I-V characteristics of biased double barrier. We show here examples of experimental curves where a shoulder at the right hand side of the resonance (see red circles) are observed . In **a** figure reported by Collins [36] (reproduced with permission of Springer Verlag), and in **b** by Zeng et al. [37] (reproduced with the permission of AIP Publishing)

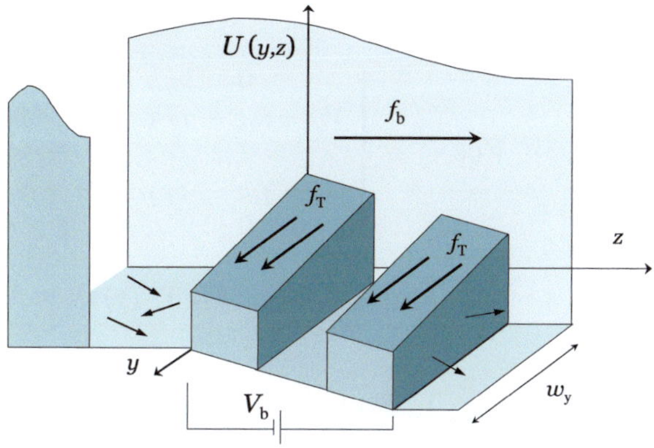

Fig. 12.5 Biased double barrier with height V_o, barrier width b, well width a and transverse width w_y. Besides the longitudinal bias field $\mathscr{E}_b$, a channel mixing transverse field $\mathscr{E}_T$ is also applied. Reproduced from [55], with the permission from AIP Publisher

12.2.1 Transverse and Longitudinal Bias. General Approach

In real systems, the transverse dimensions imply several propagating modes. It is known that for a fixed transverse width w_y of a 2D wave guide, and a given incoming electron energy E, the number of (spin-independent) propagating modes (open channels) is of the order of $w_y\sqrt{2m^*E}/\pi\hbar$. For lateral dimensions of the order of 40 nm and energies below 0.2 eV, the number of propagating modes is $\sim$5. In the presence of a transverse electric field, the propagating modes are mixed with significant consequences on the transport properties. The resonant tunneling is highly modified by the coherent interference of the propagating modes. Although the transverse wave-guide modes were mentioned to explain the peak splitting, the Schrödinger equations were taken, in general, in the 1D approximation. The analysis of electronic transport of transverse and longitudinal modes in semiconductor wave guides has been almost absent. A theoretical calculation of electron transport through a laterally confined double barrier resonant tunneling (DBRT) systems, connected to emitter and collector reservoirs via two narrow constrictions, is found in [62], where multimode transmission coefficients trough 3D-0D-3D and 2D-0D-2D structures are reported. In the presence of many open channels, the Landauer conductance depends also on the inelastic transmission amplitudes t_{ij}. The strength of the channel mixing effect can be tuned with both the longitudinal and the transverse electric field. In this way important quantities like the tunneling time and the peak-to-valley ratios can, in principle, be controlled. Though the effect on the tunneling time might be of great interest, we will concentrate here in the effects on the Landauer conductance.

To simplify the analysis we shall assume a two-dimensional system, as shown in Fig. 12.5, with transverse width w_y. The extension to 3D is straightforward. Inside the wave guide we have a biased double barrier structure with barriers height V_o, barrier-width b and well-width a. We will assume also that, besides the bias field $\mathscr{E}_b = -f_b/e$, a transverse field $\mathscr{E}_T = -f_T/e$ is applied. The 2D potential $U(y, z)$ is defined as

$$U(y, z) = V_C(y) + V(y, z) \tag{12.12}$$

where

$$V(y,z) = \begin{cases} 0 & z \leq 0 \\ V_o - f_b z - f_T y & 0 < z < b \\ -f_b z & a \leq z \leq a+b \\ U_o - f_b z - f_T y & a+b < z < a+2b \\ 0 & z \geq a+2b = L_z \end{cases}, \tag{12.13}$$

for $0 < y < w_y$. Otherwise, $U(y,z) = V_C(y) = \infty$. To solve the Schrödinger equation

$$-\frac{\hbar^2}{2m}\nabla^2\Psi(\mathbf{r}) + \big(V_C(y) + V(y,z)\big)\Psi(\mathbf{r}) = E\Psi(\mathbf{r}), \tag{12.14}$$

we use the multichannel approach outlined in [55]. After performing the variables separation, and neglecting the evanescent channels contribution, one is left with the system of coupled equations

$$\frac{d^2}{dz^2}\varphi_i(z) + \big(k^2 - k_{Ti}^2\big)\varphi_i(z) = \sum_{j=1}^{N} K_{ij}(z)\varphi_j(z) \tag{12.15}$$

where N is the number of open channels and

$$K_{ij}(z) = \frac{2m^*}{\hbar^2}\int_0^{w_y} dy\phi_i^*(y)V(y,z)\phi_j(y) \tag{12.16}$$

the coupling channels matrix. The functions $\phi_i(y)$ are the eigenfunctions of the infinite square well in the leads, with eigenvalues $k_{Tj} = \pi j/w_y$. As explained in Sect. 6.2, the channel indices i, j, characterize the propagating modes and $k_{Tj} = \pi j/w_y$ the transverse momenta. Here w_y is the transverse width, or the diameter when the transverse section is circular. For the DB potential $U(y,z)$, defined above, the coupling matrix is

$$K_{ij}(z) = \frac{2m^*}{\hbar^2}\begin{cases} -f_b z - \dfrac{f_T w_y}{2} + V_o & \text{for} \quad i = j \\ \dfrac{f_T w_y}{\pi^2}\left(\dfrac{1-\cosh(i+j)\pi}{(i+j)^2} - \dfrac{1-\cosh(i-j)\pi}{(i-j)^2}\right) & \text{for} \quad i \neq j \end{cases}, \tag{12.17}$$

Notice that in the absence of a transverse field, no channel mixing exists, and one is left with an uncoupled-channels description. But in the presence of a transverse field, we have a system of N coupled equations. To deal with these equations, one can use the transfer matrix method and the notation of [63].

We will start in the next section assuming that the transverse field is absent and no channel mixing exists, and we will then briefly consider the channel mixing case.

12.2.2 Longitudinally Biased Double Barrier

When only a longitudinal bias potential is present, we have a simple system and easier calculation of transmission coefficients of the biased DB. For a system whose Shrödinger equation is

$$-\frac{\hbar^2}{2m}\nabla^2\Psi(\mathbf{r}) + \big(V_C(x,y) + V_f(z)\big)\,\Psi(\mathbf{r}) = E\Psi(\mathbf{r}). \tag{12.18}$$

where

$$V_f(z) = \begin{cases} -f_b z & -l_i < z \le 0 \\ V_o - f_b z & 0 < z < b \\ -f_b z & b \le z \le a+b \\ V_o - f_b z & a+b < z < a+2b \\ -f_b z & a+2b < z < a+2b+l_r \end{cases}. \tag{12.19}$$

with f_b the electric force applied between $z = -l_i$ and $z = a + 2b + l_r$, and a and b the well and barrier widths. $V_C(x, y)$ is a confining hard wall potential, as in (12.13). After performing the variables separation, and neglecting the evanescent channels contribution, one is left with the system of coupled equations

$$\frac{d^2}{dz^2}\varphi_i(z) + \big(k^2 - k_{Ti}^2\big)\varphi_i(z) = \sum_{j=1}^{N} K_{ij}(z)\varphi_j(z). \tag{12.20}$$

This system of coupled equations is similar to (12.15), with slight changes. Here

$$K_{ij}(z) = \frac{2m^*}{\hbar^2}\int_0^{w_y w_x} dxdy\phi_i^*(x,y)V(y,z)\phi_j(x,y) \tag{12.21}$$

is the coupling channels matrix. The functions $\phi_i(x, y)$ are the eigenfunctions of the infinite 2D square well in the leads. In this case, the coupling matrix becomes

$$K_{ij}(z) = \frac{2m^*}{\hbar^2}(-f_b z + V_o)\delta_{i,j} \tag{12.22}$$

Notice that in the absence of a transverse field, no channel mixing exists, and one is left with soluble Airy equations

$$\frac{d^2}{dz^2}\varphi_i(z) + \big(k^2 - k_{Ti}^2\big)\varphi_i(z) = \frac{2m^*}{\hbar^2}\big(-f_b z + V_o\big)\varphi_i(z) \qquad i = 1, 2, ..., N. \tag{12.23}$$

As was explained in Sect. 5.6.1, given the solutions and their derivatives, it is possible and some times convenient to define transfer matrices $W(z_2, z_1)$, that connects vector

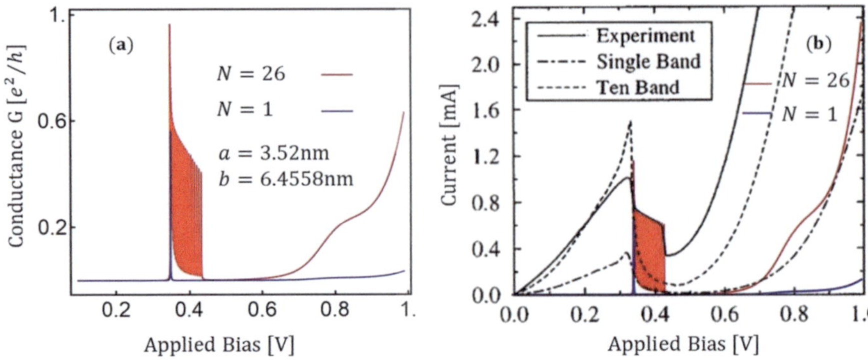

Fig. 12.6 Multichannel conductance and $I - V$ characteristic of a biased double barrier. In the right, the experimental and theoretical I–V characteristics of the DB reported in [60], reproduced with the permission from AIP Publisher. In **a** and **b**, the conductance and current calculated for a single propagating mode (blue) and for the N propagating modes. In this case, to reproduce faithfully the shoulder observed in several experiments $N = 26$. Figure reproduced from [64], with permission of John Wiley & Sons Publisher

states defined with the wave functions and their derivatives. For the DB, the N channel transfer matrix is a $2N \times 2N$ block-diagonal matrix, with block matrices on the diagonal of the form

$$W(z_2, z_1)=\begin{pmatrix} \mathrm{Ai}(z_2, k_{Ti}) & \mathrm{Bi}(z_2, k_{Ti}) \\ \mathrm{Ai}'(z_2, k_{Ti}) & \mathrm{Bi}'(z_2, k_{Ti}) \end{pmatrix} \begin{pmatrix} \mathrm{Ai}(z_1, k_{Ti}) & \mathrm{Bi}(z_1, k_{Ti}) \\ \mathrm{Ai}'(z_1, k_{Ti}) & \mathrm{Bi}'(z_1, k_{Ti}) \end{pmatrix}^{-1} \qquad (12.24)$$

where $\mathrm{Ai}(z_2, k_{Ti})$ and $\mathrm{Bi}(z_2, k_{Ti})$ are the Airy functions in channel i. After the similarity transformation is applied, see (5.166), it is easy to calculate the Landauer conductance

$$G = \frac{e^2}{h}\mathrm{Tr}tt^\dagger \qquad (12.25)$$

as well as the current

$$j = n_e \mu_e f_b L \mathrm{Tr}tt^\dagger \qquad (12.26)$$

as a function of the incoming energy E. In the last equation n_e is the charge carriers concentration, e is the electron's charge, μ_e the mobility, h the Planck's constant, and $L = a + 2b + l_i + l_r$ the distance between the contact points. In Fig. 12.6 a and b we show the calculated conductance and current for the parameters mentioned in [60] (with a and b slightly modified to fit multiples of the lattice constant). Since the number of propagating modes (or open channels), is proportional to $2w\sqrt{2m^*E}/\pi\hbar$, where w is the transverse width, it is clear the relevance of the transverse dimension in the splitting of the propagating modes, limited by the energy E. The calculation shows not only the origin of the shoulder in the negative resistance domain of a biased DB observed in the experimental reports but also the ability of the multichannel transfer matrix method to describe faithfully and to explain features that previous theoretical approaches did not fully captured. It is evident also from Fig. 12.6a that the propagating modes with higher longitudinal energy transmit at higher bias, but with lower transmission probability, a well-known effect that led to envision new devices where the symmetry of the DB, broken by the bias field, could be restored "not by applying an electric field, but by injecting minority carriers or ballistic electrons" [65].

12.3 Ballistic Electrons Through heterostructures and SLs

In the 80s of the last century, when the size of semiconductor devices reached dimensions of the mean free path of charge carriers, it became natural to think that the carriers' current is not only less diffusive, but ballistic or quasi ballistic [67,68]. Thus, new structures and devices were proposed, in which ballistic electrons pass resonantly through the quantized energy levels of symmetric double barriers or the minibands of SLs located in the base [69,70] or in the emitter region [71]. The main objectives were to increase the transmission probability and to obtain high peak-to-valley ratios, by restoring the symmetry of DBs or avoiding the loss of periodicity and the ensuing phase coherence in SLs. Among the numerous resonant tunneling transistors with SLs, we will revisit the three-terminal device, with anti-reflecting coating, studied experimentally and theoretically by Pacher et al. [66]

Figure 12.7 shows the schematic band structure of the three-terminal device with the superlattice studied by Pacher et al. The width of a $GaAs$ valley is $a = 6.5\,\mathrm{nm}$, the width of the $Al_{0.3}Ga_{0.7}As$ barrier is $b = 2.5\,\mathrm{nm}$, and the number of unit cells is $n = 6$. The width b' of the first and last barriers, which act as anti-reflection coating, is $b' = b/2$. Pacher et al calculated and measured the transmission coefficients shown in Fig. 12.8. For the calculation of the transmission coefficient of the uncoated and the coated superlattice, in the 1D one-propagating mode limit, one has first to determine the transfer matrix given by

$$M_c = M_{b'} M_n M_{b'}, \tag{12.27}$$

where M_n is the SL transfer matrix

$$M_n = \begin{pmatrix} U_n - \alpha^* U_{n-1} & \beta U_{n-1} \\ \beta^* U_{n-1} & U_n - \alpha U_{n-1} \end{pmatrix}, \tag{12.28}$$

defined in (6.25) and $M_{b'}$ the transfer matrix of a coating layer, defined as

$$M_{b'} = \begin{pmatrix} \alpha_{b'} & \beta_{b'} \\ \beta^*_{b'} & \alpha^*_{b'} \end{pmatrix}. \tag{12.29}$$

In these matrices, we have

$$\alpha = e^{ika}\left(\cosh qb + i\frac{k^2-q^2}{2qk}\sinh qb\right) \qquad \beta = -i\frac{k^2+q^2}{2qk}\sinh qb$$
$$\alpha_{b'} = e^{ika}\left(\cosh qb' + i\frac{k^2-q^2}{2qk}\sinh qb'\right) \qquad \beta_{b'} = -i\frac{k^2+q^2}{2qk}\sinh qb' \tag{12.30}$$

with $k = \sqrt{2m_w^* E/\hbar^2}$ and $q = \sqrt{2m_b^*(V_o - E)/\hbar^2}$ the wave numbers in the wells and barriers, and m_w^* and m_b^* the effective masses and $V_o = 0.23\,\text{eV}$. As mentioned before, the Chebyshev polynomials U_n are evaluated at the real part of α. After multiplying and simplifying, one can obtain the relevant matrix element to determine the transmission coefficient, i.e. the element $\alpha_c = M_{c11}$ that can be written in compact form as

$$\alpha_c = \left(\alpha_b'^2 - \beta_b'^2\right)U_n - \left(\alpha_b'^2\alpha^* - \beta_b'^2\alpha + 2\alpha_b'\beta_b'\beta\right)U_{n-1} = 1 + h_{b'}U_n + f_{b'}U_{n-1}. \tag{12.31}$$

Pacher et al. by assuming that $f_{b'}U_{n-1} = 0$, reported a simplified transmission coefficient, denoted as T_{SL}^{ARC} in the equation (1) of [66], plotted it and compared with the transmission coefficient for a SL without coating layers (see graphs in the left panel of Fig. 12.8, with dotted and continuous curves respectively). It is worth noticing that the assumption $f_{b'}U_{n-1} = 0$, is correct only for $b' \simeq b/2$. By applying a DC voltage between the emitter and the base, hot electrons were injected and transmitted through the superlattice. From the ratio I_C/I_E of the measured currents at 4.2 K, the transmission results shown in the right panel of Fig. 12.8 were obtained.

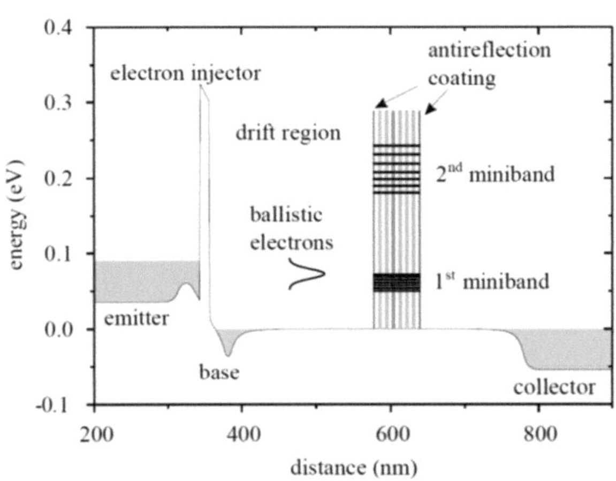

Fig. 12.7 Schematic band structure of the three-terminal resonant tunneling transistor with superlattice in the base with anti-reflection coating barriers at the ends. Horizontal lines correspond to energy levels intra-miniband. Figure reproduced from [66] with permission of Elsevier

12.4 Multichannel Conductance Through a 3D SL

Multichannel conductance calculations through 1D SLs with coupled channels were published, among others, by Ulloa et al. [72], based on self-consistent numerical calculations, and using the transfer matrix method in [55,63,73,74]. In most of the latter works a transverse electric field couples the propagating modes leading to interference and flux transition from one propagating mode to another, clearly observed in the transmission coefficients T_{ij} and conductance. In [73], 3D superlattices with 2D arrays of δ-scatterer centers, separated by homogeneous layers of width l_c, as shown in Fig. 12.9, were considered, and the transmission coefficients and conductance $G_{N,n} = (e^2/h)\mathrm{Tr}tt^\dagger$ were calculated for different values of the number of propagating modes N and different number of unit cells n. Here we will review some of the specific formulas and present new results for $N = 3$ and $n = 2, 3, 6$.

For a periodic potential

$$V_P(x, y, z) = \gamma\delta\big(z - \eta l_c\big)\sum_{\nu=1}^{N_\nu}\sum_{\mu=1}^{N_\mu}\delta\big(x - x_\nu\big)\delta\left(y - y_\mu\right) \qquad \kappa = 1, ..., n \quad (12.32)$$

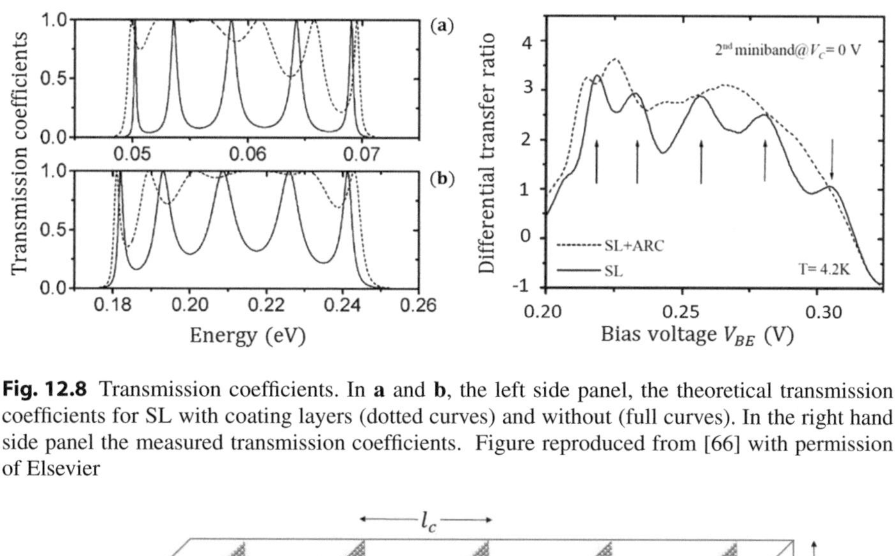

Fig. 12.8 Transmission coefficients. In **a** and **b**, the left side panel, the theoretical transmission coefficients for SL with coating layers (dotted curves) and without (full curves). In the right hand side panel the measured transmission coefficients. Figure reproduced from [66] with permission of Elsevier

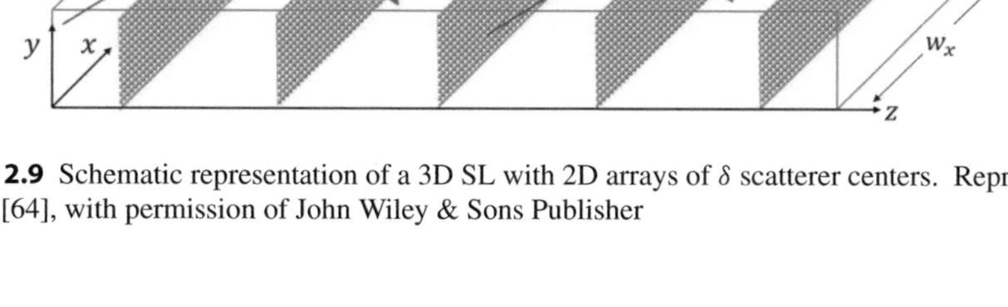

Fig. 12.9 Schematic representation of a 3D SL with 2D arrays of δ scatterer centers. Reproduced from [64], with permission of John Wiley & Sons Publisher

with confining 2D hard walls, infinite outside $\{0 \leq x \leq w_x,\ 0 \leq y \leq w_y\}$, longitudinal lattice parameter l_c and interaction strength γ, the solutions

$$\phi_i\,(x, y) = \frac{2}{\sqrt{w_x w_y}} \sum_{\{i^2=n_x^2+n_y^2\}} \sin\frac{n_x \pi x}{w_x} \sin\frac{n_y \pi y}{w_y}, \tag{12.33}$$

and energies

$$E_i = \frac{\hbar^2\pi^2}{2m^*}\left(\frac{n_x^2}{w_x^2} + \frac{n_y^2}{w_y^2}\right) \leq E \tag{12.34}$$

with E the incoming particles energy, define the transverse propagating modes labeled by the channel-index i, which is determined by the pairs of quantum numbers $n_x, n_y = 1, 2, 3, \ldots$ The functions $\phi_i\,(x, y)$ constitute a specific realization of the set of orthonormal functions, mentioned in Sect. 6.2, which allow us to obtain the channel coupling matrix

$$K_{ij} = \frac{8\pi^2 m\gamma}{h^2}\delta(z - \eta l_c)\sum_{\nu=1}^{N_\nu}\sum_{\mu=1}^{N_\mu}\phi_i^*\left(x_\nu, y_\mu\right)\phi_j\left(x_\nu, y_\mu\right) = \delta(z - \eta l_c)\,\Gamma_{ij}, \tag{12.35}$$

Replacing and integrating the Schrödinger equation upon the variable z and following the usual procedure with δ-potentials to impose the continuity conditions, one can straightforwardly determine the transfer matrix

$$M_\delta = \begin{pmatrix} \alpha_\delta & \beta_\delta \\ \beta_\delta^* & \alpha_\delta^* \end{pmatrix}. \tag{12.36}$$

with

$$\alpha_\delta = I_N + \beta_\delta, \qquad \beta_\delta = \frac{1}{2i}\begin{pmatrix} \frac{\Gamma_{11}}{k_1} & \frac{\Gamma_{12}}{k_1} & \\ \frac{\Gamma_{21}}{k_2} & \frac{\Gamma_{22}}{k_2} & \\ & & \ddots \end{pmatrix}, \qquad \text{and} \qquad \frac{\Gamma_{ij}}{\Gamma_{ji}} = \frac{k_i}{k_j} \tag{12.37}$$

To use the polynomials and invariant functions mentioned in previous sections, it is necessary to determine the eigenvalues of the $2N \times 2N$ transfer matrix M. A specific procedure to evaluate the matrix-polynomials, for this type of transfer matrix, was shown in [73], and we refer the reader to this reference.

Given the transfer matrix and applying the Theory of Finite Periodic Systems, we can evaluate the transmission coefficients $T_{i,j}$ and the conductance $G = (2e^2/h)\mathrm{Tr}tt^\dagger$ for any set of parameters. In Fig. 12.10 we show transmission coefficients and conductance for 3 propagating modes, the channels with the lowest energy, and for different number of unit cells. For this figure we consider $l_c = 20\,\text{Å}$, $w_x = 100\,\text{Å}$, $w_y = 50\,\text{Å}$, $N_u = 30$, $N_\nu = 15$, and $\gamma = -200\,\text{eV}$. For these parameters, the channel thresholds are: $E_{T1} = 0.281\,\text{eV}$; $E_{T2} = 0.449\,\text{eV}$ and

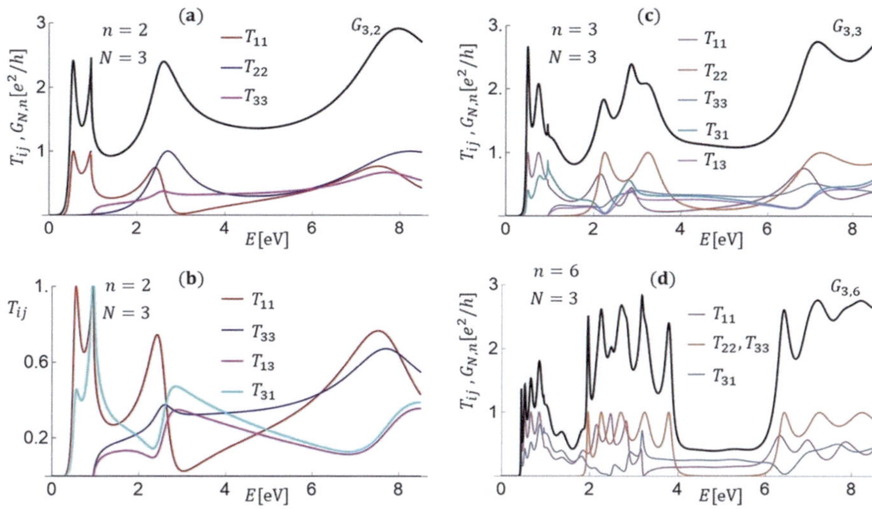

Fig. 12.10 Transmission coefficients and conductance for 3 propagating modes. The coupling effects on the transmission coefficients T_{ij} between the channels with indices $i = 1$ and $i = 3$, is evident from figures in panels (**a**) and (**b**). Notice that the transition probability T_{ij}, from channel j to channel i, rises as soon as the energy E allows to open the incoming channel j. In figures of panels **c** and **d**, transmission coefficients and conductance are plotted for $n = 3$ and for $n = 6$, respectively. Reproduced from [64], with permission of John Wiley & Sons Publisher

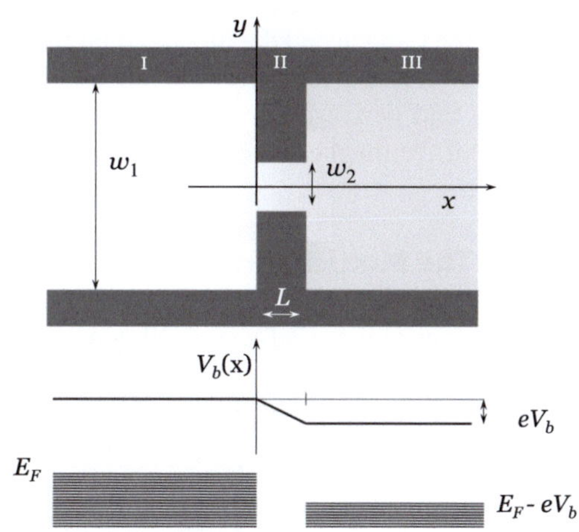

Fig. 12.11 A constriction with potential parameters w_1, w_2 and L. We show also a bias potential V_b restricted to the constriction

$E_{T3} = 0.954$ eV. In Fig. 12.10a and b, the number of unit cells is $n = 2$. In Fig. 12.10a we plot the transmission coefficient T_{ii} and the conductance G, while in Fig. 12.10b we plot the transmission coefficients T_{ij} for channels indices $i, j = 1$ and 3, with the strongest channel coupling. It is clear from this figures that the transition probability T_{ij}, from channel j to channel i, starts to rise once the energy E is enough to open the incoming channel j. In Fig. 12.10c and d we plot the transmission coefficients when the number of units cells are $n = 3$ and $n = 6$, respectively.

12.5 Quantization of the Conductance and Resistance

One of the simplest systems to observe quantization of the conductance is transport through a nanoscopic constriction, like the one shown in Fig. 12.11. The transport through a narrow ballistic constrictions in a two dimensional system of electrons, similar to that in the MOS inversion layer, has been studied extensively in the past, [75–82] and important results have been obtained. Some of these are the quantization of conductance, the absence of Hall voltages across narrow conductors, and the prediction of resonances in the longitudinal conductance. It has been suggested that wave function localization in highly doped semiconductors such as $Ga_xMa_{1-x}As$ may be responsible for oscillations in longitudinal conductance. Another system where conductance quantization was observed is the Hall bar, at low temperature and strong magnetic field.

In this section we shall briefly refer to ballistic conduction through a constriction connecting two semi-infinite regions of two-dimensional electron gas, and, to the quantum Hall effect.

The energy of the electrons can be controlled with an external potential V_g. We will calculate the conductance g, in units of $2e^2/h$, as a function of the Fermi energy E_F. Although the transport properties of the constriction as a function of the potential V_g and the magnetic field are interesting, we will concentrate here on the effects of the geometric constriction parameters.

In our reference to the quantum Hall effect (QHE), we will first review the well known relations that rise from the experimental results. Specifically, the relation between the Landau and the integral quantum numbers that define the Hall resistance plateaus. We will then review the most relevant results in the Landauer approach to the QHE and derive a new formula to describe the Hall resistance behavior as a function of the magnetic field.

12.5.1 The Multichannel Constriction Conductance

In the system of Fig. 12.11, we will consider the potential $V(x, y)$ defined as

$$V(x,y) = \begin{cases} 0 & x \leq 0 & y \leq |w_1/2| \\ V_o x & 0 < x < L & y \leq |w_2/2| \\ V_o L & x \geq L & y \leq |w_1/2| \\ \infty & \text{elsewhere} & \end{cases} . \tag{12.38}$$

In this two-dimensional system, electrons move through multiple propagation modes. In the wider regions separated by the narrow constriction, the transverse motion is quantized. The wave vectors and the corresponding wave functions are $k_{Tb} = n_b\pi/w_j$ and $\phi_b(y) = \sqrt{2/w_j}\,\sin(k_{Tb}y)$. These functions are particularly important in the analysis of the constriction and act as the sources of the multichannel or multimode process in the constriction. In fact, if the wave function is expressed as a combination of the form

$$\Psi(x, y) = \sum_{b}^{N} \phi_b(y)\varphi_b(x) \tag{12.39}$$

and substitute into the Schrödinger equation at the constriction, multiply by $\phi_b^*(y)$ and integrate, we obtain the set of coupled-channel equations

$$\frac{d^2}{dx^2}\varphi_a(x) + \left(k_{ia}^2(x) - k_{Tj}^2\right)\varphi_a(x) = \sum_{b=1}^{N} V_{ab}(z)\varphi_b(z) \tag{12.40}$$

Here N is the number of modes taken into account. When the wave numbers are real and such that $k_{ja}^2 > 0$, they are called open channels. The functions $V_{ab}(z)$ are matrix elements (of the coupling matrix) defined as

$$V_{ab}(x) = \frac{2}{w_2}\int_0^{w_2} dy V(y, x)\phi_a^*(y)\phi_b(y), \tag{12.41}$$

k_{ia} is the wave vector in channel a and in region $i = \text{I, II}$ and III. To simplify the notation we will use $i = 1, 2$ and 3:

$$k_{1a}^2 = \frac{2m^*}{\hbar^2}E_F - k_{Ta}^2 \qquad x \le 0 \tag{12.42}$$

$$k_{2a}^2 = \frac{2m^*}{\hbar^2}\left(E_F - \sum_b V_{ab}(x)\right) - k_{Ta}^2 \qquad 0 < x < L \tag{12.43}$$

$$k_{3a}^2 = \frac{2m^*}{\hbar^2}\left(E_F - eV_oL\right) - k_{Ta}^2 \qquad x \ge L. \tag{12.44}$$

Since the propagating modes, or open channels, are defined by the condition $k_{ja}^2 > 0$, the number of propagating modes for a Fermi energy E_F and transverse width w_j, is of the order of

$$N_j = w_j\sqrt{m^* E_F}/h. \tag{12.45}$$

For $w_2 \sim 10\,\text{nm}$, N_2 is of the order of 10. With the assumption that in the transverse direction we have an infinite potential well, the matrix elements $V_{ab}(x)$ are

$$V_{ab}(x) = \frac{2}{w_2}\int_0^{w_2} dy V(y, x)\phi_a^*(y)\phi_b(y) = eV_o x\delta_{ab}. \tag{12.46}$$

This means that the longitudinal electrical potential in the constriction does not produce any channel coupling. We have seen in the previous section that a transverse field does mix the propagating modes. We are going to assume that the electric potential applied to the system in Fig. 12.11 is longitudinal and restricted to the constriction. We will skip details of the calculation and quote here only the main result: the transmission amplitude.

In this system as in the other quantum systems, considered before, we have to impose continuity condition for the wave functions. After some direct algebra, we find that the matrix of transmission amplitudes is diagonal, with matrix elements

$$t_{aa} = \frac{2k_{1a}^{1/2} k_{3a}^{1/2}}{e^{ik_{1a}\Delta x_1} e^{ik_{3a}\Delta x_3}} \left[(k_{1a} + k_{3a}) \cos k_{2a} L + i \frac{\left(k_{2a}^2 + k_{1a} k_{3a}\right)}{k_{2a}} \sin k_{2a} L \right]^{-1} . \tag{12.47}$$

If we set the number of channels as N, some of them will be open and the remaining closed. As mentioned above, the open channels are those with energy smaller than the Fermi energy. The transmission amplitudes allow us also to evaluate the conductance. In fact, the conductance in units of $2e^2/h$ given by

$$g = \sum_{a,b=1}^{N} t_{ab} t_{ab}^{*} = \sum_{a=1}^{N} |t_{aa}|^2 \tag{12.48}$$

is a function potential parameters and the Fermi energy. To obtain the conductance of Fig. 12.12, we set $w_1 = 500\,\text{nm}$; $w_2 = 10\,\text{nm}$, $L = 40\,\text{nm}$ and $V_b = 0\,\text{V}$, and we plot the conductance as a function of the Fermi energy. As the energy grows, more propagation modes, i.e. open channels, contribute to the current. In addition to the quantization, we observe oscillations. To understand the relationship between the channel thresholds and the magnitude of the oscillations, we plot in Figs. 12.13 and 12.14 the conductances as functions of the Fermi energy and vary the constriction parameters in the following way.

In Fig. 12.13 we keep w_1, w_2 and the polarization potential V_b fixed and plot the conductance as a function of E_F, for three values of the constriction length L. Note that in this case, there are no changes in the positions of the thresholds, but the frequency of the conductance oscillations grows as the constriction length L increases.

In Fig. 12.14 we keep w_1, L, and the polarization potential V_b and plot the conductance as a function of E_F for three values of the constriction width w_2. In this case there is no change in the frequency of the conductance oscillations, but the thresholds change their position. As the width of the constriction increases, the threshold energies decrease.

12.5.2 The Quantum Hall Effect

We have seen in the last section an example of conductance quantization in electronic transport through a constriction, a quantization that results from the coherent superposition of multiple propagating modes. The most known and intriguing conductance quantization phenomenon, however, is the quantum Hall effect (QHE) observed in the measurement of magnetoresistance and Hall effect, first in the inversion layer of a Si MOSFET, and later at the interface of GaAs and AlGaAs layers. In these measurements, typical 'Hall bar' configurations as the one shown in Fig. 12.15 were used.

In the quantum Hall experiments, the quasi-2D concentration of electrons of the inversion layer of a Si MOSFET, attracted towards the semiconductor-oxide interface by applying a gate potential V_G, is subject to a longitudinal potential V_L and to a magnetic field $\boldsymbol{\mathscr{B}} = \mathscr{B}\hat{\mathbf{z}}$ perpendicular to the x-y plane. Using the Hall bar configuration, shown in the insets of Fig. 12.15, the Hall resistances R_{xx} and R_{xy} were measured. In panel (a), as functions of the gate potential V_G at temperature $T = 1.5$ K and $B = 15$ T, and in panel (b), as functions of the magnetic field at $T = 1.5$ K and $B = 15$ T, with the striking result that the conductivity σ_{xy} is quantized and $\rho_{xx} = \sigma_{xx} = 0$, at the plateaus of σ_{xy}. Since the effect is robust, and it has been observed also in systems with high degree of disorder, the inclussion of disorder in the explanation of features of the measured resistances became a central issue. It has been also common to suggest that the conductance G_{xy} quantization is better defined in disordered samples.

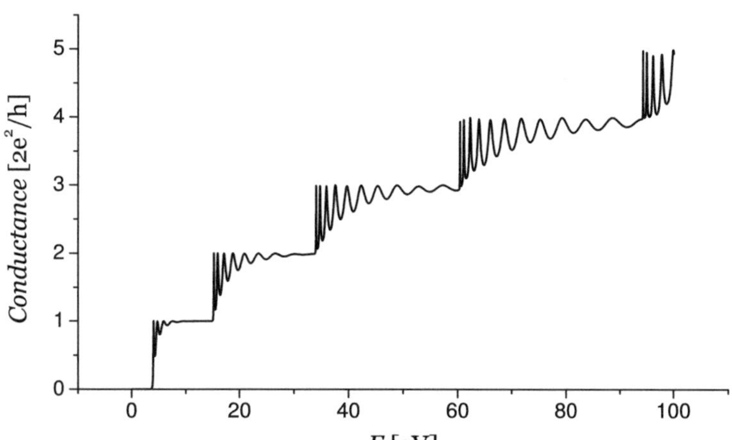

Fig. 12.12 The transmission coefficient trough a ballistic constriction as a function of the Fermi energy E. We have here the following parameters: $w_1 = 500$ nm; $w_2 = 10$ nm and $L = 40$ nm. Here the bias potential is $V_b = 0$ V

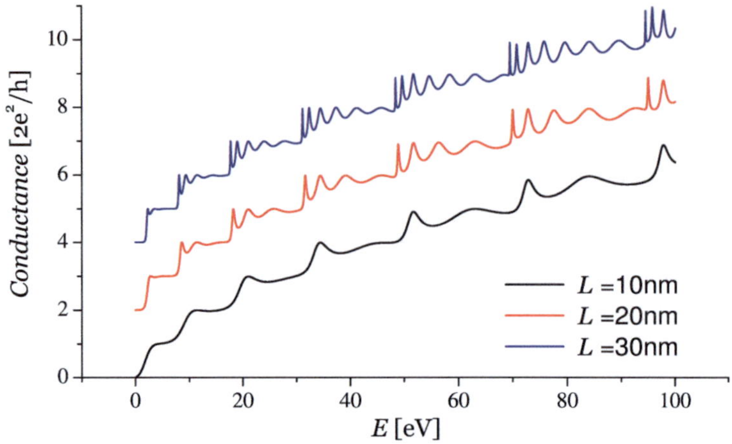

Fig. 12.13 The transmission coefficient through a ballistic constriction as a function of the Fermi energy E, for different constriction lengths L. Here $w_1 = 500\,\text{nm}$, $w_2 = 10\,\text{nm}$ and $V_b = 0\,\text{V}$

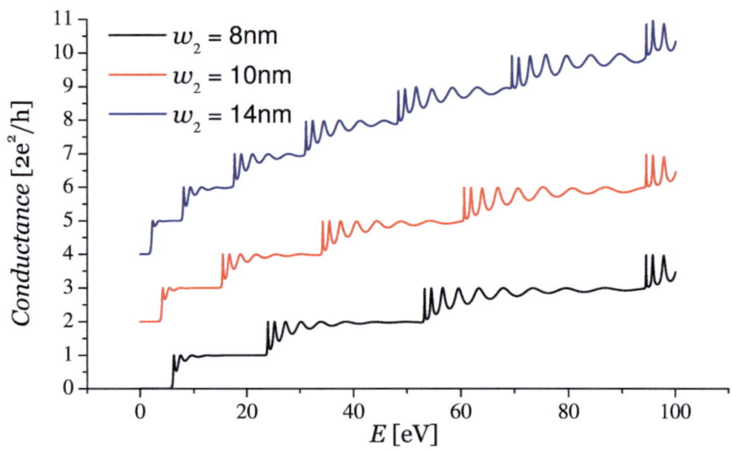

Fig. 12.14 The transmission coefficient through a ballistic constriction as a function of the Fermi energy E, for different constriction widths w_2. Here $w_1 = 500\,\text{nm}$, $L = 30\,\text{nm}$ and $V_b = 0\,\text{V}$

The explanation of this phenomenon, particularly the extension of the plateaus, suggested not only the need to analyze the system from the point of view of the quantum theory but also taking into account the role of localized states. Before we come to the Schrödinger equation for electrons in a magnetic field, let us recall some of the classical Hall effect relations introduced in Chap. 3. We found that the resistivity tensor may be written as

$$\rho = \frac{1}{\sigma_{xx}^2 + \sigma_{xy}^2} \begin{pmatrix} \sigma_{xx} & -\sigma_{xy} \\ \sigma_{xy} & \sigma_{xx} \end{pmatrix} = \begin{pmatrix} \rho_{xx} & \rho_{xy} \\ -\rho_{xy} & \rho_{xx} \end{pmatrix}. \tag{12.49}$$

Here we see that the vanishing of σ_{xx} implies also the vanishing of ρ_{xx}. We have seen also that

$$v_{qx} = \frac{q\tau_q}{m_q^*}\mathscr{E}_x + \frac{q\tau_q}{m_q^*}v_{qy}\mathscr{B}, \tag{12.50}$$

$$v_{qy} = \frac{q\tau_q}{m_q^*}\mathscr{E}_y - \frac{q\tau_q}{m_q^*}v_{qx}\mathscr{B}. \tag{12.51}$$

It is easy to show that for $q = -e$ and defining the currents $j_x = n_e e v_{ex}$ and $j_y = n_e e v_{ey}$, with n_e the electrons' concentration and Drude's conductivity $\sigma_o = e^2\tau_e n_e/m_e^*$, we obtain

$$\frac{1}{\sigma_o}\begin{pmatrix} 1 & \dfrac{e\tau_e}{m_e^*}\mathscr{B} \\ -\dfrac{e\tau_e}{m_e^*}\mathscr{B} & 1 \end{pmatrix}\begin{pmatrix} j_x \\ j_y \end{pmatrix} = \begin{pmatrix} \mathscr{E}_x \\ \mathscr{E}_y \end{pmatrix}, \tag{12.52}$$

which gives also the resistivity tensor as

$$\rho = \begin{pmatrix} \sigma_o^{-1} & \dfrac{1}{en_e}\mathscr{B} \\ -\dfrac{1}{en_e}\mathscr{B} & \sigma_o^{-1} \end{pmatrix}. \tag{12.53}$$

The behavior of the classical Hall resistance $(1/en_e)\mathscr{B}$ matches the overall trends of the Hall resistances curves in Fig. 12.15a and b, since $n_e \propto V_G$. In fact, in Sect. 10.10 we saw that, in the inversion layer of a MOS, the electron concentration grows when the gate potential is increased (see Fig. 10.21).

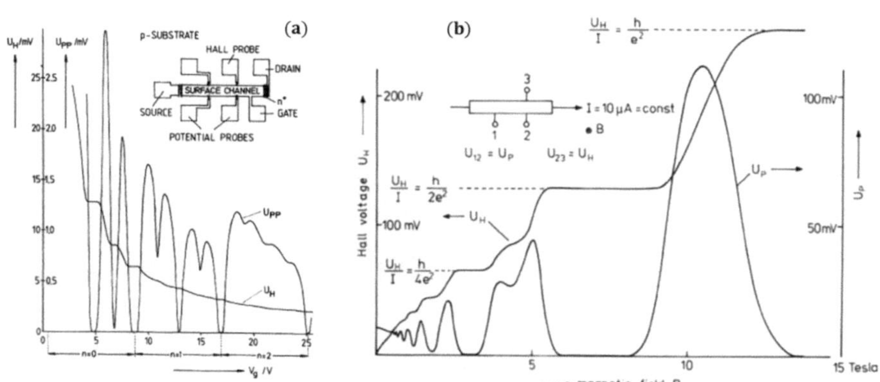

Fig. 12.15 **a** Hall voltage V_H (denoted U_H) and longitudinal potential drop V_P (denoted U_{pp}) as functions of: **a** the gate potential V_G and **b** the magnetic field $\mathscr{B}$, reported in [83] (reproduced with permission of American Physical Society) and [84] (reproduced with permission of Europhysics News). The fields were applied to electrons in the inversion layer of a Si MOSFET and the potential probes in the Hall bar configuration shown in the insets. The graphs show plateaus of the resistance $R_{xy} = U_H/I$. They also show that while R_{xy} remains constant the resistance R_{xx} vanishes

The experimental results were obtained at low temperatures ($\sim$1.5 K), and the resistance plateaus at high magnetic fields and electron concentrations. Assuming that the charge distribution is two-dimensional, some information related to the quantization has been obtained from the Schrödinger equation

$$\left(\frac{1}{2m^*}\left[-i\hbar\left(\frac{\partial}{\partial x}\hat{\mathbf{i}}+\frac{\partial}{\partial y}\hat{\mathbf{j}}\right)-e\mathbf{A}\right]^2+e\mathscr{E}_x x+\mu_B\sigma_z\mathscr{B}+V(y)\right)\psi(x,y)=E\psi(x,y), \tag{12.54}$$

which can be solved in different ways, the most common is the Landau gauge where $\mathbf{A}=-\mathscr{B}y\hat{\mathbf{x}}$ and $\nabla\cdot\mathbf{A}=0$. To simplify the analysis we will neglect the energies $e\mathscr{E}_x x$ and $\mu_B\sigma_z\mathscr{B}$, but we keep the confining potential $V(y)$, which is zero inside the sample and finite at the edges and outside the sample. In this case, the wave function is separable. Taking the function

$$\psi(x,y)=e^{-ik_x x}\eta(y)=e^{iy_o x/l_B^2}\eta(y), \tag{12.55}$$

where $\ell_B^2=\hbar/e\mathscr{B}$ is the magnetic length, and replacing in the Schrödinger equation, we have the following differential equation in the variable y

$$\frac{\partial^2\eta(y)}{\partial y^2}-\frac{1}{\ell_B^4}(y-y_o)^2\eta(y)=-\frac{2m_e^*E}{\hbar^2}\eta(y). \tag{12.56}$$

This is the well known harmonic oscillator equation, centered at $y_o=-k_x\ell_B^2$, a point that we will refer to as the guiding center moving with momentum $\hbar k_x$. With the change of variable

$$\xi=\frac{1}{\ell_B}(y-y_o), \tag{12.57}$$

the non-divergent solution is

$$\eta(\xi)=c_o 2^n e^{-\xi^2/2}H_n(\xi)=c_o 2^n e^{-m_e^*\omega_c(y-y_o)^2/2\hbar}H_n((y-y_o)/\ell_B), \tag{12.58}$$

where $H_n(\xi)$ is the Hermite polynomial of order n, being

$$n=-\frac{1}{2}+\frac{\ell_B^2 m_e^* E}{\hbar^2}, \tag{12.59}$$

an integer. In the previous equation, the cyclotron frequency (in MKS units) $\omega_c=e\mathscr{B}/m_e^*$, was also introduced. From (12.59) one has the well-known energy quantization

$$E_n=\hbar\omega_c\left(n+\frac{1}{2}\right). \tag{12.60}$$

Although this energy appeared related to a differential equation in the variable y, it should not be considered the transverse energy. In fact, for a different gauge, it can appear related to a differential equation in the variable x, or related to $r = \sqrt{x^2 + y^2}$. The effect of the repulsive potential and the strong magnetic field is to redistribute, at the edge, the quantized energy E_n into longitudinal kinetic energy and magnetic confining potential energy. We still have to impose the boundary conditions at $y = \pm Ly/2$. Because the transverse structure of Landau states is fixed a priori by the magnetic quantization, the lateral confinement does not generate new eigenmodes, but instead imposes constraints on the Hermite–Gaussian wave packet $\eta_n(y)$. The Dirichlet and Neumann boundary conditions, discretize the guiding-center positions and localize the wave packet at the nodes or local maxima of the packet.

Since this microscopic approach to the quantum Hall effect is in progress, we will refer briefly to these conditions. The Dirichlet condition,

$$\eta_n|_{y=\pm L_y/2} = 0, \tag{12.61}$$

has n zeros, corresponding to the Hermite polynomial zeros, and the Neumann condition,

$$\left.\frac{d\eta_n}{dy}\right|_{y=\pm L_y/2} = 0, \tag{12.62}$$

which implies the relation

$$H_n\left(\frac{\pm L_y/2 - y_o}{\ell_B}\right) = \frac{2n\ell_B}{\pm L_y/2 - y_o} H_{n-1}\left(\frac{\pm L_y/2 - y_o}{\ell_B}\right). \tag{12.63}$$

has $n + 1$ zeros, corresponding to the local extreme of the wave packet. The Neumann condition is physically motivated by the requirement that the normal component of the probability current vanishes at the boundary, and by the backscattering suppression of the Landauer-Büttiker formalism that will be considered below. At this point it is not yet clear which are the implications of the small difference in the number of zeros of these conditions.[1] What is evident and important is that, given the relation $k_x = -y_0/\ell_B^2$, the number of edge states, in both cases, scales with the Landau index n. For this preliminary analysis we shall refer to the approximate zeros of the Neumann boundary condition. There is no an exact closed formula for the zeros of the Hermite polynomial,[2] neither for the zeros of the Neumann condition, but a good approximation to the Neumann zeros is:

$$\frac{1}{\ell_B}\left(\pm\frac{L_y}{2} - y_{o,nj}\right) \simeq -\sqrt{2n-2} + j\frac{2\sqrt{2n-2}}{n} \qquad j = 0, 1, 2, ..., n. \tag{12.64}$$

[1] An extension of the present boundary-induced quantization approach leads to a unified description of integer and fractional quantum Hall regimes in finite systems. See P. Pereyra, arXiv:2603.04652.
[2] An approximate formula to determine the zeros x_l of a Hermite polynomial $H_n(x)$ of large n, is $x_l = \sqrt{2n-2}\cos\frac{\pi(l-1/2)}{n}$. If n is large $\cos\frac{\pi}{2n} \simeq 1$.

Here we have replaced y_o by the guiding centers $y_{o,nj}$ at the energy level n. Solving for $y_{o,nj}$ we have

$$y_{o,nj} \simeq \pm\frac{L_y}{2} - \ell_B\sqrt{2n-2}\left(1 - \frac{2j}{n}\right). \tag{12.65}$$

Therefore

$$k_{x,nj} = -\frac{y_{o,nj}}{\ell_B^2} \simeq \mp\frac{L_y}{2\ell_B^2} - \frac{\sqrt{2n-2}}{\ell_B}\left(1 - \frac{2j}{n}\right). \tag{12.66}$$

This result shows that the longitudinal velocities (along the x direction) in the upper and lower edges are different, and that the longitudinal velocity is also quantized and depends on the same quantum number n of the Landau energy level E_n. This is an important result that we will use in the last section where the quantum Hall effect is analyzed from the point of view of the Landauer-Büttiker approach. It is worth noticing that the probability to find the electrons, ($\propto |H_n(y)|^2$), is higher near the sample edges. So far, from the theoretical results obtained, we do not have yet a theoretical prediction for the Hall conductance and resistance. It is the experimental result which gives the conductance quantization from Hall resistance measurements, accurately described by the phenomenological formula $\rho_{xy} = (h/e^2)(1/\nu)$ with an accuracy of 10^{-7}. If one also assumes that $R_H = \rho_{xy}$, we have

$$\rho_{xy} = R_H = \frac{h}{e^2}\frac{1}{\nu} \qquad \nu = 1, 2, 3, ..., \tag{12.67}$$

and the conductance quantization

$$\sigma_{xy} = \frac{e^2}{h}\nu \qquad \nu = 1, 2, 3, ... \tag{12.68}$$

It is clear that these equations are based on experimental results and the phenomenological formulas that describe them. These are not yet formulas derived from physical principles. Given these relations and the Landau energy levels, we still have two important questions. What is the relation between the quantum number ν in the last equations and the quantum number n of the Landau energy levels? The second question comes from the fact that at resistance values $h/e^2\nu$ we have plateaus. How is that the energy quantization and disorder combine to keep the resistance ρ_{xy} constant, at these values, while the magnetic field varies continuously?

To establish a relation between the quantum numbers n and ν, let us first introduce some common definitions. If the cyclotron orbit area is given by ℓ_B^2, the product $n_e\ell_B^2$ corresponds to the number of electrons per cyclotron orbit. The experimental result shows that the resistance ρ_{xy}, satisfies the relation

$$\rho_{xy} = \frac{\mathscr{B}}{n_e e} = \frac{h}{e^2}\frac{1}{\nu}, \tag{12.69}$$

thus, the quantum number ν can be written as

$$\nu = n_e \frac{h}{e\mathscr{B}} = n_e \ell_B^2, \qquad \text{hence} \qquad n_e = \nu \frac{e\mathscr{B}}{h} = \nu N, \tag{12.70}$$

and corresponds to the number of electrons per magnetic orbit. The factor $N = e\mathscr{B}/h$ is known as the degeneracy factor, see [84]. This factor gives the number of magnetic orbits in the sample. Let us now analyze the meaning of the quantum number ν. It is clear from (12.70) that for $\nu = 1$, the degeneration number is equal to the number of electrons in the Hall bar, and the applied magnetic field for this configuration is

$$\mathscr{B}_1 = n_e \frac{h}{e}. \tag{12.71}$$

If the electrons concentration is large, $\mathscr{B}_1$ is also large. For $\nu = 2$, the applied magnetic field is

$$\mathscr{B}_2 = n_e \frac{h}{e} \frac{1}{2} = \frac{1}{2} \mathscr{B}_1. \tag{12.72}$$

Therefore,

$$\mathscr{B}_\nu = n_e \frac{h}{e} \frac{1}{\nu} = \frac{1}{\nu} \mathscr{B}_1. \tag{12.73}$$

It is also common to analyze the relation of ν with the magnetic flux. When the magnetic field is $\mathscr{B}_1$, the magnetic flux through the sample is

$$\Phi_1 = \mathscr{B}_1 L_x L_y = n_e L_x L_y \frac{h}{e} = N_e \frac{h}{e} = N_e \Phi_o. \tag{12.74}$$

Here $\Phi_o = h/e$ is the unit of quantum flux, called *fluxon*, while N_e is the number of electrons in the sample. From this equation, the factor N_e is also

$$N_e = \frac{\Phi_1}{\Phi_o} = N_{\phi 1} \qquad \text{and} \qquad N_e = n_e L_x L_y. \tag{12.75}$$

N_e is equal to the number of fluxons $N_{\phi 1}$ through the sample when the magnetic field is $\mathscr{B}_1$. When the magnetic field is $\mathscr{B}_\nu$, we have

$$\Phi_\nu = \mathscr{B}_\nu L_x L_y = \frac{1}{\nu} \mathscr{B}_1 L_x L_y = \frac{1}{\nu} \Phi_1 \qquad \text{thus} \qquad N_{\phi\nu} = \frac{1}{\nu} N_{\phi 1}. \tag{12.76}$$

This shows that larger ν not only means smaller magnetic fields, it also means a reduction of the number of fluxons through the sample by the factor $1/\nu$. In other words, the magnetic field necessary to accommodate all electrons in the first energy level is the largest magnetic field $\mathscr{B}_1 = n_e h/e$. When $\mathscr{B} = \mathscr{B}_2$, the number of magnetic orbits get reduced by a factor of 1/2, thus the number of energy levels needed to accommodate the N_e electrons is now 2. Therefore, when $\mathscr{B} = \mathscr{B}_\nu$ the required number of energy levels is ν, and this number corresponds to the quantum number

n_{max} of the highest of the energy levels occupied by the N_e electrons of the 2DEG in the inversion layer. The energy level $E_{n_{max}} = E_\nu$ is also the closest to the Fermi energy. The relation

$$\nu = n_{max} \tag{12.77}$$

is an important result that will be used below for the Landauer approach to the quantum Hall effect. The most intriguing topic of the quantum Hall effect has been the Hall plateaus, and the most known explanation of the Hall plateaus, observed independently of the presence or not of disorder, relies on the broadening of delocalized density of states around the Landau levels, whose eigenfunctions are extended wave functions. It is assumed that the states at the tails of the density of states are localized.[3] Given this structure of localized-delocalized states in the energy range between Landau energy levels, whose separation is $\Delta E = \hbar\omega_c = \hbar e\mathscr{B}/m_e^*$, it is clear that as the magnetic field varies, the Fermi level passes from one energy level to the next and the localized and delocalized energy domains are visited. It is assumed that while the Fermi level remains inside the localized tails, $\sigma_{xx} = 0$, and the resistance ρ_{xy} remains constant.

Before we continue with the quantum Hall effect from the point of view of the Büttiker formalism in the Landauer approach, let us make a couple of additional remarks. The quantum Hall effect that we considered here is known as the integral quantum Hall effect (IQHE) because the quantum number ν that fits the experimental results is an integer number. Soon after the discovery of the QHE, similar Hall bar experiments with 2DEG at the interface of two semiconductor layers lead to observe plateaus at resistance values that correspond to factors ν which are fractions of integers [86]. This is known as the fractional quantum Hall effect (FQHE) that was explained in terms of electro-electron and many body interactions. These complex calculations go beyond the purpose of this book.[4]

The amazing simplicity of the Hall resistance is that this quantity reduces practically to multiples of h/e^2, a simple quantity in terms of fundamental constants. The high accuracy with which the Hall resistance plateaus are measured, has been an essential argument to adopt this quantity as the new unit of resistance, named the von Klitzing or RK, 1RK$\simeq$ 25812.807 Ω.

12.5.3 The IQHE in the Landauer Approach

We will summarize the main arguments introduced by M. Bütikker in his article on the backscattering and quantum Hall effect from the point of view of the Landauer approach [88]. We will show that based on Büttiker's results and the results and relations obtained in the last section, one can derive explicit formulas for the Hall

[3] For a good explanation of this argument, see the contribution of Dung-Hai Lee in Cardona and Yu's book, [85].

[4] A good review of the integral and fractional quantum Hall effects, and of the theoretical and experimental approaches developed to understand these QHEs, is given in [87].

resistance and for the conductance, which are simple functions of the magnetic field and the Fermi energy. We will show that these results account well the main features of the experimental results.[5] In the Landauer approach it is assumed that carriers are emitted by reservoirs into current-carrying states up to its chemical potentials. Among the important properties that characterize the QHE is that the current states, as suggested by Halperin [89], are at the edges of the sample, and one of the most important results from Büttiker's analysis is that the strong magnetic field suppresses the backscattering at the edges where scattered electrons, by inelastic interactions, return moving in the same direction as before the interaction. This picture of edge states has been supported by numerous experimental articles, among them [90,91]. We have shown above that the wave numbers of the edge states are

$$k_{x,nj} \simeq \mp \frac{L_y}{2\ell_B^2} - \frac{\sqrt{2n-2}}{\ell_B}\left(1 - \frac{2j}{n}\right), \quad j = 1, 2, .., n. \tag{12.78}$$

This means that the velocities $v_{u/d,nj}$ in the "upper/lower" edge state of the energy level n are

$$v_{u/d,nj} = \frac{\hbar k_{x,nj}}{m_e^*} \simeq \mp \frac{L_y \omega_c}{2} - \frac{\sqrt{2\hbar\omega_c(n+1/2)}}{m_e^*}\left(1 - \frac{2j}{n}\right) \tag{12.79}$$

On the other hand, we have seen that the number n_{max} of energy levels below the Fermi energy is equal to the quantum number ν used to characterize the experimental resistance ρ_{xy}. Thus, the number of edge states is defined by the Fermi level and is equal to ν.

In Bütikker's analysis of the quantum Hall effect, it is shown that the resistances that characterize the QHE are four probe resistances with voltage contacts different from the current source-drain contacts, as shown in Fig. 12.17. The contact acting as current source populates the states of the sample, the edge states, with a non-equilibrium distribution. These states, due to the inelastic interactions equilibrate and become equally populated, one inelastic length away from the current contact. On the other side, due to backscattering suppression in the high magnetic field regime, each side of the sample becomes equipotential and the contacts become non-dissipative, regardless of the number of contacts. The important properties and appearance of the QHE plateaus are strongly determined by the absence of the backscattering in the strong field regime (Fig. 12.16).

To derive the Hall resistance and conductance quantization following Büttiker's approach, let us consider the sample and contacts shown in Fig. 12.17. The scattering at the contacts are characterized by the transmission coefficients T_1 and T_2 and the reflections R_1 and R_2.

Let us start defining these coefficients. In the left hand side, the reservoir with chemical potential μ_1 emits electrons into the edge states through a disorder contact

[5] See P. Pereyra, arXiv:2603.04652.

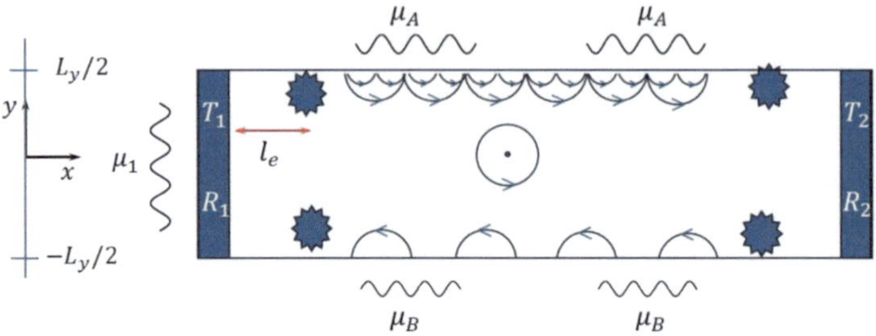

Fig. 12.16 The chemical potentials in the source-drain reservoirs and in the probe contacts in equilibrium with the edge states of the QHE system. The coefficients $T_1 ... R_2$, at the source-drain contacts, refer to the probabilities for the electrons of the edge states to transmit through the contacts into the reservoirs or get reflected by disordered contacts. The electrons fed into or reflected towards the edge states equilibrate one inelastic length l_e away from the contacts. In the presence of strong magnetic field, these electrons move in equipotential paths at the upper and lower edges with backscattering suppressed by the magnetic field, and in equilibrium with the non-dissipative probes

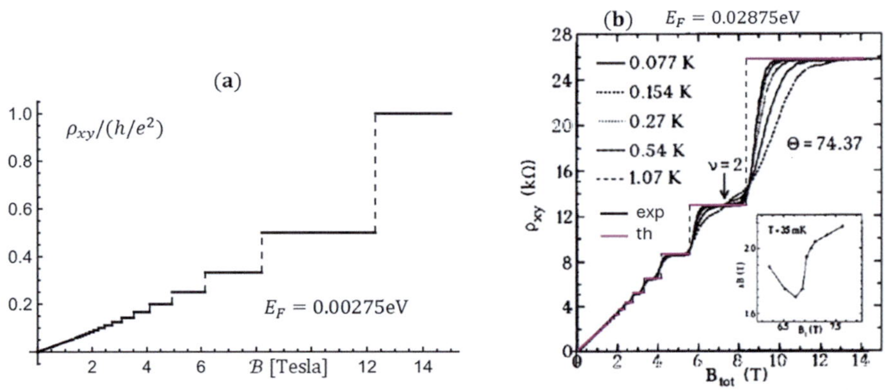

Fig. 12.17 The Hall resistance (in unit of h/e^2) predicted by (12.100), for the Fermi energy shown in the graphs. The resistance plateaus correspond to the magnetic field intervals where the magnetic-field grows without changing the number of edge states and transmission coefficients. When the magnetic field reaches a value such that the Landau level that was the second next to the total energy of the electrons $E_T = E_F + \hbar\omega/2$ becomes the upper energy level, and the number of edge states reduces from ν to $\nu - 1$. At this point the total transmission coefficient diminishes and the resistance jumps. In **b** the theoretical prediction for the QHE compared with the experimental curve at the interface of GaInAs/InP reported in [92]. The experimental figure is reproduced with permission of Elsevier

with transmission probability T'_{ij}, from channel j (in the reservoir) to edge state i. The total transmission to edge state i is then

$$T'_i = \sum_{j=1}^{M} T'_{ij}, \tag{12.80}$$

Where M is the number of incident channels. We assume that M is a large number. The total probability for the incident electrons of the reservoir to be transmitted is

$$T_1' = \sum_{i=1}^{\nu} \sum_{j=1}^{M} T_{ij}' \tag{12.81}$$

Therefore, the probability that the incident electrons are reflected is

$$R_1' = \sum_{j'=1}^{M} \sum_{j=1}^{M} R_{j'j}'. \tag{12.82}$$

Flux conservation implies that $T_1' + R_1' = M$. In a similar way, for the edge states that reach this contact coming from the other side, the probability to transmit into the reservoir (with chemical potential μ_1) is

$$T_1 = \sum_{j=1}^{M} \sum_{i=1}^{\nu} T_{ji} = T_1', \tag{12.83}$$

and the probability to be reflected is

$$R_1 = \sum_{i=1}^{\nu} \sum_{i'=1}^{\nu} R_{i'i}. \tag{12.84}$$

On this side, flux conservation implies that $T_1 + R_1 = \nu$. In a similar way, we can write the corresponding coefficients on the right hand side. For our purpose here, the probabilities for the edge states to be transmitted and reflected into and from the reservoir with chemical potential μ_2 are

$$T_2 = \sum_{j=1}^{M} \sum_{i=1}^{\nu} T_{ji} \quad \text{and} \quad R_2 = \sum_{i=1}^{\nu} \sum_{i'=1}^{\nu} R_{i'i} \quad \text{with} \quad T_2 + R_2 = \nu. \tag{12.85}$$

As shown in Fig. 12.17, the effective reservoirs with chemical potentials μ_A (upper edge) and μ_B (lower edge) represent the inelastic interactions responsible for equilibrating the edge states. These interactions begin to take place immediately after the particles pass the contacts; however, it is only after travelling approximately one inelastic scattering length that the particles are expected to have fully equilibrated. An important property from Büttiker's analysis is that once the edge states reach the equilibrium, they remain in equilibrium at the chemical potentials μ_A and μ_B along their respective paths. The relation between these effective potentials and the contact potentials μ_1 and μ_2, will be derived below. Before doing so, and before obtaining the expression for the quantum Hall resistance, it is useful to recall a general model for a current I_j transmitted to a state j from a reservoir at chemical potential μ, defined relative to a reference level μ_o. If the transmission coefficient into state j is T_j, the corresponding current is given by

$$I_j = ev_j \frac{\partial n_e}{\partial E_j}(\mu - \mu_o)T_j. \tag{12.86}$$

Replacing the density of states

$$\frac{\partial n_e}{\partial E_j} = \frac{1}{2\pi\hbar v_j}, \tag{12.87}$$

the current becomes

$$I_j = \frac{e}{h}(\mu - \mu_o)T_j. \tag{12.88}$$

We will now employ this simple relation between the current, the chemical potential and the transmission coefficient to express the currents I_1, I_2 and the total current through the sample. All chemical potentials will be defined relative to μ_2, the final results will of course be independent of this choice. The current I_1 contains two components, the current coming from the reservoir and the current reflected from the lower edge states back to the upper edge. Specifically, we write (recall that $T_{ji} = T'_{ij}$)

$$\begin{aligned} I_1 &= \sum_i^{\nu} I_{i1} + \sum_i^{\nu} I_{il} = \frac{e}{h}\sum_i^{\nu} T_i(\mu_1 - \mu_2) + \frac{e}{h}\sum_i^{\nu} R_i(\mu_B - \mu_2) \\ I_1 &= \frac{e}{h}T_1(\mu_1 - \mu_2) + \frac{e}{h}R_1(\mu_B - \mu_2). \end{aligned} \tag{12.89}$$

When this current undergoes inelastic scattering, it relaxes to an equilibrium distribution. Due to the suppression of backscattering in the presence of a strong magnetic field, electrons in the upper edge states continue propagating along the upper edge without significant mixing. Consequently, the equilibrium current can be expressed as

$$I_1 = \frac{e}{h}\sum_i^{\nu}(\mu_A - \mu_2) = \frac{e}{h}\nu(\mu_A - \mu_2). \tag{12.90}$$

On the right hand side, the current I_2 is the reflected fraction of I_1. Thus

$$I_2 = \frac{e}{h}R_2(\mu_A - \mu_2). \tag{12.91}$$

Electrons reflected by the contact at the right, move in the lower edge states, the inelastic interactions redistribute their energy within a distance of one inelastic length away from the contact, and the current reaches the equilibrium. The equilibrium current is

$$I_2 = \frac{e}{h}\nu(\mu_B - \mu_2). \tag{12.92}$$

Thus, the transmitted fraction of I_1, into the reservoir μ_2, is

$$I_t = \frac{e}{h} T_2(\mu_A - \mu_2) = I. \tag{12.93}$$

This current is also the total current I through the sample. The transmitted plus the reflected must satisfy the relation

$$I_1 = I_t + I_2 = \frac{e}{h}(T_2 + R_2)(\mu_A - \mu_2) = \frac{e}{h}\nu(\mu_A - \mu_2). \tag{12.94}$$

After some straightforward algebra, one obtains the total current I through the sample and the Hall potential $eV_H = \mu_A - \mu_B$. The reader may readily verify that these quantities are given by the following expressions:

$$I = \frac{e}{h} \frac{\nu}{\nu^2 - R_1 R_2} T_1 T_2(\mu_1 - \mu_2); \tag{12.95}$$

and

$$\mu_A - \mu_B = \frac{T_1(\nu - R_2)}{\nu^2 - R_1 R_2}(\mu_1 - \mu_2); \tag{12.96}$$

Using the flux conservation relation $T_2 = \nu - R_2$, and without even knowing the explicit functions for the transmission and reflection coefficients T_1,...R_2, we can write the Hall potential and Hall resistance as

$$\mu_A - \mu_B = \frac{h}{e} \frac{I}{\nu}, \qquad \text{thus} \qquad \rho_{xy} = \frac{h}{e^2} \frac{1}{\nu}, \tag{12.97}$$

and the Hall conductance as

$$\sigma_{xy} = \frac{e^2}{h} \nu. \tag{12.98}$$

These are precisely the phenomenological formulas given in (12.69). To conclude this section we will derive an analytical formula for the resistance ρ_{xy} as a function of the magnetic field. To this purpose, we will use the important result obtained above that the quantum number ν in the Hall resistance corresponds to the highest Landau energy level n_{max}, i.e. the quantum number of the energy level closest to the total electrons' energy E_T. Thus

$$\nu = n_{max} = \lfloor \frac{4\pi m_e^* E_T - he\mathscr{B}}{2he\mathscr{B}} \rfloor \tag{12.99}$$

where $\lfloor x \rfloor$ means the integer part of x. Let us now adopt the interpretation in which the zero-point energy of a charged particle in a magnetic field is understood as arising from its interaction not only with the applied magnetic field, but also with the fluctuating electromagnetic vacuum. In this view, the energy $\hbar\omega/2$ represents a field-dependent vacuum contribution, exchanged with the background field, while the particle is within the magnetic region,[6] in which case $E_T = E_F + \hbar\omega/2$. Replacing ν, the Hall resistance becomes

$$\rho_{xy} = \frac{h}{e^2} \frac{1}{\left\lfloor \dfrac{4\pi m_e^* E_T - he\mathscr{B}}{2he\mathscr{B}} \right\rfloor} \tag{12.100}$$

This resistance is plotted in Fig. 12.17a, in units of h/e^2 as a function of the magnetic field $\mathscr{B}$, assuming that electrons in Si have a Fermi energy $E_F = 0.00275\,\text{eV}$. In Fig. 12.17b the experimental resistance reported by S. Koch et al. in [92], for electrons at the interface of a AlGaAs/GaAs structure and the resistance predicted by (12.100) for an effective mass $m_e^* = 0.067$, and Fermi energy $E_F = 0.02875\,\text{eV}$.

Before we conclude it is important to notice that several of the properties considered and used here were part of the understanding of the IQHE, at the end of the 1980s and beginning of the 1990s. Finally, we remark that several theoretical questions related to the IQHE remain open, and research in this field is still in progress.

12.6 Problems

1. Obtain the transmission coefficient T_2 of the double barrier and plot it as a function of the barrier width for $E = 0.2\,\text{eV}$, $a = 3\,\text{nm}$ and $V_0 = 0.3\,\text{eV}$
2. Obtain the solutions of (12.11) for $a = b = 3\,\text{nm}$ and show that the transmission coefficients at these energies are resonant.
3. Consider the double barrier of Sect. 12.2.1. Show that the coupling matrix elements are effectively those given by the relations in (12.17).
4. Consider the double barrier of Sect. 12.2.2. Show that in this case, the coupling matrix elements of the previous problem reduce to those in (12.22).
5. Obtain the transfer matrix in (12.24) and calculate the conductance in (12.25) and the current for the parameters in Fig. 12.6.
6. Consider the superlattice and delta potentials array defined in (12.32). Show that the eigenfunctions in the wave guide of rectangular cross section are those in (12.33). For the energies defined in (12.34), obtain the coupling matrix elements in (12.35), the transfer matrix defined in (12.36) and (12.37), and determine the transmission coefficients shown in figure 12.10.
7. Consider the system defined in (12.38). Determine the matrix elements in (12.41) and the wave vectors in (12.42)–(12.44). Calculate the conductances in Figs. 12.12, 12.13 and 12.14.
8. Show that the experimental results of the QHE imply (12.67) and (12.68).
9. Show the second part of (12.76)
10. Show that (12.64) gives approximate zeros of Neumann condition. Plot and verify.

[6] In stochastic electrodynamics, the zero-point energy is attributed to the particle's coupling with the fluctuating electromagnetic vacuum, whose spectral density is given by $\rho(\omega) = \hbar\omega^3/2\pi^2c^3$. See [93,94].

11. Obtain the transmission and reflection coefficients $T_1, \ldots R_2$ in (12.83)–(12.85).
12. Obtain the current in (12.95) and the chemical potential difference in (12.97) and determine the Hall conductance (12.98).

References

1. D. Bohm, *Quantum Theory* (Prentice-Hall, Inc., Englewood Cliffs, New Jersey, 1951)
2. R.H. Davis, H.H. Hosack, J. Appl. Phys. **34**, 864 (1963)
3. C.A. Mead, J. Appl. Phys. **32**, 646 (1961)
4. L.L. Chang, L. Esaki, R. Tsu, Appl. Phys. Lett. **24**, 593 (1974)
5. E.E. Mendez, W.J. Wang, B. Ricco, L. Esaki, Appl. Phys. Lett. **47**, 415 (1985)
6. T.C.L.G. Sollner, W.D. Goodhue, P.E. Tannenwald, C.D. Parker, D.D Peck, Appl. Phys. Lett. **43**, 588 (1983)
7. T.C.L.G. Sollner, P.E. Tannenwald, D.D Peck, W.D. Goodhue, Appl. Phys. Lett. **45**, 1319 (1984)
8. B. Ricco, M. Ya Azbel, Phys. Rev. B **29**, 1970 (1984)
9. S. Luryi, Appl. Phys. Lett. **47**, 490 (1985)
10. A.R. Bonnefoi. R T. Collins. T.E. McGill, R.D. Burnham, F.A. Ponce, Appl. Phys. Lett. **46**, 285 (1985)
11. F. Capaso, R.A. Kiehl, J. Appl. Phys. **58**, 1366 (1985)
12. D. Coleman, J. Freeman, H. Morkoc, K. Hess, B. Streetman, M. Keever, Appl. Phys. Lett. **40**, 493 (1982); P.D. Coleman, S. Goedeke, T.J. Shewchuk, P.C. Chapin, J.M. Gering, H. Morkoc, Appl. Phys. Lett. **48**, 422 (1986)
13. D.D. Coon, H.C. Liu, Appl. Phys. Lett. **49**, 94 (1986)
14. S. Ray, P. Ruden, V. Sokolov, R. Kolbas, T. Boonstra, J. Williams, Appl. Phys. Lett. **48**, 1666 (1986)
15. S. Luryi, F. Capasso, Appl. Phys. Lett. **48**, 1347 (1986)
16. C.I. Huang, M.J. Paulus, C.A. Bozada, S.C. Dudley, K.R. Evans, C.E. Stutz, R.L. Jones, M.E. Cheney, Appl. Phys. Lett. **51**, 121 (1987)
17. E.R. Brown, T.C.L.G. Sollner, C.D. Parker, W.D. Goodhue, C.L. Chen, Appl. Phys. Lett. **55**, 1777 (1989)
18. E.R. Brown, J.R. Söderström, C.D. Parker, L.J. Mahoney, K.M. Molvar, T.C. McGill, Appl. Phys. Lett. **58**, 2291 (1991)
19. Sollner T. C. L. G., Brown E. R., Goodhue W. D. and Le H. Q., in *Physics of Quantum Electron Devices*, ed. by F. Capasso. Springer Series in Electronics and Photonics, vol. 28 (Springer, New York 1990), p. 147
20. B. Su, V.J. Goldman, M. Santos, M. Shayegan, Appl. Phys. Lett. **47**, 747 (1991)
21. H.C. Liu, D.D. Coon, Appl. Phys. Lett. **50**, 1246 (1987)
22. S. Miyasaki, Y. Ihara, M. Hirose, Phys. Rev. Lett. **59**, 125 (1987)
23. M.V. Petrychuk, A.E. Belyaev, A.M. Kurakin, S.V. Danylyuk, N. Klein, S.A. Vitusevich, Appl. Phys. Lett. **91**, 222112 (2007)
24. P.F. Bagwell, T.P.E. Broekaert, T.P. Orlando, C.G. Fonstad, J. Appl. Phys. **68**, 4634 (1990)
25. M.A. Reed, J.H. Luscombe, J.N. Randall, W.R. Frensley, R.J. Aggarwal, R.J. Matyi, T.M. Moore, A.E. Wetsel, N.A.T.O.A.S.I. Ser, Ser. B **214**, 139 (1989)
26. H.C. Liu, G.C. Aers, J. Appl. Phys. **65**, 4908 (1989)
27. S.Y. Chou et al., Appl. Phys. Lett. **52**, 1982 (1988)
28. J. Stephen et al., J. Appl. Phys. **95**, 2930 (2004)
29. S. Tsujino, N. Usamib, A. Weber, G. Mussler, V. Shushunova, D. Grützmacher, Y. Azuma, K. Nakajima, Appl. Phys. Lett. **91**, 032104 (2007)
30. M. Feiginov, C. Sydlo, O. Cojocari, P. Meissner, Appl. Phys. Lett. **99**, 133501 (2011)

31. Y.P. Li, A. Zaslavsky, D.C. Tsui, M. Santos, M. Shayegan, Phys. Rev. B **41**, 8388 (1990)
32. L.Y. Chen, C.S. Ting, Phys. Rev. B **43**, 4534 (1991)
33. M. Büttiker, Phys. Rev. B **46**, 12485 (1992)
34. V.V. Kuznetsov, E.E. Mendez, J.D. Bruno, J.T. Pham, Phys. Rev. B **58**, R10159 (1998)
35. M. Feiginov, C. Sydlo, O. Cojocari, P. Meissner, Europhys. Lett. **94**, 48007 (2011)
36. D.A. Collins, D.H. Chow, E.T. Yu, D.Z. -Y. Ting, J.R. Söderström, Y. Rajakarunanayake, T.C. McGill, Resonant tunneling in semiconductors, physics and applications, in Proceedings of a NATO Advanced Research Workshop, ed. by L.L. Chang, E.E. Mendez, C. Tejedor (Springer, 1991)
37. Y. Zeng, Ch.-I. Kuo, R. Kapadia, Ch.-Y. Hsu, A. Javey, C. Hu, J. Appl. Phys. **114**, 024502 (2013)
38. J.R. Hayes, A.F.J. Levi, W. Wiegmann, Phys. Rev. Lett. **54**, 1570 (1985)
39. A.P. Long, P.H. Beton, M.J. Kelly, Semicolld. Sci. Techno. **1**, 63 (1986)
40. T. Mori, H. Ohnishi, K. Imamura, S. Muto, N. Yokoyama, Appl. Phys. Lett. **49**, 1779 (1986)
41. Atanas Groshev, Phys. Rev. B **42**, 5895 (1990)
42. P. Gueret, N. Blanc, R. Germann, H. Rothuizen Phys. Rev. Lett. **68**, 1896 (1992)
43. M.W. Dellow, P.H. Beton, C.J.G.M. Langerak, T.J. Foster, P.C. Main, L. Eaves, M. Henini, S.P. Beaumont, C.D.W. Wilkinson, Phys. Rev. Lett. **68**, 1754 (1992)
44. T. Schmidt, M. Tewordt, R.H. Blick, R.J. Haug, D. Pfannkuche, K.V. Klitzing, A. Förster, H. Lüth, Phys. Rev. B **51**, 5570 (1995)
45. B. Su, V.J. Goldman, J.E. Cunningham, Phys. Rev. B **46**, 7644 (1992)
46. K.M. Indlekofer, J. Lange, A. Förster, H. Lüth, Phys. Rev. B **53**, 7392 (1996)
47. D.V. Averin, A.N. Kortkov, K.K. Likharev, Phys. Rev. B **44**, 6199 (1991)
48. M. Griebel, K.M. Indlekofer, A. Förster, H. Lüth, J. Appl. Phys. **84**, 6718 (1998)
49. H. Morkoc, J. Chen, U.K. Reddy, T. Henderson, S. Luryi, App. Phys. Lett. **49**, 70 (1986)
50. K.K. Choi, B.F. Levine, C.G. Bethea, J. Walker, R.J. Malik Phys. Rev. Lett. **59**, 2459 (1987)
51. F. Capasso, S. Sen, F. Beltram, A.Y. Cho, in *Physics of Quantum Electron Devices*, ed. by F. Capasso. Springer Series in Electronics and Photonics (Springer, New York, 1990) vol. 28, p. 181, and references therein
52. Esaki, L.L. Chang, Phys. Rev. Lett. **33**, 495 (1974)
53. K.K. Choi, B.F. Levine, R.J. Malik, J. Walker, C.G. Bethea, Phys. Rev. B **35**, 24172 (1987)
54. F. Capasso, K. Mohammed, A.Y. Cho, IEEE J. Quantum Electron. **22**, 1853 (1986)
55. P. Pereyra, M.G. Mendoza-Figueroa, J. Appl. Phys. **117**, 114306 (2015)
56. P. Pereyra, Phys. Rev. Lett. **84**, 1772 (2000)
57. E.R. Brown, T.C.L.G. Sollner, W.D. Goodhue, C.D. Parker, Appl. Phys. Lett. **50**, 83 (1987)
58. T.K. Woodward, T.C. McGill, R.D. Burnham, Appl. Phys. Lett. **50**, 451 (1987); T.K. Woodward, T.C. McGill, H.F. Chung, R.D. Burnham, Appl. Phys. Lett. **51**, 1542 (1987)
59. B. Su, V.J. Goldman, M. Santos, M. Shayegan, Appl. Phys. Lett. **58**, 747 (1991)
60. RCh. Bowen, G. Klimeck, R.K. Lake, W.R. Frensley, T. Moise, J. Appl. Phys. **81**, 3207 (1997)
61. Z.H. Dai, J. Ni, Y.M. Sun, W.T. Wang, Eur. Phys. J. B **60**, 439 (2007)
62. D. Csontos, H.Q. Xu, Microelectr. Eng. **51–52**, 201 (2000)
63. A. Anzaldo-Meneses, P. Pereyra, Ann. Phys. **322**, 2114 (2007)
64. P. Pereyra, Semicond. Sci. Technol. **39**, 095006 (2024)
65. F. Capasso, R. Kiehl, J. App. Phys. **58**, 1366 (1985)
66. C. Pacher, E. Gornik, Phys. Rev. B **68**, 155319 (2003)
67. M.S. Shur, L.F. Eastman, IEEE T. Electron Dev. **26**, 1677 (1979)
68. M. Heiblum, M.I. Nathan, D.C. Thomas, C.M. Knoedler, Phys. Rev. Lett. **55**, 2200 (1985)
69. A.F.J. Levi, J.R. Hayes, P.M. Platzman, W. Wiegmann, Phys. Rev. Lett. **55**, 2071 (1985)
70. F. Capasso, K. Mohammed, A.Y. Cho, IEEE J. Quantum Elect. **22**, 1853 (1986)
71. S.-Y. Cheng, P.-H. Lin, W.-C. Wang, J.-Y. Chen, W.-C. Liu, W. Lin, Electronics Lett. **33**, 98 (1997)
72. S.E. Ulloa, E. Castaño, G. Kirczenow, Phys. Rev. B **41**, R12350 (1990)
73. P. Pereyra, E. Castillo, Phys Rev. B **65**, 205120 (2002)
74. P. Pereyra, A. Anzaldo-Meneses, Microelectr. J. **36**, 419 (2005)

75. B.J. van Wees, H. van Houten, C.W.J. Beenakker, J.G. Williamson, L.P. Kouwenhoven, D. van der Marel, C.T. Foxon, Phys. Rev. Lett. **60**, 848 (1988)
76. B.J. van Wees, L.P. Kouwenhoven, H. van Houten, C.W.J. Beenakker, J.E. Mooij, C.T. Foxon, J.J. Harris, Phys. Rev. B **38**, 3625 (1988)
77. A. Szafer, A.D. Stone, Phys. Rev. Lett. **62**, 300 (1989)
78. G. Kirczenov, Phys. Rev. B **39**, 10452 (1989)
79. V. Gopar, Eur. Phys. J. B **62**, 299 (2008)
80. S. Ihnatsenka, I.V. Zozoulenko, Phys. Rev. B **79**, 255313 (2009)
81. R.A. Jalabert, W. Szewc, S. Tomsovic, D. Weinmann, Phys. Rev. Lett. **105**, 166802 (2010)
82. A.M. Lunde, A. De Martino, A. Schulz, R. Egger, K. Flensberg, New J. Phys. **11**, 023031 (2009)
83. K. von Klitzing, G. Dorda, M. Pepper, Phys. Rev. Lett. **45**, 494 (1980)
84. K. von Klitzing, Europhysics News **13**, 3 (1982)
85. P.Y. Yu, M. Cardona, *Fundamentals of Semiconductors*, 1ra edn. (Springer, Berlin, 1996)
86. D.C. Tsui, L. Störmer, A.C. Gossard, Phys. Rev. Lett. **48**, 1559 (1982)
87. T. Chakraborty, P. Pietiläinen, *The Quantum Hall Effects: Fractional and Integral*, 2nd edn (Springer, Berlin, Heidelber 1995)
88. M. Büttiker, Phys. Rev. B **38**, 9375 (1988)
89. B.I. Halperin, Phys. Rev. B **25**, 2185 (1982)
90. B.J. van Wees, E.M.M. Willems, C.J.P.M. Harmans, C.W.J. Beenakker, H. van Houten, J.G. Williamson, C.T. Foxon, J.J. Harris, Phys. Rev. Lett. **62**, 1181 (1989)
91. G. Müller, D. Weiss, S. Koch, K. von Klitzing, H. Nickel, W. Schlapp, R. Lösch, Phys. Rev. B **42**, 7633 (1990)
92. S. Koch, R.J. Haug, K. von Klitzing, M. Razeghi, Physica B **184**, 76 (1993)
93. T.H. Boyer, Phys. Rev. D **11**, 790 (1975)
94. L. de la Peña, in *Stochastic Process Applied to Physics and Other Related Fields*, ed. by B. Gomez, S.M. Moore, A. Rueda (World Scientific, Singapore, 1983)

Low-Dimensional Systems 13

Abstract

The miniaturization of electronic, optoelectronic, and sensing devices reached the nanometer scale in the 1980s, not only in the growth-direction widths of quantum wells (QWs), multiple quantum wells (MQWs), and superlattices (SLs), but also in the transverse dimensions of quasi-1D quantum wires (QWrs) and quasi-0D quantum dots (QDs). In this chapter, we explore the fundamental quantum properties that arise in these systems, features that are largely universal despite differences in fabrication methods, materials, geometries, or intended applications. To illustrate a representative application of quantum theory to QDs, modeled as artificial atoms, we revisit the calculation of excitonic energy levels and eigenfunctions in an asymmetric disk-shaped QD subjected to an external magnetic field. The quantum problem is solved, and the resulting Zeeman splitting is analyzed as a function of magnetic field strength. A crossover to pure angular momentum eigenstates is observed across all excitonic levels, independent of the dot's asymmetry or size. Explicit calculations of photoluminescence spectra are presented. We also examine quasi-1D disordered conductors and the broad applicability of the Dorokov-Mello-Pereyra-Kumar (DMPK) equation, including one of its original derivations, to underscore the importance of the transfer matrix method in the study of transport and optoelectronic properties in both periodic and disordered systems. The chapter concludes with an introductory analysis of transport in 2D graphene ribbons. The calculation of transmission coefficients as function of the interaction strength and the particle energy, shows that a gap opens, and the gap-width increases when interaction strength increases. Given the vast and rapidly expanding literature on QDs and graphene, our discussion is necessarily limited. Our aim is to provide a foundational framework for understanding their essential transport and optoelectronic characteristics.

P. Pereyra, *Contemporary Theory of Semiconductors and Devices*,
Graduate Texts in Physics, https://doi.org/10.1007/978-3-032-00296-9_13

13.1 Quantum Dots, Asymmetry and Magnetic Field Effects

Quantum dots are, generally, semiconductor structures that are able to confine quantum particles within a three-dimensional volume of nanoscopic size. These systems attracted great interest because of the variety of applications, but also as physical systems with quantum properties of artificial atoms. Despite the specific features related with the shape or the material, the theoretical analysis requires, basically, to solve the Schrödinger equation for the confined electrons or electron-hole excitations. The initial calculations were done based on the already developed approaches to describe atoms, with the appropriate adjustments to describe quantum dots (QDs) containing several electrons and excitons [1–8]. A good and early review of QDs can be found, for example in [9]. As an explicit example of the quantum theory applied to a QD, we will present here a summary of the calculation in [10] of exciton properties in asymmetric quantum dots, and in the presence of a magnetic field.

Although the numerical results are obtained using typical parameters for quantum dots formed by thickness fluctuations in narrow $Al_xGa_{1-x}As/GaAs$ quantum-well structures, similar qualitative features are expected in all quantum-dot structures, after proper scaling of the physical quantities is made (Fig. 13.1).

To explore the degree of influence of characteristic length scales in the problem, we will analyze the effect of the geometrical confinement parameters related to the quantum dot size, its asymmetry and the magnetic field. We will see that the magnetic

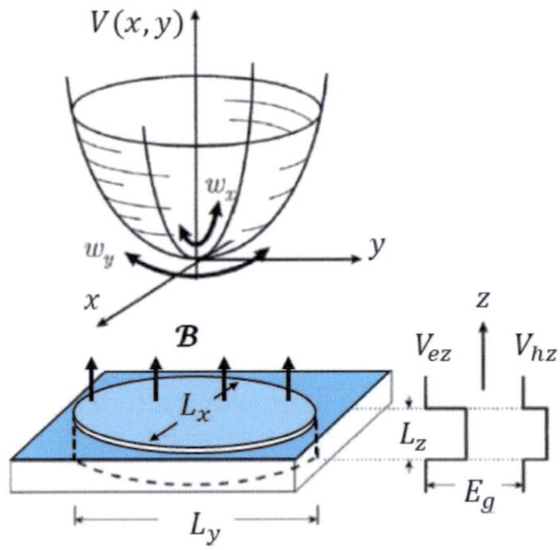

Fig. 13.1 An asymmetric and flat quantum dot, with height $L_z \ll L \simeq \sqrt{L_x L_y}$. produced as a quantum-well width fluctuation is subject to an external magnetic field $\mathscr{B}$ along the z axis. An e-h pair in the quantum dot is assumed to be confined by a lateral parabolic potential $V(x, y)$, with harmonic oscillator frequencies $w_x \propto 1/L_x^2$ and $w_y \propto 1/L_y^2$. The conduction and valence band-edge potential energies V_{ez} and V_{hz}, for electrons and holes in the quantum well of width L_z are shown. Figure reproduced from [10], with permission from American Physical Society

length $l_B = \sqrt{\hbar c / e\mathcal{B}}$,[1] the lateral extension L_x of the dot, and the effective Bohr radius for the exciton $a_B^* = \hbar^2 \varepsilon / \mu e^2$, compete with one another. Here ε is the dielectric constant of the medium, and μ is the reduced mass of the exciton.

We will show that for fixed quantum dot width $L \simeq \sqrt{L_x L_y}$ and lateral asymmetry $\eta = L_y / L_x$, different physical regimes correspond to different ratios between the characteristic length scales. At low magnetic fields, small QD sizes, and in the regime where $l_B \gg L \sim a_B^*$, all optical properties associated with excitons localized in these structures, especially photoluminescence excitation (PLE), reflect the effects of geometric asymmetry. It will be shown that by losing rotational symmetry the angular momentum is no longer conserved and then, with no good quantum number any more, the corresponding well-defined values for the ground state collapse. We will analyze this effect and show that for increasing magnetic field, the exciton and single-carrier states tend to recover a well-defined angular momentum.

The carrier confining potential in a quantum dot depends substantially on the particular method used to create this system. For quantum dots created by layer fluctuations in narrow quantum wells, it is safe to assume that the lateral confining potential is much weaker than the confinement produced by the quantum well potential. Therefore, one can reliably neglect the heavy- and light-hole mixing and consider also the approximation where the lateral motion is decoupled from the z-axis degree of freedom. At low excitation power experiments, less than one e-h pair per quantum dot is excited, on the average. It is thus appropriate to study the physics of one exciton in a quantum dot when additional confinement is produced by applying tunable external magnetic fields. This system is described by the Schrödinger equation with a Hamiltonian given by

$$\hat{H} = \hat{H}_e + \hat{H}_h + \hat{H}_{e-h} \tag{13.1}$$

with

$$\hat{H}_i = \frac{1}{2m_i^*}\left(\mathbf{p_i} \pm \frac{e\mathbf{A}_i}{c}\right)^2 + \frac{1}{m} m_i^* \left(w_x^2 x_i^2 + w_y^2 y_i^2\right) + V_{iz} \qquad i = e, h \tag{13.2}$$

where V_{iz} is the quantum-well confinement potential and

$$\hat{H}_{e-h} = -\frac{e^2}{\varepsilon |\mathbf{r}_h - \mathbf{r}_e|} \,. \tag{13.3}$$

The anisotropic harmonic confinement ($w_x \neq w_y$, in general) is understood to be the result of actual structural/geometrical constraints in addition to strain fields present in the system. The latter are especially important in the case of self-assembled dots [11–13], and in strain-defined structures [14]. Notice that this Hamiltonian can not completely be decoupled, neither in the lateral and z-axis motion nor in the center of mass and relative coordinates. The strategy that is generally followed is to first

[1] In CGS units.

make an approximate decoupling between the z and x-y coordinates, and then to transform the Hamiltonian in the x-y plane to relative and center of mass coordinates. Corrections to this approximation can be implemented systematically, as we will comment below.

In the effective mass approximation for quantum wells, holes are characterized by a pair of effective masses $m_\perp$ and $m_\parallel$, perpendicular and parallel to the quantum well. It is known that these masses, in terms of the Luttinger parameters, are $m_{h\parallel} = \gamma_1 - 2\gamma_2$, and $m_{h\perp} = \gamma_1 + \gamma_2$ [15]. For numerical calculations we will take the effective masses of $Al_{0.3}Ga_{0.7}As/GaAs$, and the corresponding dielectric constant ($\varepsilon = 13.1$ in GaAs).

To decouple the z-axis motion from the x-y plane motion, we write the Coulomb interaction as

$$\begin{aligned} \hat{H}_{e-h} &= -\frac{e^2}{\varepsilon r_{xy}} + \frac{e^2}{\varepsilon}\left(\frac{1}{r_{xy}} - \frac{1}{|\mathbf{r}_e - \mathbf{r}_h|}\right) \\ &= -\frac{e^2}{\varepsilon r_{xy}} + \Delta\hat{H}_{e-h}\,, \end{aligned} \tag{13.4}$$

and treat $\Delta\hat{H}_{e-h}$ as a small perturbation. The effective expansion parameter is proportional to $\langle |z_h - z_e| \rangle / \langle r_{xy} \rangle \lesssim L_z/L$. In typical systems as the one considered here, the expansion parameter is $\simeq 0.1$. Thus, we can neglect the perturbation contribution.

For the in-plane motion, we write the Hamiltonian $\hat{H}_{xy}$ in the center of mass and relative coordinates system, defined as usual by $\mathbf{r}_{xy} = (x_h, y_h) - (x_e, y_e)$, and $(m_h + m_e)\mathbf{R}_{xy} = m_h(x_h, y_h) + m_e(x_e, y_e)$. Here and below, the hole masses refer to $m_{h\parallel}$, for motion parallel to the quantum well. We will drop the label $\parallel$ to simplify the notation. Similarly, for the position vectors referred to the x-y plane, we omit the subindexes xy unless it is necessary to indicate. As in [16], we find convenient to use the symmetric gauge

$$\mathbf{A}_e = \frac{1}{2}\mathscr{B} \times (\mathbf{r}_e - \mathbf{r}_h) \qquad \text{and} \qquad \mathbf{A}_h = \frac{1}{2}\mathscr{B} \times (\mathbf{r}_h - \mathbf{r}_e). \tag{13.5}$$

In this gauge, the Hamiltonian $\hat{H}_{xy}$ takes the form

$$\hat{H}_{xy} = \hat{H}_{CM} + \hat{H}_{rel} + \hat{H}_c\,, \tag{13.6}$$

where

$$\hat{H}_{CM} = \frac{P^2}{2M} + \frac{M}{2}\left(w_x^2 X^2 + w_y^2 Y^2\right), \tag{13.7}$$

$$\hat{H}_{rel} = \frac{p^2}{2\mu} + \frac{1}{2}\mu\left(\widetilde{w}_x^2 x^2 + \widetilde{w}_y^2 y^2\right) + \frac{1}{2}\gamma w_{c\mu} l_z - \frac{e^2}{\varepsilon r}\,, \tag{13.8}$$

and

$$\hat{H}_c = -\frac{ie\hbar}{Mc}\mathscr{B}\left(x\frac{\partial}{\partial Y} - y\frac{\partial}{\partial X}\right). \tag{13.9}$$

Although the decoupling is not complete, the center of mass motion described by $\hat{H}_{CM}$ is (nearly) harmonic with the *same* frequencies as the constituent particles, and its motion is *independent* of the external magnetic field. This $\mathcal{B}$-independence is as having an uncharged center of mass. $\hat{H}_{rel}$ describes the relative excitonic motion. This motion depends strongly on the magnetic field. The term $\hat{H}_c$ couples the center of mass and the relative motion degrees of freedom. As will be shown below, this term is a rather weak perturbation for the structural parameters of interest, and will be treated as such. In the previous equations, we have used the usual notations $M = m_e + m_h$, $\mu = m_e m_h / M$, and defined $\gamma = (m_h - m_e)/M$, and $\widetilde{w}^2_{x(y)} = w^2_{x(y)} + w^2_{c\mu}/4$, with $w_{c\mu} = e\mathcal{B}/\mu c$.

The magnetic field effects are decomposed into two types of contributions: those producing the so-called *diamagnetic* shift, associated with the magnetic dependence on the effective parabolic confinement in (13.8); and the *Zeeman* splitting induced by the Hamiltonian term that depends on the angular momentum $\hat{l}_z$. The spin Zeeman splitting is small and will be neglected here. As mentioned before, we can deal with the rotational symmetry breaking in terms of the asymmetry ratio $\eta = L_y/L_x$ or, equivalently, in terms of the potential curvatures $w_x = \hbar/\mu L_x^2$, and $w_y = \hbar/\mu L_y^2$ (assumed here the same for electrons and holes).

13.1.1 Exciton Characteristics

To calculate the exciton eigenstates, we use the basis $|n_x n_y\rangle$ of the 2D harmonic oscillator, the scaled frequencies $\widetilde{w}_x$ and $\widetilde{w}_y$, defined above, and we will follow a numerical matrix diagonalization scheme. Taking $\hat{H}_c$ as a perturbation, we concentrate first on the relative-motion Hamiltonian whose elements are

$$
\begin{aligned}
\left\langle n'_x n'_y \middle| \hat{H}_{rel} \middle| n_x n_y \right\rangle &= \hbar\widetilde{w}_x \left(n_x + \frac{1}{2}\right) + \hbar\widetilde{w}_y \left(n_y + \frac{1}{2}\right) - \left\langle n'_x n'_y \middle| \frac{e^2}{\varepsilon r} \middle| n_x n_y \right\rangle \\
&+ i\frac{1}{4}\gamma\hbar w_{c\mu} \left\langle n'_x n'_y \right| \left(a_x^+ a_y^+ - a_x a_y\right) \left(\widetilde{\eta} - \frac{1}{\widetilde{\eta}}\right) |n_x n_y\rangle \\
&+ i\frac{1}{4}\gamma\hbar w_{c\mu} \left\langle n'_x n'_y \right| \left(a_x a_y^+ - a_x^+ a_y\right) \left(\widetilde{\eta} + \frac{1}{\widetilde{\eta}}\right) |n_x n_y\rangle .
\end{aligned}
\tag{13.10}
$$

Here the effective asymmetry factor $\widetilde{\eta} = \sqrt{\widetilde{w}_x/\widetilde{w}_y}$ depends on the magnetic field and reduces to η for $\mathcal{B} = 0$) and $\widetilde{\eta}(\mathcal{B} \gg 1) \to 1$, recovering effectively the circular symmetry. The operators $a_{x(y)}$ and $a^+_{x(y)}$ annihilate and create quantum excitations with energy $\hbar\widetilde{w}_{x(y)}$. To obtain the low-lying energy eigenvalues as functions of the magnetic field, dot size and asymmetry, we diagonalize the matrix H_{rel}. For $\mathcal{B} = 0$, this reduces, in the symmetric case to models discussed in [4,5]. The last two terms in (13.10) are responsible for Zeeman splitting and depend on $\mathcal{B}$ through $w_{c\mu}$ and $\widetilde{\eta}$. Notice that for equal particle masses, these terms vanish ($\gamma = 0$). Since the center of mass motion has no magnetic field dependence, the entire magnetic field evolution

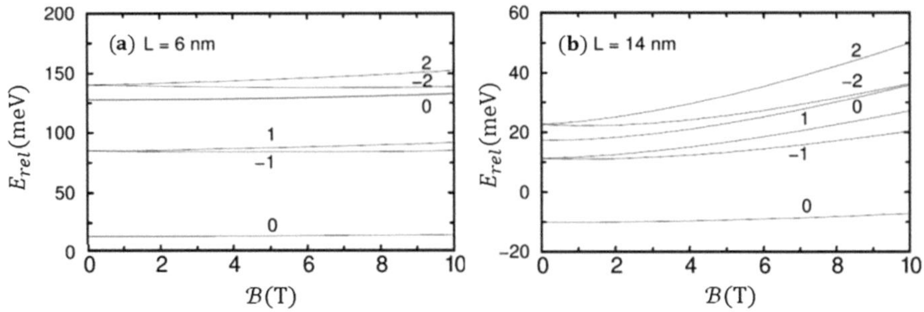

Fig. 13.2 Six lowest-energy states of the exciton relative-coordinate component, E_{rel}. Labels indicate the quantized value of $\langle l_z \rangle / \hbar$ of each state in these circular dots ($\eta = 1$) with different sizes. Figure reproduced from [10], with permission from American Physical Society

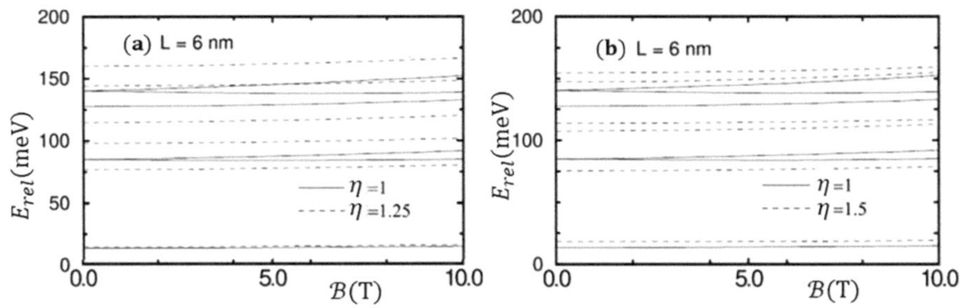

Fig. 13.3 Relative-motion eigenstates for $L = 6$ nm and different asymmetries. Solid lines are results for the circular dot. Notice how larger η in (**b**) yields larger splitting. For any $\eta \neq 1$, the zero-field degeneracies are lost. Figure reproduced from [10], with permission from American Physical Society

of the exciton level structure comes from H_{rel}. Figure 13.2 shows typical results of the low lying relative-motion exciton states, for symmetric quantum dots with sizes $L = 6$ and 14 nm. The states are labeled with the corresponding values of the angular momentum, $\hat{l}_z$; a good quantum number in the symmetric ($\eta = 1$) cases. The level ordering, Zeeman splitting, and overall magnetic-field dependence is reminiscent of atomic levels, with the addition of the diamagnetic shift via $\widetilde{w}$, which is more pronounced for larger dots. One important feature is the energy difference between the ground and the first excited states. This difference is proportional to the binding energy, E_{cb}, and is directly measurable in PLE experiments. We see that this quantity decreases as a function of L, and it is a good indication of quantum dot size.

In Fig. 13.3, we show similar evolution for quantum dots with the same length, $L = \sqrt{L_x L_y} = 6$ nm, but with different ratios $\eta = L_y / L_x$. To exhibit the asymmetry effects, the symmetric and asymmetric low-lying relative-motion energies are plotted with solid and dotted lines, respectively. We notice a general blue-shift of the ground state. This blue-shift tends to disappear as the magnetic field increases (although slowly for the parameter values considered here). Notice that the asymmetry breaks the level degeneracies at $\mathscr{B} = 0$, as shown in Fig. 13.3. The splits are already visible for $\eta = 1.25$, in the second and third excited states, and would be evident in PLE experiments.

13.1.2 Coulomb and Magnetic Field Interactions and Size Quantization

As mentioned above, an interesting property to analyze is the competition between the Coulomb interaction and the size quantization. This can be done in terms of the exciton-binding energy E_{cb}, a concept that was addressed in Sect. 1.5.2, and in the Appendix A as a function of dimensionality, for dimensionality varying from 2D to 3D. Here the interest is in exciton binding energy for a 0D system, and in the presence of a magnetic field. The exciton binding energy can be defined as the difference between the ground state energies calculated with and without the Coulomb-interaction term, i.e. as

$$E_{cb} = \left\langle \phi_o \left| \hat{H}_o \right| \phi_o \right\rangle - \left\langle \Phi_0 \left| \hat{H}_{rel} \right| \Phi_0 \right\rangle . \tag{13.11}$$

In this equation $\hat{H}_o = \hat{H}_{rel} - e^2/\varepsilon r$. The states $|\phi_o\rangle$ and $|\Phi_0\rangle$ are the ground eigenstates of $\hat{H}_o$ and $\hat{H}_{rel}$, respectively.

In the size-quantization regime, $L \ll a_B^*$, the binding energy at zero and small fields, $E_{cb}(\mathscr{B} \approx 0)$, depends strongly on the quantum dot sizes, and is larger for smaller dots. In this limit, the Coulomb interaction is a small perturbation to the confinement energy. On the other hand, in the Coulomb regime, for $a_B^* \ll L$ (corresponding to dot sizes larger than $\simeq$20 nm in GaAs, for example), the zero-field binding energy $E_{cb}(\mathscr{B} \approx 0)$ becomes independent of the quantum dot size. The quantum dot boundary effect is weaker and the exciton behaves nearly as if it would not feel the boundary. These two extreme regimes provide the framework to understand the behavior of excitons for all intermediate sizes, where both the Coulomb interactions and the confinement energy compare in magnitude.

Since the magnetic field enhances the *e-h* pair confinement, it is clear that the Coulomb and magnetic effects in the Hamiltonian compete also with each other as the field and/or size change. The quantum dot sizes at which E_{cb} become size-independent are smaller for $\mathscr{B} \neq 0$ than for $\mathscr{B} = 0$. In the limit of strong magnetic fields, the binding energy varies linearly with $\mathscr{B}$, and the magnetic confinement is dominant over the structural or geometrical factors. Increasing $\mathscr{B}$, the dot size effectively diminishes. It is interesting to notice that as the quantum dot size increases, the center of mass (COM) energy and its associated quantum of energy $\hbar w$ become smaller, and a number of low-lying exciton states corresponding to excitations of the COM motion appear between the ground state and the first relative-motion excited state. This effect is clearly seen in the susceptibility calculations and can also explain the equidistant peaks in photoluminescence experiments observed by Kash et al. [17], and by Lipsanen et al. [18]. We will illustrate this behavior further below with a few sample results.

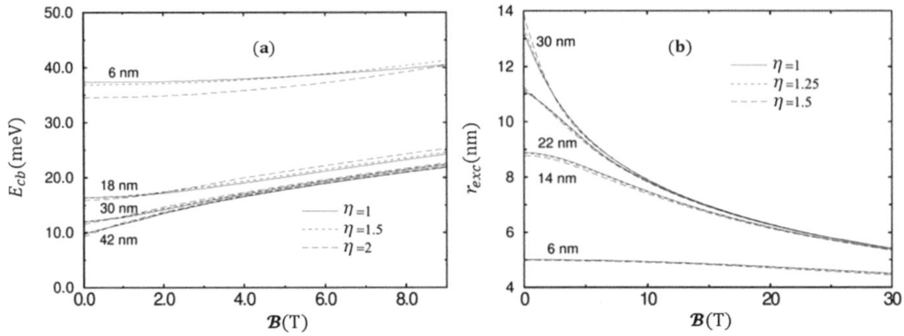

Fig. 13.4 Exciton binding energy (**a**) and exciton size (**b**). In **a** the exciton binding energy as a function of magnetic field for various dot sizes and asymmetries. For larger dots, the binding energy increases almost linearly with the field. In **b** the exciton 'size' for the ground state and different dot sizes and asymmetries. As field increases, the effective exciton size decreases. Figure reproduced from [10], with permission from American Physical Society

In Fig. 13.4 the binding energies E_{cb} are plotted for different quantum dot sizes and different asymmetry ratios η. The upper three curves correspond to $L = 6\,\text{nm}$, with $\eta = L_y/L_x = 1, 1.5$, and 2. It is clear from these three curves that one of the effects of asymmetry is to lower the binding energy at $\mathscr{B} = 0$, with respect to the symmetric case. The second set of three curves is for $L = 18\,\text{nm}$. The size L_m for which the binding energy is somewhat close to its asymptotic value, depends on the magnetic field. For $\mathscr{B} = 0$, the asymptotic limit is reached when $L_m \sim 30\,\text{nm}$, while for $\mathscr{B} = 27\,\text{T}$, it is reached when $L_m \sim 15\,\text{nm}$. At the same time, given the rather strong magnetic fields, the asymmetry effects disappear, and the binding energy depends mostly on $\mathscr{B}$, as l_B controls the effective confinement geometry.

Having the eigenvectors, $|\Phi\rangle = \sum_{n_x n_y} c_{n_x n_y} \left|n_x n_y\right\rangle$, where the coefficients $c_{n_x n_y}$ are the result of diagonalizing the matrix H_{rel}, it is possible to calculate other quantities of physical significance and experimental relevance. Among these, a quantity closely related with the size regimes, is the mean electron-hole separation for each state,

$$r_{exc} = \left(\left\langle r_{eh}^2 \right\rangle\right)^{1/2} = \left[\sum_{j=x,y} \frac{\hbar}{2\mu\widetilde{w}_j} \langle\Phi| \left(a_j^+ a_j^+ + a_j a_j + a_j a_j^+ + a_j^+ a_j\right) |\Phi\rangle \right]^{1/2} . \tag{13.12}$$

The values of r_{exc} give an idea of the binding of the electron and hole, and of the effective exciton size. In Fig. 13.4, the behavior of the exciton size of the ground state is shown as a function of the magnetic field. This figure shows a fast drop as $\mathscr{B}$ increases, especially for larger dots. It is clear, but surprising, that moderate asymmetries do not influence much neither E_{cb} nor r_{exc}, as they represent perhaps quantities that average over the entire dot domain.

On the other hand, excited states of the relative-coordinates show clear Zeeman splitting, in addition to the diamagnetic shifts. One can then explore the dot asymmetries effects on the excited states. Figure 13.5 illustrates such effects, by

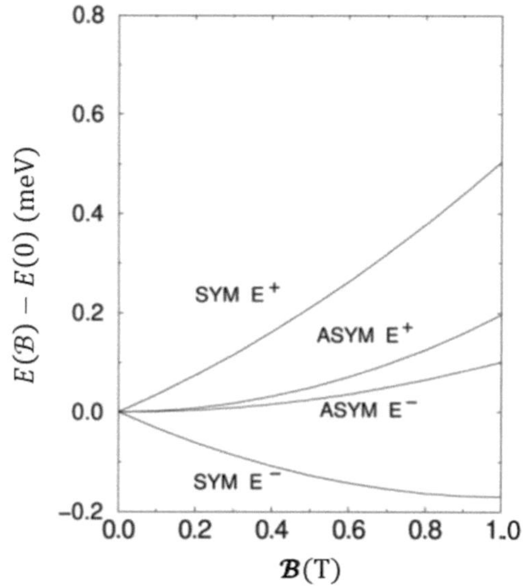

Fig. 13.5 Illustration of the orbital Zeeman level splitting for a symmetric (SYM, $\eta = 1$) and asymmetric (ASYM, $\eta = 1.25$) quantum dot with $L = 18$ nm. Here the lowest two excited states of the exciton, E^+ and E^-, are shown. Notice that the asymmetry strongly suppresses their linear splitting. Figure reproduced from [10], with permission from American Physical Society

showing the first pair of excited states (E^+ and E^-) as functions of the magnetic field, for symmetric ($\eta = 1$, SYM) and asymmetric ($\eta = 1.25$, ASYM) quantum dots with $L = \sqrt{L_x L_y} = 18\,\text{nm}$. The energies $E^\pm$ correspond to $\langle l_z \rangle = \pm\hbar$ states in Fig. 13.2b for the SYM case (but for a larger L in Fig. 13.5). The energies are shifted by $E^\pm(B = 0)$, to emphasize the net $\mathscr{B}$-dependent shift. In the SYM case, $E^\pm$ split apart linearly from each other (due to the Zeeman term) up to $\mathscr{B} \simeq 0.2\,\text{T}$, commensurate with their well-defined angular momentum, although the splitting is later taken over by diamagnetic effects. For $\eta = 1.25$, the ASYM levels hardly split for small fields, since their effective angular momentum has basically collapsed to zero (as we will show below), and only the quadratic diamagnetic shift remains. This strong suppression of the Zeeman splitting provides then a strong signature of the asymmetry of the dot, which is directly accessible spectroscopically, and can in principle even be used to quantify η.

In order to illustrate the effects of magnetic field and size confinement competition, a quantity sensitive to rotational symmetry breaking is the expectation value of the $z-$component of the angular momentum, given by

$$\left\langle \hat{l}_z \right\rangle = \frac{i\hbar}{2} \langle \Phi | \left(\left(a_x^+ a_y^+ - a_x a_y \right) \left(\widetilde{\eta} - \frac{1}{\widetilde{\eta}} \right) + \left(a_x a_y^+ - a_x^+ a_y \right) \left(\widetilde{\eta} + \frac{1}{\widetilde{\eta}} \right) \right) | \Phi \rangle \, . \tag{13.13}$$

The calculation of this quantity is useful to assign angular momentum values and to identify angular momentum transitions. Although the angular momentum is not a good quantum number when the rotational symmetry is broken, the angular momen-

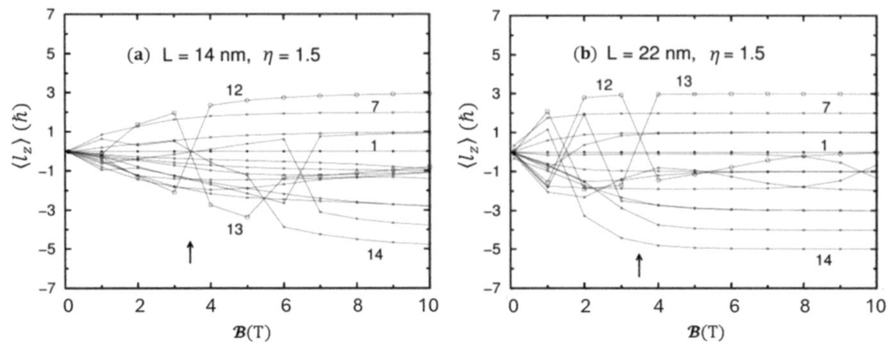

Fig. 13.6 Plot of the expectation values of $\hat{l}_z$ (in units of $\hbar$) for the first fifteen relative-motion levels of asymmetric quantum dots of different sizes. The points are the calculated values and the solid lines are just guidelines to follow the evolution of $\langle \hat{l}_z \rangle$. In this plot, $\langle \hat{l}_z \rangle$ evolves continuously in some levels while in others, it experiences abrupt changes due to energy level crossings. For example, in the ground state (label 1) and the 7th energy level, there are no energy level crossings thus no abrupt changes in $\langle \hat{l}_z \rangle$. In the 13th energy level, $\langle \hat{l}_z \rangle$ changes because this level crosses with the 12th and the 14th, as explained in the text. Notice that the strong suppression of $\langle \hat{l}_z \rangle$ caused by the asymmetry is lifted for high magnetic fields, as the circular symmetry is recovered. Figure reproduced from [10], with permission from American Physical Society

tum calculations show that as one reaches the magnetic regime at strong magnetic fields, the exciton states become close to pure angular momentum states, in essence restoring rotational symmetry to the system.

To further illustrate the effects of asymmetry on the dot levels, one can see the evolution of the expectation value of $\hat{l}_z$. Figure 13.6 shows typical examples of $\langle \hat{l}_z \rangle$ for different relative-coordinate states of the exciton. The points are the calculated values of $\langle \hat{l}_z \rangle$ and the solid lines are just guidelines. In general the z-component of the angular momentum of the ith energy level changes when it crosses with the $(i+1)$th or the $(i-1)$th energy level. For example, the $z-$component of the angular momenta of the 12th and the 13th energy levels (in Fig. 13.6a) change somewhere between $B \simeq 3$ and 4 T (see the arrow). The crossed lines indicate that the energy levels cross each other and the angular momentum of the 12th energy level passes from about $-2\hbar$ to about $2.5\hbar$, correspondingly, the angular momentum of the 13th energy level experiences the opposite change. The lowest energy level (label 1) has always $\langle \hat{l}_z \rangle = 0\hbar$. Both panels show that the quantized $\hat{l}_z$ values for circularly symmetric dots ($\eta = 1$) are strongly affected by the asymmetry, regardless of the dot size or even the η value. As L grows, however, the suppression of $\langle \hat{l}_z \rangle$ weakens and gives rise to finite but not quantized values of the angular momentum. Asymptotically, stronger $\mathcal{B}$ fields reinstate rotational symmetry to the system and therefore z-component angular momentum conservation, so that in fact, $\langle \hat{l}_z \rangle$ recovers well-quantized values in this region, regardless of L and η values (of course, smaller L and/or larger η values require stronger fields for this regime to be reached).[2]

[2] One should notice that sudden crossings are indicative of the $\mathcal{B}$-dispersion of the levels, since the negative dispersion forces rearrangements of level orderings (used as index in the plottings).

Since different η values produce different degree of $\langle \hat{l}_z \rangle$ suppression, one can conversely use this feature (measurable in PLE experiments, as explained in and by Fig. 13.5) to evaluate the degree of asymmetry of the quantum dots. The suppression, together with light-polarization dependent shifts [19], complement the analysis of geometrical effects on the spectral response of the system. Notice further that as states recover well-defined $\langle \hat{l}_z \rangle \neq 0$ values, they also decrease their overall oscillator strength. This effect, included in the calculation of the PLE figures give interesting $\mathscr{B}$-dependent lineshapes of the PLE data. Detailed analysis of this type of experiments serve as additional probes of effective confinement geometries.

For a direct comparison with high spatial and energy resolution experiments reporting well defined photo- and magneto-luminescence peaks for single quantum dots, one can calculate the excitonic contributions to the optical susceptibility of the system,

$$\chi(w) = \sum_l \left|\langle 0\,|P|\,1\rangle_l\right|^2 (\hbar w - E_l - i\hbar\Gamma)^{-1} , \tag{13.14}$$

where $\langle 0\,|P|\,1\rangle_l$ is the dipole matrix element between the excitonic state l (of one e-h pair) and the vacuum [5]. E_l is the *total* energy, i.e. the gap energy E_g plus the center of mass and the z-motion component, as well as the relative motion energy in the l-state. Physically, the optical susceptibility, $\chi(w)$, as was explained in Chap. 7, describes the possible radiative transitions or exciton recombination processes, whose strengths are determined by the dipole matrix elements. A phenomenological Lorentzian weight factor or level broadening factor of width Γ is assumed. This width mimics the finite spectral lifetimes and possible instrument resolution of PL and PLE experiments. One should note that the transition selection rules obtained in the presence of magnetic field are formally quite similar to the $\mathscr{B} = 0$ results in [5], since they are also expressed by

$$|\langle 0|P|1\rangle_l|^2 = |p_{cv}|^2 \, |\Phi_{rel}(\mathbf{r} = 0)|^2 \left| \int \Psi_{COM}(\mathbf{R}) \, d^2\mathbf{R} \right|^2 , \tag{13.15}$$

where p_{cv} is the interband matrix element, near the Γ $(k = 0)$ point of these cubic materials, and explicit analytical expressions for the other factors are obtainable in a straightforward way [5]. These factors give no exact suppression of transitions in the relative coordinate (or 'hard' selection rules), but do impose the restriction that COM states accessible via PLE should have even parity (i.e., with both COM harmonic oscillator indexes N_X and N_Y even).

From the calculations described above, and as a synthesis of all the effects, we can obtain the linear susceptibility spectra defined in Eq. (13.14), and compare with experimental photo- and magneto-luminescence excitation measurements. Notice, incidentally, that using the parameters mentioned in [19], one can be very close to the experimental values both in photon energy as in the peak separation for the system discussed in that reference, for $L \simeq 18$ nm. In Fig. 13.7, the susceptibility is plotted for several values of $\mathscr{B}$, as functions of the photon energy, assuming a Lorentzian

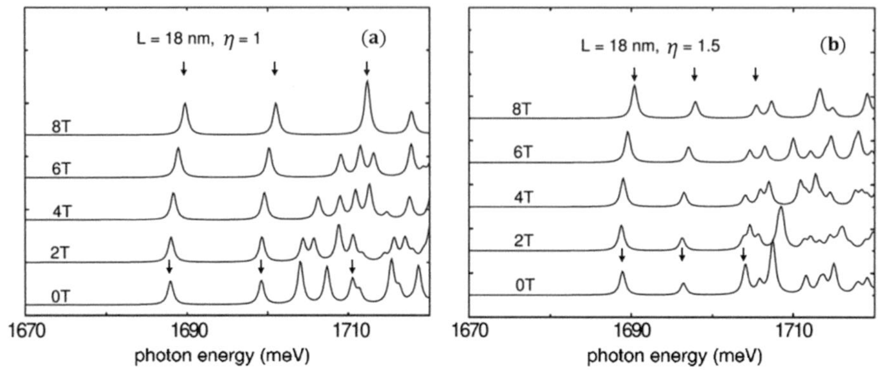

Fig. 13.7 **a** Exciton contribution to the optical response of a large circular quantum dot, $L \simeq 18\,\text{nm}$. Arrows indicate the center of mass replicas. Here, $\hbar w_x = \hbar w_y = 5.7\,\text{meV}$. In **b** $\eta = 1.5$; notice that the first center of mass replica is only 7.6 meV from the ground state, as $\hbar w_y = 3.8\,\text{meV}$ and $\hbar w_x = 8.5\,\text{meV}$. Figure reproduced from [10], with permission from American Physical Society

width of the order of $\Gamma = 0.1\,\text{meV}$. These figures show the anticipated diamagnetic shift, clear Zeeman splitting, and replicas.

Notice that as $\eta = 1.5$, the first COM replica moves closer.[3] The lowest two excited states of the exciton, in this case non-degenerate, split further as the field grows, at $\mathscr{B} = 2\text{T}$ the additional splitting is less than $\sim 0.2\,\text{meV}$. This suppressed Zeeman splitting is what we referred to in connection with Fig. 13.5.

Finally, the results of perturbation theory can be used to explore the contribution of the coordinate coupling term $\hat{H}_c$. From the form of the H_c term, proportional to the COM momentum components, and given that the COM states have definite parity along each direction, one can easily see that the first order contribution in perturbation theory vanishes, $\langle kN|\hat{H}_c|kN\rangle = 0$, for any set of quantum numbers. Therefore, the first non-vanishing energy correction for a given state E^0_{kN} (where N is a short-hand notation for N_X and N_Y) is given by (for the *non-symmetric* dot case),

$$\Delta E^{(2)}_{kN} = \sum_{k'N' \neq kN} \frac{\left|\langle k'N'|\, \hat{H}_c \,|kN\rangle\right|^2}{E^0_{kN} - E^0_{k'N'}}, \tag{13.16}$$

which is proportional to $\mathscr{B}^2$. Consequently, for moderate values of the magnetic field, the effect is small, proving that the decoupling scheme presented here is, indeed, amply justified.[4]

[3] COM replicas are readily identified since they mimic the field dependence of the ground state.
[4] Additional analysis on this field can be seen, for example, in [20,21].

13.2 Quantum Wires

Quasi-1D systems or quantum wires (QWrs) are systems where particles have one degree of freedom, one more than particles in quantum dots. In quantum wires the transverse dimension reaches the nanometer limit and their motion in the transverse direction is quantized. As for the quantum dots, there is an overwhelming specialized literature, review articles and books. Therefore, we shall introduce only the basic formalism. We will distinguish the homogeneous from the disordered quasi-1D systems. It is important also to distinguish the 1D approximation from quasi-1D systems. In this book we have frequently referred to *the 1D approximation of 3D systems*, which are generally systems large in the transverse plane, such that the physics is insensitive to the lateral boundaries, and invariant under transversal translations. In this case the physics is described by functions of one variable, generally the variable z, known also as of the growing direction. In these systems, at variance with the QWrs or quasi-1D systems, the effects of lateral boundaries can be neglected. Since the transverse dimensions are large, the energy levels are too close and the quantized energies in the growing direction are highly degenerate. In QWrs the transverse dimensions are small, of a few nanometers, and the boundaries too close. As a consequence, the transverse motion is quantized, and the longitudinal motion of the particle is strongly determined by the transverse quantization. The motion in QWrs is in some respects similar to the propagation of electromagnetic waves in wave guides.

Let us assume that QWr is a semiconductor and that the particles motion can be desicribed within the independent-particle and effective-mass approximations. We will assume that the potential can be written as

$$V(\mathbf{r}) = V_T(x, y) + V_L(z) \tag{13.17}$$

where $V_T(x, y)$ is essentially a confining, finite or infinite, 1D or 2D quantum well. In this case, the Schrödinger equation is separable and the quantized transverse wave vectors, for transverse dimensions w_x and w_y, are given by

$$\mathbf{k}_T = \frac{n_x \pi}{w_x}\hat{\mathbf{x}} + \frac{n_y \pi}{w_y}\hat{\mathbf{y}} \tag{13.18}$$

The energy consumed in the transverse motion is then

$$E_{T,n_x n_y} = \frac{\hbar^2 \pi^2}{2m^*}\left(\frac{n_x^2}{w_x^2} + \frac{n_y^2}{w_y^2}\right) \tag{13.19}$$

All pairs of integers (n_x, n_y) define the physical states or physical channels allowed to the particle. The concept of physical channel has been adopted from nuclear physics. When the particle energy, is such that $E_F \geq E_{T,n_x n_y}$, the physical channel defined by the pair (n_x, n_y) is an open channel, otherwise it is a closed channel. In open channels, the motion in the growing direction z is a propagating mode. The lowest energy for non evanescent motion in the physical channel (n_x, n_y) is the threshold energy $E_{th,n_x n_y}$ of channel (n_x, n_y). Notice that the threshold energy $E_{th,n_x n_y}$ and $E_{T,n_x n_y}$ are equal, i.e.

$$E_{th,n_x n_y} = E_{T,n_x n_y} = \frac{\hbar^2 \pi^2}{2m^*}\left(\frac{n_x^2}{w_x^2} + \frac{n_y^2}{w_y^2}\right) \tag{13.20}$$

If the particle is an electron in GaAs, with effective mass $m_e^* = 0.067 m_e$ and the transverse QWr widths are $w_T = w_x = w_y$. The minimum particle energy E_{Fm} for an open channel is one of the following threshold energies: $E_{th,10}$ or $E_{th,01}$, for $(n_x, n_y) = (1, 0)$ *or* $(0, 1)$. These thresholds, in eV, are

$$E_{Tm} = \frac{\hbar^2 \pi^2}{2m^* w_T^2} = 5.6\left(\frac{1}{w_T^2}\right) = \begin{cases} 5.6\,\text{eV if } w_T = 1\,\text{nm} \\ 0.6\,\text{eV if } w_T = 3\,\text{nm} \end{cases} \tag{13.21}$$

This shows that electrons in too narrow QWrs may not have enough energy for the transverse motion, in which case the wave function is an evanescent mode. It was already mentioned in Chap. 6, that the propagating modes, or open channels, are determined by the non evanescent modes, with energies larger than the threshold energies.

The QWr can be an homogeneous medium, a heterostructure, a quasi-1D disorder conductor, or any other kind of system. In any case, to describe the physics in this systems we will refer to the physical channels just defined. For QWrs used as light emitters, an important parameter is the transverse width w_T. Generally the first energy level of the QW in the conduction band corresponds to an eigenfunction with parity even φ_e, thus, the transition with lowest energy allowed by the selection rules, will be to the second energy level in the QW of the valence band that corresponds to an eigenfunction with parity odd φ_o. Assuming the effective masses m_e^* and m_h^* of electron and hole in GaAs, $n_1^2 = n_{xc}^2 = 1$, $n_2^2 = n_{xv}^2 = 4$, the emitted photon energy will be given by

$$\hbar\omega \simeq \frac{\hbar^2 \pi^2}{2 w_T^2}\left(\frac{n_1^2}{m_e^*} + \frac{n_2^2}{m_h^*}\right) + E_g \tag{13.22}$$

and will correspond to $\lambda \sim 720\,\text{nm}$ when $w_T \sim 10\,\text{nm}$. Other wave lengths require different cross section. A large variety of other QWr structures and heterostructures have been considered for different kinds of purposes. In Chap. 9, we referred to an axial heterostructure for solar cells.

To study quantum wire heterostructures, one can apply the Bagwell's formalism introduced in Chap. 6 to decouple the transverse from the longitudinal variables, and apply then the one-dimensional systems formalism, taking into account that the minimum of the longitudinal energies are the threshold energies. One must also take into account whether the system is open to the leads, or bounded for optical transitions purpose. When the transverse potential is constant, there is no channel mixing, however, when the transverse potential is a function of the transverse coordinates, there will be a mixing of channels and a coupling-channel matrix will be obtained. In the former case, of no channel mixing, one has a similar problem to the one considered in Sect. 12.5.1. If on the other side the channels are mixed, one has a problem similar to the one considered in Sect. 12.2.1.

Another important class of systems are the quasi-1D disordered systems, where both mass particles and light might propagate. We will present here the basic arguments that made possible the derivation of the diffusion equation that determines the evolution of the probability distribution of the transfer matrices that carry the physical information through the quasi-1D system. A diffusion equation known as the DMPK equation.[5]

13.3 Theory of Quasi-1D Disordered Conductors

In the original approaches to describe systems sensitive to impurity configurations and boundary conditions, it was common to develop numerical simulations of disordered scatterers, or to model complex formalisms based on microscopic properties plus statistical laws for individual scatterers. These formalisms were used to evaluate averages of quantities of the whole conductor, with final results that do not depend on the details of the microscopic quantities. For this problem, as for the compound nucleus reactions theory, an alternative formalism, compatible with Landauer's scattering theory, was pushed forward by Pier A. Mello. In Mello's approach the idea is to determine directly the statistical distribution of transfer matrices of the whole conductor [22–24].

In previous sections we have introduced and derived transfer matrices for one and more propagating modes. We have seen that the physical symmetries and conservation principles that are present in the Hamiltonian, reflect in the explicit representation and symmetries of the transfer matrices. We have seen also that the close and direct relation between the transfer matrix (TM) elements and the scattering amplitudes, makes the transfer matrices a useful quantity to embrace and carry in a precise and simple manner the whole complexity of physics and symmetries throughout the multiple scattering process that occur inside the physical system. We have shown that in the theory of finite periodic systems we could determine the transfer matrix of the whole system knowing the transfer matrix of a unit cell. This property, the relation between the unit cell and the whole system transfer matrix, that characterize the periodic systems, will also appear in the DMPK theory, but in a different kind of relation between the mathematical entity related to the whole disordered system and the entity related to the building block. In disordered systems, we obtain the probability density $p_L(M)$ for the TMs of the whole system based on the probability density $p_{\delta L}(\delta M)$ of the building block.

Several properties of TMs were already met in previous sections of this book. For most of the systems consider so far it was possible to express the TMs elements as explicit functions of the growing coordinate z. This is not possible in disordered

[5] Or Dorokov–Mello–Pereyra–Kumar equation.

systems. In fact if the length L is larger than the mean free path ℓ, there is no way to keep track of the functional relation between the wave function phase and the coordinate z. The transfer matrices, however, keep the physical symmetries. For this reason, the transfer matrices of disordered systems, connect coefficients rather than the products of coefficients and wave functions.

Before we present the derivation of the multichannel disordered conductors diffusion equation, we will introduce some properties related with the transfer matrices and the general relation between the probability density $p_L(M)$ and the probability density $p_{\delta L}(\delta M)$ of an infinitesimal system, relation that will be transformed into the diffusion equation whose solution is $p_L(M)$.

13.3.1 Density Probability of the Whole and Infinitesimal Systems

The final goal is to obtain the differential probability

$$dp_L(M) = p_L(M)d\mu(M) \tag{13.23}$$

where $p_L(M)$ is the density probability and $d\mu(M)$ is the volume element in the space of matrices. This quantity, derived in the Appendix J, is

$$d\mathsf{V} = C_\beta \prod_{a<b}^{N} |\lambda_a - \lambda_b|^\beta \prod_a d\lambda_a d\mu(u_1)d\mu(u_2)d\mu(v_1)d\mu(v_2), \tag{13.24}$$

with $\beta = 1, 2$ and 3 for the orthogonal, unitary and symplectic universality classes, respectively. Universality classes are determined by the symmetry of the Hamiltonian in matrix representation, which fix the algebraic structure of the matrices used to diagonalize it. Systems in the same universality class exhibit the same spectral statistics. In the derivation presented in the next section, we will consider transfer matrices of the orthogonal universality class, of time-reversal invariant systems, of dimension $2N \times 2N$, being N the number of propagating modes in the leads to which, we assume, the disordered sample is connected.

When we add to a system of length L and probability density $p_L(M)$, a system of length δL, small but still macroscopic, and probability density $p_{\delta L}(\delta M)$, we obtain a system of length $L' = L + \delta L$, described by the transfer matrix $M' = \delta M M$ and the probability density $p_{L'}(M')$. These probabilities are such that for a given function $f(M)$, we can determine the averages

$$\langle f(M)\rangle_L = \int f(M)p_L(M)d\mu(M) \tag{13.25}$$

$$\langle f(M')\rangle_{L'} = \int f(M')p_{L'}(M')d\mu(M') \tag{13.26}$$

Since $M' = \delta M M$, we can write the last equation as

$$\langle f(M')\rangle_{L'} = \int f(\delta M M)p_L(M)d\mu(M)p_{\delta L}(\delta M)d\mu(\delta M) \tag{13.27}$$

which we can also write as

$$\langle f(M')\rangle_{L'} = \int p_{\delta L}(\delta M)d\mu(\delta M) \int f(\delta M M)p_L(M)d\mu(M). \tag{13.28}$$

Here we recall that $M = \delta M^{-1}M'$ and the invariance of the volume element $d\mu M$ under a transformation, i.e. that the transformation $M \to M' = \delta M\, M$ leaves $d\mu(M) = d\mu(M')$. Using this property, we write the last equation as follows

$$\langle f(M')\rangle_{L'} = \int f(M')d\mu(M') \int p_{\delta L}(\delta M)p_L(\delta M^{-1}M')d\mu(\delta M). \tag{13.29}$$

which compared with (13.26) leads to the convolution relation

$$p_{L'}(M') = \int p_L(\delta M^{-1}M')p_{\delta L}(\delta M)d\mu(\delta M). \tag{13.30}$$

or

$$p_{L'}(M') = \int p_L(M)p_{\delta L}(\delta M)d\mu(\delta M). \tag{13.31}$$

This relation shows on side that the probability density of the building block acts as a reproductive density. It has been shown in [24] that among all possible differential probability densities that fulfill this requirement, that also maximizes the Shannon's information entropy, the following probability density was proposed

$$p_{\delta L}(\delta M) = exp\Big[\widetilde{\alpha} - \frac{\widetilde{\beta}}{\delta L}\mathrm{Tr}\left(\delta M^{\dagger}\delta M\right)\Big] \tag{13.32}$$

where $\widetilde{\alpha}$ and $\widetilde{\beta}$ are Lagrange multipliers that are fixed with the normalization of $p_{\delta L}(\delta M)$ and the requirement

$$D = \frac{1}{4N}\lim_{\delta L \to 0}\langle \mathrm{Tr}\,(\delta M^{\dagger}\delta M - I)\rangle = \frac{1}{\alpha\ell} \tag{13.33}$$

13.3.2 The Focker–Planck (DMPK) Equation

A useful representation of a transfer matrix is the Bargmann representation. In this representation the TM for a time reversal invariant system (the orthogonal universality class) can be written as[6]

$$\begin{pmatrix} u & 0 \\ 0 & u^* \end{pmatrix}\begin{pmatrix} \sqrt{1+\lambda} & \sqrt{\lambda} \\ \sqrt{\lambda} & \sqrt{1+\lambda} \end{pmatrix}\begin{pmatrix} v & 0 \\ 0 & v^* \end{pmatrix} = U\Lambda V, \tag{13.34}$$

[6] See P. Pereyra, *J. Math. Phys.* **36**, 1166 (1995).

13.3.2.1 Important Definitions and Properties

For the derivation presented here, it is convenient to use the Bargmann representation and the following notation:

$$\delta M_{\delta L} = \begin{pmatrix} u' & 0 \\ 0 & u'^* \end{pmatrix} \begin{pmatrix} \sqrt{1+\lambda'} & \sqrt{\lambda'} \\ \sqrt{\lambda'} & \sqrt{1+\lambda'} \end{pmatrix} \begin{pmatrix} v' & 0 \\ 0 & v'^* \end{pmatrix} \equiv M' \tag{13.35}$$

$$M_L = \begin{pmatrix} u'' & 0 \\ 0 & u''^* \end{pmatrix} \begin{pmatrix} \sqrt{1+\lambda''} & \sqrt{\lambda''} \\ \sqrt{\lambda''} & \sqrt{1+\lambda''} \end{pmatrix} \begin{pmatrix} v'' & 0 \\ 0 & v''^* \end{pmatrix} \equiv M'' \tag{13.36}$$

$$M_{L+\delta L} = \begin{pmatrix} u & 0 \\ 0 & u^* \end{pmatrix} \begin{pmatrix} \sqrt{1+\lambda} & \sqrt{\lambda} \\ \sqrt{\lambda} & \sqrt{1+\lambda} \end{pmatrix} \begin{pmatrix} v & 0 \\ 0 & v^* \end{pmatrix} \equiv M \tag{13.37}$$

It is also convenient to consider the following properties:

1. the flux conservation requirement

$$M^\dagger \Sigma_z M = \Sigma_z, \tag{13.38}$$

where Σ_z is the Pauli's σ_z with 1 replaced by the unit matrix I_N of dimension $N \times N$. This means that

$$\Sigma_z M^\dagger \Sigma_z M = \mathrm{I}_{2N} \quad \Rightarrow \quad M^{-1} = \Sigma_z M^\dagger \Sigma_z \quad \Rightarrow \quad (M^\dagger)^{-1} = \Sigma_z M \Sigma_z; \tag{13.39}$$

2. since

$$M = M'M'' \quad \rightarrow \quad M'' = (M')^{-1} M; \tag{13.40}$$

3. for each matrix M we have the following property

$$M^\dagger M = \mathrm{I}_{2N} + K = \mathrm{I}_{2N} + \begin{pmatrix} 2v^\dagger \lambda v & 2v^\dagger \sqrt{\lambda}\sqrt{1+\lambda} v^* \\ 2v^\mathsf{T} \sqrt{\lambda}\sqrt{1+\lambda} v & v^\mathsf{T} \lambda v^* \end{pmatrix}, \tag{13.41}$$

thus

$$M M^\dagger = \mathrm{I}_{2N} + \tilde{K} = \mathrm{I}_{2N} + \begin{pmatrix} 2u\lambda u^\dagger & 2u\sqrt{\lambda}\sqrt{1+\lambda} u^\mathsf{T} \\ 2u^* \sqrt{\lambda}\sqrt{1+\lambda} u^\dagger & 2u^* \lambda u^\mathsf{T} \end{pmatrix}; \tag{13.42}$$

4. from the previous representations, it is clear that the $(1, 1)$ blocks have the structure

$$(M M^\dagger)_{1,1} = \mathrm{I}_N + 2u\lambda u^\dagger = h \tag{13.43}$$

Therefore, the probability $p_{\delta L}(M')$ depends on the eigenvalues $\lambda_a^{'}$ of $(M'M'^\dagger)_{1,1} = \mathrm{I}_N + 2u'\lambda' u'^\dagger = h'$ and the probability density $p_L(M'')$ on the eigenvalues λ'' of $(M M^\dagger)_{1,1} = \mathrm{I}_N + 2u''\lambda'' u''^\dagger = h''$

Notice that since

$$M = M'M'' \quad \rightarrow \quad M'' = (M')^{-1}M = \Sigma_z(M')^\dagger\Sigma_z M, \tag{13.44}$$

we have

$$\begin{aligned}(M'')^\dagger M'' &= M^\dagger\Sigma_z M'\Sigma_z\Sigma_z M'^\dagger\Sigma_z M = M^\dagger\Sigma_z M'M'^\dagger\Sigma_z M \\ &= M^\dagger M + M^\dagger\Sigma_z\tilde{K}'\Sigma_z M\end{aligned} \tag{13.45}$$

The block (1, 1) of this of this matrix equation can be written as

$$h'' = h + (M^\dagger\Sigma_z\tilde{K}'\Sigma_z M)_{1,1} = h + \delta h \tag{13.46}$$

The last term will be treated as a perturbation, which is justified because $p_{\delta L}(M')$, in the convolution relation, concentrates λ', hence $\tilde{K}'$, near the origin. We can then write

$$\lambda'' = \mathrm{v}^\dagger(\lambda + \delta\lambda)\mathrm{v} \tag{13.47}$$

This matrix, defined in (13.46) and (13.47), is explicitly given as

$$\begin{aligned}\lambda + \delta\lambda = \lambda &+ \left(1+\lambda\right)^{1/2} v v'^\dagger \lambda' v' v^\dagger \left(1+\lambda\right)^{1/2} \\ &- \left(1+\lambda\right)^{1/2} v v'^\dagger \left[\lambda'(1+\lambda')\right]^{1/2} v'^* v^\mathsf{T} \lambda^{1/2} \\ &- \lambda^{1/2} v^* v'^\mathsf{T} \left[\lambda'(1+\lambda')\right]^{1/2} v' v^\dagger \left(1+\lambda\right)^{1/2} + \lambda^{1/2} v^* v'^\mathsf{T} \lambda' v'^* v^\mathsf{T} \lambda^{1/2}\end{aligned} \tag{13.48}$$

Notice that the matrix v' appears always multiplied by $v^\dagger$. Since the measure is invariant under multiplication with a matrix $v^\dagger$, we can write

$$\begin{aligned}\delta\lambda = &\left(1+\lambda\right)^{1/2} v'^\dagger \left(\lambda' v' (1+\lambda)^{1/2} - (\lambda'(1+\lambda'))^{1/2} v'^* \lambda^{1/2}\right) \\ &- \lambda^{1/2} v'^\mathsf{T} \left((\lambda'(1+\lambda'))^{1/2} v'(1+\lambda) - \lambda' v'^* \lambda^{1/2}\right)\end{aligned} \tag{13.49}$$

Notice also that this matrix is not necessarily a diagonal matrix. The eigenvalues of $\lambda + \delta\lambda$ define those of λ''. They are related as

$$\lambda''_a = \lambda_a + \delta\lambda_a \quad \text{with} \quad a = 1, \ldots, N \tag{13.50}$$

Using the perturbation theory to order λ', the eigenvalues $\delta\lambda_a$ can be written as

$$\begin{aligned}\delta\lambda_a &= \left(\delta\lambda\right)_{aa} + \sum_{b(\neq a)} \frac{\left(\delta\lambda\right)_{ab}\left(\delta\lambda\right)_{ba}}{\lambda_a - \lambda_b} \\ &= \left(\delta\lambda_a\right)^{(1)} + \left(\delta\lambda_a\right)^{(2)} + \ldots\end{aligned} \tag{13.51}$$

13.3.2.2 Derivation of the DMPK Equation

With the definitions and relations introduced before, the convolution relation in (13.32) becomes

$$p_{L+\delta L}(\lambda) = \int p_L(\lambda + \delta\lambda) p_{\delta L}(\lambda') d\mu(M') = \langle p_L(\lambda + \delta\lambda)\rangle_{\delta L} \tag{13.52}$$

Here the density $p_L(\lambda + \delta\lambda)$ depends upon the N eigenvalues λ_a'' which are also the eigenvalues of $\lambda + \delta\lambda$. If we develop the function $p_L(\lambda + \delta\lambda)$ around λ as

$$p_L(\lambda + \delta\lambda) = p_L(\lambda) + \sum_a \frac{\partial p_L(\lambda)}{\partial \lambda_a}\delta\lambda_a + \frac{1}{2}\sum_{ab}\frac{\partial^2 p_L(\lambda)}{\partial\lambda_a\partial\lambda_b}\delta\lambda_a\delta\lambda_b + \ldots \tag{13.53}$$

and replace in the integral, we have

$$p_{L+\delta L}(\lambda) = p_L(\lambda) + \sum_a \frac{\partial p_L(\lambda)}{\partial \lambda_a}\langle\delta\lambda_a\rangle_{\delta L} + \frac{1}{2}\sum_{ab}\frac{\partial^2 p_L(\lambda)}{\partial\lambda_a\partial\lambda_b}\langle\delta\lambda_a\delta\lambda_b\rangle_{\delta L} + \ldots \tag{13.54}$$

Here we need to evaluate some averages $\langle\ \rangle_{\delta L}$ assuming that λ_k' and v_{ab}' are statistically independent, and using the well-known averages

$$\langle v_{ad} v_{cd}\rangle_0 = \langle v_{ab}^* v_{cd}^*\rangle = 0 \tag{13.55}$$

$$\langle v_{ab}^* v_{cd}\rangle_0 = \frac{1}{N}\delta_{ac}\delta_{bd} \qquad \langle v_{cb}^* v_{cb}^* v_{da} v_{db}\rangle_0 = \frac{1+\delta_{ab}}{N(N+1)}\delta_{cd} \tag{13.56}$$

$$\langle \mathrm{Tr}(\lambda')\rangle_{\delta L} = DN\delta L \qquad \langle\lambda_{a_1}' \ldots \lambda_{a_k}'\rangle_{\delta L} \sim (\delta L)^k \tag{13.57}$$

To first order in δL we have (see [24])

$$\langle(\delta\lambda_a)^{(1)}\rangle_{\delta L} = \langle\delta_{aa}\rangle_{\delta L} = \sum_c (1+2\lambda_a)\langle\lambda_c' v_{ca}'^* v_{ca}'\rangle_{\delta L} = (1+2\lambda_a)\frac{1}{N}\langle\sum_c \lambda_c'\rangle_{\delta L} \tag{13.58}$$

$$= (1+2\lambda_a) D\delta L \tag{13.59}$$

Similarly

$$\langle(\delta\lambda_a)^{(2)}\rangle_{\delta L} = \sum_{b(\neq a)} \frac{\langle(\delta\lambda)_{ab}(\delta\lambda)_{ba}\rangle}{\lambda_a - \lambda_b} = \frac{D\delta L}{N+1}\sum_{b(\neq a)}\frac{\lambda_a + \lambda_b + 2\lambda_a\lambda_b}{\lambda_a - \lambda_b} \tag{13.60}$$

and

$$\langle\delta\lambda_a\delta\lambda_b\rangle_{\delta L} \approx \langle\delta\lambda_{aa}\delta\lambda_{bb}\rangle_{\delta L} = \frac{4}{N+1}\lambda_a(1+\lambda_a)\delta_{ab} D\delta L \tag{13.61}$$

On the other hand, one can also develop the function $p_{L+\delta L}(\lambda)$ as powers of δL, around L, as

$$p_{L+\delta L}(\lambda) = p_L(\lambda) + \frac{\partial p_L}{\partial L}\delta L \tag{13.62}$$

Replacing the previous results we end up with

$$\begin{aligned}\frac{\partial p_L}{\partial L} = D\Big[\sum_a \frac{\partial p_L(\lambda)}{\partial \lambda_a}(1+2\lambda_a) + \frac{1}{N+1}\sum_{a\neq b}\frac{\lambda_a+\lambda_b+2\lambda_a\lambda_b}{\lambda_a-\lambda_b}\frac{\partial p_L(\lambda)}{\partial \lambda_a} \\ + \frac{2}{N+1}\sum_a \frac{\partial^2 p_L(\lambda)}{\partial \lambda_a^2}\big(\lambda_a(1+\lambda_a)\big)\Big]\end{aligned} \tag{13.63}$$

This equation can still be simplified by using the identities

$$\sum_{b\neq a}\frac{\lambda_a+\lambda_b+2\lambda_a\lambda_b}{\lambda_a-\lambda_b} = -(N-1)(1+2\lambda_a) + 2\lambda_a(1+\lambda_a)\sum_{b\neq a}\frac{1}{\lambda_a-\lambda_b} \tag{13.64}$$

and

$$\frac{\partial J(\lambda)}{\partial \lambda_a} = \sum_{b\neq a}\frac{1}{\lambda_a-\lambda_b} \tag{13.65}$$

where $J(\lambda) = \prod_{a<b}^{N}|\lambda_a-\lambda_b|$ is the Jacobian from the invariant measure. Replacing we end up with a more compact form of the equation

$$\frac{\partial p_L}{\partial L} = \frac{2D}{N+1}\frac{1}{J(\lambda)}\sum_a^N \frac{\partial}{\partial \lambda_a}\Big[\lambda_a(1+\lambda_a)J(\lambda)\frac{\partial p_L(\lambda)}{\partial \lambda_a}\Big] \tag{13.66}$$

The most known form of the equation is written for the probability density

$$w_L(\lambda) = p_L(\lambda)J(\lambda) \tag{13.67}$$

in which case the diffusion equation becomes

$$\alpha\ell\frac{\partial w_L}{\partial L} = \frac{2}{N+1}\sum_a^N \frac{\partial}{\partial \lambda_a}\Big[\lambda_a(1+\lambda_a)J(\lambda)\frac{\partial}{\partial \lambda_a}\frac{w_L(\lambda)}{J(\lambda)}\Big] \tag{13.68}$$

with the initial condition

$$w_L(\lambda) = \delta(\lambda). \tag{13.69}$$

In the strictly 1D, one-channel limit, the DMPK equation is [22,25]

$$\ell\frac{\partial w_L}{\partial L} = \frac{\partial}{\partial \lambda}\Big[\lambda(1+\lambda)\frac{\partial w_L}{\partial \lambda}\Big] \tag{13.70}$$

known as the Melnikov equation [26].

13.3.2.3 Some Multichannel Transport Results of Interest

The DMPK equation has been used in a variety of explicit calculations, and deeply analyzed in [23]. We will end our reference to this theory quoting some of basic and important results. The average of a physical quantity $F(\lambda)$, in a quasi-1D disordered system described by the probability density $p_L(\lambda)$, is defined as

$$\langle F\rangle = \int F(\lambda)p_L(\lambda)d\mu(M) = \int F(\lambda)w_L(\lambda)d^N\lambda \tag{13.71}$$

It was shown in [24] that the evolution of this quantity as function of the system size is determined by the evolution equation

$$\begin{aligned}\alpha\ell\frac{\partial\langle F\rangle}{\partial L} &= \frac{2}{N+1}\Big\langle\sum_a^N\Big[\lambda_a(1+\lambda_a)\frac{\partial^2 F}{\partial\lambda_a^2} + (1+2\lambda_a)\frac{\partial F}{\partial\lambda_a}\Big]\\ &+\frac{1}{2}\sum_{a\neq b}\frac{1}{\lambda_a-\lambda_b}\Big[\lambda_a(1+\lambda_a)\frac{\partial F}{\partial\lambda_a} - \lambda_b(1+\lambda_b)\frac{\partial F}{\partial\lambda_b}\Big]\Big\rangle\end{aligned} \tag{13.72}$$

Two of the important results in transport theory are the transmission and reflexion coefficients to a given channel, say, to channel a. In this approach, they read

$$\mathcal{T}_a = (tt^\dagger)_{aa} = \sum_b\frac{|u_{ab}|^2}{1+\lambda_b} \tag{13.73}$$

$$\mathcal{R}_a = (tt^\dagger)_{aa} = \sum_b|v_{ba}|^2\frac{\lambda_b}{1+\lambda_b} \tag{13.74}$$

with $\mathcal{T}_a + \mathcal{R}_a = 1$. The total transmission and reflection coefficients are

$$\mathcal{T} = \sum_a\mathcal{T}_a \quad\text{and}\quad \mathcal{R} = \sum_a\mathcal{R}_a \tag{13.75}$$

It has been found [24] that in the diffusive regime $\ell \ll L$ and $N >> 1$ the average transmission probability is

$$\langle\mathcal{T}\rangle_L \simeq \frac{N\alpha\ell}{\alpha\ell+L} \tag{13.76}$$

and

$$\langle\mathcal{R}\rangle_L \simeq \frac{N}{2}(e^{2L/\alpha\ell}-1) \tag{13.77}$$

Other interesting applications [27] and extensions [28,29] have been published in numerous articles. A review of those developments goes beyond the purpose of this book.

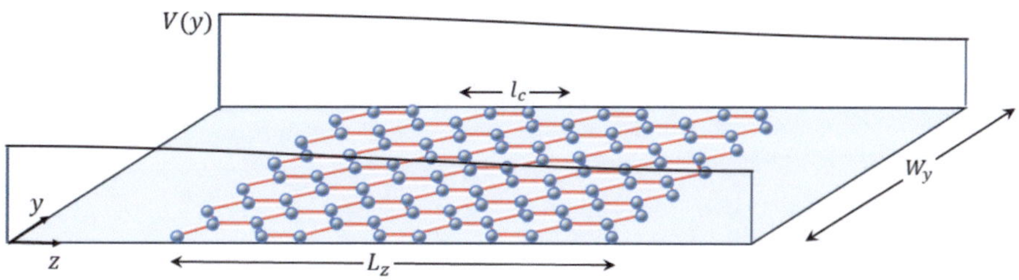

Fig. 13.8 A 2D graphene ribbon confined by an infinite square well

13.4 Graphene, the Opening of a Gap

The rise of excitement on the graphene field has been as astonishing as it has been in QHE, QDots, and spintronics. The potential applications of graphene attracted more of the enthusiasm and achievements. Soon after the isolation of graphene flakes the theoretical analysis recovered the Wallace theory of 1947, where a variational, tight-binding and Bloch-like functions were considered [30]. In Wallace theory a linear dispersion relation and zero gap at the hexagon corners of the Brillouin zones were reported. Since then the physics of this 2D system did not change much but got prestigious names, the linear response are now known as Dirac cone and the hexagonal corners, the Dirac points. In the abundant experimental work, some striking properties have been obtained. Among them, relations between the cyclotron mass and the electrons concentration that led to think on the massless Dirac-like quasi-particles, and the anomalous Hall effect observed in quantum Hall experiments. The literature on this system, and the standard tight-binding approach is overwhelming. Since most of the approaches have much in common with the original Wallace approach and results, and are well known, we shall consider the Graphene from a different perspective. We will address the gap issue from the point of view of the allowed and forbidden energy domains. We will study the electronic transport problem through a periodic structure of scattering centers, located in the corners of a finite 2-D hexagonal array, with graphene parameters. We will assume that the honeycomb-like array has a length $L_z = nl_c$ and a width W_y. We will also assume that this array is inside a confining potential $V(y)$, which is zero for $0 < y < W_y|$, and ∞ for $y \leq 0$ and $y \geq W_y|$, and assume also that this confining potential extends to the left and right of the sample (as a wave guide, see Fig. 13.8), where the incoming and outgoing electrons move. The results presented here are part of the work done in [31]. Here we will discuss only the arm-chair configuration and we will show that a gap, proportional to the strength of effective repulsion interaction, appears naturally in this system.

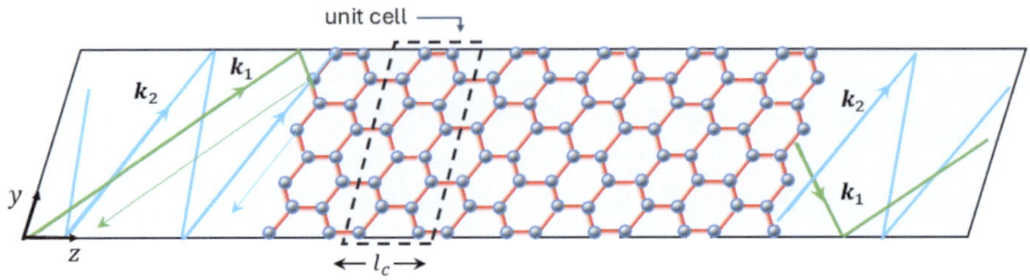

Fig. 13.9 A graphene stripe to determine the unit-cell transfer matrix, and the leads at the left and right hand sides of the sample where the transverse motion of the incoming, reflected and transmitted particles is quantized

13.4.1 The TM Approach for an Arm-Chair Graphene Ribbon

An important input in the transfer matrix approach of open and closed systems is the unit-cell transfer matrix. The unit-cell that we require to obtain this matrix, and to build the transfer matrix of the whole scattering system, not necessarily corresponds to the unit cell that builds the periodic structure. For the graphene nano-ribbon shown in the Fig. 13.9, the unit cell is the stripe of hexagons limited by the dotted frame. To obtain the transfer matrix M of this unit-cell we need to solve the Schrödinger equation in the stripe. Let us consider the Hamiltonian

$$H = -\frac{\hbar^2}{2m_e^*}\left(\frac{\partial^2}{\partial y^2} + \frac{\partial^2}{\partial z^2}\right) + V_c(y) + V_p(y, z) \tag{13.78}$$

where m_e^* is the electrons effective mass, $V_c(y)$ is the confining infinite square well and $V_p(y, z)$ models the interactions with the hexagonal array of scattering centers with the symmetry of the graphene sample. The unit cell of the periodic potential is defined as

$$V_p(y, z) = \sum_{m,l}^{N_y,4} \gamma\delta\big(y - y_m^l\big)\delta\big(z - z_l\big) \tag{13.79}$$

The pairs (y_μ^l, z_l) are the coordinates of the scattering centers, N_y is the number of hexagons along the y direction, and γ the strength of the interaction. The index l is attached to coordinate y_μ^l to recall that they, and the quantities defined with them, depend on the coordinate z_l. To describe the electronic transport of particles through the scattering configuration shown in Fig. 13.9, it is convenient to define, first, the propagating modes in the leads, i.e. at the left and right hand sides of the sample. It was argued in Sect. 6.2 and at the beginning of the last section, that for a given Fermi energy, E_F, the propagating modes (moving to the right and to the left) are defined as those whose wave number is such that

$$0 < |\mathbf{k}_i| \leq \left|\sqrt{\frac{2m_e^* E_F}{\hbar^2}}\right| \tag{13.80}$$

The eigenfunctions $\varphi_i(y)$ along the y axis are

$$\varphi_i(y) = \sqrt{\frac{2}{W_y}} \sin\frac{n_i \pi}{W_y}, \tag{13.81}$$

and following Bagwell [32], these functions can be used as a basis to express the wave function $\psi(y, z)$, inside the sample, as the combination

$$\psi\,(y, z) = \sum_i \varphi_i(y)\phi_i(z). \tag{13.82}$$

If we introduce this function into the Schrödinger equation, multiply by $\varphi_i^*(y)$ and integrate, we obtain the system of coupled equations

$$\frac{d^2}{dz^2}\phi_i(z) + \left(k^2 - k_{Ti}^2\right)\phi_i(z) = \sum_j K_{ij,l}(z)\,\phi_j(z) \qquad \text{with} \qquad i = 1, 2, ... \qquad \text{and} \qquad l = 1, ..., 4 \tag{13.83}$$

Here $k = \sqrt{2mE}/\hbar, k_{Ti} = \sqrt{2m_e^* E_{Ti}}/\hbar$ and $K_{ij,l}(z)$ the elements of the coupling matrix

$$K_{ij,l}(z) = \frac{2m}{\hbar^2}\int \varphi_i^*(y)V_P(y, z)\varphi_j(y)dy = \Gamma_{ij,l}\delta(z - z_l) \qquad \text{with} \qquad l = 1, ..., 4 \tag{13.84}$$

Here

$$\Gamma_{ij,l} = \frac{4m_e\gamma}{c^2\hbar^2 W_y}\sum_{\mu,\nu} \sin\left(\frac{n_i\pi}{W_y}y_\mu^l\right)\sin\left(\frac{n_j\pi}{W_y}y_\nu^l\right). \tag{13.85}$$

As mentioned before, there is no physical need to consider the evanescent channels. Thus the number of coupled equations that will be considered is $N = s\mathcal{N}$, which is equal to the number of open channels $\mathcal{N}$ times s, the number of spin projections, when the spin of the particle should be taken into account. Notice that one has to solve the Schrödinger equation for each of the four delta potentials $\delta(z - z_l)$ and determine a transfer matrix $M_{\delta,l}$ to connect the wave functions at the left and right sides of z_l, each transfer matrix with the following time-reversal-invariant structure

$$M_{\delta,l} = \begin{pmatrix} \alpha_{\delta,l} & \beta_{\delta,l} \\ \beta_{\delta,l}^* & \alpha_{\delta,l}`* \end{pmatrix} \qquad \text{where} \qquad \alpha_{\delta,l} = I_2 + \beta_{\delta,l},$$

and

$$\beta_{\delta,l} = \frac{1}{2i}\begin{pmatrix} \dfrac{\Gamma_{11,l}}{k_1} & \dfrac{\Gamma_{12,l}}{k_1} \\ \dfrac{\Gamma_{21,l}}{k_2} & \dfrac{\Gamma_{22,l}}{k_2} \end{pmatrix} \tag{13.86}$$

Time reversal invariance requires that $\Gamma_{12,lk_2} = \Gamma_{21,lk_1}$. In this model, this property is satisfied. To propagate the state vectors from z_l to z_{l+1} one needs also the transfer matrices

$$m_1 = \begin{pmatrix} e^{ik_1a/2} & 0 & 0 & 0 \\ 0 & e^{ik_2a/2} & 0 & 0 \\ 0 & 0 & e^{-ik_1a/2} & 0 \\ 0 & 0 & 0 & e^{-ik_2a/2} \end{pmatrix}, \qquad m_2 = \begin{pmatrix} e^{ik_1a} & 0 & 0 & 0 \\ 0 & e^{ik_2a} & 0 & 0 \\ 0 & 0 & e^{-ik_1a} & 0 \\ 0 & 0 & 0 & e^{-ik_2a} \end{pmatrix}. \tag{13.87}$$

where $a = 1.42\text{Å}$ is the interatomic distance in graphene. Since $M_{\delta,4} = M_{\delta,1}$ and $M_{\delta,3} = M_{\delta,2}$, the unit-cell transfer matrix is obtained from

$$M = m_1 M_{\delta,1} m_1 M_{\delta,2} m_2 M_{\delta,2} m_1 M_{\delta,1} m_1. \tag{13.88}$$

Given this transfer matrix, all we need is to apply the closed formulas of the theory of finite periodic systems to obtain, first, the transfer matrix of the whole system, and to obtain the transmission coefficients, the corresponding connection between the transfer matrix elements and the scattering matrices. In [33] it was shown that since the transfer matrix of the n unit-cells system is

$$M_n = M^n = \begin{pmatrix} \alpha & \beta \\ \gamma & \delta \end{pmatrix}^n = \begin{pmatrix} \alpha_n & \beta_n \\ \gamma_n & \delta_n \end{pmatrix}, \tag{13.89}$$

the matrix elements, or matrix blocks for more than one propagating mode, of the whole system, are given by the product of polynomials and the single-cell transfer matrix elements, i.e. expressions similar to those introduced in Chap. 6. The resulting connections between the transfer matrix blocks are

$$\alpha_n = p_n - \gamma^{-1}\delta\gamma\, p_{n-1} \quad \text{and} \quad \beta_n = \beta\, p_{n-1} \tag{13.90}$$

with the matrix polynomials p_n determined by the matrix-recurrence relation

$$p_n - (\beta^{-1}\alpha\beta)p_{n-1} + (\delta\beta^{-1}\alpha\beta - \gamma\beta)p_{n-2} = 0 \tag{13.91}$$

whose solutions are given in [33,34]. These polynomials, in the one propagating mode case, reduce to the Chebyshev polynomials. The connection between the transfer matrix blocks and the scattering amplitudes is direct, and given by

$$t = (\alpha^\dagger)^{-1} \quad \text{and} \quad r = (\alpha^*)^{-1}\beta^*. \tag{13.92}$$

In this approach, it is easy to calculate the transmission coefficients. Since

$$t = \frac{1}{\det \alpha_n^\dagger} \begin{pmatrix} \alpha_{n,22} & -\alpha_{n,21} \\ -\alpha_{n,12} & \alpha_{n,11} \end{pmatrix}^* \tag{13.93}$$

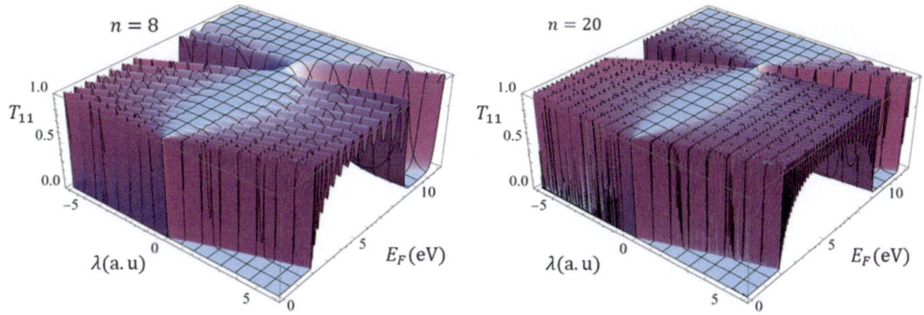

Fig. 13.10 Transmission coefficient as a function of the strength and Fermi energy for $n = 8$ and $n = 20$. The only difference between the graphs at the left and right is the number of unit cells n. For larger n the bandstructure gets better defined. The band structure is sensitive to λ. For $\lambda = 0$, there is no band structure, no gap, and the sample is transparent ($T_{11} = 1$). For $\lambda > 0$, the gap grows as a linear function of λ (see lower corner). For $\lambda < 0$, the first gap appear at too large values of the energy

the transmission coefficient is

$$T = tt^{\dagger} = \frac{1}{|\det \alpha_n|^2} \begin{pmatrix} |\alpha_{n,22}|^2 + |\alpha_{n,21}|^2 & -\alpha_{n,21}\alpha^*_{n,22} - \alpha_{n,11}\alpha^*_{n,12} \\ -\alpha^*_{n,21}\alpha_{n,22} - \alpha^*_{n,11}\alpha_{n,12} & |\alpha_{n,11}|^2 + |\alpha_{n,12}|^2 \end{pmatrix} \tag{13.94}$$

For the particular configuration shown in Fig. 13.9, which is zig-zag at the left and right borders and arm-chair at the lateral borders, we obtain the transmission coefficients T_{11} shown in Fig. 13.10, for $n = 8$ (left), $n = 20$ (right) and $N_y = 20$. It is clear from this figures that zero gap is possible only for $\lambda = 0$, which implies no interaction and complete transmission. We see that for an effective repulsive interaction, i.e. for $\lambda > 0$, the gap opens up as a function of the strength λ. In this approach we have only one parameter, which is λ. This parameter can be fixed with experimental measurements of the gap. There is also the possibility, within the TFPS, to determine λ from first principles.[7] Fixing λ one can predict not only the channels mixing, calculating T_{ij}, but also the conductance $G \propto \mathrm{Tr}tt^{\dagger}$, and other properties, including the anomalous Hall effect in the presence of a magnetic field.

References

1. A.L. Efros, A.L. Efros, F. Tekh, Poluprovodn **16**, 1209 (1982) [Sov. Phys. Semicond. **16**, 772 (1982)]; G.W. Bryant, Phys. Rev. B **37**, 8763 (1988); **41**, 1243 (1990)
2. L. Banyai, Y.Z. Hu, M. Lindberg, S.W. Koch, Phys. Rev. B **38**, 8142 (1988); T. Takagahara, *ibid.* **39**, 10 206 (1989); S.V. Nair, S. Sinha, K.C. Rustagi, *ibid.* **35**, 4098 (1987)
3. Y.Z. Hu, M. Lindberg, S.W. Koch, Phys. Rev. B **42**, 1713 (1990)

[7] This is a work on progress.

4. W. Que, Phys. Rev. B **45**, 11 036 (1992)
5. J. Song, S.E. Ulloa, Phys. Rev. B **52**, 9015 (1995)
6. V. Halonen, T. Chakraborty, P. Pietiläinen, Phys. Rev. B **45**, 5980 (1992)
7. U. Bockelmann, Phys. Rev. B **48**, 17 637 (1993); **50**, 17 271 (1994); U. Bockelmann, Ph. Roussignol, A. Filoramo, W. Heller, G. Abstreiter, K. Brunner, G. Böhm, G. Weimann, Phys. Rev. Lett. **76**, 3622 (1996)
8. A.O. Govorov, A.V. Chaplik, Zh. E' ksp. Teor. Fiz. **99**, 1853 (1991) [Sov. Phys. JETP **72**, 1037 (1991)]
9. L. Jacak, Hawrylak, A. Wójs, *Quantm Dots* (Springer, Berlin Heidelberg, 1998)
10. P. Pereyra, S.E. Ulloa, Phys. Rev. B **61**, 2128 (2000)
11. J.-Y. Marzin, J.-M. Gerard, A. Izrael, D. Barrier, G. Bastard, Phys. Rev. Lett. **73**, 716 (1994)
12. R. Leon, P.M. Petroff, D. Leonard, S. Fafard, Science **267**, 1966 (1995)
13. M. Grundmann, J. Christen, N.N. Ledentsov, J. Böhrer, D. Bimberg, S.S. Ruvimov, P. Werner, U. Richter, U. Gösele, J. Heydenreich, V.M. Ustinov, AYu. Egorov, A.E. Zhukov, P.S. Kop'ev, Zh.I. Alferov, Phys. Rev. Lett. **74**, 4043 (1995)
14. R. Rinaldi, P.V. Giugno, R. Cingolani, H. Lipsanen, M. Sopanen, J. Tulkki, J. Ahopelto, Phys. Rev. Lett. **77**, 342 (1996)
15. P.Y. Yu, M. Cardona, *Fundamentals of Semiconductors* 1ra Ed. (Springer, Berlin, 1996)
16. V. Halonen, T. Chakraborty, P. Pietiläinen, Phys. Rev. B **45**, 5980 (1992)
17. K. Kash, D.D. Mahoney, B.P. Van der Gaag, A.S. Gozdz, J.P. Tarbisson, L.T. Florez, J. Vac. Sci. Technol. B **10**, 2030 (1993)
18. H. Lipsanen, M. Sopanen, J. Ahopelto, Phys. Rev. B **51**, 13868 (1995)
19. D. Gammon, E.S. Snow, B.V. Shanabrook, D.S. Katzer, D. Park, Phys. Rev. Lett. **76**, 3005 (1996)
20. C. Simserides, A. Zora, G.P. Triberis, *Low-Dimensional Carriers Under In-plane Magnetic Field: Novel Phenomena* (Nove Science Publisher, N.Y., 2010)
21. J.Y. Song, S.E. Ulloea, Phys. Rev. B **63**, 125302 (2001)
22. P.A. Mello, Phys. Rev. B **35**, 1082 (1987)
23. P.A. Mello, N. Kumar, *Quantum Transport in Mesoscopic Systems. Complexity and Statistical Fluctuations* (Oxford University Press, Oxford, 2004)
24. P.A. Mello, P. Pereyra, N. Kumar, Ann. Phys. **181**, 290 (1988)
25. N. Kumar, Phys. Rev. B **31**, 5513 (1985)
26. V.I. Mel'nikov, Pis'maZh. Ekps. Teor. Fiz. **32**, 244(1980); V.I. Mel'nikov, JETP Lett. **32**, 225(1980)
27. V.A. Gopar, M. Martinez, P.A. Mello, Phys. Rev. B **50**, 2502 (1994)
28. K.A. Muttalib, V.A. Gopar, Phys. Rev. B **66**, 115318 (2002)
29. L.S. Froufe-Pérez, M. Yepez, P.A. Mello, J.J. Sáenz, Phys. Rev. E **75**, 031113 (2007)
30. P.R. Wallace, Phys. Rev. **71**, 622 (1947)
31. C.J. Avila, *Estructura de Bandas y Conductancia del Grafeno PTI* (UAM, Mexico, 2020)
32. P.F. Bagwell, T.P.E. Broekaert, T.P. Orlando, C.G. Fonstad, J. Appl. Phys. **68**, 4634 (1990)
33. P. Pereyra, Phys. Rev. Lett. **80** 2677 (1998); P. Pereyra, J. Phys. A **31**, 4521 (1998)
34. P. Pereyra, E. Castillo, Phys Rev. B **65**, 205120 (2002)

Tunneling Time and Wave Packets Evolution

14

Abstract

In this chapter, we explore a foudational topic of significant contemporary interest: the tunneling time. Following the experimental measurement of tunneling times for photons through optical superlattices, [1,2] there has been growing interest in understanding the tunneling time of massive particles. A central theoretical challenge has been how to define the transit time, the velocity and the wave function to describe a quantum particle, within the limits imposed by the uncertainty principle. In this chapter, we summarize several tunneling-time calculations for particles and Gaussian wave packets, both with and without mass, using the closed-form expressions of the scattering amplitudes derived from the theory of finite periodic systems (TFPS), introduced in Chap. 6. We include the first theoretical prediction of this theory, which matches the experimental results within an error margin of 10^{-16} s.

14.1 Wave Packets and Tunneling Time

We have seen in chap. 5 that dealing with the free particle wave function, one way to fix the normalization of the linearly independent solutions $\varphi^+(x) = Ae^{ikx}$ and $\varphi^-(x) = Ae^{-ikx}$ is by assuming that the current density is equal to 1, in that case $A = 1/\sqrt{\hbar k/m}$. We mentioned also that including the temporal part of the wave function, a particle that moves to the right is described by $\psi(x,t) = A\varphi^+(x)e^{-i\omega t}$, while a particle that moves to the left by $\phi(x,t) = A\varphi^-(x)e^{-i\omega t}$. Since the free particle is not a confined particle, there is no quantization of the energy. If the energy of the particle is fixed, the wave number is $k_p = \sqrt{2mE_p/\hbar^2}$, the momentum $p = \hbar k_p$ and the frequency is $\omega_p = E_p/\hbar$. We use the subscript p to refer to the energy, momentum and particle frequency. In quantum theory, questions like: Which is the particle's position? Which is the particle speed? Do not have trivial answers, and many times lead to questionable assertions.

P. Pereyra, *Contemporary Theory of Semiconductors and Devices*,
Graduate Texts in Physics, https://doi.org/10.1007/978-3-032-00296-9_14

If the particle is described by the wave function $\psi(x,t) = Ae^{-i(kx-\omega t)}$, one can imagine that it moves with the speed of the constant phase points in the wave. This idea can be used to determine the speed of the particle. If the phase is constant, we have:

$$\frac{d(kx-\omega t)}{dt} = 0 \qquad \Rightarrow \qquad v_g = \frac{dx}{dt} = \frac{d\omega}{dk} = \frac{\hbar k}{m}. \tag{14.1}$$

This speed is known as the optical wave group velocity. If we accept the idea that the particle motion accompanies the constant phase motion, the group velocity will be the one that describes the particle motion. This result reiterates the relationship $p = mv_g = \hbar k$, which is an essential relation in quantum theory. The group velocity differs from the phase velocity $v = \omega/k$ which is also defined an a well-known quantity in optics.

The probabilistic interpretation of wave function states that the probability of finding the particle at x and time t is $|\psi(x,t)|^2$. For the particle described by a plane wave moving to the right or left, the probability of finding it at point x is

$$|\psi(x,t)|^2 = |A|^2, \tag{14.2}$$

which means a uniform probability distribution, independent of x and t. If instead of a plane wave, the particle is described by a superposition of wave functions, say, of the form

$$\psi(x,t) = a\,\varphi^+(x)e^{-i\omega_p t} + b\,\varphi^+(x)e^{-i\omega_p t}, \tag{14.3}$$

the probability of finding the particle at point x will be

$$|\psi(x,t)|^2 = |A|^2\left(|a|^2 + |b|^2 + 2\Re\left[a\,b^*\,e^{i2(k_p x-\omega_p t)}\right]\right), \tag{14.4}$$

In the particular case of $a = b$, the wave function becomes

$$\psi(x,t) = 2a\cos kxe^{-i\omega t}. \tag{14.5}$$

The real part of this function oscillates in space with wave number k and in time with angular frequency ω. The probability of finding the particle at point x, at time t, is

$$|\psi(x,t)|^2 = 4|a|^2\cos^2 kx, \tag{14.6}$$

independent of time but periodic as function of x. This is no longer constant. Although this function is still defined from $-\infty$ to ∞, it provides insight into how one could obtain a more localized wave function associated with an actual quantum particle. Indeed, since a linear superposition of wave functions is also a solution of the Schrödinger equation, it has long been recognized that, with an appropriate superposition of functions, the probability $|\psi(x,t)|^2$ can eventually be 'localized'. In fact,

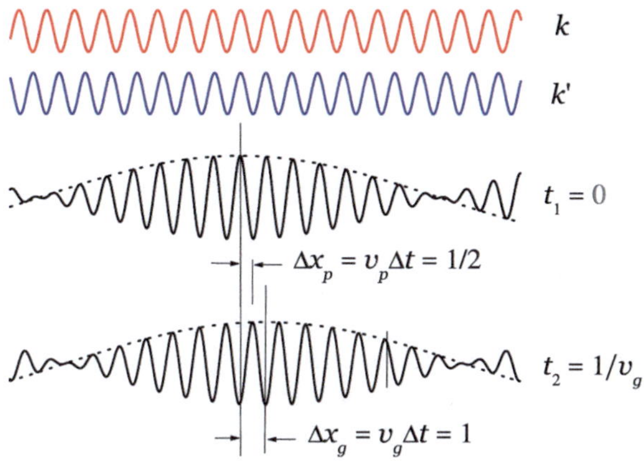

Fig. 14.1 The real parts of two wave functions, with wave numbers $k = k_0 - \delta k$ and $k' = k_0 + \delta k$, and snapshots of their superposition at $t_1 = 0$ and at $t_2 = 1/v_g$. During the time interval $\Delta t = t_2 - t_1 = 1/v_g$, a constant-phase point of the internal waves moves a distance Δx_p and the maximum of the modulation envelope moves a distance Δx_g

if we consider the superposition of two waves[1] describing free particles moving in the same direction like

$$\psi_2(x,t) = a\, e^{i(kx-\omega t)} + a\, e^{i(k'x-\omega' t)}, \tag{14.7}$$

such that

$$\begin{aligned} k &= k_0 - \delta k & k' &= k_0 + \delta k \\ \omega &\simeq \omega_0 - \delta\omega & \omega' &\simeq \omega_0 + \delta\omega, \end{aligned} \tag{14.8}$$

with $\delta\omega = \hbar k_0 \delta k/m$, the two-component wave function in (14.7) can be written as

$$\psi_2(x,t) = 2a\cos(\delta k\, x - \delta\omega\, t) e^{i(k_0 x - \omega_0 t)}. \tag{14.9}$$

This is also a product of two wave functions: one with frequency ω_0 and wave number k_0 and the other a wave function with frequency $\delta\omega_0$ and wave number δk_0. These functions are plotted in Fig. 14.1. When $\delta k_0 \ll k_0$, the wave function $\cos(\delta k\, x - \delta\omega\, t)$ behaves as a modulation of the amplitude of the internal oscillations described by $e^{i(k_0 x - \omega_0 t)}$. These waves have different physical meaning and move with different velocities. The constant-phase condition applied to $(k_0 x - \omega_0 t)$ provides the phase velocity

$$v_p = \frac{dx}{dt} = \frac{\omega_0}{k_0}, \tag{14.10}$$

[1] This simple example is frequently used in optical physics to introduce phase and group velocities.

while applied to $(\delta k\, x - \delta\omega\, t)$ gives the group velocity[2]

$$v_g = \frac{dx}{dt} = \frac{\delta\omega}{\delta k} \simeq \frac{\hbar k_0}{m}, \tag{14.12}$$

which, as mentioned above, reiterates the relation $p = mv = \hbar k$. In Fig. 14.1 we plot the real parts of the wave components together with snapshots of their superposition at $t_1 = 0$ and at $t_2 = 1/v_g = m/\hbar k_0$. During the period of time $\Delta t = t_2 - t_1$, a point of constant phase of the internal oscillations travels with the phase velocity a distance $\Delta x_p = v_p \Delta t = 1/2$ while the modulation envelope moves a distance $\Delta x_g = v_g \Delta t$. This shows that the modulating function propagates with the group velocity $v_g = \hbar k_0/m$. The same thing happens in the superposition of more than two components, with wave numbers $k_0 \pm k$ symmetric with respect to k_0. A superposition that builds a wave packet (WP). The overall wave envelope will move with the group velocity. The phase velocity $\omega_0/k_0 = \hbar k_0/2m$ of the internal oscillations is approximately the same as the phase velocities ω/k and ω'/k' of the wave components.

We will consider now a superposition of functions like

$$\Psi(x,t) = \int_{k_p-k_a}^{k_p+k_a} A(k) e^{i(kx-\omega t)} dk. \tag{14.13}$$

In this superposition, pairs of wave functions with wave numbers k in the neighborhood of k_p and, amplitude $A(k)$ each, add to form a wave packet. The shape of the packet relies strongly on $A(k)$. One of the most frequently used amplitude is the Gaussian function centered at k_p, with dispersion (or standard deviation) σ_k. In this case, we have

$$\Psi(x,t) = \frac{1}{\sqrt{2\pi}\sigma_k} \int_{-\infty}^{\infty} e^{-(k-k_p)^2/2\sigma_k^2} e^{i(kx-\omega t)} dk. \tag{14.14}$$

Let us see if a superposition like this can be used to describe a free particle with wave number k_p and speed $v = \hbar k_p/m$. To visualize this function, we need to perform the integral upon the variable k. To this purpose it is convenient to write the exponent as follows

$$-\frac{(k-k_p)^2}{2\sigma_k^2} + i\left(kx - \frac{\hbar k^2 t}{2m}\right) = -a\,(k-\kappa)^2 + a\,\kappa^2 - \frac{k_p^2}{2\sigma_k^2}, \tag{14.15}$$

[2] In the limit $\delta k \to 0$ we have

$$v_g = \frac{d\omega}{dk}. \tag{14.11}$$

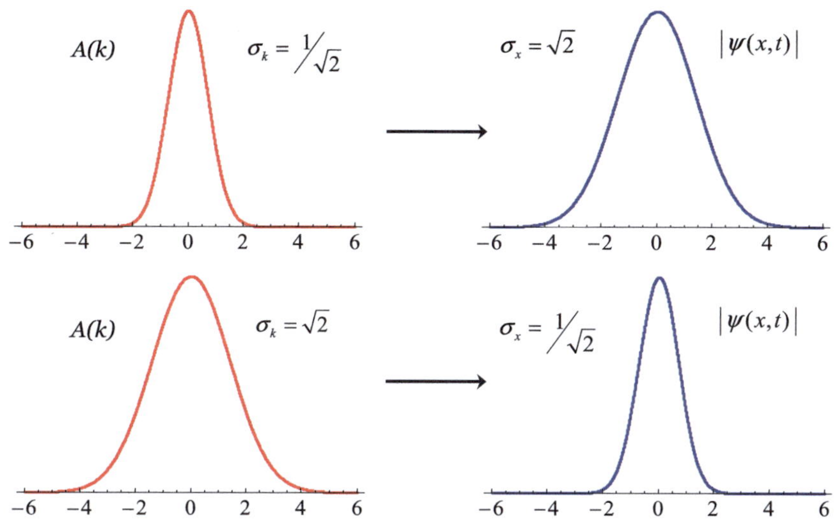

Fig. 14.2 The Gaussian amplitude $A(k)$ in the superposition (14.14), produces a Gaussian function whose variance σ_x^2 is roughly the inverse of the variance σ_k^2 in $A(k)$. This is usually referred as the uncertainty relation of position and linear momentum $\Delta x \simeq 1/\Delta k$

with

$$a = \frac{m + i\hbar\sigma_k^2 t}{2m\,\sigma_k^2} \qquad \text{and} \qquad \kappa = \frac{im\sigma_k^2(x - ik_p/\sigma_k^2)}{m + i\hbar\sigma_k^2 t}, \tag{14.16}$$

the integration on the variable k is direct and the packet takes the form

$$\Psi(x,t) = \frac{1}{\sigma_x\sigma_k} e^{\frac{i(k_p x - w_p t)}{\sigma_k^2\sigma_x^2}} e^{-\frac{x^2}{2\sigma_x^2}}. \tag{14.17}$$

This is also a Gaussian function with

$$\sigma_x^2 = \frac{m + i\hbar\sigma_k^2 t}{m\sigma_k^2} = \frac{1 + i\frac{\hbar t}{m}\sigma_k^2}{\sigma_k^2}. \tag{14.18}$$

As expected, the procedure of building up a wave packet, effectively concentrates the probability $|\Psi(x,t)|^2$ of finding the particle around the Gaussian centroid. If the wave packet describes a particle, the peak position will coincide on the average with the particle position, with a dispersion Δx. In Fig. 14.2 we plot the functions $A(k)$ and $|\Psi(x,t)|$ for the standard deviations indicated there. It is evident from these plots that $\sigma_x = \Delta x \simeq 1/\Delta k = 1/\sigma_k$. This relation is consistent with (14.18). We will return to this subject some lines below.

Does the wave packet move with the speed of a free particle? To visualize this, we follow the space time evolution of the wave packet. By observing it at two times, say

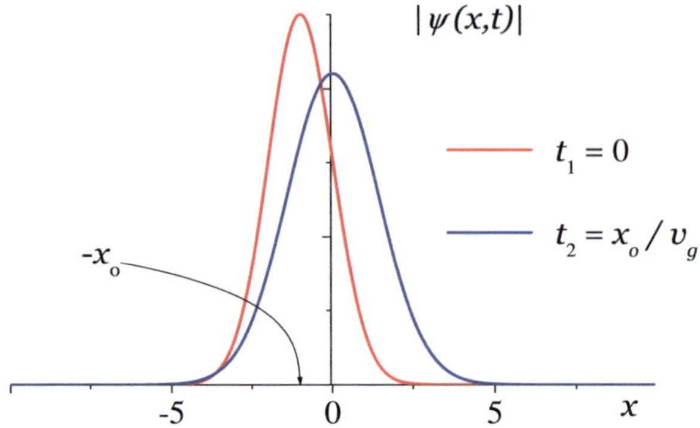

Fig. 14.3 Snapshots of the Gaussian wave packet at times $t_1 = 0$ and $t_2 = x_0/v$, defined as the instants at which a free particle with speed $v = \hbar k_0/m$ passes the positions $-x_o$ and the origin. Since the wave-packet peak also moves from $-x_0$ at t_1 and reaches the origin t_2, we conclude that, the wave-packet group velocity coincides with the speed of the free particle

t_1 and t_2, we can obtain the speed of the wave packet peak. In Fig. 14.3 we present results for this "experiment". We start with the wave function

$$\Psi(x,t) = \frac{1}{\sigma_x \sigma_k} e^{\frac{i(k_p(x+x_0)-w_p t)}{\sigma_k^2 \sigma_x^2}} e^{-\frac{(x+x_0)^2}{2\sigma_x^2}}, \tag{14.19}$$

located at $x = -x_0$ when $t = 0$. We plot the wave packet at $t_1 = 0$ s and at $t_2 = x_0/v$, with $v = \hbar k_p/m$ the speed of the free particle. If the wave-packet peak moves with a group velocity $v_g = v$, at $t = t_2$ it must be at $x = 0$. The snapshots show that this is precisely what happens. Hence, wave packets can be used to describe free particles. The wave-packet peak describes the average particle position, and the standard deviation of the wave packet describes the dispersion of the particle position.

A widely discussed problem in quantum physics, is the time that a particle takes to pass through a region with a potential, such as the potential barrier. In classical physics the procedure to calculate the transit time is well defined. In quantum physics, the tunneling time has been a controversial topic since the 1930s, more precisely since MacColl, in 1932 [3], first addressed the question of tunneling time through a potential barrier. As is well-known now, one of the striking quantum properties is the tunneling effect, that is, the possibility that quantum particles can pass through a classically forbidden potential region, even when the particle energy is lower than the height of the potential barrier. While in classical physics it is natural to consider the transit time when the particle goes from one point to another, it was not clear how to calculate the transit time for a quantum particle, in particular how long it takes to tunnel through a barrier? One of the first answers was the phase time or Wigner time [4], which we will introduce below. There were other proposals [4–10]. The interested reader can look at the review articles, for example at [7]. As mentioned in the introduction, in recent years accurate measurements of the transit time of

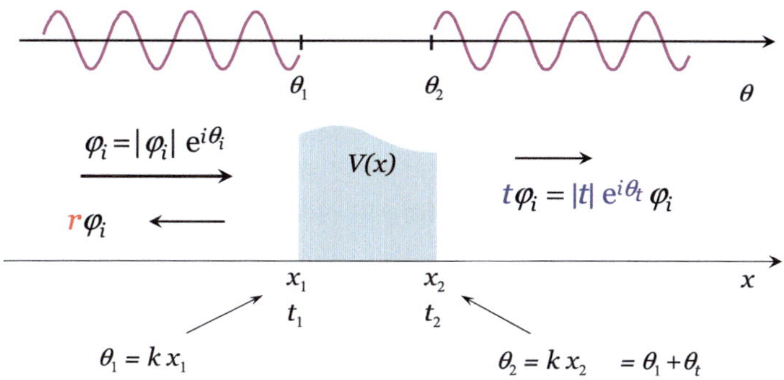

Fig. 14.4 The phase evolution when a quantum particle moves through a potential region, and the phases of the incoming and transmitted wave functions

photons and electromagnetic pulses have been reported. It has also been shown that the tunneling, or delay time, in the quantum theory and in the electromagnetic theory is correctly described by the so-called phase time [11].

How is the phase time defined? We have seen in different examples that the transmitted wave function $\varphi_t(x_2)$ and the incident wave function $\varphi_i(x_1)$ have among them a very simple relation. In fact, if the points x_1 and x_2 are just at the left and right of a potential region, say a potential barrier like in Fig. 14.4, there exist a barrier transfer matrix such that

$$\begin{pmatrix} a_t e^{ikx_2} \\ b_3 e^{-ikx_2} \end{pmatrix} = \begin{pmatrix} \alpha_b & \beta_b \\ \beta_b^* & \alpha_b^* \end{pmatrix} \begin{pmatrix} a_i e^{ikx_1} \\ b_r e^{-ikx_1} \end{pmatrix}. \tag{14.20}$$

In terms of the reflection and transmission amplitudes, r_b and t_b, this relation reads

$$\begin{pmatrix} a_t e^{i\theta_2} \\ b_3 e^{-i\theta_2} \end{pmatrix} = \begin{pmatrix} 1/t_b^* & -r_b^*/t_b^* \\ -r_b/t_b & 1/t_b \end{pmatrix} \begin{pmatrix} a_i e^{i\theta_1} \\ b_r e^{-i\theta_1} \end{pmatrix}. \tag{14.21}$$

If the incident particle comes only from the left, $b_3 = 0$, the matrix relation implies the following equations

$$b_r e^{-ikx_1} = r_b\, a_i e^{ikx_1} \tag{14.22}$$

$$a_t e^{ikx_2} = \frac{1}{t_b^*} a_i e^{ikx_1} - \frac{r_b^*}{t_b^*} b_r e^{-ikx_1}. \tag{14.23}$$

Combining these equations and taking into account that $1 - r_b r_b^* = t_b t_b^*$, we obtain

$$a_t e^{ikx_2} = t_b a_i e^{ikx_1} = |t_b| e^{i\theta_t} a_i e^{ikx_1}, \tag{14.24}$$

which is a known result that we have used to obtain the transmission coefficient and the transmitted wave function. In the last equation we have written the transmission amplitude in terms of its absolute value $|t_b|$ and its phase θ_t. This allows us to observe the following important relation between phases

$$kx_2 = kx_1 + \theta_t, \qquad \theta_t = k(x_2 - x_1). \tag{14.25}$$

If we take the frequency derivative of the transmission amplitude phase θ_t, we have

$$\frac{\partial \theta_t}{\partial \omega} = \frac{\partial k}{\partial \omega}(x_2 - x_1) = \frac{1}{v_g}(x_2 - x_1). \tag{14.26}$$

If, on the other hand, the elapsed time while the particle moves from x_1 to x_2 is $(x_2 - x_1)/v_g$, the time we are seeking, known as the *phase time*, is

$$\tau = \frac{\partial \theta_t}{\partial \omega}. \tag{14.27}$$

This time has been evaluated for photon transmission problems and electromagnetic pulses through optical media. The theoretical calculations of the phase time [11] describe correctly the experimental results.[3] When the transmission is by quantum tunneling, the transit time is known as the *tunneling time*. Consequently, one of the possible ways to calculate the elapsed time to tunnel is the phase time τ, (we do not use the letter t here to avoid confusion with the transmission amplitude) which can also be written as

$$\tau = \hbar \frac{\partial \theta_t}{\partial E} \tag{14.28}$$

14.2 Experimental Tunneling Times

As mentioned before, the interest in the tunneling, or transit, time of quantum particles through potential barriers grew in the 1990s, after the experiments of Steinberg et al. at Berkeley [1] and Spielmann's group in Vienna [2]. These groups performed ingenious and compelling experiments to measure photon transit times and pulse propagation through silica and titanium oxide superlattices. These are examples of how semiconductor superlattices were also used for basic research in physics. The experiments of Steinberg and of Spielmann put, practically, an end to the variety of models and methods proposed to calculate and measure tunneling time. The measured transit times are of the order of 1 fs, and the errors are tenths of fs, that is, $\sim 10^{-16}$ s. The numerical calculations by Steinberg et al. and by Spielmann et al. used to compare their results with the proposed models tended to support the phase time, although the numerical error was greater than the experimental one. The application of the TFPS and the analytical and exact formulas, valid regardless of the

[3] See for example [12].

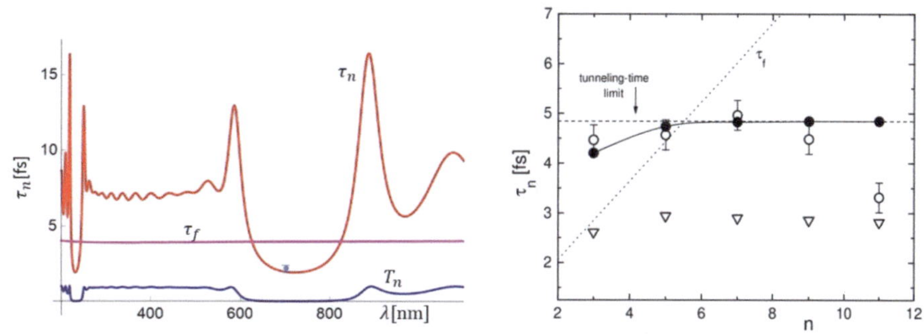

Fig. 14.5 Tunneling times through the superlattices $(HL)^nH$ and $(LH)^nL$ with high (H) and low (L) refractive indices for wavelengths in the photonic bandgaps. In **a** the times measured by Steinberg et al. in the superlattice $(HL)^nH$ for $n = 5$. In the left graph, the tunneling time τ_n, transmission coefficient T_n and τ_f, are shown. τ_f is the time that would pass by if the medium is empty and the photon speed is c. In **b** the tunneling times (open circles) reported by Spielmann et al. through $(LH)^nL$, for different values of n. The black circles are the TFPS predictions. Figure reproduced from [11] with permission of the American Physical Society

size of the superlattices, allowed to calculate the phase of the transmission amplitude, exactly, and to show [11] that Wigner's phase time [4] correctly describes the observations as seen in Fig. 14.5, where the measured times by Steinberg et al. and by Spielmann et al., together with the theoretical prediction, that falls within the experimental error, are shown. In the following section we will briefly comment on the TFPS applied to the calculation of the tunneling time in these experiments, and in Sect. 14.4 study the transport and transit time of Gaussian packets of electrons and of fields through semiconductor superlattices.

14.3 Transit Time Through Superlattices

If we write the transmission amplitude of a superlattice in terms of its real and imaginary parts, that is, if

$$t_n = t_{nr} + i t_{ni} \tag{14.29}$$

and we substitute this amplitude in the definition of tunneling time in (14.28), we have

$$\tau_n = |t_n|^{-2} \left(t_{nr} \partial t_{ni}/\partial\omega - t_{ni} \partial t_{nr}/\partial\omega\right). \tag{14.30}$$

With the general expressions that were introduced in Chap. 6, the general, closed formula for the evaluation of this time is

$$\tau_n = \frac{\hbar}{(U_n - \alpha_r U_{n-1})^2 + (\alpha_i U_{n-1})^2} \left(A_r \frac{d\alpha_r}{dE} + A_i \frac{d\alpha_i}{dE} \right) \tag{14.31}$$

with

$$A_r = \frac{\alpha_i}{1 - \alpha_r^2} \left((\alpha_r U_{n-1} + n U_{n-2})\, U_n - \left(n + \alpha_r^2\right) U_{n-1}^2 \right) \tag{14.32}$$

and

$$A_i = (U_n - \alpha_r U_{n-1})U_{n-1} \tag{14.33}$$

These formulas were first reported in [11]. In the next section we present the theoretical predictions for the tunneling time and compare it with the time spent by electron and photon wave packets propagating through superlattices with velocities dictated by thir respective dynamical laws.

14.3.1 Transit Time of Photons Through Optical Superlattice

For the specific superlattice in which the tunneling time was measured, the elements of the unit cell transfer matrix are

$$\begin{aligned} \alpha &= \frac{e^{ik_H d_H}}{4n_H n_L}\left(e^{ik_L d_L}(n_H + n_L)^2 - e^{-ik_L d_L}(n_H - n_L)^2\right) \\ \beta &= i\frac{n_H^2 - n_L^2}{2n_H n_L}\sin k_L d_L = i\beta_i \end{aligned} \tag{14.34}$$

For a more precise calculation, the transfer matrix of the additional layer H or L of the substrate, whose length is l_s, should be included. The tunneling times obtained are shown in Fig. 14.5. In the left panel of this figure, we show the tunnel time reported by Steinberg et al. through the SL $air(HL)^n H(substrateL)air$, for $n = 5$, along with the calculated tunneling times τ_n and τ_f, and the transmission coefficient T_n. The curve labeled with τ_f is the time that would pass by if the medium is empty and the particle moves with velocity c. In the right panel, we have the tunneling times reported by Spielmann et al. (open circles) through superlattices $air(substrateL)(HL)^n air$ of titanium oxide (H) and silica (L) for different values of n. The black circles are the phase time predictions calculated from (14.31) [11]. These predictions fall within the error bars, which are of the order of 10^{-16} s. The open triangles are the phase times calculated numerically by Spielmann et al.

14.3.2 Tunneling Time of Electrons Through Superlattice

To our knowledge, there is no direct measurement of electron tunneling time, it has been an issue of interest to understand mechanisms related to transport or optical transitions in multiple quantum wells structures. The tunneling time of electrons, with a given energy E and momentum $p = \hbar k$, through a superlattice $(A^{1/2}BA^{1/2})^n$ contained between two thick cladding layers, was calculated in [11], and tunneling times of wave packets peaked at a given particle momentum $p_p = \hbar k_p$, as in (14.14) were reported in [13], for mass particles, and for wave packets in [12]. We will refer to the latter two in the next section. We will here briefly refer to the tunneling time of electrons with a given momentum $\hbar k$. In this case, (14.31) can directly be applied. All

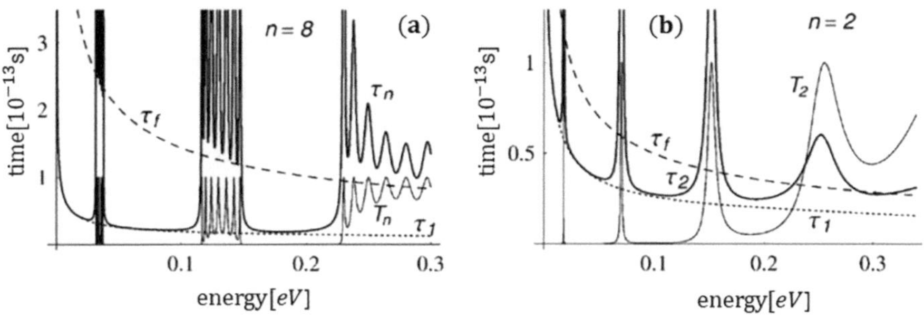

Fig. 14.6 Electron tunneling times through the $(GaAs^{1/2}Al_xGa_{1-x}AsGaAs^{1/2})^n$ superlattice with $x = 0.03$ for $n = 8$ in (**a**) and $n = 2$ in (**b**). Here τ_n is the tunneling time through the whole superlattice, τ_1 is the tunneling time through a single cell, and τ_f is the time that a free particle needs to move a distance equal to the superlattice length $L = n\ell_c$, with the speed $v_e = \hbar k_p/m_A^*$. Figure reproduced from [11] with permission from the American Physical Society

we need is the single-cell $A^{1/2}BA^{1/2}$ transfer matrix, or just its well-known matrix element

$$\alpha = e^{ikd_A}\left(\cosh\kappa d_B + i\frac{k^2-\kappa^2}{2k\kappa}\sinh\kappa d_B\right), \qquad (14.35)$$

where $k = \sqrt{2m_A^*E/\hbar^2}$ and $\kappa = \sqrt{2m_B^*(V_0 - E)/\hbar^2}$.

In Fig. 14.6 the transmission coefficient T_n and tunneling times τ_n, τ_1 and τ_f (in units of 10^{-13} s) are plotted for a superlattice where A is $GaAs$, B is $Al_xGa_{1-x}As$ with $x = 0.03$, $n = 8$, $d_A = 10\,\text{nm}$, $d_B = 3\,\text{nm}$, $m_A^* = 0.067$, $m_B^* = 0.1$ and $GaAs$ cladding layers. In this figure, τ_n is the tunneling time through the whole superlattice, τ_1 is the tunneling time through a single cell, and τ_f is the time that a free particle needs to move a distance equal to the superlattice length $L = n\ell_c$, with the speed $v_e = \hbar k_p/m_A^*$.

As can be seen in Fig. 14.6, for SL (with $n = 8$ or double barrier structure with $n = 2$), the tunneling times of electrons are two orders of magnitude larger than the photon tunneling times.

In the next section we calculate again the electron and photon tunneling times representing them as wave packets.

14.4 Space-Time Evolution of Wave Packets in Superlattices

As mentioned above, a recognized way to study the transport of quantum particles through a superlattice is by studying the evolution of a wave packet built as the superposition of wave functions for particles with mass and electromagnetic fields for photons. In this section, we will present the tunneling times of electron and photon Gaussian wave packets that were briefly reported in [14] and studied with more detail in [12,13], using the transfer matrix method introduced in the previous section and in Chap. 6. In the case of electrons we will consider the energy of the

Gaussian centroid coinciding with a resonance of the transmission coefficient, while for the electromagnetic case, the wavelength of the centroid will be in the transparent region, in one case, and in a relatively opaque region in the other.

We know that if we have a finite periodic system with n cells of length l_c, the wave function of a particle, in vector representation, can be written as

$$\Phi(z) \;=\; \begin{pmatrix} \psi_a(z) \\ \psi_b(z) \end{pmatrix} \tag{14.36}$$

where $\psi_a(z)$ and $\psi_b(z)$ are functions that describe particles moving to the right (or exponentially increasing) and to the left (or exponentially decreasing wave functions). The state vectors $\Phi(z)$ and $\Phi(z')$, at arbitrary points z and z', are related using the transfer matrix

$$M(z, z') \;=\; \begin{pmatrix} \alpha & \beta \\ \gamma & \delta \end{pmatrix}. \tag{14.37}$$

In the approximation of one propagating mode, α , β , γ and δ are complex scalar functions of the potential parameters and of z and z'. As explained in Chap. 6, owing to the multiplicative property of the transfer matrices, the state vector $\Phi_{j+1}(z)$ at any point z in the $(j+1)$th unit cell of the superlattice can be written as

$$\Phi_{j+1}(z) \;=\; M(z, jl_c)\,[M(l_c, 0)]^j\,\Phi_1(0)\;,\; j = 0, 1, \dots, (n-1) \tag{14.38}$$

where $[M(l_c, 0)]^j$ is the transfer matrix for j unit cells, i.e. from $z = 0$ to $z = z_j = jl_c$, and $M(z, jl_c)$ is the transfer matrix that connects the state vector at z_j with the state vector at $z > z_j$. For square barrier potentials of height V_0 and width b separated by potential wells of width a (so that $l_c = a + b$), the transfer matrix of a unit cell $M(l_c, 0)$ is

$$M(l_c, 0) \;\equiv\; \begin{pmatrix} \alpha & \beta \\ \beta^* & \alpha^* \end{pmatrix} \;=\; \begin{pmatrix} \alpha_b e^{ika} & \beta_b \\ \beta_b^* & \alpha_b^* e^{-ika} \end{pmatrix} \tag{14.39}$$

where

$$\alpha_b = \cosh(\kappa b) - \frac{i}{2}\left(\frac{\kappa}{k} - \frac{k}{\kappa}\right)\sinh(\kappa b), \quad \text{and} \quad \beta_b = -\frac{i}{2}\left(\frac{\kappa}{k} + \frac{k}{\kappa}\right)\sinh(\kappa b), \tag{14.40}$$

with $k = \sqrt{2mE/\hbar^2}$ and $\kappa = \sqrt{2m(V_0 - E)/\hbar^2}$. On the other hand, the transfer matrix of j cells $[M(l_c, 0)]^j$ is

$$[M(l_c, 0)]^j = \begin{pmatrix} \alpha & \beta \\ \beta^* & \alpha^* \end{pmatrix}^j \;\equiv\; \begin{pmatrix} \alpha_j & \beta_j \\ \beta_j^* & \alpha_j^* \end{pmatrix} \tag{14.41}$$

where $\alpha_j = U_j(\alpha_R) - \alpha^* U_{j-1}(\alpha_R)$, $\beta_j = \beta U j - 1(\alpha_R)$. Here $U_j(\alpha_R)$ is the Chebyshev polynomial of the second class, evaluated in the real part of α.

We write the wave function on the left of the superlattice as

$$\psi(z) = A\left(e^{ikz} + r_n e^{-ikz}\right). \tag{14.42}$$

where the reflection amplitude is $r_n(k) = -\beta_n^*/\alpha_n^*$. As explained in Chap. 6, the wave function at z can be written as

$$\psi_{j+1}(z) = A\left[(\alpha_p + \gamma_p)\left(\alpha_j - \beta_j \frac{\beta_n^*}{\alpha_n^*}\right) + (\beta_p + \delta_p)\left(\beta_j^* - \alpha_j^* \frac{\beta_n^*}{\alpha_n^*}\right)\right] \tag{14.43}$$

where α_p , β_p , γ_p , δ_p are the elements of $M(z, jl_c)$.

Let us now consider the Gaussian packet that at $t = 0$ is located in z_0. The wave packet at time $t > 0$ in point z, is given by

$$\Psi(z,t) = \int dk\, e^{-\gamma(k-k_0)^2} e^{ikz_0} \psi(z,k) e^{-i\omega(k)t} \tag{14.44}$$

where $\omega(k) = \hbar k^2/2m$. Here we write $\psi(z,k)$ for the function $\psi(z)$ in (14.42). Within the superlattice $\psi(z,k)$ becomes the function $\psi_{j+1}(z)$, and outside, in the transmitted region, it becomes $\psi(z,k) = \frac{1}{\alpha_n^*} e^{ik(z-n\ell_c)}$.

By performing Gaussian integration, the superlattice tunneling time $\tau(k_0)$, i.e. the time required by the centroid of the wave packet to traverse the superlattice, is

$$\tau(k_0) = -\frac{1}{v_g}\left[a + \left(\frac{d\theta_{\alpha nc}(k)}{dk}\right)_{k_0}\right] \tag{14.45}$$

where $\theta_{\alpha nc}$ is the phase of $\alpha_n^* = |\alpha_n| e^{i\theta_{\alpha nc}}$ and $v_g = (d\omega/dk)_{k_0}$ is the group speed. When instead of the transmission amplitude phase one considers the phase of the reflection amplitude, we have the reflection time

$$\tau_d(k_0) = -\frac{1}{v_g}\left[a - \left(\frac{d\theta_{rn}(k)}{dk}\right)_{k_0}\right] \tag{14.46}$$

where $\theta_{rn} = \pi/2 - \theta_{\alpha nc}$ is the phase of $r_n = |r_n| e^{i\theta_{rn}}$. Therefore, the tunneling time $\tau(k_0)$ is the same as the reflection time $\tau_d(k_0)$. This is the tunneling time derived in [11], which agrees well with experimental results [1,2].

14.4.1 Results for Electron Waves Packet

In Fig. 14.7, we have a time series calculations for a Gaussian packet of electron waves with centroid at $E_0 = 0.075$ eV, which is at the center of the gap (the forbidden band) of the superlattice $(AB)^n = GaAs(Al_{0.3}Ga_{0.7}As/GaAs)^6$, with unit-cell length equal to $\ell_c = 130$ Å and effective masses $m_A^* = 0.067$, $m_B^* = 0.1$. The width of the

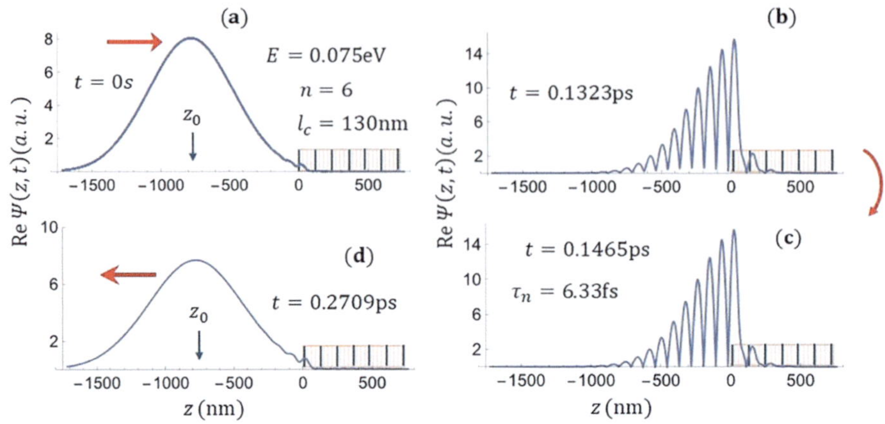

Fig. 14.7 Snapshots of a Gaussian packet of electron waves with frequencies in the gap, with transmission probability approximately equal to zero. The width of the packet is defined by $\gamma = 12.133$ Å^2. The superlattice is $(AB)^n = GaAs(Al_{0.3}Ga_{0.7}As/GaAs)^6$, with unit-cell length equal to $\ell_c = 130$ Å and effective masses $m^*_A = 0.067$ and $m^*_B = 0.1$. At $t = 0s$, the maximum of the packet (the centroid) is at $z_0 = -6l_c = -780$ nm. In panel **b** $t = (z_0 + a/2)/v_g$=0.1323 ps, the maximum of the packet reaches the first barrier of the superlattice, and at (**c**), for $t = (z_0 + a/2)/v_g + \tau_n$ = 0.1386 ps, the centroid of the packet is, to a larger fraction, reflected presumably to the point where it was in panel (**b**). In panel **d** the elapsed time is $\Delta t = (2z_0 + a)/v_g + \tau_n$=0.2709ps. If the tunneling time is indeed τ =6.33fs, the packet centroid should be back in the initial position z_0

Gaussian WP is defined here with $\gamma = 12.133$ Å^2. The tunneling time for the centroid of the wave packet, with energy $E_0 = 0.075$ eV, is $\tau_n = 6.33 \times 10^{-15}$ s. As shown in [15], and can be seen in Fig. . 14.7, the reflected packet, as well as the transmitted packet (not shown here), preserve the shape of the incident packet. This is possible when the packet width is smaller than the gap width, and also because almost all components have the same tunneling time.

At the initial instant, $t = 0s$, the maximum of the wave packet is at $z_0 = -6l_c = -780$ nm, the length of the superlattice is also $L = 6\ell_c = 780$ nm. In (**b**) $t = (z_0 + a/2)/v_g$=0.1323ps. At this instant the maximum of the packet reaches the first barrier of the superlattice. In (**c**), we have the packet when $t = (z_0 + a/2)/v_g + \tau_n$ =0.1386ps. At this time the centroid of the packet, if the tunneling time prediction is correct, should have been transmitted or reflected. The centroid of the packet should return to the point where it was on panel (**b**). This is certainly what happens, thus, the predicted tunneling time is correct. If it were incorrect, the centroid will be somewhere else. In panels (**b**) and (**c**) the interference between the arriving and returning components is strong. In panel (**d**) the wave packet should have, theoretically, been returned to the initial point. The elapsed time for this round is $\Delta t = (2z_0 + a)/v_g + \tau_n$=0.2709ps. If the tunneling time considered here is indeed τ_n =6.33 fs, the centroid of the Gaussian packet should be at its initial position z_0, as it is effectively seen in the figure.

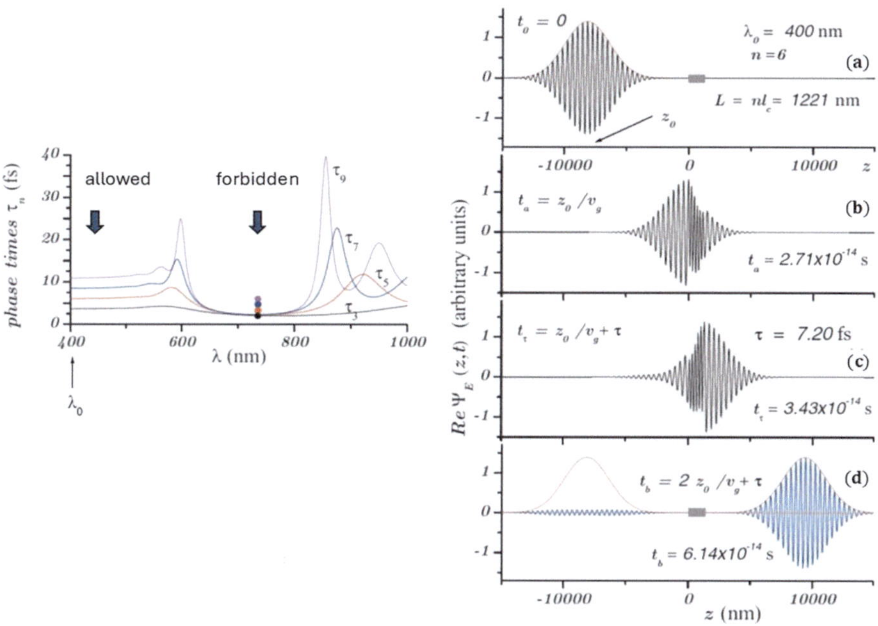

Fig. 14.8 Transmission of a Gaussian packet of electromagnetic waves defined in the allowed band of the optical superlattice $(L/2HL/2)^n$ defined in the text. In this case the centroid wavelength is $\lambda_0 = 400$ Å. In the left panel, the phase time of the optical superlattice (OSL), as a function of the wave length λ, for different values of n. In panel **a**, the centroid of the WP localized at $z_0 = -40\ell_c$ when $t = 0$. In **b**, the WP reaches the OSL at $t_a = z_0/v_g = 2.71 \times 10^{-14}$ s, and in **c** the wave packet when $t_\tau = z_0/v_g + \tau = 3.43 \times 10^{-14}$ s. At this time the centroid of the transmitted WP starts to appear on the other side of the OSL, with a tunneling time $\tau = 7.2 \times 10^{-15}$ s. In **d**, the wave packet at $t_b = 2z_0/v_g + \tau = 6.14 \times 10^{-14}$ s, when the reflected WP should return and appear at a distance z_0 from the right edge of the OSL, if the predicted tunneling time is correct. Figure reproduced from [12], with permission of the American Physical Society

14.4.2 Results for Packets of Electromagnetic Waves

The same method can be applied to the case of electromagnetic waves packets, but here we will present only the results to avoid unnecessary repetitions. We will consider a superlattice $(\text{LH})^n$ made of silica L and titanium oxide H with $l_L = 124.46$ nm, $l_H = 79.06$ nm, and refractive indices $n_L = 1.41$ and $n_H = 2.22$, such that $n_L l_L = n_H l_H$. Systems like this were used by Steinberg et al. [1] and by Spielmann et al. [2].

We will consider two examples of electromagnetic WPs moving across the optical superlattices $(L/2HL/2)^n$ made of silica L and titanium oxide H. The wave packet in one case is defined in the allowed band of the optical superlattice (OSL), hence it will be almost completely transmitted, and in the other case, in the gap, thus almost completely reflected.

In Fig. 14.8 we show the snapshots of a Gaussian electromagnetic wave packets, whose centroid wavelength is equal to $\lambda_0 = 400$ nm, at times $t = 0$, $t = t_a$, $t = t_\tau$,

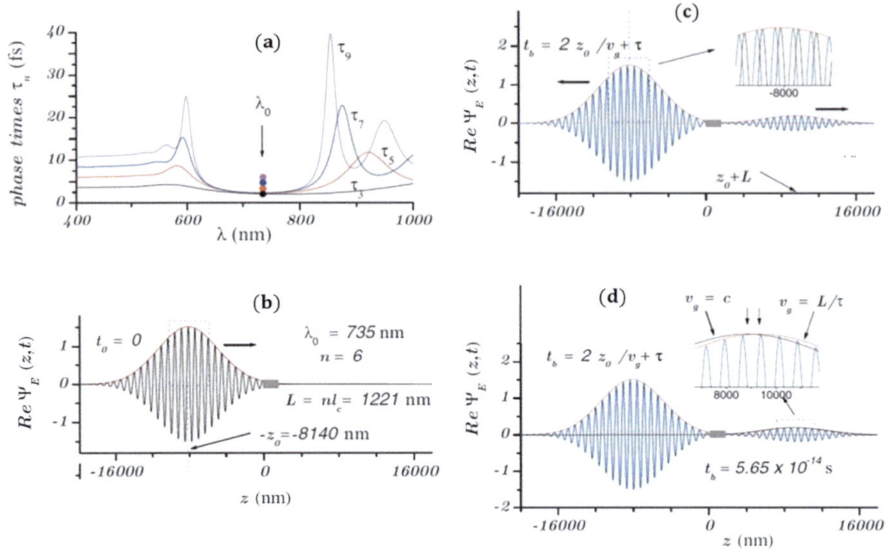

Fig. 14.9 Reflected Gaussian wave packet defined in the gap of the optical superlattice $(L/2HL/2)^n$. In panel **a** we show again the phase time as a function of the wave length λ. The OSL is the same as in Fig. 14.8. In panel *b* the WP with its centroid located at $z_0 = -40\ell_c$, when t =0s. In panels *c* and *d* we show the reflected WP, with different insets, both at $t_r = 2z0/v_g + \tau$ with $\tau = 2.32$ fs, when presumably the WP returns to its original position at $t = 0$. In the inset of panel **c** the WP and the gaussian envelope at $t = 0$ are plotted with the WP at t_r. This inset shows that the WP returned exactly at the original position. The red Gaussian curve in the inset of panel **d** is drawn assuming that the group velocity is $v_g = L/\tau$ while the black curve is drawn assuming that $v_g = c$. These figures show that the actual time spent in the OSL by the WP, defined with solutions of Maxwell equations, coincides with the phase time τ. Figure reproduced from [12], with permission of the American Physical Society

and $t = t_b$. As can be seen in Fig. 14.8, the OSL is transparent for $\lambda_0 = 400$ nm and wavelengths around this value. The transmission coefficients are close to 1 and the phase times of the order of 7.2 fs. In panel (**a**) of Fig. 14.8, we have the approaching wave packet, centered at $z_0 = -40\ell_c = -8140$ nm, at the instant $t = 0$. In (**b**), the WP at $t_a = z_0/v_g = 2.71 \times 10^{-14}$ s, and in (**c**) the wave packet at $t_\tau = z_0/v_g + \tau = 3.43 \times 10^{-14}$ s, when presumably the centroid of the WP should appear on the other side of the OSL, if the predicted tunneling time $\tau = 7.2 \times 10^{-15}$ s , is correct. In (**d**), the wave packet at $t_b = 2z_0/v_g + \tau = 6.14 \times 10^{-14}$ s, when the reflected WP should return to z_0, and the transmitted should appear at a distance z_0 from the right edge of the OSL. The figure shows that the phase time predictions are correct.

In Fig. 14.9 we show a Gaussian wave packet defined in the gap of the optical superlattice $(L/2HL/2)^n$. In this case the wavelength of the centroid is λ_0 =735 nm. Panel (**a**) on the left, shows the phase time of the OSL as a function of the wavelength λ. The OSL is the same as in Fig. 14.8. In panel **b**, the centroid is located at $z_0 = -40\ell_c$ when t =0s. In panels **c** and **d** we plot the wave packet at $t_r = 2z0/v_g + \tau$ where $\tau = 2.32$ fs. The difference between panels (**c**) and (**d**) are their insets. When $t = t_r$, the wave packet, presumably, returns to its original position where it was at $t = 0$. In the inset of panel (**c**), the packet and the Gaussian envelope are plotted when $t = 0$ and when $t = t_r$. This inset shows that the packet that returns at time t_r coincides precisely with the WP at the original position. This means that the time $t_r = 2t_a + \tau$, which includes the phase time τ, is correct. The red Gaussian in the inset of panel (**d**) will be the envelope of both packets if the group velocity is $v_g = L/\tau$, and the black curve will be the envelope if $v_g = c$. These figures show that the correct envelope is the red one, that is, the time used by the package, with solutions of Maxwell's equations, coincides with the predicted phase time τ.

References

1. A.M. Steinberg, P.G. Kwiat, R. Chiao, Phys. Rev. Lett. **71**, 708 (1993)
2. C. Spielmann, R. Szipöcs, A. Stingl, F. Krausz, Phys. Rev. Lett. **73**, 2308 (1994)
3. L.A. MacColl, Phys. Rev. **40**, 621 (1932)
4. E.P. Wigner, Phys. Rev. **98**, 145 (1955)
5. D. Bohm, *Quantum Theory* (Prentice-Hall, New York, 1951)
6. M. Büttiker, R. Landauer, Phys. Rev. Lett. **49**, 1739 (1982)
7. E.H. Hauge, J.A. Støvneng, Rev. Mod. Phys. **61**, 917 (1989)
8. A. Ranfagni, D. Mugnai, P. Fabeni, G.P. Pazzi, Appl. Phys. Lett. **58**, 774 (1991)
9. A. Enders, G. Nimtz, J. Phys. I (France) **2**, 1693 (1992)
10. R. Landauer, Th. Martin, Rev. Mod. Phys. **66**, 217 (1994)
11. P. Pereyra, Phys. Rev. Lett. **84**, 1772 (2000)
12. P. Pereyra, H.P. Simanjuntak, Phys. Rev. E **75**, 056604 (2007)
13. H.P. Simanjuntak, P. Pereyra, Phys. Rev. B **67**, 045301 (2003)
14. H.P. Simanjuntak, P. Pereyra, Phys. Stat. Sol. C **2**, 3108 (2005)
15. P. Pereyra, Ann. Phys. **320**, 1 (2005)

Semiconductor Spintronics and Magnetic Superlattices

15

Abstract

The intrinsic angular momentum known as the spin is a property of quantum particles, as important as the electric charge. Particles such as electrons, protons, light holes, etc., whose spin has in a given direction, say z, a projection equal to a half-odd-integer of $\hbar$, are fermions and obey the Pauli exclusion principle and those particles as photons whose spin projection is an integer multiple of $\hbar$, are bosons. In the previous chapters, the physical property that has allowed us to distinguish and define concentrations of electrons and holes, to establish currents and to modify them with the action of fields, particularly of the electric field, was the electric charge. Electrons and holes, in addition to charge, have also spin which, likewise charge, has two possible projections $s_z = \pm\hbar/2$, projections that are recognized when a magnetic field interacts with the spin of the particle. This interaction is proportional to $\mathbf{s} \cdot \mathbf{B}$, and the gradient of this energy allows to detect and distinguish one spin orientation from another, as in the Stern–Gerlach experiment. The idea of using electronic spin in addition to charge has been highly appealing and has sparked a great deal of research work mainly oriented to develop specific devices. Simultaneously, this has also become a basic physics topic [1]. The injection and extraction of spin-polarized electrons has been one of the important issues in the transition to the applications step in spintronics. When the conductivity mismatch for the spin injection from a ferromagnetic metal into a diffusive semiconductor was revealed as a fundamental obstacle [2], the interest in dilute magnetic semiconductors grew up [3–15]. Further theoretical calculations, based on the standard drift-diffusion theory, suggested the possibility that tunnel-contact resistance might play a relevant role to increase the efficiency of ferromagnetic/nonmagnetic (F/N) spin injectors [11,16]. This results uphold the important role of the quantum mechanical tunneling process in the F/N spin injectors interface and the importance of quantum mechanical calculations to understand the behavior of the spin-polarized devices. In the next sections, we will outline the basic concepts introduced to describe and characterize the spintronic devices. We will start with a brief review of the widely used semiclassical

P. Pereyra, *Contemporary Theory of Semiconductors and Devices*,
Graduate Texts in Physics, https://doi.org/10.1007/978-3-032-00296-9_15

approach to spin transport in polarized semiconductors, [17] and in the second part we will present a summary of the magnetic superlattice approach [18,19]. The semiclassical approach has from the beginning been the mainstream research approach, and the low spin-injection efficiency a crucial topic. We will concentrate a bit more on this topic.

15.1 The Semiclassical Approach to Semiconductor Spintronics

The discovery of ferromagnetism in III-V and II-IV Mn-based semiconductors provided the possibility to inject, manipulate and detect spin-polarized carriers by purely electrical means. Particular interest was paid on the fabrication and physical properties of the compound $Ga_{1-x}Mn_xAs$ [20–23]. The spin-polarized electrons detected by Johnston-Halperin et al. [24] and by Kohda et al. [25], using a p^+-$(Ga, Mn)As/n^+$-$GaAs$, led to recognize there an Esaki diode structure as spin injector. This fact triggered a huge experimental and theoretical research activities to try to understand the electronic and magnetic properties of this system [26–29].

Despite the open questions concerning the band structure of the $(Ga_{1-x}, Mn_x)As$ and the various models and first principle calculations performed to elucidate whether the free carriers (holes) reside in an impurity band, detached from the host valence band, or in the valence band itself, an important number of physical properties have been clarified. It is now well established that: the exchange interaction of itinerant holes (impurity-band or valence-band holes) and the magnetic impurities [29,30] lead to a ferromagnetic behavior of the dilute magnetic semiconductor $(Ga_{1-x}, Mn_x)As$, with a split of the spin-bands, induced by the spin-exchange interaction, of the order of a few tenths of eV [31,32]; the phase coherence length is of the order of 100 nm while the mean free path is rather small [33]. For magnetic impurity concentrations of the order of 6%, an Esaki barrier is built up at the magnetic/nonmagnetic interface. Due to this barrier, an important issue in the $Ga_{1-x}Mn_xAs/n^+$-$GaAs$ structure, is the spin-injection efficiency.

The phenomenological drift-diffusion theory and the two-component model introduced by Mott [34] in 1936 (used later by Campbell [35], van Son et al. [36], Valet et al. [37], Hershfield et al. [38], Schmidt [2] and Rashba [16], among others), have been widely used to study issues like the spin accumulation and spin injection efficiency in magnetic/nonmagnetic structures. Within this formalism, it was shown that the spin-injection efficiency, from ferromagnetic to normal (non-magnetic) semiconductor, can be written as (see [24,39,40])

$$P_j = \frac{R_c P_\Sigma + R_F P_{\sigma F}}{R_c + R_N + R_F}, \tag{15.1}$$

where $P_{\sigma F}$ is the conductivity spin polarization in the magnetic side, defined as

$$P_{\sigma F} = \frac{\sigma_{\uparrow F} - \sigma_{\downarrow F}}{\sigma_{\uparrow F} + \sigma_{\downarrow F}}, \tag{15.2}$$

R_F and R_N are the magnetic and non-magnetic resistances, and

$$R_c = \frac{\Sigma_\uparrow + \Sigma_\downarrow}{4\Sigma_\uparrow \Sigma_\downarrow}, \qquad P_\Sigma = \frac{\Sigma_\uparrow - \Sigma_\downarrow}{\Sigma_\uparrow + \Sigma_\downarrow}. \tag{15.3}$$

the contact resistance and conductance spin polarization, with $\Sigma_\uparrow$ and $\Sigma_\downarrow$ the spin up and spin down surface conductances. Equation (15.1) makes clear the importance of the contact resistance, hence the need of quantum-mechanical calculations of tunneling probabilities of spin-polarized carriers through the space-charge region.

Within few years, the phenomenological theory of diffusion and drift and the two-component model introduced by Mott in 1936 were extended to define useful concepts such as spin accumulation and spin injection efficiency, as well as physical variables, such as the spin-polarized currents. The literature on spin polarization and magnetization phenomena is vast [40,41]. As mentioned before, here we restrict ourself on introducing the basic concepts of the semiclassical two-component model applied to highly-doped inhomogeneous semiconductors. We will address the spin injection efficiency, and we will present a summary of the magnetic superlattice formalism as well as of some relevant topics of spin dynamics.

15.2 Two-Spin Model, Drift-Diffusion Theory and Tunneling

In this section we will derive, among other useful quantities, the spin injection efficiency P_j introduced in (15.1) of the last section. Let us first recall some of the basic definitions of the standard model, extensively discussed in the review article of Fabian et al. [40]. To describe a Fermi gas in a highly-doped inhomogeneous, ferromagnetic or nonmagnetic, semiconductor in the presence of an electric field $-\nabla\phi$, the phenomenological drift-diffusion equations (within the two-spin model) are written as

$$\mathbf{j}_\uparrow = \sigma_\uparrow \mathbf{E} - D_\uparrow \nabla n_\uparrow, \tag{15.4}$$

$$\mathbf{j}_\downarrow = \sigma_\downarrow \mathbf{E} - D_\downarrow \nabla n_\downarrow. \tag{15.5}$$

where $\sigma_{\uparrow(\downarrow)} = e^2 \mathscr{N}_{\uparrow(\downarrow)} D_{\uparrow(\downarrow)}$ is the spin-up (down) conductivity, $\mathscr{N}_{\uparrow(\downarrow)}$ and $D_{\uparrow(\downarrow)}$ the density of states and the spin diffusion constant, and $n_{\uparrow(\downarrow)}$ the spin-up (down) density. It is clear that for the nonmagnetic media $\sigma_\uparrow = \sigma_\downarrow$. If the rate at which spin-up scatters into spin-down is denoted by $1/\tau_{\uparrow\downarrow}$ and the rate at which spin-down scatters into spin-up by $1/\tau_{\downarrow\uparrow}$, the continuity condition implies

$$\nabla \cdot \mathbf{j}_\uparrow = -\frac{n_\uparrow}{\tau_{\uparrow\downarrow}} + \frac{n_\downarrow}{\tau_{\downarrow\uparrow}}, \tag{15.6}$$

$$\nabla \cdot \mathbf{j}_\downarrow = -\frac{n_\downarrow}{\tau_{\downarrow\uparrow}} + \frac{n_\uparrow}{\tau_{\uparrow\downarrow}}. \tag{15.7}$$

It is well-known that the spin-up (down) Landauer conductance [38], written in terms of the spin-up (down) transmission coefficients, depends on the spin-dependent electrochemical potentials. Hence, we will collect some relevant results of the standard drift-diffusion model, written in terms of the electrochemical potentials $\mu_\uparrow$ and $\mu_\downarrow$, generally defined by

$$n_\uparrow(x) = n_{\uparrow 0} - e\mathcal{N}_\uparrow\big(\mu_\uparrow(x) + e\phi(x)\big), \tag{15.8}$$

$$n_\downarrow(x) = n_{\downarrow 0} - e\mathcal{N}_\downarrow\big(\mu_\downarrow(x) + e\phi(x)\big), \tag{15.9}$$

where $n_{\uparrow 0}$ and $n_{\downarrow 0}$ are the equilibrium densities and $\phi(x)$ the electric potential. Using these relations, the drift diffusion equations become

$$\mathbf{j}_\uparrow = \frac{\sigma_\uparrow}{e}\nabla\mu_\uparrow, \tag{15.10}$$

$$\mathbf{j}_\downarrow = \frac{\sigma_\downarrow}{e}\nabla\mu_\downarrow. \tag{15.11}$$

while the continuity equations take the form

$$\nabla\cdot\mathbf{j}_\uparrow = e\frac{\mathcal{N}_\uparrow}{\tau_{\uparrow\downarrow}}\mu_\uparrow - e\frac{\mathcal{N}_\downarrow}{\tau_{\downarrow\uparrow}}\mu_\downarrow + e^2\phi\left(\frac{\mathcal{N}_\uparrow}{\tau_{\uparrow\downarrow}} - \frac{\mathcal{N}_\downarrow}{\tau_{\downarrow\uparrow}}\right), \tag{15.12}$$

$$\nabla\cdot\mathbf{j}_\downarrow = e\frac{\mathcal{N}_\downarrow}{\tau_{\downarrow\uparrow}}\mu_\downarrow - e\frac{\mathcal{N}_\uparrow}{\tau_{\uparrow\downarrow}}\mu_\uparrow - e^2\phi\left(\frac{\mathcal{N}_\uparrow}{\tau_{\uparrow\downarrow}} - \frac{\mathcal{N}_\downarrow}{\tau_{\downarrow\uparrow}}\right). \tag{15.13}$$

Invariance of the spin flip rate under a uniform shift of the chemical potential leads to the detailed balance relation

$$\frac{\mathcal{N}_\uparrow}{\tau_{\uparrow\downarrow}} = \frac{\mathcal{N}_\downarrow}{\tau_{\downarrow\uparrow}}. \tag{15.14}$$

Thus

$$\nabla\cdot\mathbf{j}_\uparrow = e\frac{\mathcal{N}_\uparrow}{\tau_{\uparrow\downarrow}}(\mu_\uparrow - \mu_\downarrow), \tag{15.15}$$

$$\nabla\cdot\mathbf{j}_\downarrow = e\frac{\mathcal{N}_\downarrow}{\tau_{\downarrow\uparrow}}(\mu_\downarrow - \mu_\uparrow). \tag{15.16}$$

The analysis of the spin effects becomes simpler when the drift diffusion equations and transport properties are written in terms of the following definitions

$$\sigma = \sigma_\uparrow + \sigma_\downarrow\,, \qquad \sigma_s = \sigma_\uparrow - \sigma_\downarrow, \tag{15.17}$$

$$\mathcal{N} = \mathcal{N}_\uparrow + \mathcal{N}_\downarrow\,, \qquad \mathcal{N}_s = \mathcal{N}_\uparrow - \mathcal{N}_\downarrow, \tag{15.18}$$

and

$$\mu = \frac{1}{2}\left(\mu_\uparrow + \mu_\downarrow\right), \qquad \mu_s = \frac{1}{2}\left(\mu_\uparrow - \mu_\downarrow\right). \tag{15.19}$$

We will use these definitions and the previous equations to express, in the next subsection, the current spin polarization and the non-equilibrium or spin accumulation in terms of charge and spin eigenmodes of the chemical potential, which will be then useful to express these quantities in terms of the Esaki-barrier transmission coefficients. We follow here the arguments and notation in [38,40].

15.2.1 Eigenmodes of the Electrochemical Potential

Combining (15.10), (15.11), with the continuity (15.15) and (15.16), and the detailed balance relation, one can easily obtain the following coupled differential equations

$$\nabla^2 \mu_\uparrow = \frac{1}{D_\uparrow \tau_{\uparrow\downarrow}} \mu_\uparrow - \frac{1}{D_\uparrow \tau_{\uparrow\downarrow}} \mu_\downarrow, \tag{15.20}$$

$$\nabla^2 \mu_\downarrow = -\frac{1}{D_\downarrow \tau_{\downarrow\uparrow}} \mu_\uparrow - \frac{1}{D_\downarrow \tau_{\downarrow\uparrow}} \mu_\downarrow, \tag{15.21}$$

which can be written in matrix form as

$$\nabla^2 \begin{pmatrix} \mu_\uparrow \\ \mu_\downarrow \end{pmatrix} = \begin{pmatrix} \dfrac{1}{D_\uparrow \tau_{\uparrow\downarrow}} & -\dfrac{1}{D_\uparrow \tau_{\uparrow\downarrow}} \\ -\dfrac{1}{D_\downarrow \tau_{\downarrow\uparrow}} & \dfrac{1}{D_\downarrow \tau_{\downarrow\uparrow}} \end{pmatrix} \begin{pmatrix} \mu_\uparrow \\ \mu_\downarrow \end{pmatrix}. \tag{15.22}$$

To uncouple these equations we follow [38]. The (2×2)-matrix on the right side can easily be diagonalized. In fact, using $a = (D_\uparrow \tau_{\uparrow\downarrow})^{-1}$ and $b = (D_\downarrow \tau_{\downarrow\uparrow})^{-1}$, one has

$$\begin{pmatrix} a & -a \\ -b & b \end{pmatrix} = \begin{pmatrix} 1 & a \\ 1 & -b \end{pmatrix} \begin{pmatrix} 0 & 0 \\ 0 & a+b \end{pmatrix} \begin{pmatrix} 1 & a \\ 1 & -b \end{pmatrix}^{-1}. \tag{15.23}$$

Multiplying (15.22) by the inverse of the eigenvectors matrix and dropping the factors $1/(a+b)$, the last equation becomes

$$\begin{pmatrix} b & a \\ 1 & -1 \end{pmatrix} \nabla^2 \begin{pmatrix} \mu_\uparrow \\ \mu_\downarrow \end{pmatrix} = \begin{pmatrix} 0 & 0 \\ a+b & -(a+b) \end{pmatrix} \begin{pmatrix} \mu_\uparrow \\ \mu_\downarrow \end{pmatrix}, \tag{15.24}$$

or, in the form

$$\nabla^2 \begin{pmatrix} b\mu_\uparrow + a\mu_\downarrow \\ \mu_\uparrow - \mu_\downarrow \end{pmatrix} = \begin{pmatrix} 0 \\ (a+b)(\mu_\uparrow - \mu_\downarrow) \end{pmatrix}, \tag{15.25}$$

which is equivalent to the following equations named of charge and spin modes:

$$\nabla^2 \left(\frac{\mu_\uparrow}{D_\downarrow \tau_{\downarrow\uparrow}} + \frac{\mu_\downarrow}{D_\uparrow \tau_{\uparrow\downarrow}} \right) = 0 \tag{15.26}$$

and

$$\nabla^2 (\mu_\uparrow - \mu_\downarrow) = \left(\frac{1}{D_\uparrow \tau_{\uparrow\downarrow}} + \frac{1}{D_\downarrow \tau_{\downarrow\uparrow}} \right) (\mu_\uparrow - \mu_\downarrow). \tag{15.27}$$

In fact, since

$$\nabla\left(\frac{\mu_\uparrow}{D_\downarrow\tau_{\downarrow\uparrow}}+\frac{\mu_\downarrow}{D_\uparrow\tau_{\uparrow\downarrow}}\right)=\frac{ej}{\sigma_\uparrow D_\downarrow\tau_{\downarrow\uparrow}}, \tag{15.28}$$

Equation (15.26) is equivalent to $\nabla\cdot\mathbf{j}=0$. Therefore, for currents along the x-axis only, one has

$$\frac{\mu_\uparrow}{D_\downarrow\tau_{\downarrow\uparrow}}+\frac{\mu_\downarrow}{D_\uparrow\tau_{\uparrow\downarrow}}=\frac{ejx}{\sigma_\uparrow D_\downarrow\tau_{\downarrow\uparrow}}+C. \tag{15.29}$$

Equation (15.27), on the other hand, governs the space evolution of the spin mode and the spin accumulation. To recognize this property, we establish the relation between μ_s and the spin accumulation. Combining (15.8) and (15.9), and making use of the local charge neutrality ($n=n_0$) and defining the non-equilibrium or spin accumulation

$$\delta s=\frac{n_s-n_{s0}}{-e} \tag{15.30}$$

results

$$\mathcal{N}(\mu+e\phi)+\mathcal{N}_s\mu_s=0 \tag{15.31}$$

and

$$\delta s=\mathcal{N}\mu_s+\mathcal{N}_s(\mu+e\phi)=\frac{\mathcal{N}^2-\mathcal{N}_s^2}{\mathcal{N}}\mu_s, \tag{15.32}$$

that gives the relation

$$\delta s=\frac{4\mathcal{N}_\uparrow\mathcal{N}_\downarrow}{\mathcal{N}}\mu_s, \tag{15.33}$$

between the electrochemical potential difference μ_s and the spin accumulation (see [40]).

Using the spin-up and spin-down diffusion lengths $L_{s\uparrow}^2=D_\uparrow\tau_{\uparrow\downarrow}$ and $L_{s\downarrow}^2=D_\downarrow\tau_{\downarrow\uparrow}$, it is common to define the spin diffusion length

$$\frac{1}{L_s^2}=\left(\frac{1}{L_{s\uparrow}^2}+\frac{1}{L_{s\downarrow}^2}\right). \tag{15.34}$$

Thus

$$\nabla^2(\mu_\uparrow-\mu_\downarrow)=\frac{1}{L_s^2}(\mu_\uparrow-\mu_\downarrow). \tag{15.35}$$

This equation has been studied for different F/N structures and an important number of results, related with the space evolution of the spin accumulation, were published (see for example [2,16,17,36,38]). Before specifying any particular structure, it is worth writing the general solution for the spin mode $\mu_s(x) = (\mu_\uparrow(x) - \mu_\downarrow(x))/2$ in the form

$$\mu_s(x) = \frac{1}{2}\Big(\mu_\uparrow(x) - \mu_\downarrow(x)\Big) = Ae^{x/L_s} + Be^{-x/L_s}. \tag{15.36}$$

Combining this and the linear solution in (15.29), one obtains

$$\mu_\uparrow(x) = \frac{ejx}{\sigma} + C\frac{\sigma_\uparrow\sigma_\downarrow\tau_{\uparrow\downarrow}}{\sigma e^2 \mathcal{N}_\uparrow} + \frac{2\sigma_\downarrow}{\sigma}\left(Ae^{x/L_s} + Be^{-x/L_s}\right), \tag{15.37}$$

$$\mu_\downarrow(x) = \frac{ejx}{\sigma} + C\frac{\sigma_\uparrow\sigma_\downarrow\tau_{\downarrow\uparrow}}{\sigma e^2 \mathcal{N}_\downarrow} - \frac{2\sigma_\uparrow}{\sigma}\left(Ae^{x/L_s} + Be^{-x/L_s}\right). \tag{15.38}$$

If we have a ferromagnetic semiconductor of length L_F in contact with a non magnetic semiconductor of length L_N, the general solutions of the respective second order differential equations are

$$\mu_{sF}(x) = A_F e^{x/L_s} + B_F e^{-x/L_{sF}}, \quad L_F < x < 0, \tag{15.39}$$

$$\mu_{sN}(x) = A_N e^{x/L_s} + B_N e^{-x/L_{sN}}, \quad 0 < x < L_N. \tag{15.40}$$

In the particular case of semi-infinite ($L_F >> L_{sF}$) ferromagnetic and semi-infinite ($L_N >> L_{sN}$) nonmagnetic media, with boundary conditions $\mu_{sF}(-\infty) = 0$ and $\mu_{sN}(\infty) = 0$, the spin mode of the electrochemical potential in the ferromagnetic and normal sides is written as

$$\mu_{sF}(x) = \frac{1}{2}\big(\mu_{\uparrow F} - \mu_{\downarrow F}\big) = A_F e^{x/L_{sF}}, \quad x < 0, \tag{15.41}$$

$$\mu_{sN}(x) = \frac{1}{2}\big(\mu_{\uparrow N} - \mu_{\downarrow N}\big) = B_N e^{-x/L_{sN}}, \quad 0 < x. \tag{15.42}$$

These are well-known results. Our purpose in the next subsection is to express the spin injection efficiency in terms of the contact spin transmission coefficients.

15.2.2 Spin Injection and Transmission Coefficients

Before writing the spin injection in the ferromagnetic, nonmagnetic and contact regions, it is convenient to use again the definitions introduced before to write the charge and spin currents in the following form [17]

$$\mathbf{j} = \mathbf{j}_\uparrow + \mathbf{j}_\downarrow = \sigma\nabla\mu + \sigma_s\nabla\mu_s, \tag{15.43}$$

$$\mathbf{j}_s = \mathbf{j}_\uparrow - \mathbf{j}_\downarrow = \sigma_s\nabla\mu + \sigma\nabla\mu_s. \tag{15.44}$$

For a current along the x-axis perpendicular to the interface, an important relation that can be obtained between the current spin polarization $P_j(x) = (j_\uparrow - j_\downarrow)/j$, the conductivity spin polarization $P_\sigma = (\sigma_\uparrow - \sigma_\downarrow)/\sigma$ and the electrochemical spin mode μ_s. Solving for $\nabla\mu$ in the first equation and substituting for this in the second, we have

$$\begin{aligned} P_j(x) = \frac{j_s}{j} &= \frac{\sigma_\uparrow - \sigma_\downarrow}{\sigma} + \frac{4}{j}\frac{\sigma_\uparrow\sigma_\downarrow}{\sigma}\nabla\mu_s \\ &= P_\sigma + \frac{4}{j}\frac{\sigma_\uparrow\sigma_\downarrow}{\sigma}\nabla\mu_s(x). \end{aligned} \tag{15.45}$$

Knowing the space evolution of the electrochemical potential $\mu_s(z)$, one can determine the current spin polarization at any point of a magnetic or nonmagnetic medium. These are also well-known solutions and have been used to obtain the current spin polarization, or device spin injection efficiency P_j in (15.1). We will use (15.45) and the Landauer formula to obtain the device spin injection efficiency. Writing the resistance of a ferromagnetic conductor with unit cross section and length L as

$$R = \left(\frac{1}{\sigma_\uparrow} + \frac{1}{\sigma_\downarrow}\right)L, \tag{15.46}$$

the current spin polarization for the ferromagnetic medium at the left side of the ferromagnetic/nonmagnetic junction, can be written as (see [40])

$$\begin{aligned} P_{jF}(x) &= \frac{\sigma_\uparrow - \sigma_\downarrow}{\sigma} + \frac{4}{j}\frac{\sigma_\uparrow\sigma_\downarrow}{\sigma}\nabla\mu_{sF} \\ &= P_{\sigma F} + \frac{1}{j}\frac{1}{R_F}\mu_{sF}(x), \end{aligned} \tag{15.47}$$

with R_F the effective resistance. In the same way one has

$$P_{jN}(x) = -\frac{1}{j}\frac{1}{R_N}\mu_{sN}(x). \tag{15.48}$$

for the nonmagnetic side. For the Esaki barrier, characterized by spin-dependent conductances $G_\uparrow$ and $G_\downarrow$, we have

$$\frac{4}{j}\frac{\sigma_\uparrow\sigma_\downarrow}{\sigma}\nabla\mu_s(x) \rightarrow \frac{4}{j}\frac{G_\uparrow G_\downarrow}{G_\uparrow + G_\downarrow}\nabla\mu_s(x)\cdot\Delta x \tag{15.49}$$

$$= \frac{4}{jR_c}\Big(\mu_{sN}(x_n) - \mu_{sF}(-x_p)\Big). \tag{15.50}$$

Applying (15.45) and taking into account the last relation, we have for the current spin polarization in the contact the following expression

$$P_{jc} = \frac{G_\uparrow - G_\downarrow}{G_\uparrow + G_\downarrow} + \frac{4}{jR_c}\Big(\mu_{sN}(x_n) - \mu_{sF}(-x_p)\Big). \tag{15.51}$$

Notice that using (15.47) and (15.48) to evaluate $\mu_{sF}(-x_p)$ and $\mu_{sN}(x_n)$ in the ferromagnetic and nonmagnetic sides, and taking into account that $P_{jF}(-x_p) = P_{jc} = P_{jN}(x_n) = P_j$ the device spin injection efficiency becomes

$$P_j = \frac{R_c P_G + R_F P_{\sigma F}}{R_c + R_N + R_F}. \tag{15.52}$$

This equation led, as mentioned above, to observe that whenever $R_c >> R_N, R_F$ the efficiency is controlled basically by the contact resistance. The spin efficiency is then written as [40]

$$P_j \simeq P_G = \frac{G_\uparrow - G_\downarrow}{G_\uparrow + G_\downarrow}. \tag{15.53}$$

If we use the Landauer formula

$$G_\sigma = \frac{e^2}{\hbar} \sum_{\sigma'=\uparrow,\downarrow} T_{\sigma\sigma'} \tag{15.54}$$

and neglect the spin-flip transitions in the tunneling region, i.e. if $T_{\sigma\sigma'} << T_{\sigma\sigma}, T_{\sigma'\sigma'}$ (which is justified because the depletion layer thicknesses is much smaller than the spin-flip length L_{sf}), we can write the spin efficiency as

$$P_j \simeq P_T = \frac{T_{\uparrow\uparrow} - T_{\downarrow\downarrow}}{T_{\uparrow\uparrow} + T_{\downarrow\downarrow}}. \tag{15.55}$$

which means that large spin injection efficiency can be expected when $T_{\uparrow\uparrow}$ and $T_{\downarrow\downarrow}$ are rather different. High spin efficiency does not guaranty large spin injection. If we have a double barrier systems, like in the spin sensing system, we have a resonant tunneling transmission. The resonant tunneling has been used to determine the impurity band structure.

Other quantities like the spin accumulation and the electrochemical potential that drops at the contact, will be written in terms of this contact transmission coefficients. We will discuss this we will discuss this in the next subsection. We are interested in the spin tunneling process across the Esaki potential and we will discuss this topic in Sect. 15.3. The spin transport problem, outside the Esaki potential, will be studied ahead with the magnetic superlattices.

15.2.3 Spin Accumulation and the Electrochemical Potential Drop

To evaluate the spin accumulation $\delta s(x)$ at the ferromagnetic and nonmagnetic sides using (15.33) we need first to define the electrochemical potentials, which in the semi-infinite ($L_F >> L_{sF}$) ferromagnetic and semi-infinite ($L_N >> L_{sN}$) nonmagnetic media cases, and (15.47) and (15.48), are given by

$$\mu_{sF}(x) = (P_j - P_\sigma) j R_F e^{(x+x_p)/L_{sF}}, \qquad x < -x_p, \tag{15.56}$$

$$\mu_{sN}(x) = -P_j j R_N e^{-(x-x_n)/L_{sN}}, \qquad x_n < x, \tag{15.57}$$

where the spin efficiency $P_j \simeq P_T$ of (15.55) is assumed. The spin accumulation on the ferromagnetic and magnetic sides are then given by

$$\delta s_F(x) = (P_T - P_\sigma) \frac{4 j R_F \mathcal{N}_\uparrow \mathcal{N}_\downarrow}{\mathcal{N}_{\mathcal{F}}} e^{(x+x_p)/L_{sF}} \tag{15.58}$$

$$\delta s_N(x) = -\frac{T_{\uparrow\uparrow} - T_{\downarrow\downarrow}}{T_{\uparrow\uparrow} + T_{\downarrow\downarrow}} j R_N \mathcal{N}_N e^{-(x-x_n)/L_{sN}}. \tag{15.59}$$

Other quantities of interest are the electrochemical potential drop at the contact

$$\mu_N(x_n) - \mu_F(-x_p) = \frac{j}{G} - P_T \Delta\mu_s, \tag{15.60}$$

where

$$\Delta\mu_s = \mu_{sN}(x_n) - \mu_{sF}(-x_p) \tag{15.61}$$

$$= j\big(R_F P_\sigma - (R_N + R_F) P_T\big), \tag{15.62}$$

and the nonequilibrium spin resistance

$$\delta R \simeq R_F \big(P_\sigma - P_T\big)^2 + R_N P_T^2. \tag{15.63}$$

To evaluate all these quantities, we need the spin up and spin down transmission coefficients, as well as the current j. We will be concerned here only with transmission probabilities across the contact, independent of the density of states and occupation probabilities at the ferromagnetic and nonmagnetic sides.

15.3 Exchange Energy and Spin Injection

Injecting, manipulating and detecting spin polarization have been the main objectives of the semiconductor spintronics [42]. Different proposals have been published by countless research groups [40]. Spin injection and detection represent two important stages of the problem and are different processes from spin manipulation. In the following sections we will deal with the dynamics, and to a certain extent, the manipulation of spin through a magnetic superlattice.

Despite great advances in this field, the goal of producing an all-electric, all-semiconductor and above all efficient detection and injection device remains quite elusive. This problem has slowed the development of real devices to control and manipulate the spin filter and spin reversal mechanisms.

The theoretical suggestion that higher contact resistance R_c can increase the efficiency of ferromagnetic/nonmagnetic (F/N) spin injectors if [11, 16, 44, 45] promoted research on ferromagnetic semiconductors. To study the contact resistance as a function of the exchange energy, the non-equilibrium spin splitting, the impurity concentrations, and the bias potential, accurate quantum calculations are required. Interband tunneling calculations across Schottky barriers, based on a drift-diffusion transport model, were carried out by Smith et al. [11], Fabian et al. [46] studied the spin injection through magnetic p-n junctions within the standard theory of semiconductors. A microscopic calculation of spin transport through a Schottky (triangular) barrier was carried out by Osipov et al. [47]. Nevertheless, none of these works include a detailed calculation of spin-selective I-V characteristics (IVC).

In the standard drift-diffusion theory, the contact resistance is a kind of abstract entity, defined by the boundary conditions at z = 0. However, as shown in Fig. 15.1a, where the profiles of the band edges at the junction $p^+Ga_{0.95}Mn_{0.05}As/Al_{0.36}Ga_{0.64}As/n^+GaAs$ are plotted, a barrier forms at this interface, an Esaki barrier similar to the Schottky barrier, with classical return points at z_L and z_r. This barrier has size and shape which depend, among other factors, on the exchange energy, the impurity concentrations, and the bias potential. The quantum nature of this contact has been ignored by semiclassical approaches, which generally limit themselves to recognizing it as a resistance with some numerical value.

The Esaki barrier at the interface is a quantum structure that determines the transmission coefficients, and as will be shown below, together with the exchange energy determine the low-bias features observed in the I-V characteristics. In Fig. 15.2a and b, we have the I-V characteristics obtained by Shiogai et al. [48] and by Holmberg et al. [49], respectively. The plateau and negative resistance around 0.4 V are features that suggest that spin injection and detection should be studied as tunneling processes, in which electrons come from spin bands separated by the exchange energy (~0.4 eV) and thus pass through different Esaki barriers. To understand the IV characteristics of the spins ↑ and ↓ ('up' and 'down') through a dynamic bar-

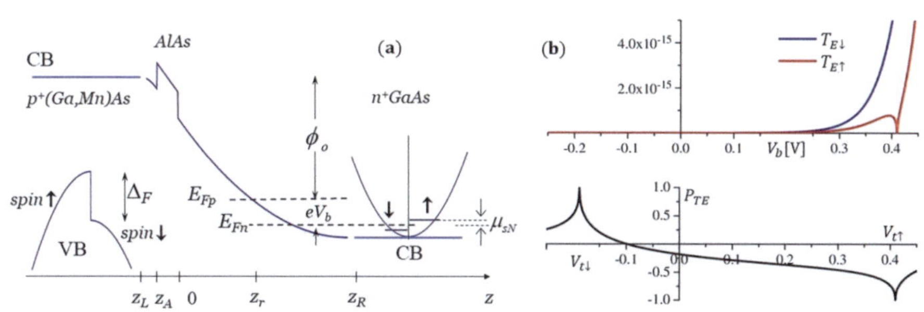

Fig. 15.1 In **a**, the band structure of a polarized Esaki barrier consisting of a ferromagnetic semiconductor, a stop layer and a non-magnetic semiconductor, with displacement Δ_F in the valence band of the p side, and on the n side an electrochemical spin accumulation potential μ_{sN}. In **b**, the spin-up and spin-down transmission coefficients, when $\Delta_F = 0.6$ eV and E_{Fp} is 0.112 eV below E_p^0. The cancellation of $T_{E\sigma}$ at the threshold potentials $V_{t\uparrow}$ and $V_{t\downarrow}$ gives rise to the spin polarization transmission seen in the graph below. Figure reproduced from [43] with permission of the American Physical Society

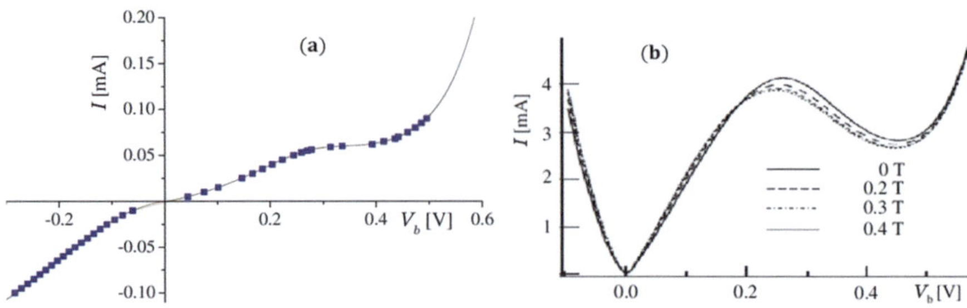

Fig. 15.2 I-V characteristic in spin injectors across the Ga(Mn)As/GaAs junction. In **a** the I-V characteristics measured in [48] for the spin injector $p^+Ga_{0.95}Mn_{0.05}As/n^+GaAs$ with stop layer. In **b** effects of the magnetic field on the I-V characteristic of the Ga(Mn)As/GaAs structure measured in [49]. Figures reproduced with permission of the American Physical Society and from EDP Sciences-Europhysics Letters, respectively

rier, one has to consider, in addition to the potential profile of the semiconductor junction pn, the shift of the Fermi levels determined by the strong exchange energy interaction and by the impurity states in the gap.

For the accurate calculation of the transmission coefficients, we use the transfer matrix method (TMM) in the WKB approximation (details can be seen in [50,51]). In reverse (injection) and forward (extraction) configuration, the corresponding 4×4 transfer matrices fulfill the relations

$$\Phi(z_r) = M_I(z_r, z_L)\Phi(z_L) = \begin{pmatrix} \alpha_I & \beta_I \\ \gamma_I & \delta_I \end{pmatrix} \Phi(z_L) \tag{15.64}$$

and

$$\Phi(z_L) = M_E(z_L, z_r)\Phi(z_L) = \begin{pmatrix} \alpha_E & \beta_E \\ \gamma_E & \delta_E \end{pmatrix} \Phi(z_L). \tag{15.65}$$

The state vectors

$$\Phi(z) = \left(\varphi_\uparrow^+, \varphi_\downarrow^+, \varphi_\uparrow^-, \varphi_\downarrow^-\right)^T, \tag{15.66}$$

in the propagating modes representation, are written in terms of the propagating wave functions (σ stands for spin up $\uparrow$ or spin down $\downarrow$)

$$\varphi_\sigma^+(z) = \frac{a_\sigma}{\sqrt{k_\sigma}} e^{ik_\sigma z}\chi_\sigma \quad \text{and} \quad \varphi_\sigma^- = \frac{b_\sigma}{\sqrt{k_\sigma}} e^{-ik_\sigma z}\chi_\sigma, \tag{15.67}$$

and the evanescent functions

$$\varphi_\sigma^+(z) = \frac{a_\sigma}{\sqrt{q_\sigma}} e^{q_\sigma z}\chi_\sigma \quad \text{and} \quad \varphi_\sigma^- = \frac{b_\sigma}{\sqrt{q_\sigma}} e^{-q_\sigma z}\chi_\sigma. \tag{15.68}$$

Here $k_{\uparrow,\downarrow} = \sqrt{\frac{2m^*}{\hbar^2}(E_{Fi} - U_{\uparrow,\downarrow}(x, V_b))}$ and $q_{\uparrow,\downarrow} = \sqrt{\frac{2m^*}{\hbar^2}(U_{\uparrow,\downarrow}(x, V_b) - E_{Fi})}$, are the corresponding wave numbers with $E_{Fi} = E_{Fp}$ and E_{Fn} the quasi-Fermi energies. Given the transfer matrices and the relations with the scattering amplitudes, $t_I = (\delta_I^\dagger)^{-1}$ and $t_E = (\delta_E^\dagger)^{-1}$, one can obtain the transmission coefficients

$$T_{\sigma,\sigma'} = |t_{\sigma,\sigma'}|^2 = |(\delta^\dagger)^{-1}_{\sigma,\sigma'}|^2. \tag{15.69}$$

and the spin polarized extraction probability, defined as

$$P_{TE} = \frac{T_{E\uparrow} - T_{E\downarrow}}{T_{E\uparrow} + T_\downarrow}, \qquad \text{with} \qquad T_{E\sigma} = \sum_{\sigma'} T_{E\sigma,\sigma'} \tag{15.70}$$

These quantities are plotted in the upper and lower panels of Fig. 15.1b, respectively. The current spin polarization, spin injection efficiency and the spin accumulation defined as

$$P_j \simeq \frac{j_\uparrow - j_\downarrow}{j_\uparrow + j_\downarrow}, \qquad P_{jQ} \simeq \frac{j_{Q\uparrow} - j_{Q\downarrow}}{j_{Q\uparrow} + j_{Q\downarrow}} \qquad \text{and} \qquad \delta\mu_s \propto -P_j, \tag{15.71}$$

where $j_\sigma = j_{I\sigma} + j_{E\sigma}$, and $Q = I, E$ (for injection and extraction), have similar behavior and, they are also very sensitive to the exchange energy and the threshold potentials $V_{t\uparrow}$ and $V_{t\downarrow}$.

The currents are evaluated, as usual, from

$$j_\sigma^Q = \mathscr{G} \int \mathfrak{D}_{N\sigma}(E) \mathfrak{D}_{F\sigma}(E - eV_b) T_{\sigma\sigma}^Q(E, V_b) f_F(E) g_N(E) dE$$

where $\mathscr{G}$ is a geometrical factor, $\mathfrak{D}_{N\sigma}$ and $\mathfrak{D}_{F\sigma}$ are the densities of states (DOS) in the non-magnetic and ferromagnetic sides, and $f(E)$ and $g(E) = 1 - f(E)$ are the occupation probabilities. Since in experiments the temperature is $\sim$4 K, the occupation probabilities factor becomes $\delta(E - E_F)$. The currents shown in Fig. 15.3 are calculated by assuming that in the ferromagnetic side the DOS is $\mathfrak{D}_{F\sigma} \propto$

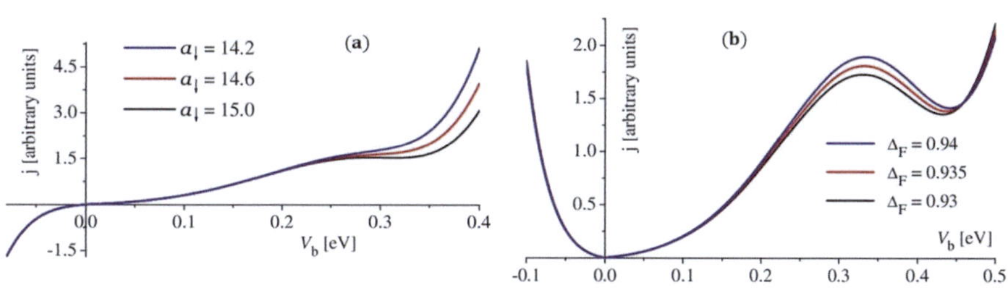

Fig. 15.3 **a** The effect of the parameter p on $\mathfrak{D}_{F\sigma}$, in the neighborhood of the valence band edges. **b** Total current through the Esaki barrier for the same parameters of Fig. 15.1, and the densities of states explained in the text. Figure reproduced from [43] with permission of the American Physica Society

$e^{-a_\sigma \epsilon_\sigma}(a_\sigma \epsilon_\sigma)^{p/2}$ for $\epsilon_\sigma = E - E_{Fp} - \sigma \Delta_F/2 > 0$, and $\mathfrak{D}_{F\sigma} \propto (-\epsilon_\sigma)^{q/2}$ with $q = 1$ for $\epsilon_\sigma < -\Delta_F$, $q = p$ for $-\Delta_F < \epsilon_\sigma < 0$, being a_σ and p parameters chosen to fit the DOS's obtained numerically by Turek et al. [17], and $\mathfrak{D}_{N\sigma} = \mathfrak{D}_{K\sigma} \propto (E - E_{0\sigma})^{p/2}$ (for $p = 1$). Here $\mathfrak{D}_{K\sigma}$ is the Kane's y function [52]

$$y(E) = \int_{-\infty}^{E} (E - E_{0\sigma} - \xi)^{p/2} \mathfrak{F}(\xi) d\xi. \tag{15.72}$$

As can be seen in Fig. 15.2, currents account for the experimental behavior. These results show also that in the presence of exchange, the spin splitting and the DOS are strongly correlated quantities, and define the shape and low-bias I-V characteristics of the F/N structures.

15.3.1 Homogeneous Magnetic Superlattices

The transfer matrix method and the TFPS approach to spintronics is a different approach to conventional spintronic frameworks, which are mostly treated using the semi-classical methods outlined in the previous sections. In this section, we briefly review the formalism developed to study the electronic transport and spin manipulation in magnetic superlattices.

The aim to control spin carriers transport through layered magnetic structures, has long been a topic of much interest [13,54,55]. Motivated by this goal, a line of research was conducted to investigate various aspects of spin $1/2$ carrier interactions with magnetic superlattices under conditions of phase-coherent periodic magnetic fields [18,53,56]. This work explored phenomena such as Shubnikov-de Haas oscillations, the Landé g factor [56], Fano resonances and the magnetic-field-tilting angle as a significant mechanism for the spin transitions in magnetic superlattices. Leveraging these spin-transport effects, the research proposed simple and efficient designs for spin-filter and spin-inverter devices [53].

For practical reasons, a simple system was conceived, where only the external field changes periodically, that is, the same semiconductor under an external magnetic field whose amplitude varies periodically [18]. We refer to this superlatice as the homogeneous magnetic superlattice (HMSL), and represent it as $(L_F L_H)^n$, where L_F stands for a stripe free of field and L_H for a stripe under an external magnetic field. As shown in Fig. 15.4, the HMSL is basically a quasi two-dimensional semiconductor wave guide subject to an external magnetic field, which is blocked on stripes that alternate along the growing direction z. As is well-known, the spin 1/2 electrons, moving in those regions subject to an external magnetic field $\mathbf{B}$, are described by

$$\left[\frac{1}{2m^*}\left(\mathbf{p} - \frac{e}{c}\mathbf{A}\right)^2 + \left(V(y,z) - E_f\right)\right]\Psi(\mathbf{r}) + \left(\frac{1}{2} g\mu_B \boldsymbol{\sigma} \cdot \mathbf{B} + V_R + V_D\right)\Psi(\mathbf{r}) = \mathbf{0} \tag{15.73}$$

where $V(x, y)$ is a lateral confining potential, V_R and V_D are the Rashba and Dresselhaus spin-orbit interaction. If we have a tilted magnetic field, say $\mathbf{B} =$

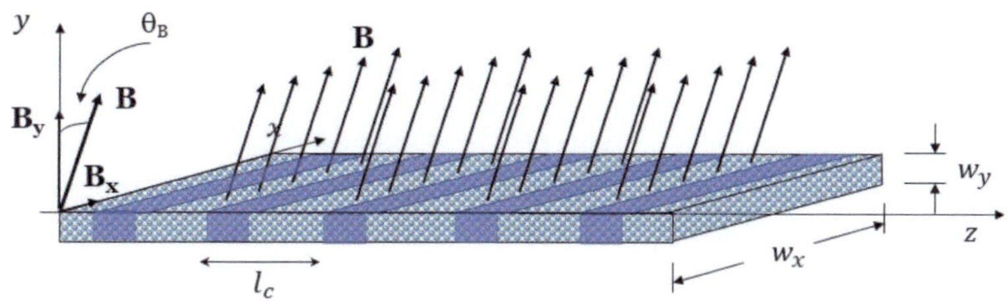

Fig. 15.4 A homogeneous magnetic superlattice where a tilted external magnetic field **B**, blocked in alternate (darker) stripes, acts also only on quasi-twodimensional stripes of thickness d and transverse widths w_x and w_y. Figure reproduced with permission [53] 2008, Microelectronics Journal

$B\,(\sin\theta_B, \cos\theta_B, 0) = B\hat{\mathbf{B}}$, the spin precession and spin transitions induced by the Rashba and Dresselhaus interactions are negligible compared with those induced by the Zeeman interaction. Thus, for this review we drop those terms. It was shown in [18] that after some analytical calculation, one is left with the equation

$$\left(\frac{d^2}{dz^2} - \frac{z^2\cos\theta_B}{l_B^4} + k_z^2 - \frac{g}{2l_B^2}\boldsymbol{\sigma}\cdot\hat{\mathbf{B}}\right) Z(z) = 0, \tag{15.74}$$

whose solutions are the hypergeometric matrix functions [57]

$$\mathscr{A}_\sigma\,(z) = {}_1\mathrm{F}_1\left(-\frac{\mathbf{b}_\sigma}{2}; \frac{1}{2}; \frac{\cos\theta_B}{l_B^2}z^2\right) e^{-z^2\cos\theta_B/2l_B^2}, \tag{15.75}$$

and

$$\mathscr{B}_\sigma\,(z) = z\,{}_1\mathrm{F}_1\left(\frac{\mathbf{12} - \mathbf{b}_\sigma}{2}; \frac{3}{2}; \frac{\cos\theta_B}{l_B^2}z^2\right) e^{-z^2\cos\theta_B/2l_B^2}. \tag{15.76}$$

In these functions, the matrix function $\mathbf{b}_\sigma$ is

$$\mathbf{b}_\sigma = \frac{l_B^2}{2\cos\theta_B}\left(k_z^2 - \frac{g}{2l_B^2}\boldsymbol{\sigma}\cdot\hat{\mathbf{B}} - \mathbf{1}\right), \tag{15.77}$$

with $l_B = \sqrt{\hbar c/Be}$ the magnetic length. and k_z the longitudinal wave number. We will show here examples of distinct spin-projection transmission coefficients. To better visualize the effect of the spin-dependent transmission coefficients, we will plot the space-time evolution of Gaussian pulses of spin 1/2 electrons. To this purpose we need to calculate the wave packet (WP)

$$\psi_s\,(z,t) = \int_{-\infty}^{\infty} dk e^{-\gamma(k-k_0)^2} e^{ikz_0} \phi_s\,(z,k)\, e^{-iwt} \tag{15.78}$$

with centroid at $t_o = 0$ located at the position $-z_o$. The subindex s refers to the spin projection. For the evaluation of this integral, one requires to know analytically the wave functions $\phi_s(z,k)$, inside and outside the SL, based on the solutions $\mathscr{A}_\sigma(z)$ and $\mathscr{B}_\sigma(z)$, the transfer matrix method and the TFPS. To determine the transfer matrix $M(z, z_0)$, that connects the state vector at z_0 with the state vector at any point z inside and outside the superlattice, one can evaluate the scattering amplitudes

$$\mathbf{t}_n = \begin{pmatrix} t_{\uparrow\uparrow} & t_{\uparrow\downarrow} \\ t_{\downarrow\uparrow} & t_{\downarrow\downarrow} \end{pmatrix} \quad \text{and} \quad \mathbf{r}_n = \begin{pmatrix} r_{\uparrow\uparrow} & r_{\uparrow\downarrow} \\ r_{\downarrow\uparrow} & r_{\downarrow\downarrow} \end{pmatrix} \tag{15.79}$$

and the wave functions $\phi_\uparrow(z,k)$, $\phi_\downarrow(z,k)$ inside and outside the superlattice.

In this approach, schematically reviewed here, one can obtain different type of effects by varying the SL parameters and the magnetic field orientation. The behavior of the spin-up and spin-down transmission coefficients, T_{uu} and T_{dd}, respectively, is an important guide to visualize in advance the outcome of desired effects for the WP through the HMSL. We will illustrate the HMSL acting as a spin filter, the tilting angle effect and the HMSL acting as spin inverter.

15.3.1.1 The HMSL as a Spin Filter

In Fig. 15.5 we have the transmission coefficients T_{uu} and T_{dd}, on the left, and the spin wave packets on the right. The HMSL is a quasi-2D wave guide with widths $w_x = 50\,\text{nm}$ and $w_y = 3\,\text{nm}$, under an external magnetic field with $B = 0.18\,\text{T}$ acting on stripes of width $L_H = 10\,\text{nm}$, unit-cell width $l_c = 130\,\text{nm}$, and $n = 6$ unit cells. We consider first that $\theta_B = 0$. In this case, no spin$\uparrow \rightarrow$ spin$\downarrow$ transitions occur, i.e. $T_{ud} = T_{du} = 0$. This implies well defined band structures for the spin$\uparrow$ and the spin$\downarrow$ transmission coefficients, as shown in Fig. 15.5a. In this example, an unpolarized electrons WP is prepared with centroid and wave-packet components entirely in the gap of T_d and in the resonant region of T_u. We expect that the HMSL will work as a spin filter for the spin$\downarrow$ electrons. Indeed, the snapshots in Fig. 15.5b, at $t = 0$ and at $t = 2z_0/v_g + \tau_\sigma$, show the filter effect. At $t = 0$, the spin$\uparrow$ and spin$\downarrow$ Gaussian wave packets with centroids located at $z_0 = -40l_c$, moving towards the HMSL. The centroids of these packets reach the left edge of the HMSL at $t_a = x_o/v_g$, with a group velocity v_g. In the lower frame we plot the spin packets at $t = 2t_a + \tau_u$ and at $t = 2t_a + \tau_d$, when presumably they should be back at the starting point, having spent a time τ_u or τ_d, respectively, in the HMSL. As expected, from the transmission coefficient behavior, the spin$\uparrow$ wave-packet components are partially reflected and partially transmitted, while those of the spin$\downarrow$ packet are completely reflected.

It is interesting to notice that, even though a shape distortion is seen for the spin-up wave packet, the transmitted Gaussian wave-packets are in the position that one would expect them, if the phase-time prediction were correct and the same for most of the spin$\uparrow$ Gaussian components. The tails of the Gaussian in Fig. 15.5b transmit less and have also different tunneling times, while the spin$\downarrow$ components, all have the same transmission coefficient and the same tunneling time, see lower panel in Fig. 15.5b. For the centroid of the spin$\uparrow$ electrons, the phase time predicts a tunneling time $\tau_u = 1.995\,10^{-12}\,\text{s}$, while for spin$\downarrow$ electrons, it predicts $\tau_d = 3.483\,10^{-13}\,\text{s}$. These are quite different times and they are also rather different from the time $\tau_f \simeq$

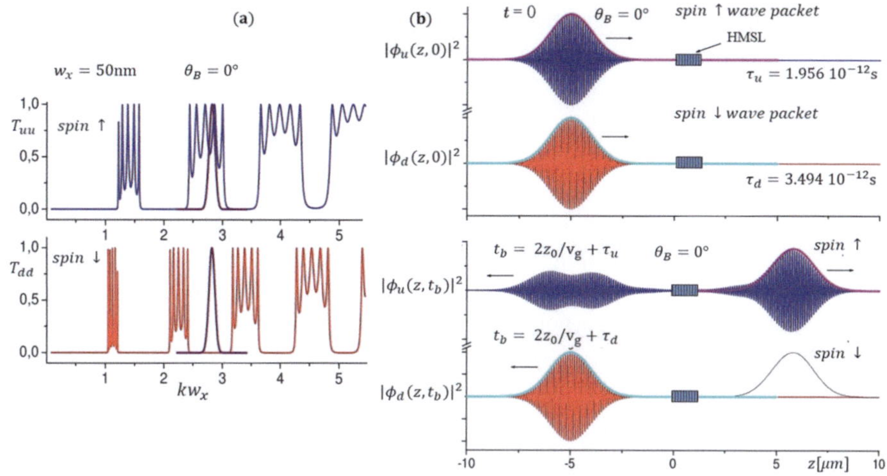

Fig. 15.5 Spin-up and spin-down transmission coefficients and Gaussian wave packets through a HMSL. In **a** the spin-up and spin-down transmission coefficients for $\theta_B = 0$ on a HMSL with $w_x = 50$ nm and $w_y = 3$ nm, $l_c = 130$ nm, $n = 6$, and $B = 0.18$T. The Gaussian curves in the gap of T_{dd} and the allowed region of T_{uu} show the position and shape of the unpolarized spin wave packet that will move through the HMSL. In the upper frame of (**b**), the spin↑ and spin↓ wave packets at $t = 0$ and $-z_0 = -40l_c$, moving towards the HMSL. In the lower frames the spin↑ and spin↓ packets at $t = 2t_a + \tau_u$ and $t = 2t_a + \tau_d$, respectively, when presumably, both packets should be back at $-z_o$ (a distance z_o from the HMSL). It is clear from these graphs *the filter effect* manifest by the absence of transmitted spin-down wave packet components. The Gaussian curves centered at $-z_o$ and at $L + z_o$ are plotted as reference. Here $L = nl_c = 0.78$ µm is the HMSL length. Figure reproduced with permission [53] 2008, Microelectronics Journal

$1.625\ 10^{-12}$ s that they would spent if they were to move (the same distance) in a region free of magnetic fields. This example shows clearly that the system behaves as a spin↓ filter and the phase time predictions are correct.

15.3.1.2 Spin Mixing and Space-Time Evolution When $\theta_B = 40°$

We consider again the unpolarized wave packet of the previous example but now with the magnetic field in the HMSL tilted with $\theta_B = 40°$. In Fig. 15.6a the transmission coefficients T_{uu} and T_{dd} are shown. Now, the reflection coefficient R_{dd} (which was $\simeq$1) reduces to $\simeq$0.7 and the transmission coefficients are smaller than 1, as shown in Fig. 15.6a. An important amount of flux passes from the spin↑ to the spin↓ packet, and vice versa. The spin↑ → spin↓ transitions are induced by the in-plane field component $\mathbf{B_y}$. In addition, the band structures experience small red an blue shifts, respectively. The snapshots in Fig. 15.6b show the wave-packets at $t = 2t_a + \tau_u$ and at $t = 2t_a + \tau_d$, when their centroids are back at $-z_o$ or transmitted at $nl_c + z_o$. It is worth noticing that the wave packet distortion is more evident in this example than in the previous one. This is also compatible with the phase time prediction. In the lower panel of Fig. 15.6b, we observe a kind of spin↓ wave-packet hastening, as the reflected and transmitted packets appear beyond the position expected if they were to move in the HMSL with the centroid's velocity. This can be explained because the phase time of an important fraction of wave-packet components is smaller than the centroid's phase time. Those components leave the HMSL a bit earlier. Because of the spin transitions, there are now reflected and transmitted wave packets for spin↑ and for spin↓ electrons.

15.3.1.3 Spin Inversion

In the previous examples we had at $t = 0$ an unpolarized Gaussian wave packet defined in a spin↓ gap. The spin mixing made it possible to have a small fraction of spin↓ electrons transmitted. We will see now that a spin inversion is also possible. In fact, if at $t = 0$ we have a polarized Gaussian wave packet, say a spin↓ WP in

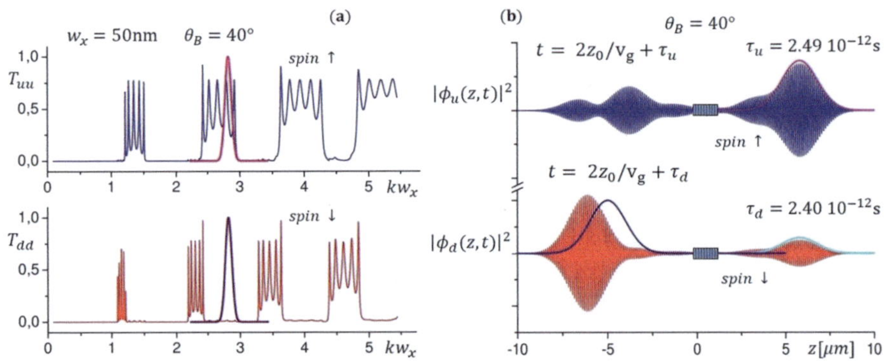

Fig. 15.6 Transmitted and reflected spin↑ and spin↓ wave packets when the magnetic field tilting angle is $\theta_B = 40$. The incoming packets are the same as in Fig. 15.5. The spin↑ and spin↓ wave packets are plotted for $t = 2t_a + \tau_u$ and $t = 2t_a + \tau_d$, respectively. Although the centroid phase time predictions, describes well the transmitted spin-packet tunneling time, it does not represent the reflected spin-packet tunneling time because of the high wave-packet distortion. The Gaussian curves centered at $-x_o$ and at $l_s + x_o$ are plotted for reference. Figure reproduced with permission [53] 2008, Microelectronics Journal

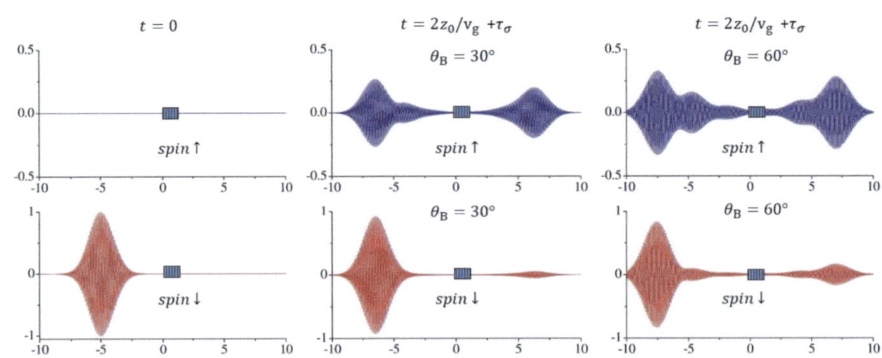

Fig. 15.7 Tilted magnetic field effect. In these graphs a completely polarized WP is sent at $t = 0$ towards a HMSL. The reflected and transmitted WP remains polarized when $\theta_B = 0°$, but a strong spin mixing is observed when $\theta_B = 60°$

lower graph of Fig. 15.6a. Tilting the field, as we have just seen, and as shown in Fig. 15.7 an important amount of electrons with spin↑ are transmitted through the HMSL. This fraction grows with the tilting angle θ_B. Eventually the population of transmitted spin↑ electrons is equal and even larger than the population of spin↓ electrons. All one needs to complete the spin inversion is a second step of spin filter, as described in 15.3.1.1.

References

1. I. Zutić, J. Fabian, S. Das Sarma, Rev. Mod. Phys. **76**, 323 (2004)
2. G. Schmidt, D. Ferrand, L.W. Molenkamp, A.T. Filip, B.J. van Wees, Phys. Rev. B **62**, R4790 (2000)
3. A.T. Filip, B.H. Hoving, F.J. Jedema et al., Phys. Rev. B **62**, 9996 (2000)
4. Y. Guo, H. Wang, B.L. Gu et al., J. Appl. Phys. **88**, 6614 (2000)
5. D. Ferrand, A. Wasiela, S. Tatarenko et al., Sol. Stat. Com. **119**, 237 (2001)
6. S. Das Sarma , J. Fabian, X. Hu, I. Zutić, Solid State Commun. **119**, 207 (2001)
7. S. Sanvito, P. Ordejon, N.A. Hill, Phys. Rev. B **63**, 165206 (2001)
8. T. Dietl, Physica E-Low-Dimens. Syst. Nanostruct. **10**, 120 (2001)
9. A.T. Filip, F.J. Jedema, B.J. van Wees et al., Physica E-Low-Dimens. Syst. Nanostruct. **10**, 478 (2001)
10. J.A.C. Bland, A. Hirohata, C.M. Guertler et al., J. Appl. Phys. **89**, 6740 (2001)
11. D.L. Smith, R.N. Silver, Phys. Rev. B **64**, 045323 (2001)
12. C.M. Hu, Matsuyama. T Phys. Rev. Lett. **87**, 066803 (2001)
13. M. Oestreich, M. Bender, J. Hubner et al., Sem. Sci. Tech. **17**, 285 (2002)
14. J. De Boeck, W. Van Roy, J. Das et al., Sem. Sci. Tech. **17**, 342 (2002)
15. B.T. Jonker, Y.D. Park, B.R. Bennett, H.D. Cheong, G. Kioseoglou, A. Petrou, Phys. Rev. B **62**, 8180 (2000)
16. E.I. Rashba, Phys. Rev. B **62**, R16267 (2000)
17. M. Turek, J. Siewert, J. Fabian, Phys. Rev. B **78**, 085211 (2008)
18. J.L. Cardoso, P. Pereyra, A. Anzaldo-Meneses, Phys. Rev. B **63**, 153301 (2001)
19. P. Pereyra, V.G. Ibarra-Sierra, J.L. Cardoso, Microelectronics J. **40**, 779 (2009)
20. A. Oiwa, Y. Mitsumori, R. Moriya, T. Supinski, H. Munekata, Phys. Rev. Lett. **88**, 137202 (2002)
21. A. Van Esch, L. Van Bockstal, J. De Boeck, G. Verbanck, A.S. van Steenbergen, P.J. Wellmann, B. Grietens, R. Bogaerts, F. Herlach, G. Borghs, Phys. Rev. B **56**, 13 103 (1997)
22. F. Matsukura, D. Chiba, T. Omiya, E. Abe, T. Dietl, Y. Ohno, K. Ohtani, H. Ohno, Physica E **12**, 351 (2002)
23. H. Ohno, Science **281**, 951 (1998)
24. E. Johnston-Halperin, D. Lofgreen, R.K. Kawakami, D.K. Young, L. Coldren, A.C. Gossard, D.D. Awschalom, Phys. Rev. B **65**, 041306 (2002)
25. M. Kohda, Y. Ohno, K. Takamura, F. Matsukura, H. Ohno, Jpn. J. Appl. Phys., Part 2 **40**, L1274 (2001)
26. Y. Ohno, D.K. Young, B. Beschoten, F. Matsukura, H. Ohno, D.D. Awschalom, Nature (London) **402**, 790 (1999)
27. T. Dietl, H. Ohno, F. Matsukura, J. Cibert, D. Ferrand, Science **287**, 1019 (2000)
28. M. Ciorga et al., Phys. Rev. B **79**, 165321 (2009)
29. K. Sato, L. Bergqvist, J. Kudrnovský, P.H. Dederichs, O. Eriksson, I. Turek, B. Sanyal, G. Bouzerar, H. Katayama-Yoshida, V.A. Dinh, T. Fukushima, H. Kizaki, R. Zeller, Rev. Mod. Phys. **82**, 1633 (2010)
30. T. Jungwirth, J. Sinova, J. Masek, J. Kucera, A.H. MacDonald, Rev. Mod. Phys. **78**, 809 (2006)

31. H. Saito, A. Yamamoto, S. Yuasa, K. Ando, App. Phys. Lett. **92**, L192512 (2008)
32. G. Bouzerar, Eur. Phys. Lett. **79**, 57007 (2007). G. Bouzerar, R. Bouzerar, O. Cepas, Phys. Rev. B **76**, 144419 (2007)
33. K. Wagner, D. Neumaier, M. Reinwald, W. Wegscheider, D. Weiss, Phys. Rev. Lett. **97**, 56803 (2006)
34. N.F. Mott, Proc. R. Soc. London, Ser. A **124**, 425 1929. N.F. Mott, Proc. Roy. Soc. London Ser. A **153**, 699 (1936), and **156**, 368 (1936), and Adv. Phys. **13**, 325 (1964)
35. A. Fert, I.A. Campbell, J. Phys. (Paris), Colloq. **32**, C 1-46 (1971)
36. P.C. van Son, H. van Kempen, P. Wyder, Phys. Rev. Lett. **58**, 2271 (1987)
37. T. Valet, A. Fert, Phys. Rev. B **48**, 7099 (1993)
38. S. Hershfield, H.L. Zhao, Phys. Rev. B **56**, 3296 (1997)
39. A. Fert, S.F. Lee, Phys. Rev. B **53**, 6554 (1996)
40. J. Fabian, A. Matos-Abiague, C. Ertlera, P. Stano, I. Zutić, Act. Phys. Slov. **57**, 565 (2007)
41. H.P. Simajuntak, *Macroscopic Quantum Phenomena in Spintronics* (Nova Science Publishers, Inc., New York, 2019)
42. S. Datta, B. Das, Appl. Phys. Lett. **56**, 665 (1990)
43. P. Pereyra, D. Weiss, Physica B **455**, 96 (2014)
44. A. Fert, H. Jaffres, Phys. Rev. B **64**, 184420 (2001); A. Fert, H. Jaffres, J. Appl. Phys. **91**, 811 (2002)
45. S. Takahashi, S. Maekawa, Phys. Rev. B **67**, 052409 (2003)
46. J. Fabian, I. Zutić, S. Das Sarma, Phys. Rev. B **66**, 165301 (2002)
47. V.V. Osipov, A.M. Bratkovsky, Phys. Rev. B **72**, 115322 (2005)
48. J. Shiogai, M. Ciorga, M. Utz, D. Schuh, M. Kohda, D. Bougeard, T. Nojima, J. Nitta, D. Weiss, Phys. Rev. B **89**, R081307 (2014)
49. H. Holmberg, N. Lebedeva, S. Novikov, J. Ikonen, P. Kuivalainen, M. Malfait, V.V. Moshchalkov, Europhys. Lett. **71**, 811 (2005)
50. P. Pereyra, D. Weiss, Phys. Rev. B **90**, 245310 (2014)
51. P. Pereyra, D. Weiss, Proc. SPIE **9931**, 993124 (2016). https://doi.org/10.1117/12.2237995
52. E.O. Kane, Phys. Rev. **131**, 79 (1963)
53. J.L. Cardoso, P. Pereyra, Europhys. Lett. **83**, 38001 (2008)
54. S. Seikin, M. Shen, M.-C. Cheng, Y. Privman, J. Appl. Phys. **94**, 1769 (2003)
55. D.K. Young, J.A. Gupta, E. Johnston-Halperin, R. Epstein, Y. Kato, D.D. Awschalom, Semicond. Sci. Technol. **17**, 275 (2002)
56. J.L. Cardoso, P. Pereyra, Physica Status Solidi (c) **2**, 3082 (2005); J.L. Cardoso, P. Pereyra, Physica Status Solidi (c) **2**, 3153 (2005)
57. P. Pereyra, E. Castillo, Phys. Rev. B **65**, 205120 (2002)

Variational Asymmetric Binding Energy

A

Given the quasi-two-dimensional Schrödinger equation with a Coulomb potential and azimuthal quantum number $m = 0$, written as

$$\left[-\frac{\hbar^2}{2\mu_\|} \frac{1}{\rho} \frac{d}{d\rho} \rho \frac{d}{d\rho} - \frac{e^2}{\epsilon\sqrt{\rho^2 + z^2}} \right] \phi_n(\rho, z) = -E_n^{eh} \phi_n(\rho, z) \tag{A.1}$$

where ρ and z are the in-plane $\rho_e - \rho_h$ and perpendicular $z_e - z_h$ electron-hole distances, and $\mu_\|$ the in-plane relative e-h effective mass, it is convenient to define the dimensionless parameters

$$u = \frac{\rho}{a_0}, \qquad \zeta = \frac{z_e - z_h}{a_0}, \qquad w(\zeta) = \frac{E_n^{eh}}{R_{eh}} \qquad \text{and} \qquad G(u; \zeta) = a_0 \phi_n(u; \zeta), \tag{A.2}$$

where

$$a_0 = \frac{\epsilon\hbar^2}{\mu_\| e^2} \qquad \text{and} \qquad R_{eh} = \frac{\mu_\| e^4}{2\epsilon^2\hbar^2}, \tag{A.3}$$

are the Bohr's radio and the effective Rydberg energy. With these definitions one has the dimensionless differential equation

$$\left(\frac{1}{u} \frac{d}{du} u \frac{d}{du} + \frac{2}{\sqrt{u^2 + \zeta^2}} \right) G(u, \zeta) = w(\zeta) G(u, \zeta). \tag{A.4}$$

Leavitt and Little, [1] by using a trial wave function defined with the correct results in the limits $\zeta = 0$ and $\zeta >> u$, were able to obtain a wave function and the ground state binding energy. In fact, knowing the exact solutions: $G(u; 0) = (1/u)e^{-2u}$ with $w(0) = 4$, for $\zeta = 0$, and $G(u; \zeta >> 1) = \sqrt{2}e^{-u^2/2\zeta^{3/2}}/u\zeta^3$ with $w(\zeta) = 2/\zeta - 2/\zeta^{3/2}$, for $\zeta >> u$, the trial function

$$G_\lambda(u, \zeta) = N e^{-\lambda(\sqrt{u^2+\zeta^2}-\zeta)} \tag{A.5}$$

P. Pereyra, *Contemporary Theory of Semiconductors and Devices*, Graduate Texts in Physics, https://doi.org/10.1007/978-3-032-00296-9

was suggested, with N a normalization constant and λ a variational parameter chosen as $\lambda = 2/(1+2\sqrt{\zeta})$, in such a way that as $\zeta \to 0$ the variational parameter $\lambda \to 2$, and for $\zeta >> 1$, in which case $\lambda \to 1/\sqrt{\zeta}$.

Since the first excited energy and the exact solutions are also known in these limits: $w_1(0) = 4/(3+2m)^2$ for $\zeta = 0$ and $w_1(\zeta) = 2/\zeta(1-(3+m)/\sqrt{\zeta})$ for $\zeta >> u$, we propose the trial wave function

$$G_{\lambda_k}(u,\zeta) = Ne^{-\lambda_k(\sqrt{u^2+\zeta^2}-\zeta)}\left[1-2\lambda_k(\sqrt{u^2+\zeta^2}-\zeta)\right]^k \tag{A.6}$$

with

$$\lambda_k = \frac{2}{2k+1+m+2\sqrt{\zeta}} = \frac{1}{\eta_k}. \tag{A.7}$$

being η_k the energy quantum number. The expectation values $\langle w_{\lambda_k}(\zeta)\rangle$ obtained from [2]

$$\langle w_{\lambda_k}(\zeta)\rangle = \frac{\int_0^\infty G^*_{\lambda_k}(u,\zeta)\hat{H}_{2D}G_{\lambda_k}(u,\zeta)udu}{\int_0^\infty |G_{\lambda_k}(u,\zeta)|^2 udu}, \tag{A.8}$$

become, for $k = 0$

$$\langle w_{\lambda_0}(\zeta)\rangle = -\lambda_0^2 + \frac{4\lambda_0 + 4\zeta^2\lambda_0^4}{1+2\sqrt{\zeta}}e^{2\zeta\lambda_0}\Gamma(0, 2\zeta\lambda_0), \tag{A.9}$$

and for $k = 1$

$$\begin{aligned}\langle w_{\lambda_1}(\zeta)\rangle = &-4e^{2\zeta\lambda_1}\zeta^2\lambda_1^4(3+2\zeta\lambda_1)\Gamma(0,2\zeta\lambda_1)\\ &+\frac{4\lambda_1 - \lambda_1^2\Big(3+2\zeta\lambda_1(5+2\zeta\lambda_1(5+2\zeta\lambda_1))\Big)}{3+2\zeta\lambda_1}.\end{aligned} \tag{A.10}$$

Notice that for $\zeta = 0$ and $m = 0$, these (dimensionless) binding energies become $\langle w_{\lambda_0}(0)\rangle = -\lambda_0^2 + 4\lambda_0 = 4$ and $\langle w_{\lambda_1}(0)\rangle = (-3\lambda_1^2 + 4\lambda_1)/3 = 4/9$. In Fig. A.1 we plot the expectation value of the dimensionless exciton binding energies in the ground and first exited states. The ratio between these energies changes from a factor of 9 in the 2D limit to a factor that tends to 3 as $\zeta \to 1$, in the 3D limit.

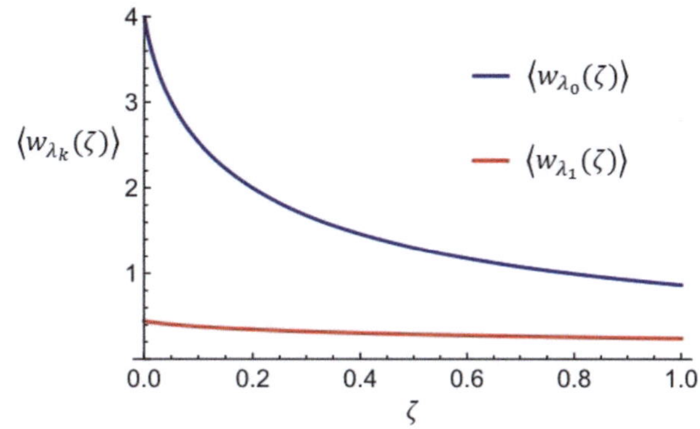

Fig. A.1 Dimensionless exciton binding energies as function of ζ, for the ground and the first excited states. The binding energy decreases as the exciton dimension grows from $D = 2$ ($\zeta = 0$) to $D = 3$ ($\zeta = 1$)

Recombination-Generation Currents. The SRH Model

B

Recombination processes and generation of carriers in the transition zone, mediated by the so-called deep levels, which are located in the gap, contribute to the emission phenomenon through electron and hole capture mechanisms, as shown in Fig. B.1b.

Shockley, Read, and Hall determined the rate at which concentrations n and p change in the transition zone. We will assume that the system is connected to a battery with potential difference V. In general, we should have $n(z)$ and $p(z)$ with $-w_p < z < w_n$. Since the concentration, especially the minority concentration, changes from one end of the transition zone to the other, the analysis can be done by assuming concentrations n and p, such that: if $z = -w_p$, the concentrations are $n = n(-w_p) = n_{pV} = n_p e^{eV/k_BT}$ and $p = p(-w_p) = p_p$; likewise if $z = w_n$, the concentrations are $n = n_n$ and $p = p_{nV} = p_n e^{eV/k_BT}$.

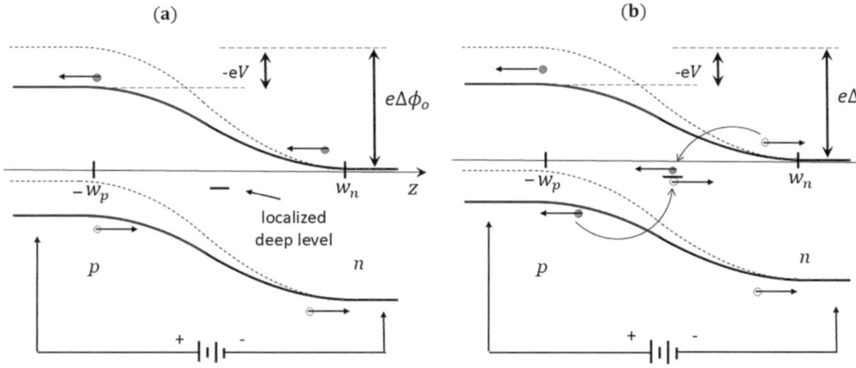

Fig. B.1 Non-radiative recombination mediated by deep levels, localized and capable of capturing both electrons and holes. The electrons and holes that are in the transition zone are captured at the deep level that is in the gap. In **b** the capture of the hole is equivalent to the emission of an electron to the valence band, while the capture of the electron is equivalent to the emission of a hole to the conduction band

P. Pereyra, *Contemporary Theory of Semiconductors and Devices*, Graduate Texts in Physics, https://doi.org/10.1007/978-3-032-00296-9

Eventhough the majority concentrations do not change much with the external field, they can be modified in the presence of capture and emission processes. If the concentration of deep levels is N_d, and the energy level of the states is E_d, we denote by $f(E_d)$ the probability that the state of energy E_d is occupied. When the electron and the hole move with speeds $v_e \simeq v_h$ of the order of the thermal speed v_{th} defined by the relation

$$\frac{1}{2}m^* v_{th} = \frac{3}{2}k_B T, \tag{B.1}$$

and the effective cross sections for the capture of electrons and holes are σ_e and σ_h, respectively, the coefficients can be defined with the capture parameters $c_n = \sigma_e v_{th}$ and $c_p = \sigma_h v_{th}$. When an electron is captured, an emission coefficient e_n can be defined for the deep level. Likewise, if a hole is captured, an emission coefficient e_p is defined. With these coefficients, the deep level occupation probabilities and the deep level concentration N_d, the capture and emission rates of electrons and holes can be defined. For electrons we have

$$r_e^c = N_d c_n (1 - f(E_d))n \qquad \text{and} \qquad r_e^e = N_d e_n f(E_d). \tag{B.2}$$

While the rate of capture does depend on the concentration of electrons n, the rate of emission does not. We will see that the emission coefficients do depend on an effective concentration of electrons (or holes). If these rates are not equal, the rate at which the population n changes is (to simplify the notation we will write f_d instead of $f(E_d)$)

$$\left.\frac{\partial n}{\partial t}\right|_{recomb} = -\big[N_d c_n (1 - f_d)n - N_d e_n f_d)\big]. \tag{B.3}$$

But if $r_e^c = r_e^e$, the system is in a stationary condition. In this case, if $f_d \to f_0$ and $n \to n_o = n_i e^{(E_F - E_{F_i})}$, with E_{F_i} the intrinsic Fermi level $\simeq E_g/2$, the relationship can be obtained as

$$\frac{e_n}{c_n} = n_o \frac{1 - f_0}{f_0} = n_0 e^{(E_d - E_F)/k_B T} \simeq n_i e^{(E_d - E_{F_i})/k_B T} = n_d. \tag{B.4}$$

Likewise, the rate of change of the population p is

$$\left.\frac{\partial p}{\partial t}\right|_{recomb} = -\big[N_d c_p p f_d - N_d e_p (1 - f_d)\big]. \tag{B.5}$$

When $r_h^c = r_h^e$ and also assuming that in this case $f_d \to f_0$ while $p \to p_0 = p_i e^{(E_{F_i} - E_F)}$, we have

$$\frac{e_p}{c_p} = p_0 \frac{f_0}{1 - f_0} = p_0 e^{-(E_d - E_F)/k_B T} \simeq n_i e^{-(E_d - E_{F_i})/k_B T} = p_d. \tag{B.6}$$

With these relations, and introducing the times of capture of an electron $\tau_{n0} = 1/c_n N_d$ and capture of a hole $\tau_{p0} = 1/c_p N_d$, the changes in the populations n and p can be written as

$$\left.\frac{\partial n}{\partial t}\right|_{recomb} = -\frac{1}{\tau_{n0}}\big[(1 - f_d)n - n_d f_d\big], \tag{B.7}$$

and

$$\left.\frac{\partial p}{\partial t}\right|_{recomb} = -\frac{1}{\tau_{p0}}\big[p f_d - n_d(1 - f_d)\big]. \tag{B.8}$$

When the rates of change of n and p are not equal, the occupation probability f_d would have to be obtained by solving the equation

$$\left.\frac{\partial n}{\partial t}\right|_{recomb} - \left.\frac{\partial p}{\partial t}\right|_{recomb} = -N_d \frac{\partial f_d}{\partial t}, \tag{B.9}$$

but, if they are equal, we have

$$f_d\left[\frac{n + n_d}{\tau_{n0}} + \frac{p + p_d}{\tau_{p0}}\right] = \frac{n}{\tau_{n0}} + \frac{p_d}{\tau_{p0}}, \tag{B.10}$$

and f_d will be given by the expression

$$f_d = \frac{n\tau_{p0} + p_d\tau_{n0}}{(n + n_d)\tau_{p0} + (p + p_d)\tau_{n0}}. \tag{B.11}$$

With this, the non-radiative recombination rate, which is equal to the rate at which the population n or p changes, can be expressed as

$$R_{rc} = \frac{np - n_i^2}{\tau_{p0}(n + n_d) + \tau_{n0}(p + p_d)}. \tag{B.12}$$

This rate of change in populations may be important because the population that does not contribute to j_n and j_p, contributes to R_{rc}. The previous expression is simplified a bit if $E_d \simeq E_F$, i.e. the deep levels are more or less close to the center of the gap, and $n \gg n_d \simeq n_i$, $p \gg p_d = n_i$ and $\tau_{n0} \simeq \tau_{p0} \simeq \tau_{rc}$. Then

$$R_{rc} = \frac{np}{\tau_{rc}(n + p)}. \tag{B.13}$$

As said before, if $z = -w_p$, $n = n_p e^{eV/k_B T}$ and $p = p_p$. This means that

$$np = n_p p_p e^{eV/k_B T} = n_i^2 e^{eV/k_B T}. \tag{B.14}$$

The same result is obtained if in $z = w_n, n = n_n$ and $p = p_n e^{eV/k_BT}$. For this reason, in the transition zone, we can approximate the concentrations n and p, on average, as

$$n \simeq p = n_i e^{eV/2k_BT}. \tag{B.15}$$

Consequently, the nonradiative recombination rate of the transition zone will be

$$R_{rc} = \frac{n_i}{2\tau_{rc}} e^{eV/2k_BT}. \tag{B.16}$$

The current associated with the recombination processes in the transition zone whose length is $W = w_p + w_n$, through an area A, can be written as

$$I_R(V) = eWAR_{rc} = \frac{eWAn_i}{2\tau_{nr}} e^{eV/2k_BT} = I_{R0} e^{eV/2k_BT}. \tag{B.17}$$

This expression suggests that when $V = 0$, a current I_{R0} remains. In reality it must be canceled and we have to add a term of spontaneous generation of charge carriers that is associated with a current $I_G = -I_R(0)$ that cancels the spontaneous recombination current I_{R0}. This contribution corresponds to the component $-n_d p_d$ neglected above. Consequently, we have a current related to the recombination-generation given by

$$I_{gr}(V) = I_{R0} e^{eV/2k_BT} - IR(0) = I_{R0}\left(e^{eV/2kBT} - 1\right). \tag{B.18}$$

Therefore, the injection current (of the majority carriers) plus the current associated with non-radiative recombination-generation processes, will be

$$I(V) = I_0(e^{eV/k_BT} - 1) + I_{R0}\left(e^{eV/2k_BT} - 1\right). \tag{B.19}$$

The contribution of the second term is greater if the potential values are small, and the first term, associated with injection currents, contributes more for large potentials. It is customary to express the two contributions in a single expression:

$$I(V) = I_0'\left(e^{eV/mk_BT} - 1\right), \tag{B.20}$$

where the factor m known as the ideality factor of the diode has been introduced. In high quality diodes $m \to 1$.

The Rectangular Potential Well

C

A solvable example that was initially widely used to illustrate the application of the Schrödinger equation, and in recent decades, as the light-emitting potential model, is the finite rectangular well shown in Fig. C.1. A system like this can be obtained when three semiconductor layers are grown, and the edges of the conduction bands of the first and third layers have higher potential energy than in the second layer, for example, in the heterostructure $Ga_{1-x}Al_xAs$/Ga $As/Ga_{1-x}Al_xAs$. The potential energy profile is effectively a potential well, mostly known as quantum well (QW). If the energy of the particle is less than the finite height of the barriers in the well, the potential confines the particle and the particle's energy is quantized. We will consider here the potential defined by the function

$$V(x) = \begin{cases} V_0 & x \leq 0, \quad x \geq a \\ 0 & 0 < x < a \end{cases} . \tag{C.1}$$

Since the potential is sectionally constant, effective masses are also sectionally constant. The Schrödinger equation of the QW, like that of the barrier, can be solved for each layer. The energy of the particle can be less than or greater than V_0. Here we will review only the case of energy less than V_0. The case of energy greater than V_0 can be seen in quantum physics texts, for example in [3].

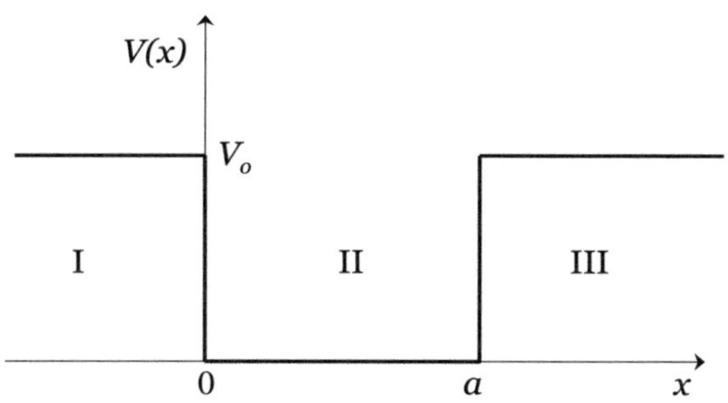

Fig. C.1 Potential well with finite depth V_0 and width a

P. Pereyra, *Contemporary Theory of Semiconductors and Devices*,
Graduate Texts in Physics, https://doi.org/10.1007/978-3-032-00296-9

If $0 < E < V_0$, the solutions in regions I, II and III (see Fig. C.1) are, respectively,

$$\begin{aligned} \varphi_{\mathrm{I}}(x) &= a_1 e^{qx} + b_1 e^{-qx} && x \leq 0, \\ \varphi_{\mathrm{II}}(x) &= a_2 e^{ikx} + b_2 e^{-ikx} && 0 < x < a, \\ \varphi_{\mathrm{III}}(x) &= a_3 e^{qx} + b_3 e^{-qx} && a \leq x. \end{aligned} \tag{C.2}$$

Here $q = \sqrt{2m_b^*(V_0 - E)/\hbar^2}$ and $k = \sqrt{2m_w^* E/\hbar^2}$ with m_b^* the effective mass in the barriers, and m_w^* the effective mass in the wells. The solutions in regions I and III are the combination of functions that increase and decrease exponentially and in II combinations of oscillating functions. When the barriers are wide, the condition of finiteness requires that the coefficients b_1 and a_3 cancel out. As we did in the barrier case, we will temporarily preserve all the coefficients and make the coefficients b_1 and a_3 zero later. Unlike the infinite well, in which the solutions in regions I and III cancel out, in the finite well the solutions are exponential functions which penetrate the walls, as will be seen below. It is clear that the penetration length in the barriers reduces as the height V_o of the walls increase or the energy E of the particle decreases.

The quantum well wave function must be continuous from $x = -\infty$ to $x = \infty$. Since we have layered solutions, we have to impose continuity conditions between the functions $\varphi_{\mathrm{I}}(x)$, $\varphi_{\mathrm{II}}(x)$ and $\varphi_{\mathrm{II}}(x)$, $\varphi_{\mathrm{III}}(x)$, as well as their first derivatives at $x = 0$ and $x = a$. The continuity conditions at $x = 0$ give rise to equations

$$\begin{aligned} a_1 + b_1 &= a_2 + b_2, \\ q(a_1 - b_1) &= ik(a_2 - b_2), \end{aligned} \tag{C.3}$$

and the conditions of continuity at $x = a$, to equations

$$\begin{aligned} a_3 e^{qa} + b_3 e^{-qa} &= a_2 e^{ika} + b_2 e^{-ika}, \\ q(a_3 e^{qa} - b_3 e^{-qa}) &= ik(a_2 e^{ika} - b_2 e^{-ika}). \end{aligned} \tag{C.4}$$

In matrix representation these equations are expressed as

$$\begin{pmatrix} a_2 \\ b_2 \end{pmatrix} = \frac{1}{2k} \begin{pmatrix} k - iq & k + iq \\ k + iq & k - iq \end{pmatrix} \begin{pmatrix} a_1 \\ b_1 \end{pmatrix} = M_i(0^+, 0^-) \begin{pmatrix} a_1 \\ b_1 \end{pmatrix}. \tag{C.5}$$

and

$$\begin{pmatrix} a_3 e^{qa} \\ b_3 e^{-qa} \end{pmatrix} = \frac{1}{2q} \begin{pmatrix} q + ik & q - ik \\ q - ik & q + ik \end{pmatrix} \begin{pmatrix} a_2 e^{ika} \\ b_2 e^{-ika} \end{pmatrix} = M_d(a^+, a^-) \begin{pmatrix} a_2 e^{ika} \\ b_2 e^{-ika} \end{pmatrix}. \tag{C.6}$$

It can be verified that the matrix $M_i(0^+, 0^-)$ is the inverse of the transition matrix $M_d(a^+, a^-)$ that relates the wave vectors at the well-barrier interface in $x = a$.

To relate the state vector at $x = 0^+$ to the state vector at $x = a^-$, we need to propagate the wave vector from the left to the right of the well. It is easy to see that

$$\begin{pmatrix} a_2 e^{ika} \\ b_2 e^{-ika} \end{pmatrix} = \begin{pmatrix} e^{ika} & 0 \\ 0 & e^{-ika} \end{pmatrix} \begin{pmatrix} a_2 \\ b_2 \end{pmatrix} = M_t(a^-, 0^+) \begin{pmatrix} a_2 \\ b_2 \end{pmatrix}, \tag{C.7}$$

where the transfer matrix $M_t(a^-, 0^+)$ that propagates the wave vector from one point to another of the well arises naturally. The leftmost and rightmost wave vectors of the well only differ in their phases. For this reason the transfer matrix in (C.7) only modifies the phase. With the matrix $M_t(a^-, 0^+)$ we have the necessary relations to connect the wave vector in region III with the wave vector in region II. Combining (C.5), (C.6) and (C.7) we obtain the relationship

$$\begin{pmatrix} a_3 e^{qa} \\ b_3 e^{-qa} \end{pmatrix} = \frac{1}{4qk} \begin{pmatrix} q+ik & q-ik \\ q-ik & q+ik \end{pmatrix} \begin{pmatrix} e^{ika} & 0 \\ 0 & e^{-ika} \end{pmatrix} \begin{pmatrix} k-iq & k+iq \\ k+iq & k-iq \end{pmatrix} \begin{pmatrix} a_1 \\ b_1 \end{pmatrix}, \tag{C.8}$$

between the state vectors at $x = 0^-$ and at $x = a^+$, to the left and right of the well. To this transfer matrix we denote it as $M_a(a^+,0^-)$. After multiplying and simplifying we have the rectangular-well transfer matrix

$$M_a(a^+, 0^-) = \begin{pmatrix} \cos ka + \frac{q^2-k^2}{2qk} \sin ka & -\frac{k^2+q^2}{2qk} \sin ka \\ \frac{k^2+q^2}{2qk} \sin ka & \cos ka - \frac{q^2-k^2}{2qk} \sin ka \end{pmatrix} \equiv \begin{pmatrix} \alpha_a & \beta_a \\ -\beta_a & \delta_a \end{pmatrix}, \tag{C.9}$$

and, consequently

$$\begin{pmatrix} a_3 e^{qa} \\ b_3 e^{-qa} \end{pmatrix} = \begin{pmatrix} \cos ka + \frac{q^2-k^2}{2qk} \sin ka & -\frac{k^2+q^2}{2qk} \sin ka \\ \frac{k^2+q^2}{2qk} \sin ka & \cos ka - \frac{q^2-k^2}{2qk} \sin ka \end{pmatrix} \begin{pmatrix} a_1 \\ b_1 \end{pmatrix}. \tag{C.10}$$

The structure of the transfer matrix is determined in general by physical properties and symmetries, such as the principle of conservation of flux and the principle of invariance under time inversion. It can be seen that when the energy $E > V_0$, then $q \to i\kappa$, and the structure of the transfer matrices related to the aforementioned principles, mentioned for example in sect. 5.5, 6.3.2, and Eq. (13.38), is recovered. It is worth recalling that the temporal reversibility imposes a slightly more general condition on the transfer matrices. As can be seen in Appendix E, temporal reversibility requires that the transfer matrices M satisfy the condition

$$M \begin{pmatrix} 0 & 1 \\ 1 & 0 \end{pmatrix} M^* = \begin{pmatrix} 0 & 1 \\ 1 & 0 \end{pmatrix}. \tag{C.11}$$

It can be verified that the matrix $M_a(a^+,0^-)$ satisfies this temporal reversibility condition. Let us now consider the finiteness of the solutions. As said before, the solutions

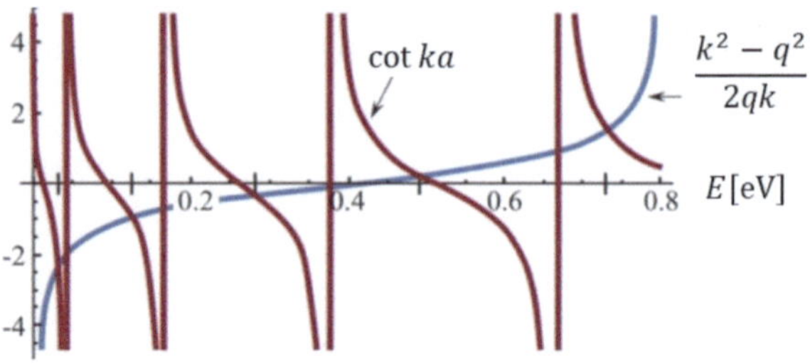

Fig. C.2 graphical method to determine the energy eigenvalues from (C.14)

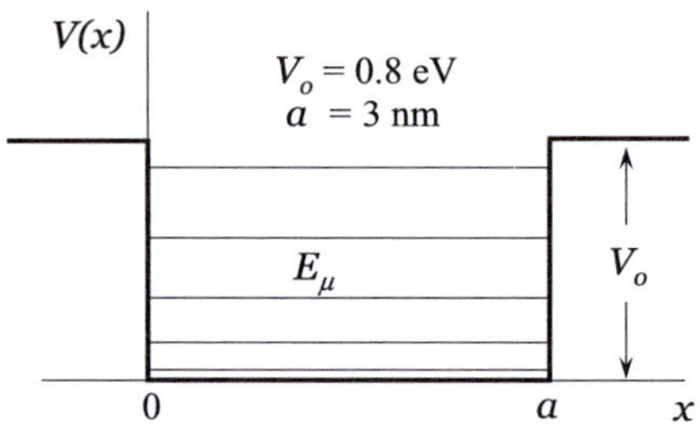

Fig. C.3 Eigenvalues E_m for a quantum well which width is $a = 3\,\text{nm}$ and depth $V_0 = 0.8\,\text{eV}$

are finite if $b_1 = a_3 = 0$. Substituting these values into the previous equation, we obtain

$$\begin{pmatrix} 0 \\ b_3 e^{-qa} \end{pmatrix} = \begin{pmatrix} \alpha_a & \beta_a \\ -\beta_a & \delta_a \end{pmatrix} \begin{pmatrix} a_1 \\ 0 \end{pmatrix} \tag{C.12}$$

from which we easily deduce the equation

$$a_1 \left(\cos ka + \frac{q^2 - k^2}{2qk} \sin ka \right) = 0. \tag{C.13}$$

Since $a_1 \neq 0$, it must hold the equality

$$\cot ka = \frac{k^2 - q^2}{2qk} \tag{C.14}$$

This equation is not satisfied for every value of the energy. It holds only for a discrete set of values E. Those are the eigenvalues of the potential well and (C.14) is *the eigenvalue equation of the finite well*. One way to solve this transcendent equation is to plot separately $\cot ka$ and $(k^2 - q^2)/2qk$ as functions of energy. In Fig. C.2 we precisely do that. The intersections of the two functions occur at the eigenvalues E_μ. It is also possible to solve the equation numerically to obtain the eigenvalues. The energy levels E_μ are indicated in Fig. C.3 for electrons in a well of width $a = 3\,\text{nm}$ and depth $V_0 = 0.8\,\text{eV}$. Equation (C.14) provides all the eigenvalues that in common textbooks are obtained from the solution of two implicit equations, generally, one associated with the odd functions and the other with even functions.

What other information can be obtained from the relation

$$\begin{pmatrix} 0 \\ b_3 e^{-qa} \end{pmatrix} = \begin{pmatrix} \alpha_a & \beta_a \\ -\beta_a & \delta_a \end{pmatrix} \begin{pmatrix} a_1 \\ 0 \end{pmatrix}? \tag{C.15}$$

Another implicit relationship in this equation is

$$b_3 = -e^{qa} \beta_a a_1. \tag{C.16}$$

This means that the coefficients b_3 and a_1 are not independent. Using this relationship, the wave functions in regions I and III must be written in the form

$$\varphi_{\mathrm{I}}(x) = a_1 e^{qx} \qquad x < 0, \tag{C.17}$$

$$\varphi_{\mathrm{III}}(x) = a_1 \frac{k^2 + q^2}{2qk} \sin ka \; e^{-q(x-a)}, \qquad x > a. \tag{C.18}$$

Both are functions that decrease exponentially as the walls of the well are penetrated. If we return to (C.5) with $b_1 = 0$, we obtain the coefficients

$$a_2 = \frac{k - iq}{2k} a_1, \tag{C.19}$$

$$b_2 = \frac{k + iq}{2k} a_1, \tag{C.20}$$

which when substituted into $\varphi_{\mathrm{II}}(x)$ leaves us with the function

$$\varphi_{\mathrm{II}}(x) = a_1 \left[\cos kx + \frac{q}{k} \sin kx \right]. \tag{C.21}$$

In this way, the solutions in the three regions were obtained in terms of a single coefficient. The coefficient a_1 is fixed when the eigenfunctions are normalized. Once the eigenvalues $\{E_\mu\}$ are set, and the functions of regions I, II and III, are given in the joined form

$$\varphi_p(x, E) = a_1 \begin{cases} e^{qx} & x < 0 \\ \cos kx + \frac{q}{k} \sin kx & 0 < x < \\ \frac{k^2 + q^2}{2qk} \sin ka \; e^{-q(x-a)} & x > a \end{cases}, \tag{C.22}$$

we can obtain the eigenfunctions $\varphi_\mu(x)$ of the well by evaluating the function $\varphi_p(x, E)$ of (C.22) for $E = E_\mu$, that is:

$$\varphi_\mu(x) = \varphi_p(x, E_\mu). \tag{C.23}$$

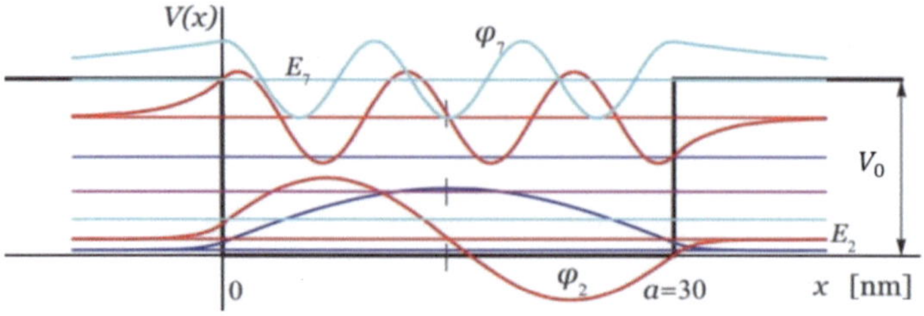

Fig. C.4 Eigenvalues E_i and the corresponding eigenfunctions φ_i. In this well $V_0 = 0.23\,\text{eV}$ and the width is $a = 30\,\text{nm}$. These parameters are different from those in Fig. C.3, in which the depth of the well is greater but the width is smaller. It is evident that by increasing the width of the well, the levels of energy fall and its density increases. The parity of the eigenfunctions is defined with respect to the center of the well. The even eigenfunctions alternate with the odd ones

In Fig. C.4 we plot the eigenfunctions for a well in which $a = 30\,\text{nm}$ and $V_0 = 0.23\,\text{eV}$. The solutions of the three regions are real, consequently the current density in each of the regions is zero. This means that the flow of particles that partially penetrates the barriers, also leaves them. With the eigenfunctions and eigenvalues, the optical susceptibility can be calculated, and the optical response of the quantum well obtained.

Interaction Representation

D

We will present here an alternative procedure frequently used to solve the time-dependent equation

$$i\hbar\frac{\partial\psi}{\partial t} = \widehat{H}\psi. \tag{D.1}$$

A procedure appropriate when the Hamiltonian can be written as

$$\widehat{H} = \widehat{H}_o + \widehat{V}(t), \tag{D.2}$$

where $\widehat{H}_o$ describes the system in equilibrium and $\widehat{V}(t)$ a perturbation, which, not necessarily, depends explicitly on time. This method is based on the assumption that the time evolution of the wave function $\psi(\mathbf{r}, t)$ is determined by the time-dependent potential $\widehat{V}(t)$. The purpose is to obtain the function $\psi(\mathbf{r}, t)$ to first order in $\widehat{V}(t)$, and expectation values

If we define the time-dependent function

$$\widetilde{\psi}(\mathbf{r}, t) = e^{i\widehat{H}_o t/\hbar}\psi(\mathbf{r}, t), \tag{D.3}$$

such that at $t = 0$

$$\widetilde{\psi}(\mathbf{r}, 0) = \psi(\mathbf{r}, 0), \tag{D.4}$$

it is clear that (to simplify the notation we show only the time dependence)

$$\begin{aligned}\frac{\partial\widetilde{\psi}(t)}{\partial t} &= \frac{i}{\hbar}\widehat{H}_o e^{i\widehat{H}_o t/\hbar}\,\psi(t) + e^{i\widehat{H}_o t/\hbar}\,\frac{\partial\psi(t)}{\partial t}\\ &= \frac{i}{\hbar}e^{i\widehat{H}_o t/\hbar}\widehat{H}_o\,\psi(t) + e^{i\widehat{H}_o t/\hbar}\left(-\frac{i}{\hbar}\widehat{H}\psi(t)\right)\\ &= -\frac{i}{\hbar}\left(e^{i\widehat{H}_o t/\hbar}\widehat{V}(t)e^{-i\widehat{H}_o t/\hbar}\right)\widetilde{\psi}(t).\end{aligned} \tag{D.5}$$

P. Pereyra, *Contemporary Theory of Semiconductors and Devices*,
Graduate Texts in Physics, https://doi.org/10.1007/978-3-032-00296-9

If we define the operator

$$\widehat{V}'(t) = e^{i\widehat{H}_o t/\hbar}\widehat{V}(t)e^{-i\widehat{H}_o t/\hbar}, \tag{D.6}$$

the third equation in (D.5) can be written as

$$i\hbar\frac{\partial\widetilde{\psi}(t)}{\partial t} = \widehat{V}'(t)\widetilde{\psi}(t), \tag{D.7}$$

This means that

$$\widetilde{\psi}(t) = e^{-i\widehat{V}'(t)(t-t_0)/\hbar}\widetilde{\psi}(t_0) \equiv \widehat{U}(t, t_0)\ \widetilde{\psi}(t_0). \tag{D.8}$$

We will try to obtain a simpler representation of the unitary operator $\widehat{U}(t, t_0)$. If we replace the function $\widetilde{\psi}(t)$ in (D.7) by $\widehat{U}(t, t_0)\ \widetilde{\psi}(t_0)$ we have the first-order differential equation

$$i\hbar\frac{d\widehat{U}(t, t_0)}{dt} = \widehat{V}'(t)\widehat{U}(t, t_0), \tag{D.9}$$

which, after integration from t_0 to t, transforms into

$$\widehat{U}(t, t_0) - \widehat{U}(t_0, t_0) = -\frac{i}{\hbar}\int_{t_0}^{t}\widehat{V}'(t_1)\widehat{U}(t_1, t_0)dt_1. \tag{D.10}$$

From the definition of the operators $\widehat{V}'(t_1)$ and $\widehat{U}(t, t_0)$, it is easy to see that the last equation can be written as

$$\widehat{U}(t, t_0) = 1 - \frac{i}{\hbar}\int_{t_0}^{t} e^{i\widehat{H}_o t_1/\hbar}\widehat{V}(t_1)e^{-i\widehat{H}_o t_1/\hbar}\widehat{U}(t_1, t_0)dt_1, \tag{D.11}$$

which is a time evolution operator where both parts of the Hamiltonian, $\widehat{H}_o$ and $\widehat{V}(t)$ appear explicitly. This is an integral equation for $\widehat{U}(t, t_0)$ which is solved iteratively. At each stage of the iterative process one has, in principle, a better approximation. If we start the iterative process with $\widehat{U}^{(0)}(t, t_0) = 1$), we obtain the first order solution

$$\widehat{U}^{(1)}(t, t_0) = 1 - \frac{i}{\hbar}\int_{t_0}^{t} e^{i\widehat{H}_o t_1/\hbar}\widehat{V}(t_1)e^{-i\widehat{H}_o t_1/\hbar}dt_1. \tag{D.12}$$

In the second step of the iterative procedure we introduce this operator in (D.11), to obtain the second order solution

$$\widehat{U}^{(2)}(t,t_0) = 1 - \frac{i}{\hbar}\int_{t_0}^{t} e^{i\widehat{H}_o t_1/\hbar}\widehat{V}(t_1)e^{-i\widehat{H}_o t_1/\hbar}dt_1 \\ \left(1 - \frac{i}{\hbar}\int_{t_0}^{t_1} e^{i\widehat{H}_o t_2/\hbar}\widehat{V}(t_2)e^{-i\widehat{H}_o t_2/\hbar}dt_2\right), \tag{D.13}$$

that can be written as

$$\widehat{U}^{(2)}(t,t_0) = 1 - \frac{i}{\hbar}\int_{t_0}^{t} e^{i\widehat{H}_o t_1/\hbar}\widehat{V}(t_1)e^{-i\widehat{H}_o t_1/\hbar}dt_1 \\ - \frac{1}{\hbar^2}\int_{t_0}^{t} dt_1 \int_{t_0}^{t_1} dt_2\, e^{i\widehat{H}_o t_1/\hbar}\widehat{V}(t_1)e^{i\widehat{H}_o(t_2-t_1)/\hbar}\widehat{V}(t_2)e^{-i\widehat{H}_o t_2/\hbar}. \tag{D.14}$$

This procedure continues to higher orders if necessary. Given this evolution operator, we write the wave function $\psi(\mathbf{r},t)$, for example to second order, as

$$\psi(\mathbf{r},t) = e^{-i\widehat{H}_o t/\hbar}\left(1 + \widehat{U}^{(1)}(t,t_0) + \widehat{U}^{(2)}(t,t_0)\right)\psi(\mathbf{r},0), \tag{D.15}$$

with $\widehat{U}^{(1)}(t,t_0)$ and $\widehat{U}^{(2)}(t,t_0)$ as defined above.

Given the wave function $\psi(\mathbf{r},t)$ one can easily calculate the expectation value of an operator $\widehat{Q}$. The expectation value to first order in $\widehat{V}$ is

$$\langle\psi^*(\mathbf{r},t)|\widehat{Q}|\psi(\mathbf{r},t)\rangle = \langle\psi^*(\mathbf{r},0)\big(1 + \widehat{U}^{(1)\dagger}(t,t_0)\big)|\widehat{Q}|\big(1 + \widehat{U}^{(1)}(t,t_0)\big)\psi(\mathbf{r},0)\rangle \tag{D.16}$$

This can be written in compact form, to first order in the interaction $\widehat{V}'(t)$, as

$$\langle\widehat{Q}\rangle_t = \langle\widehat{Q}\rangle_0 - \frac{i}{\hbar}\left\langle\int_{t_0}^{t}[\widehat{Q},\widehat{V}'(t_1)]dt_1\right\rangle_0 \tag{D.17}$$

Temporal Reversibility

E

We shall briefly discuss the time reversal invariance property. In the first part of this book we introduced the scattering and transfer matrices S and M as useful mathematical entities to embody, in a matrix array, all the physical information described by the corresponding Schrödinger or Maxwell equations, plus continuity and physical requirements. It is known that, in a scattering process like the one shown in Fig. E.1, the matrices S and M satisfy the relations

$$\begin{pmatrix} \varphi_{ol}(x_1) \\ \varphi_{or}(x_2) \end{pmatrix} = S \begin{pmatrix} \varphi_{il}(x_1) \\ \varphi_{ir}(x_2) \end{pmatrix}; \qquad \begin{pmatrix} \varphi_{or}(x_2) \\ \varphi_{ir}(x_2) \end{pmatrix} = M \begin{pmatrix} \varphi_{il}(x_1) \\ \varphi_{ol}(x_1) \end{pmatrix}. \tag{E.1}$$

The matrix S relates the incident waves to the outgoing ones; while the matrix M connects the wave vector at one point x_1 with the wave vector at another x_2. Before referring to the reversal of time, we will take the complex conjugate of the previous relations, that is:

$$\begin{pmatrix} \varphi^*_{ol}(x_1) \\ \varphi^*_{or}(x_2) \end{pmatrix} = S^* \begin{pmatrix} \varphi^*_{il}(x_1) \\ \varphi^*_{ir}(x_2) \end{pmatrix}; \qquad \begin{pmatrix} \varphi^*_{or}(x_2) \\ \varphi^*_{ir}(x_2) \end{pmatrix} = M^* \begin{pmatrix} \varphi^*_{il}(x_1) \\ \varphi^*_{ol}(x_1) \end{pmatrix}. \tag{E.2}$$

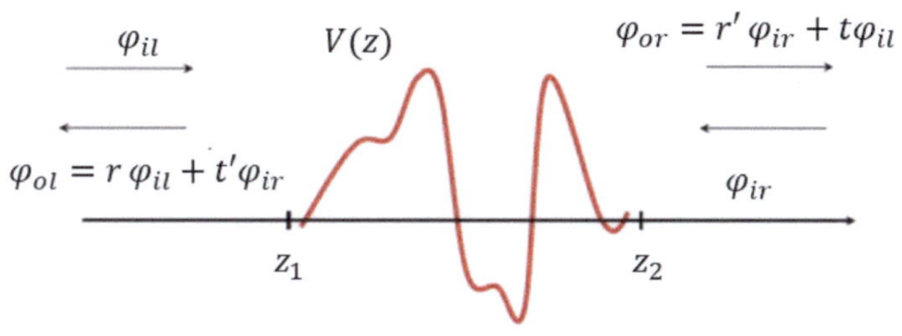

Fig. E.1 The wave functions incident, reflected and transmitted by a scattering potential $V(x)$

P. Pereyra, *Contemporary Theory of Semiconductors and Devices*,
Graduate Texts in Physics, https://doi.org/10.1007/978-3-032-00296-9

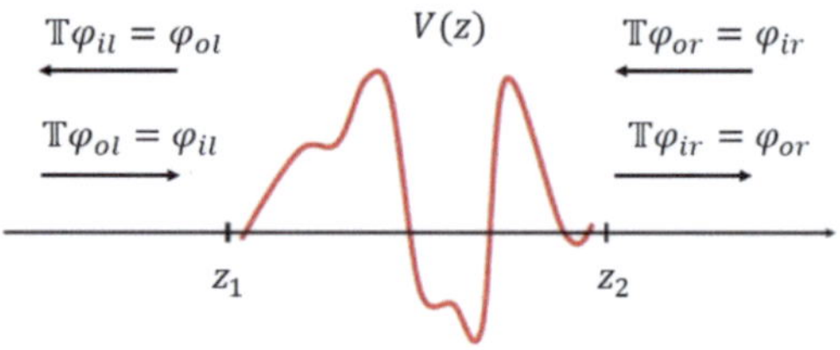

Fig. E.2 Under the action of the time reversal operator $\mathbb{T}$ the incident, reflected and transmitted wave functions change their direction

Now suppose an operation $\mathbb{T}$ that changes the sign of time. When this operation is applied, physical variables that depend linearly on time, such as speed and wave number $k = p/\hbar$, change sign. Under this time reversal operation, the particles that were moving to the right move to the left and vice versa. A wave function like $\varphi(x) = e^{ikx}$, under the action of $\mathbb{T}$, transforms as follows:

$$\mathbb{T}e^{ikx} = e^{-ikx}, \qquad \text{that means,} \qquad \mathbb{T}\varphi(x) = \varphi^*(x). \tag{E.3}$$

If the operator $\mathbb{T}$ acts on the system in the Fig. E.1, we will have:

$$\mathbb{T}\varphi_{il}(x_1) = \varphi^*_{il}(x_1) = \phi_{ol}(x_1) \qquad \mathbb{T}\varphi_{ir}(x_2) = \varphi^*_{ir}(x_2) = \phi_{or}(x_2), \tag{E.4}$$

$$\mathbb{T}\varphi_{ol}(x_1) = \varphi^*_{ol}(x_1) = \phi_{il}(x_1) \qquad \mathbb{T}\varphi_{or}(x_2) = \varphi^*_{or}(x_2) = \phi_{ir}(x_2). \tag{E.5}$$

As seen in Fig. E.2, the wave functions that were incident in Fig. E.1, become outgoing wave functions, and vice versa. If the system is invariant under time inversion, the same matrices S and M, which appear in (E.1), must be valid in the system transformed by $\mathbb{T}$, i.e., the matrices S and M must fulfil the following relations

$$\begin{pmatrix} \phi_{ol}(x_1) \\ \phi_{or}(x_2) \end{pmatrix} = S \begin{pmatrix} \phi_{il}(x_1) \\ \phi_{ir}(x_2) \end{pmatrix} \qquad \begin{pmatrix} \phi_{or}(x_2) \\ \phi_{ir}(x_2) \end{pmatrix} = M \begin{pmatrix} \phi_{il}(x_1) \\ \phi_{ol}(x_1) \end{pmatrix}. \tag{E.6}$$

These relations, rewritten in terms of the functions φ, take the form

$$\begin{pmatrix} \varphi^*_{il}(x_1) \\ \varphi^*_{ir}(x_2) \end{pmatrix} = S \begin{pmatrix} \varphi^*_{ol}(x_1) \\ \varphi^*_{or}(x_2) \end{pmatrix} \qquad \begin{pmatrix} \varphi^*_{ir}(x_2) \\ \varphi^*_{or}(x_2) \end{pmatrix} = M \begin{pmatrix} \varphi^*_{ol}(x_1) \\ \varphi^*_{il}(x_1) \end{pmatrix}. \tag{E.7}$$

If we multiply the first by S^{-1}, and if we also observe that

$$\begin{pmatrix} \varphi^*_{ir}(x_2) \\ \varphi^*_{or}(x_2) \end{pmatrix} = \begin{pmatrix} 0 & 1 \\ 1 & 0 \end{pmatrix} \begin{pmatrix} \varphi^*_{or}(x_2) \\ \varphi^*_{ir}(x_2) \end{pmatrix} \tag{E.8}$$

and

$$\begin{pmatrix} \varphi^*_{ol}(x_1) \\ \varphi^*_{il}(x_1) \end{pmatrix} = \begin{pmatrix} 0 & 1 \\ 1 & 0 \end{pmatrix} \begin{pmatrix} \varphi^*_{il}(x_1) \\ \varphi^*_{ol}(x_1) \end{pmatrix}, \tag{E.9}$$

it is evident, after comparing with (E.2), that

$$S^{-1} = S^* \quad \text{and} \quad \begin{pmatrix} 0 & 1 \\ 1 & 0 \end{pmatrix} M \begin{pmatrix} 0 & 1 \\ 1 & 0 \end{pmatrix} = M^*. \tag{E.10}$$

Concerning the transfer matrix, the last equation also means that

$$\begin{pmatrix} \delta & \gamma \\ \beta & \alpha \end{pmatrix} = \begin{pmatrix} \alpha^* & \beta^* \\ \gamma^* & \delta^* \end{pmatrix}. \tag{E.11}$$

We can than conclude that in time-reversible systems, the transfer matrices have the structure

$$M = \begin{pmatrix} \alpha & \beta \\ \beta^* & \alpha^* \end{pmatrix}. \tag{E.12}$$

The Landauer Conductance

F

We referred to the Landauer conductance in Sect. 5.6.3 and derived the Landauer formula

$$G = \frac{e^2}{2\pi\hbar}\frac{T}{R}, \tag{F.1}$$

in the one propagating mode limit. This formula was generalized for more than one propagating mode, among others, by Anderson [4], Azbel [5] and Büttiker et al. [6], who obtained the multichannel conductance

$$G = \frac{e^2}{2\pi\hbar}\frac{\mathrm{Tr}\, tt^\dagger \sum_a v_a^{-1}}{\sum_b v_b^{-1} R_{bb}} \tag{F.2}$$

which is, generally, assumed corresponds to the conductance measured between two charge reservoirs. A factor 2 must be added when the system is spin degenerated. We shall derive here this formula following a reasoning and charge concentration arguments similar to those used in Sect. 5.6.3 for the one propagating mode case.

F.1 The Landauer Conductance in Systems with N Propagating Modes

We will assume now that the number of propagating modes is N. Assume also that the incoming particles come from the left side. In this N-channels system, the scattering amplitudes r and t are matrices of dimension $N \times N$. Our purpose now is to obtain the conductance G.

Because of flux conservation, the current density at the left of the scattering system is the same as on the right side. On the right hand side the transmitted current density, in channel a, is

P. Pereyra, *Contemporary Theory of Semiconductors and Devices*,
Graduate Texts in Physics, https://doi.org/10.1007/978-3-032-00296-9

$$\vec{j}_{ta}(z) = \sum_{b,c} \frac{i\hbar}{2m} \left(\varphi_{ic} t^T_{ca} \frac{\partial}{\partial z} t^*_{ab} \varphi^*_{ib} - \varphi^*_{ic} t^\dagger_{ca} \frac{\partial}{\partial z} t_{ab} \varphi_{ib} \right) \hat{\mathbf{z}} \tag{F.3}$$

which can be written as,

$$j_{ta}(z) = \frac{\hbar}{2m} \sum_{b,c} k_b \left(\varphi_{ic} t^T_{ca} t^*_{ab} \varphi^*_{ib} + \varphi^*_{ic} t^\dagger_{ca} t_{ab} \varphi_{ib} \right), \tag{F.4}$$

It is easy to show that the first and the second terms, on the right hand side, are equal, thus

$$j_{ta}(z) = \frac{\hbar}{m} \sum_{b,c} k_b \varphi_{ic} \varphi^*_{ib} t_{ac} t^*_{ab}. \tag{F.5}$$

If we integrate upon x, y and z, with cross-section $A = \int dxdy$, and take into account that the wave functions in the leads satisfy the normalization condition

$$\int \varphi^*_{ic}(z) \varphi_{ib}(z) dz = \frac{\delta_{bc}}{\hbar k_b / m}, \tag{F.6}$$

the current in channel a takes the form

$$I_a = e \sum_{b,c} \delta_{cb} t_{ac} t^*_{ab} = e \sum_b t_{ab} t^*_{ab}, \tag{F.7}$$

and the total current becomes

$$I = e \sum_{ab} t_{ab} t^*_{ab} = e\mathrm{Tr}\, t t^\dagger. \tag{F.8}$$

On the other hand, as explained in Sect. 5.6.3, the concentration of particles on the left side, i.e., incoming plus reflected particles, is different than the concentration of the transmitted particles on the right hand side, unless all incoming particles pass through the scattering sample. On the left side the concentration is

$$n_{la}(z) = \sum_{b,c} \varphi^*_{ic} (I_N + r)^\dagger_{ca} (I_N + r)_{ab} \varphi_{ib}, \tag{F.9}$$

and on the right is

$$n_{ra}(z) = \sum_{b,c} \varphi^*_{ic} t^\dagger_{ca} t_{ab} \varphi_{ib}. \tag{F.10}$$

After integration upon z, the difference in particles concentration, in channel a, can be written as

$$\delta n_a = n_{la} - n_{ra} = \sum_b \frac{1}{v_b} \left[(I_N + r)_{ab} (I_N + r)^\dagger_{ba} - t_{ab} t^\dagger_{ba} \right]. \tag{F.11}$$

We recall now that the change in particle concentration can be written as

$$\delta n_a = \frac{dn_a}{dE}\delta E = \mathscr{D}_a(E)e\delta V, \tag{F.12}$$

where $\mathscr{D}_a$ is the level density, which was obtained in the 1D limit equal to $1/\pi\hbar v_a$, thus

$$\delta n_a = \frac{1}{\pi\hbar v_a}e\delta V, \tag{F.13}$$

and the total particle concentration difference is

$$\sum_a \delta n_a = \sum_a \frac{1}{\pi\hbar v_a}e\delta V. \tag{F.14}$$

We can then write

$$\delta V = \frac{\pi\hbar \sum_{ab} v_b^{-1}\left[(I_N + r)_{ab}(I_N + r)^{\dagger}_{ba} - t_{ab}t^{\dagger}_{ba}\right]}{e\sum_a v_a^{-1}}, \tag{F.15}$$

which in a more compact form is

$$\delta V = \frac{\pi\hbar \sum_b 2\, v_b^{-1}\big(R_{bb} + \Re e[r_b]\big)}{e\sum_a v_a^{-1}} \tag{F.16}$$

where $R_{bb} = \sum_a r^*_{ba}r_{ab} = 1 - T_{bb}$. Since $\Re e[r_b] \simeq \mathscr{O}(1/N)$ we have

$$\delta V = \frac{\pi\hbar \sum_b 2\, v_b^{-1} R_{bb}}{e\sum_a v_a^{-1}}. \tag{F.17}$$

Therefore, if we replace this potential and the current I into the conductance

$$G = \frac{I}{\delta V} \tag{F.18}$$

we have the well known [5, 6] multichannel conductance

$$G = \frac{e^2}{2\pi\hbar}\frac{\mathrm{Tr}\, tt^{\dagger}\sum_a v_a^{-1}}{\sum_b v_b^{-1} R_{bb}}. \tag{F.19}$$

A factor 2 must be added when the system is spin degenerated. A simplified version of this formula can be obtained when all particles move with almost the same velocity.

If in addition the number of channels is large,[1] $T_{bb} \simeq 1/N$, hence $R_{bb} \simeq 1$. In that case, the conductance becomes

$$G = \frac{e^2}{2\pi\hbar} \mathrm{Tr}\, t t^\dagger \tag{F.20}$$

This result has been originally derived from Kubo's formula.

With these formulae one can evaluate the conductance for a fixed value of the incoming energy E. In real systems the incoming fluxes have an energy dispersion of the order of the thermal energy k_BT and the states are filled according to the Fermi distribution

$$f(E) = \frac{1}{e^{\beta(E-\mu)} + 1} \tag{F.21}$$

[1] It was shown in [7] that $\langle T_{ab} \rangle = \frac{1}{N^2} \langle T \rangle$.

Time Distribution of Transition Probabilities

G

In Chap. 7, we referred to the particle-field interaction and derived some expressions related to the emission and absorption transition probabilities induced by this interaction. In the first part of Chap. 7 using the regular formalism, in time-dependent perturbation theory, we derived the amplitude and the transition probability $|a_{fi}(t)|^2$ from an excited state with energy E_i and lifetime τ_i to a final state with energy E_f. By multiplying with the density of accessible states to photons, we obtained the transition probability per unit of frequency as

$$\mathscr{P}_{fi}(\omega, t) = \frac{e^2(n_{ph}+1)\omega}{6m^2\pi^2\hbar\epsilon_0 c^3} \frac{|_{\rm o}\langle\varphi_f|p_z|\varphi_i\rangle_{\rm o}|^2}{\Omega^2+\gamma_i^2} \left[1+e^{-2t\gamma_i} - 2e^{-t\gamma_i}\cos\Omega t\right], \tag{G.1}$$

where $\Omega = \omega - (E_i - E_f)/\hbar$ and $\gamma_i = 1/\tau_i$. We will present here the alternative procedure mentioned in Chap. 7, to obtain a time-resolved transition probability per unit of time and also to obtain an important factor that gives the time distribution of optical, and of other kind of physical and chemical transitions. See [8]. We will see that this distribution describes with high accuracy different ultra-fast experiments.

G.1 Time Distribution of the Transition Probability

To obtain the probability per unit time, we will start from the transition probability per unit frequency $\mathscr{P}_{fi}(\omega, t)$ in (7.48) and take the derivative with respect to time t. In this way, we have the transition probability per unit frequency and per unit of time:

$$\mathscr{Q}_{fi}(\omega, t) = \frac{\partial \mathscr{P}_{fi}(\omega, t)}{\partial t}. \tag{G.2}$$

P. Pereyra, *Contemporary Theory of Semiconductors and Devices*,
Graduate Texts in Physics, https://doi.org/10.1007/978-3-032-00296-9

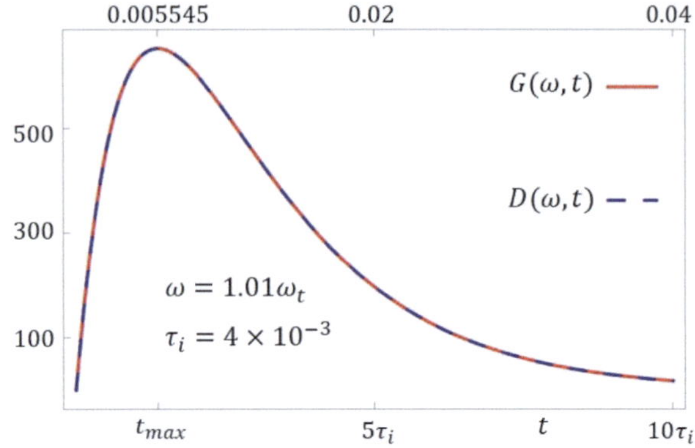

Fig. G.1 The temporal distribution functions $G(\omega, t)$ and $D(\omega, t)$. The exact function $G(\omega, t)$ and the approximate function $D(\omega, t)$ describe, in practically the same way, the temporal distribution of emission probabilities. The maximum occurs when $t = t_{max} = -2\tau_i \ln 0.5 \simeq 1.4\tau_i$

Explicitly, this functions is

$$\mathcal{Q}_{fi}(\omega, t) = \frac{e^2\hbar(n_{ph}+1)}{6m^2\pi^2\epsilon_0 c^3} |_o\langle\varphi_f|p_z|\varphi_i\rangle_o|^2 \frac{2\omega e^{-2t\gamma_i}(-\gamma_i + e^{t\gamma_i}(\gamma_i \cos\Omega t + \Omega \sin\Omega t))}{(E_f - E_i + \hbar\omega)^2 + \Gamma_i^2}. \tag{G.3}$$

Here we have defined the energy $\Gamma_i = \hbar\gamma_i = \hbar/\tau_i$, which is a parameter that, generally, is considered as related to the broadening of the energy level E_i, more specifically to the width of the resonance at E_i, and via the uncertainty relation to the lifetime τ_i.

The next step is to integrate $\mathcal{Q}_{fi}$ over the frequency ω. Before integrating it is convenient to define the function

$$G(\omega, t) = \frac{2\omega e^{-2t\gamma_i}\left(-\gamma_i + e^{t\gamma_i}(\gamma_i \cos\Omega t + \Omega \sin\Omega t)\right)}{(E_f - E_i + \hbar\omega)^2 + \Gamma_i^2}, \tag{G.4}$$

which contains all the dependence on t and ω. With this definition the probability $\mathcal{Q}_{fi}$ is written as

$$\mathcal{Q}_{fi}(\omega, t) = \frac{e^2\hbar(n_{ph}+1)}{6m^2\pi^2\epsilon_0 c^3} |_o\langle\varphi_f|p_z|\varphi_i\rangle_o|^2 \, G(\omega, t). \tag{G.5}$$

The function $G(\omega, t)$ is a relatively complex function of t, but due to the resonant character as a function of ω, it is possible to obtain an excellent approximation that not only facilitates the analysis but also reveals the temporal distribution of the transition probabilities.

The resonant behavior of the ω-dependent part is an important characteristic that concentrates the function $G(\omega, t)$ at the resonant frequency $\omega_t = (E_i - E_f)/\hbar$. At this frequency, the function $G(\omega, t)$, as seen in the Fig. G.1, is approximated perfectly well by the function

$$D(\omega, t) = \frac{2\omega\gamma_i e^{-2t\gamma_i}(-1 + e^{t\gamma_i})}{(E_f - E_i + \hbar\omega)^2 + \Gamma_i^2}, \tag{G.6}$$

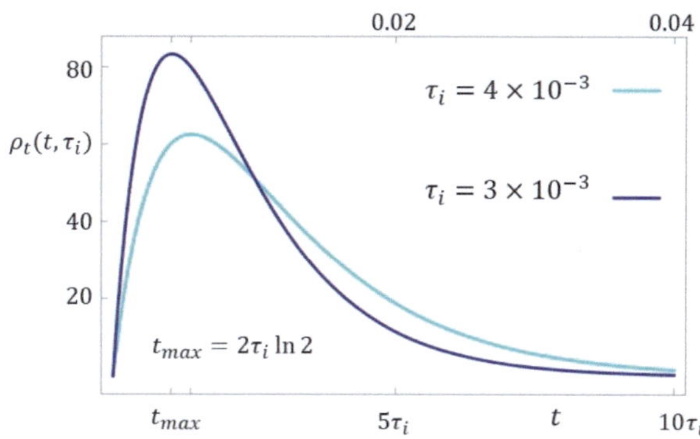

Fig. G.2 Time distribution of the emission probability for two values of τ_i. The distribution ρ_t shows the behavior with time of the emission probability from the energy level E_i. For $t < \tau_i$ it grows almost linearly, reaches a maximum when $t = t_{max} = 2\tau_i \ln 2$, and decays exponentially at larger times

which can be written as a product of two distributions, one of frequencies and another of time. That is:

$$D(\omega, t) = \rho_w(\omega, \tau_i)\rho_t(t, \tau_i), \tag{G.7}$$

where

$$\rho_w(\omega, \tau_i) = \frac{\omega}{(E_f - E_i + \hbar\omega)^2 + \Gamma_i^2} \quad \text{and} \quad \rho_t(t, \tau_i) = 2\gamma_i e^{-2t\gamma_i}\left(-1 + e^{t\gamma_i}\right). \tag{G.8}$$

The first function is basically a Lorentzian. If it is plotted as a function of ω it peaks at $\omega = \omega_t = (E_i - E_f)/\hbar$. The second function is the temporal analogue of the frequency distribution of Planck for the radiated field, emitted in the optical transitions that take place in the walls of the black body. The distribution $\rho_t(t, \tau_i)$ describes the time distribution of emission from level i. Figure G.2 shows the distribution $\rho_t(t, \tau_i)$ for two values of τ_i. It is interesting to see that, in this distribution, the lifetime τ_i of the excited state plays for the time distribution, the same role as the temperature for the Planck frequency distribution. In fact, both distributions refer to two aspects of the same physical phenomenon, the transitions in the emitters of the walls of the black body not only emit photons with different frequencies, the elapsed time from excitation to emission has a distribution, the distribution $\rho_t(t, \tau_i)$. In the last 30 years, ultra-fast experiments have been carried out which account for the temporal evolution of the mechanisms behind a great diversity of physical, chemical and biological phenomena. In Fig. G.3, some experimental results are shown that are described quite precisely by the distribution $\rho_t(t, \tau_i)$ superimposed on the figures with solid colored curves. Unlike previous theoretical fits that only account for the decreasing tail with a decreasing exponential, the function $\rho_t(t, \tau_i)$ also describes the initial growth. This topic is discussed with more detail in [8].

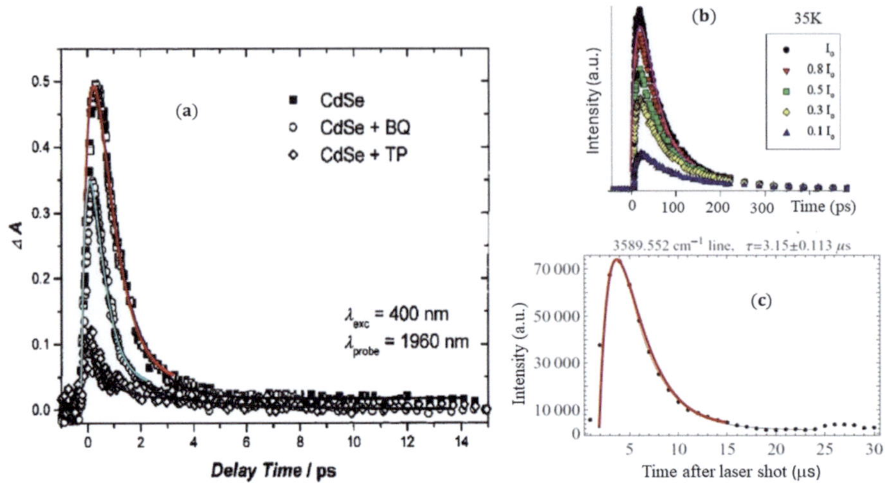

Fig. G.3 Experimental curves and fittings with the time distribution $\rho_t(t, \tau_i)$. In **a** the relaxation of transient absorption observed at 1960 nm, in the near infrared spectra for CdSe nanoparticles, reported by Burda et al. [9]. Figure reproduced with permission of the American Chemical Society. In **b**, time-resolved PL spectra, observed for Ag lines by Civiš et al. [10] Figure reproduced with permission of American Physical Society. The authors fit the temporal decay with an exponential function (black curves). In **c** ultrafast PL carrier relaxation in InN thin films [11]. Figure reproduced with permission of AIP. The continuous curves are fittings with the one parameter-function $\varrho(t, \tau_i)$

The next objective is to obtain the transition rate $W_{fi}(t)$ from the probability $\mathcal{Q}_{fi}(\omega, t)$ by integration over ω.

G.2 Time-Dependent Transition Probability Per Unit Time W_{fi}

To obtain the emission probability per unit time $W_{fi}(t)$, we have to integrate $\mathcal{Q}_{fi}(\omega, t)$ with respect to ω. The integral over the frequency of the factor in $\mathcal{Q}_{fi}(\omega, t)$ that depends on the variable ω gives

$$\int \frac{\omega d\omega}{(E_f - E_i + \hbar\omega)^2 + \Gamma_i^2} = \frac{1}{\hbar^2} \int \frac{\omega d\omega}{(\omega_t - \omega)^2 + \gamma_i^2} \simeq \frac{\pi}{\hbar^2 \gamma_i} \omega_t . \tag{G.9}$$

Here π/γ_i corresponds to the normalization constant of the Lorentzian. Consequently, the probability of transition per unit of time is

$$W_{fi}(t) = \frac{e^2(n_{ph}+1)}{6m^2\pi\hbar\epsilon_0 c^3} \, |{}_o\langle\varphi_f\,|p_z|\varphi_i\rangle_o|^2 \, \frac{\omega_t}{\gamma_i} \, \rho_t(t, \tau_i). \tag{G.10}$$

This is essentially the time-dependent Fermi golden rule. To remove the constant γ_i, which appears by the integration in ω and which multiplies in ρ_t, we redefine the distribution as $\rho_t(t, \tau_i) = \gamma_i \varrho(t, \tau_i)$, and we write the transition rate in the form

$$W_{fi}(t) = \frac{e^2(n_{ph}+1)\hbar\omega_t}{6m^2\pi\hbar^2\epsilon_0 c^3} \, |{}_o\langle\varphi_f|p_z|\varphi_i\rangle_o|^2 \, \varrho(t, \tau_i)$$

with

$$\varrho(t, \tau_i) = 2e^{-2t\gamma_i}\left(-1 + e^{t\gamma_i}\right) \tag{G.11}$$

To identify the Einstein coefficients, we rewrite this probability as

$$W_{fi}(t) = \frac{e^2\pi}{6\epsilon_0\hbar^2 m^2\omega_t^2} \, |{}_o\langle\varphi_f|p_z|\varphi_i\rangle_o|^2 \, \varrho(t, \tau_i)\frac{\hbar\omega_t^3}{\pi^2 c^3}(n_{ph}+1) \equiv B^i_f(t)\wp_{ph}(\omega, T). \tag{G.12}$$

This gives the time-dependent Einstein coefficient

$$B^i_f(t) = \frac{e^2\pi}{6\epsilon_0\hbar^2 m^2\omega_t^2} \, |{}_o\langle\varphi_f|p_z|\varphi_i\rangle_o|^2 \, \varrho(t, \tau_i) = B^i_f e^{-2t\gamma_i}(-1 + e^{t\gamma_i}). \tag{G.13}$$

The factor B^i_f in the right hand side of the last equation is the time-independent Einstein coefficient obtained in Sect. 7.4.3. Likewise, the time-dependent spontaneous emission coefficient is

$$A_{sp}(t) = \frac{e^2\omega_t}{6\epsilon_0\hbar m^2\pi c^3} \, |{}_o\langle\varphi_f|p_z|\varphi_i\rangle_o|^2 \, \varrho(t, \tau_i) = A_{sp}e^{-2t\gamma_i}(-1 + e^{t\gamma_i}). \tag{G.14}$$

The MOS in the Hauser–Littlejohn Model

H

As mentioned in Chap. 10 the first attempts to obtain an analytical expression for the charge in the inversion layer of a MOS, as a function of the gate potential and the semiconductor parameters, started right after the device was envisioned. W. Shockley (WS) and later Kingston and Neustadter (KN) were the first to model the charge density in the impurified and biased semiconductor layer. In Shockley's approach, the Poisson equation reads

$$\frac{d^2\phi}{dz^2} = -\frac{q(p(z) - n(z) + \delta N)}{\epsilon_s}, \tag{H.1}$$

where $\delta N = N_D - N_A$ is the effective doping charge concentration, when donor and acceptor impurities are completely ionized, and the free charge distributions $n(z)$ and $p(z)$ are given by

$$n(z) = n_i e^{(e\phi(z)-\mu_n)/k_BT}, \tag{H.2}$$

and

$$p(z) = p_i e^{(\mu_p-e\phi(z))/k_BT}, \tag{H.3}$$

with μ_n and μ_p the quasi Fermi energy levels and $e\phi(z)$ the electrostatic potential energy. Since neither the WS nor the KN models could be analytically solved, Hauser and Litlejohn (HL) suggested the approximate model presented in Sect. 10.7.3. The purpose in this appendix is to show that the alternative method introduced in 10.8.1, gives for the HL model the same result as by direct integration. Let us start recalling the *ansatz* that was introduced in 10.8.1 in its integral representation

$$-\phi_T \int_{\rho_i(\phi(z_b))}^{\rho_i(\phi(z))} d\rho_i(\phi(z)) = \frac{1}{2}\int_{u_E(\phi(z_b))}^{u_E(\phi(z))} du_E. \tag{H.4}$$

P. Pereyra, *Contemporary Theory of Semiconductors and Devices*, Graduate Texts in Physics, https://doi.org/10.1007/978-3-032-00296-9

It was shown the assuming that $\mathscr{E} = -d\phi(z)/dz\hat{\mathbf{z}}$, one obtains

$$\frac{\epsilon_s}{2}\left(\frac{d\phi(z)}{dz}\right)^2 - \frac{\epsilon_s}{2}\mathscr{E}_b^2 = -\phi_T\Big(\rho_i\big(\phi(z)\big) - \rho_i\big(\phi(z_b)\big)\Big). \tag{H.5}$$

It is clear that in the HL model, when

$$\rho_i(\phi(z)) = \rho_H(z) = -eN_0 e^{e\phi(z)/k_B T}, \tag{H.6}$$

the previous equation becomes

$$\frac{1}{2}\left(\frac{d\phi(z)}{dz}\right)^2 = \frac{eN_0}{\epsilon_s}\phi_T\left(e^{\phi(z)/\phi_T} - 1\right) + \frac{1}{2}\mathscr{E}_b^2. \tag{H.7}$$

This is identical to (10.54), as expected. We will now follow the procedure presented in Sect. 10.8.2. Using the Poisson equation, we write the last equation as the nonlinear second order differential equation

$$\phi_T\frac{d^2\phi(z)}{dz^2} = \frac{1}{2}\left(\frac{d\phi(z)}{dz}\right)^2 - \frac{1}{2}\xi_L^2, \tag{H.8}$$

with

$$\xi_L^2 = \mathscr{E}_b^2 - \frac{2eN_0\phi_T}{\epsilon_s}. \tag{H.9}$$

Notice that (H.8) is similar to (10.62). As mentioned before, (H.8) is a general, non-specific, equation that makes the nonlinear Poisson equation solvable. Different models imply different parameters ξ. These parameters, as mentioned before, carry the information on the charge density and also on the specific threshold and inversion population conditions. Following the procedure presented in Sect. 10.8.2, it is easy to see that the solution to (H.8) is

$$\phi(z) = \phi_0 - 2\phi_T \ln\left(\cosh\frac{\xi_L(z - z_0)}{2\phi_T} - \frac{f_0}{\xi_L}\sinh\frac{\xi_L(z - z_0)}{2\phi_T}\right). \tag{H.10}$$

For the boundary conditions at $z_0 = z_b$ where $\phi_0 = 0$ and $f_0 = -\mathscr{E}_b$, our solution for the model proposed by Hauser–Littlejohn [12] is

$$\phi(z) = -2\phi_T \ln\left(\cosh\frac{\xi_L(z - z_b)}{2\phi_T} + \frac{\mathscr{E}_b}{\xi_L}\sinh\frac{\xi_L(z - z_b)}{2\phi_T}\right). \tag{H.11}$$

This electric potential is a different representation of the solution of the Poisson equation and must coincide with (10.55), otherwise the differential equations would describe different physics. One way of checking the identity is to plot the electric potential $\phi(z)$ alongside the potential $\phi_H(z)$. In Fig. H.1 the function in (H.10) is plotted with solid curves and the potential $\phi_H(z)$ in (10.55) is plotted with dashed

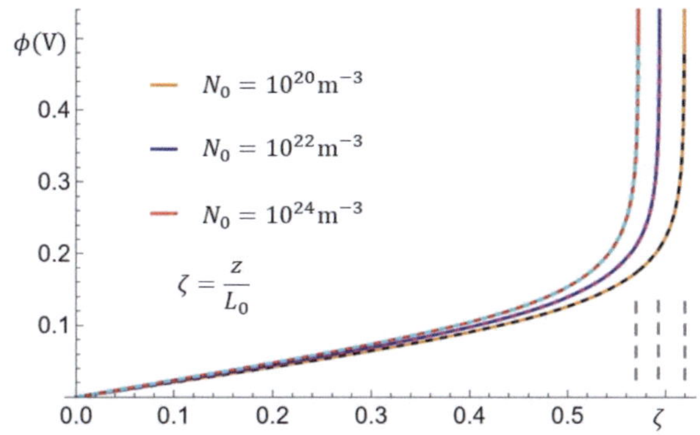

Fig. H.1 The electric potential in the HL model and using the method presented in Sect. 10.8.2. In these graphs the inversion point is fixed at $z_b = 0$. The potential function $\phi(z)$ in (H.11) is plotted with solid curves and the function $\phi_H(z)$ in (10.55) is plotted with dashed curves, as functions of the dimensionless parameter $\zeta = z/L_0$, with the "distance normalization constant" $L_0^2 = \epsilon k_B T/e^2 N_0$, introduced by Hauser–Littlejohn. The graphs are for three values of the concentration N_0. This figure corresponds to Fig 2 in [12]. Reproduced from [13], with permission of John Wiley & Sons Publisher

curves. As in Fig. 2 of [12], we plot these potentials as functions of the dimensionless parameter $\zeta = z/L_0$, where L_0 is the "distance normalization constant" introduced by Hauser and Littlejohn and defined as $L_0^2 = \epsilon k_B T/e^2 N_0$. The agreement is perfect. One can also verify analytically that one solution can be transformed into the other. Schematically this transformation goes as follows. If

$$A = \frac{(z - z_b)\sqrt{b}}{2\phi_T}, \qquad B = \tanh^{-1}\sqrt{\frac{a}{b} + 1}, \tag{H.12}$$

the function in (10.55) becomes

$$\phi_H(z) = \phi_T \ln\frac{b}{a} + \phi_T \ln\left[\frac{-1}{\cosh^2(A + B)}\right]. \tag{H.13}$$

Since

$$\cosh(A + B) = \cosh B\Big(\cosh A + \Big(\frac{a + b}{b}\Big)^{1/2} \sinh A\Big), \tag{H.14}$$

and

$$\tanh^2 B = \frac{a + b}{b}, \qquad \text{thus} \qquad \cosh^2 B = -\frac{b}{a}, \tag{H.15}$$

and using the definitions in (10.56) and (H.9), one ends up with the function $\phi(z)$ in (H.11).

In the graphs of Fig. H.1, the inversion point is fixed at $z_b = 0$. This is not a convenient presentation because, on first side, the oxide-semiconductor interface is, in each case, at different position (indicated by vertical dashed gray lines), and also, the conduction band is also at different levels. Notice that rotating this graph 180°, displacing the oxide-semiconductor interfaces such that all coincide, and using the conduction band levels to align them, the graphs in Fig. H.1 will appear similar to the potential curves in Fig. 10.20, in the inversion layer.

Notice also that the electric potential $\phi(z)$ in (H.10) can also be written in terms of the boundary conditions at the surface, and in this case $\phi_0 = \phi_s$, $f_0 = -\mathscr{E}_s$ and

$$\phi(z) = \phi_s - 2\phi_T \ln\left(\cosh\frac{\xi_L z}{2\phi_T} + \frac{\mathscr{E}_s}{\xi_L}\sinh\frac{\xi_L z}{2\phi_T}\right) \tag{H.16}$$

with the oxide-semiconductor interface at $z_0 = 0$. This equation is also formally similar to the electric potential that was obtained for the Shockley charge density. The difference between them is basically in the parameter ξ.

Drift-Diffusion Currents in MOSFETs Based on the Pao-Sah Model

I

I.1 The Double Gate JL MOSFET

We will briefly refer here to the double-gate MOSFET. Consider the double-gate device sketched in Fig. I.1, where the oxide layers are at a distance t_{Si} and the front and back gate voltages V_{Gf} and V_{Gb} are assumed to have opposite polarizations. In this case, holes accumulate near the back oxide-semiconductor interface and electrons near the front oxide-semiconductor. Before the source-drain potential U_D is applied, we can write the electric potential in the inversion layers in the 1D approximation as

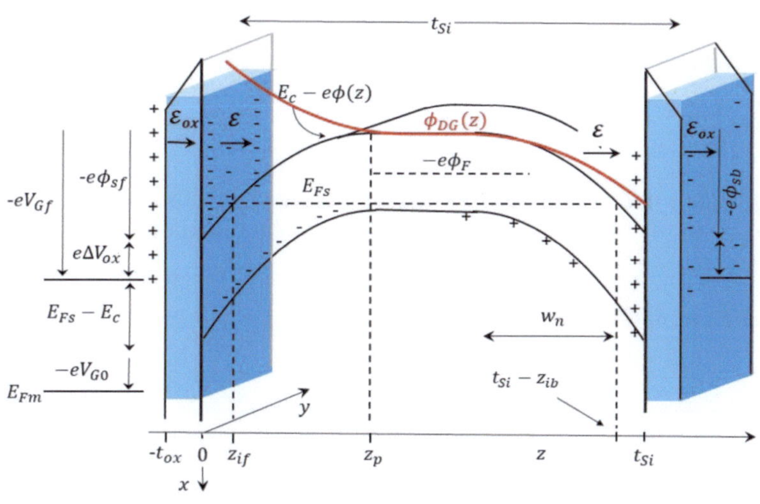

Fig. I.1 A double gate structure as a generalization of the single gate MOS of Fig. 1.1. We assume here that the back gate potential is negative and attracts holes. The potential distribution $\phi_{DG}(z)$ (red curve) is schematically shown. Reproduced from [13], with permission of John Wiley & Sons Publisher

P. Pereyra, *Contemporary Theory of Semiconductors and Devices*, Graduate Texts in Physics, https://doi.org/10.1007/978-3-032-00296-9

$$\phi_{iDG}(z) = \phi_{if}(z) + \phi_{ib}(z). \tag{I.1}$$

Assuming that the charge densities are described by the Kingston–Neustadter-like distributions, the potential distributions in the front and back inversion layers, respectively, can be written as

$$\phi_{if}(z) = \phi_{sf} - 2\phi_T \ln\left(\cosh\frac{\xi_K z}{2\phi_T} + \frac{\mathscr{E}_{Kfs}}{\xi_K}\sinh\frac{\xi_K z}{2\phi_T}\right), \tag{I.2}$$

and

$$\phi_{ib}(z) = -\phi_{sb} + 2\phi_T \ln\left(\cosh\frac{\xi_K z'}{2\phi_T} + \frac{\mathscr{E}_{Kbs}}{\xi_K}\sinh\frac{\xi_K z'}{2\phi_T}\right). \tag{I.3}$$

Here $z' = t_{Si} - z$, ϕ_{sf} and ϕ_{sb} are the front and back surface potentials, and $\mathscr{E}_{Kfs}$ and $\mathscr{E}_{Kbs}$ are the front and back surface electric fields. In the symmetric case with $t_{oxf} = t_{osb}$ and $V_{Gf} = -V_{Gb}$, the potential function $\phi_{DG}(z)$ is asymmetric, as sketched in the Fig. I.1 (red curve), and the potential energy $q\phi_{iDG}(z)$ is symmetric, as shown in Fig. I.2a, where a silicon layer $t_{Si} \simeq 40$ nm is assumed, which for an impurity concentration $N_A = 1 \times 10^{18}\,\text{cm}^{-3}$ is $\simeq (1/2)(w_{pd} + w_{nd} + z_{if} + z_{ib})$, where w_{pd} and w_{nd} are the depletion layers widths and z_{if} and z_{ib} the inversion layer widths.

Given the potential distribution $\phi_{iDG}(z)$, we can, using again the Pao-Sah's [14] formula, obtain the analytical expression for the drift-diffusion current in the DG JL MOSFET channel as a function of the relevant parameters. In the symmetric case, the total current in the DG JL MOSFET channels can be written as

$$I_{DG} = \frac{eD_n W_x n_i}{L}\left[\int_0^{z_{if}} e^{\phi_{if}(z)/\phi_T}dz + \int_{z_{if}}^{t_{Si}/2} e^{\phi_{pd}(z)/\phi_T}dz + \int_{t_{Si}/2}^{t_{Si}-z_{ib}} e^{-\phi_{nd}(z)/\phi_T}dz + \int_{t_{Si}-z_{ib}}^{t_{Si}} e^{-\phi_{ib}(z)/\phi_T}dz\right] \times \int_0^{U_D} e^{(-\eta(y)-\phi_{Fs})/\phi_T}d\eta.$$

All integrals in this equation can be performed analytically, and for $z_{if} + z_{ib} < t_{Si} < w_p + w_n + z_{if} + z_{ib}$, and $z_{if} = z_{ib}$, the current becomes

$$\begin{aligned} I_{DG} &= \frac{eD_n w_x n_i}{L}\left(1 - e^{-U_D/\phi_T}\right)\left[\frac{4\phi_T}{\xi}e^{(\phi_s - V_{Gu})/2\phi_T}\sinh\frac{\xi z_i}{2\phi_T}\right. \\ &\quad \left. + \frac{\sqrt{\pi\epsilon_s\phi_T}}{\sqrt{eN_d}}e^{-\phi_F/2\phi_T}\left(\text{Erf}\left[\frac{\sqrt{eN_d}w_p}{2\sqrt{\epsilon_s\phi_T}}\right] + \text{Erf}\left[\frac{\sqrt{eN_d}(t_{Si} - w_p - 2z_{if})}{2\sqrt{\epsilon_s\phi_T}}\right]\right)\right] \\ &= I_{DGi} + I_{DGd}. \end{aligned} \tag{I.4}$$

It can easily be seen that the current in the DG JL MOSFET is practically twice the charge and current in the single gate MOSFET channel. In fact, it is exactly twice when $t_{Si} = 2w_p + 2z_{if}$. In Fig. I.2b we plot the drift-diffusion current of (I.4) as a function of the gate potential, for three different values of the source-drain

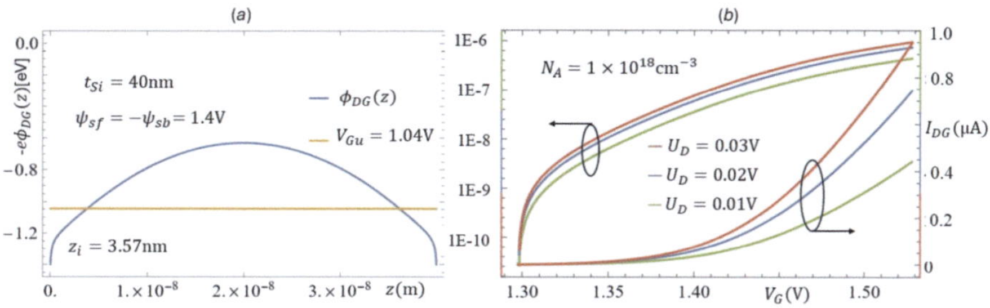

Fig. I.2 Electric potential and drift-diffusion current in a symmetric double-gate JL MOSFET. In **a** the electric potential for a DG MOS where the silicon thickness t_{Si} is smaller than the expected depletion plus inversion layers width $w_p + w_n + z_{if} + z_{ib}$. Some parameters are assumed equal to those in [15], i.e. $L = 11\,\mu$m, $W_x = 10\,\mu$m. However, nowhere in [15] are the impurity concentration and mobilities given. In this reference, the silicon layer width t_{Si} is assumed equal to 8 nm, slightly larger than the sum of the inversion layer widths for $N_A = 1 \times 10^{18}\,\text{cm}^{-3}$. Here we assume that $t_{Si} \simeq 40$ nm. Reproduced from [13], with permission of John Wiley & Sons Publisher

voltage U_D. For this graph we assume that the impurity concentration is $N_A = 1 \times 10^{18}\,\text{cm}^{-3}$, the channel length is $L = 11\,\mu$m, the gate width is $W_x = 10\,\mu$m, and the mobility is assumed equal to $40\,\text{cm}^2/\text{Vs}$. Some of these parameters are the same as those used in figure 7 of [15], where a simplified charge distribution is assumed, however we cannot compare the model with that reference because some parameters, such as the impurities concentration and mobilities, are not given there.

It is clear that the analytical expression for the current in (I.4) can also be used for asymmetric DG JL MOSFETs. The general formulas for the inversion layer width, surface electric field and effective concentration of ionized impurity atoms, introduced for the single gate MOS are also valid with the appropriate changes for the double gate MOSFET. The ability to evaluate these quantities and particularly the widths of the inversion layers is, for a given impurity concentration, an important input in the design of MOSFETs and in the definition of some of the device parameters, such as the silicon layer width t_{Si}.

The Volume Element in the Space of Transfer Matrices

J

Here we will derive the volume element in the space of pseudounitary transfer matrices, i.e. in the unitary universality class where the TRI symmetry is broken. We will follow, basically, the same procedure as in [16]. To simplify the notation we will use the following notation for the transfer matrix of the unitary universality class

$$\begin{pmatrix} u_1 & 0 \\ 0 & u_2 \end{pmatrix} \begin{pmatrix} \sqrt{1+\lambda} & \sqrt{\lambda} \\ \sqrt{\lambda} & \sqrt{1+\lambda} \end{pmatrix} \begin{pmatrix} v_1 & 0 \\ 0 & v_2 \end{pmatrix} = U \Lambda V, \tag{J.1}$$

The arc element in the space of transfer matrices is generally written [17] as

$$ds^2 = Tr\left(\Sigma_z dM_{\dagger} \Sigma_z dM\right) = \sum_{i,j}^{N} g_{ij}(x) dx_i dx_j, \tag{J.2}$$

with a volume element

$$d\mathsf{V} = |\det g(x)|^{1/2} \prod dx_i, \tag{J.3}$$

that remains invariant under a transformation like

$$M_u \mapsto M_o M_u \qquad\qquad M_u \mapsto M_u M_o \tag{J.4}$$

where M_o is a fixed transfer matrix of the spU(2N, $\mathscr{C}$) group. To determine the volume element, the first task is to write ds^2 as the right hand side of equation (J.2), i.e. in terms of the independent parameters. From the Bargmann polar representation, we have

$$dM_u = U\left(U^{\dagger} dU \Lambda + d\Lambda + \Lambda dV V^{\dagger}\right) V \tag{J.5}$$

P. Pereyra, *Contemporary Theory of Semiconductors and Devices*,
Graduate Texts in Physics, https://doi.org/10.1007/978-3-032-00296-9

and

$$dM_u^\dagger = V^\dagger\big(\Lambda dU^\dagger U + d\Lambda + V dV^\dagger \Lambda\big)U^\dagger \tag{J.6}$$

If we use the relations $U\Sigma_z U^\dagger = \Sigma_z$, $V\Sigma_z V^\dagger = \Sigma_z$ and define the matrices

$$\delta U = U^\dagger dU \qquad \text{and} \qquad \delta V = dV V^\dagger \tag{J.7}$$

the arc element becomes

$$ds^2 = Tr\big(\Sigma_z\big(\Lambda\delta U^\dagger + d\Lambda + \delta V^\dagger \Lambda\big)\Sigma_z\big(\delta U\Lambda + d\Lambda + \Lambda\delta V\big)\big) \tag{J.8}$$

Since

$$\Sigma_z\Lambda\Sigma_z\Lambda = I_{2N} \tag{J.9}$$

$$\Sigma_z\Lambda\Sigma_z d\Lambda = \frac{d\lambda}{2\sqrt{\lambda(1+\lambda)}}\begin{pmatrix} 0 & I_N \\ I_N & 0 \end{pmatrix} = -\Sigma_z d\Lambda\Sigma_z\Lambda \tag{J.10}$$

$$\Sigma_z d\Lambda\Sigma_z d\Lambda = \frac{d\lambda^2}{4\lambda(1+\lambda)}\begin{pmatrix} -I_N & 0 \\ 0 & -I_N \end{pmatrix} \tag{J.11}$$

and

$$\Sigma_z\Lambda\Sigma_z\delta U\Lambda\,\delta U = \begin{pmatrix} \sqrt{1+\lambda} & -\sqrt{\lambda} \\ -\sqrt{\lambda} & \sqrt{1+\lambda} \end{pmatrix}\begin{pmatrix} \delta u_1 & 0 \\ 0 & \delta u_2 \end{pmatrix}\begin{pmatrix} \sqrt{1+\lambda} & \sqrt{\lambda} \\ \sqrt{\lambda} & \sqrt{1+\lambda} \end{pmatrix}\begin{pmatrix} \delta v_1 & 0 \\ 0 & \delta v_2 \end{pmatrix} \tag{J.12}$$

we can write the arc element as a trace of $N \times N$ matrices in the form

$$\begin{aligned} ds^2 = \mathrm{Tr}\big[\delta u_1^2 + \delta u_2^2 + \delta v_1^2 + \delta v_2^2 - \frac{d\lambda^2}{2\lambda(1+\lambda)} + 2\sqrt{1+\lambda}\,\delta u_1\sqrt{1+\lambda}\,\delta v_1 + \\ 2\sqrt{1+\lambda}\,\delta u_2\sqrt{1+\lambda}\,\delta v_2 - 2\sqrt{\lambda}\,\delta u_2\sqrt{\lambda}\,\delta v_1 - 2\sqrt{\lambda}\,\delta u_1\sqrt{\lambda}\,\delta v_2\big]. \end{aligned} \tag{J.13}$$

we can now write the anti-Hermitian matrices $\delta u_i = -\delta u_i^\dagger$ and $\delta v_i = -\delta v_i^\dagger$ as

$$\delta u_i = \delta a_i + i\,\delta s_i \qquad \text{and} \qquad \delta v_i = \delta a_i' + i\,\delta s_i' \tag{J.14}$$

with δa_i and δs_i real antisymmetric and symmetric matrices. After some algebra we have

$$\begin{aligned} ds^2 = \mathrm{Tr}\Big[&\sum_{i=1}^{2}(\delta a_i^2 - \delta s_i^2) + \sum_{i=1}^{2}(\delta a_i'^2 - \delta s_i'^2) \\ &+2\sum_{i=1}^{2}\big(\sqrt{1+\lambda}\,\delta a_i\sqrt{1+\lambda}\,\delta a_i' - \sqrt{1+\lambda}\,\delta s_i\sqrt{1+\lambda}\,\delta s_i'\big) \\ &-2\sum_{i\neq j=1}^{2}\big(\sqrt{\lambda}\,\delta a_i\sqrt{\lambda}\,\delta a_j' - \sqrt{\lambda}\,\delta s_i\sqrt{\lambda}\,\delta s_j'\big) - \frac{d\lambda^2}{2\lambda(1+\lambda)}\Big]. \end{aligned} \tag{J.15}$$

In terms of the matrix elements, where the diagonal and off diagonal elements are taken separately, the arc element transforms into

$$
\begin{aligned}
ds^2 = -\sum_{a=1}^{N}\Big[&\frac{d\lambda_a^2}{2\lambda_a(1+\lambda_a)} + (\delta s_1)^2_{aa} + (\delta s_1')^2_{aa} + (\delta s_2)^2_{aa} + (\delta s_2')^2_{aa} \\
&-2\lambda_a(\delta s_1)_{aa}(\delta s_2')_{aa} - 2\lambda_a(\delta s_2)_{aa}(\delta s_1')_{aa} + 2(1+\lambda_a)(\delta s_1)_{aa}(\delta s_1')_{aa} \\
&+2(1+\lambda_a)(\delta s_2)_{aa}(\delta s_2')_{aa}\Big] \\
&-2\sum_{a<b}^{N}\Big[\sum_{i=1}^{2}\Big[(\delta a_i)^2_{ab} + (\delta a_i')^2_{ab} + (\delta s_i)^2_{ab} + (\delta s_i')^2_{ab} \\
&+2\sqrt{1+\lambda_a}\,(\delta a_i)_{ab}\sqrt{1+\lambda_b}\,(\delta a_i')_{ab} + 2\sqrt{1+\lambda_a}\,(\delta s_i)_{ab}\sqrt{1+\lambda_b}\,(\delta s_i')_{ab}\Big] \\
&-\sum_{i\neq j=1}^{2}\Big[2\sqrt{\lambda_a}\,(\delta a_i)_{ab}\sqrt{\lambda_b}\,(\delta a_j')_{ab} + 2\sqrt{\lambda_a}\,(\delta s_i)_{ab}\sqrt{\lambda_b}\,(\delta s_j')_{ab}\Big]\Big].
\end{aligned}
\tag{J.16}
$$

If we order the Bargmann parameters as follows

$$
\begin{aligned}
&d\lambda_1 ... d\lambda_N (\delta s_1)_{11}(\delta s_2)_{11}(\delta s_1')_{11}(\delta s_2')_{11} ... (\delta s_1)_{NN}(\delta s_2)_{NN}(\delta s_1')_{NN}(\delta s_2')_{NN} \\
&(\delta s_1)_{12}(\delta s_2)_{12}(\delta s_1')_{12}(\delta s_2')_{12} ... (\delta s_1)_{N-1,N}(\delta s_2)_{N-1,N}(\delta s_1')_{N-1,N}(\delta s_2')_{N-1,N} \\
&(\delta a_1)_{12}(\delta a_2)_{12}(\delta a_1')_{12}(\delta a_2')_{12} ... (\delta a_1)_{N-1,N}(\delta a_2)_{N-1,N}(\delta a_1')_{N-1,N}(\delta a_2')_{N-1,N},
\end{aligned}
$$

the metric tensor has a block structure, with blocks of dimensions 1×1 and 4×4. It can also be noticed that the coefficients of the off diagonal elements $(\delta s_1)_{ab}$ of the symmetric matrix are the same as the coefficients $(\delta a_1)_{ab}$ of the anti-symmetric matrix. Thus

$$
\det g(\lambda, a, ..., s') \propto \prod_{a}^{N} \frac{1}{\lambda_a(1+\lambda_a)} \begin{vmatrix} 1 & 0 & (1+\lambda_a) & -\lambda_a \\ 0 & 1 & -\lambda_a & (1+\lambda_a) \\ (1+\lambda_a) & -\lambda_a & 1 & 0 \\ -\lambda_a & (1+\lambda_a) & 0 & 1 \end{vmatrix}
$$

$$
\prod_{a<b}^{N} \begin{vmatrix} 1 & 0 & \sqrt{(1+\lambda_a)(1+\lambda_b)} & -\sqrt{\lambda_a\lambda_b} \\ 0 & 1 & -\sqrt{\lambda_a\lambda_b} & \sqrt{(1+\lambda_a)(1+\lambda_b)} \\ \sqrt{(1+\lambda_a)(1+\lambda_b)} & -\sqrt{\lambda_a\lambda_b} & 1 & 0 \\ -\sqrt{\lambda_a\lambda_b} & \sqrt{(1+\lambda_a)(1+\lambda_b)} & 0 & 1 \end{vmatrix}^2,
$$

which we denote as

$$
\det g(\lambda, a, ..., s') \propto \prod_{a}^{N} \frac{1}{\lambda_a(1+\lambda_a)} \det M_{\lambda_a} \prod_{a<b}^{N} \Big[\det M_{\lambda_a,\lambda_b}\Big]^2. \tag{J.17}
$$

Using the identity

$$\begin{vmatrix} A & B \\ C & D \end{vmatrix} = \det A \det D \det(1 - D^{-1}CA^{-1}B), \tag{J.18}$$

it is easy to verify that

$$\det M_{\lambda_a} = \left[2\lambda_a(1+\lambda_a)\right]^2 \begin{vmatrix} -1 & 1 \\ 1 & -1 \end{vmatrix} = 0 \tag{J.19}$$

This is a consequence of the N redundant parameters in the Bargmann representation of the pseudo unitary transfer matrices. To solve this problem it is common to introduce the matrix elements

$$\begin{aligned} \delta(\xi_1)_{aa} &= (\delta s_1)_{aa} + (\delta s_1')_{aa} \\ \delta(\xi_2)_{aa} &= (\delta s_2)_{aa} + (\delta s_2')_{aa} \\ \delta(\xi_3)_{aa} &= (\delta s_1)_{aa} - (\delta s_2)_{aa} \end{aligned} \tag{J.20}$$

such that the first sum in the trace of equation (J.16) becomes

$$\begin{aligned} -\sum_{a=1}^{N}\Big[&\frac{d\lambda_a^2}{2\lambda_a(1+\lambda_a)} + (\delta\xi_1)_{aa}^2 + (\delta\xi_2)_{aa}^2 - 2\lambda_a(\delta\xi_3)_{aa}^2 \\ &+2\lambda_a(\delta\xi_1)_{aa}(\delta\xi_3)_{aa} - 2\lambda_a(\delta\xi_2)_{aa}(\delta\xi_3)_{aa}\Big] \end{aligned} \tag{J.21}$$

in which case

$$\det M_{\lambda_a} \Rightarrow \begin{vmatrix} -1 & 0 & -\lambda_a \\ 0 & -1 & \lambda_a \\ -\lambda_a & \lambda_a & 2\lambda_a \end{vmatrix} = 2\lambda_a(1+\lambda_a) \tag{J.22}$$

It is easy to show, using (J.18), that the second determinant in (J.17) reduces to

$$\det M_{\lambda_a,\lambda_b} \propto \begin{vmatrix} -\lambda_a - \lambda_b - 2\lambda_a\lambda_b & 2\sqrt{\lambda_a(1+\lambda_a)}\sqrt{\lambda_b(1+\lambda_b)} \\ 2\sqrt{\lambda_a(1+\lambda_a)}\sqrt{\lambda_b(1+\lambda_b)} & -\lambda_a - \lambda_b - 2\lambda_a\lambda_b \end{vmatrix}. \tag{J.23}$$

Hence

$$\det M_{\lambda_a,\lambda_b} \propto (\lambda_a - \lambda_b)^2. \tag{J.24}$$

With these results, we finally have

$$\det g(\lambda, a, ..., s') \propto \prod_{a<b}^{N}(\lambda_a - \lambda_b)^4 \tag{J.25}$$

Therefore

$$d\mathsf{V} \propto \prod_{a<b}^{N} |\lambda_a - \lambda_b|^2 \prod_a d\lambda_a \prod_a (\delta s_1)_{aa} (\delta\xi_1)_{aa} (\delta\xi_2)_{aa} (\delta\xi_3)_{aa}$$
$$\prod_i \prod_{a<b} (\delta a_i)_{ab} (\delta s_i)_{ab} (\delta a_i')_{ab} (\delta s_i')_{ab} \tag{J.26}$$

One can transform back the variables $(\delta s_1)_{aa}$, $(\delta\xi_1)_{aa}$, $(\delta\xi_2)_{aa}$ and $(\delta\xi_3)_{aa}$ into the original ones $(\delta s_1)_{aa}$, $(\delta s_1')_{aa}$, $(\delta s_2)_{aa}$ and $(\delta s_2')_{aa}$. Since the Jacobian of this transformation is 1, the volume element in the space of transfer matrices of the unitary universality class becomes

$$d\mathsf{V} = C_u \prod_{a<b}^{N} |\lambda_a - \lambda_b|^2 \prod_a d\lambda_a d\mu(u_1) d\mu(u_2) d\mu(v_1) d\mu(v_2), \tag{J.27}$$

with $d\mu(u) = \prod_a (\delta s)_{aa} \prod_{a<b} (\delta a)_{ab} (\delta s)_{ab}$ the Haar measure of the unitary matrix u. For the orthogonal and symplectic universality classes, the volume elements are similar. It is common to write

$$d\mathsf{V} = C_\beta \prod_{a<b}^{N} |\lambda_a - \lambda_b|^\beta \prod_a d\lambda_a d\mu(u_1) d\mu(u_2) d\mu(v_1) d\mu(v_2), \tag{J.28}$$

with $\beta = 1, 2$ and 3 for the orthogonal, the unitary and the symplectic universality classes, respectively.

References

1. R.P. Leavitt, J.W. Little, Phys. Rev. B **42**, 11774 (1990)
2. P. Pereyra, Ann. Phys. **397**, 159 (2018)
3. P. Pereyra, *Fundamentals of Quantum Physics* (Springer, Berlin, 2011)
4. P.W. Anderson, Phys. Rev. B **22**, 3519 (1980)
5. M.. Ya.. Azbel, J. Phys. C **14**, L225 (1981)
6. M. Büttiker, Y. Imry, R. Landauer, S. Pinhas, Phys. Rev. B **31**, 6207 (1985)
7. P.A. Mello, N. Kumar, *Quantum Transport in Mesoscopic Systems. Complexity and Statistical Fluctuations* (Oxford University Press, Oxford, 2004)
8. P. Pereyra, J. Opt. **26**, 075501 (2024)
9. C. Burda, S. Link, M. Mohamed, M. El-Sayed, J. Phys. Chem. B **105**, 12286 (2001)
10. S. Civiš, I. Matulková, J. Cihelka, P. Kubelik, K. Kawaguchi, V.E. Chernov, Phys. Rev. A **82**, 022502 (2010)
11. D.-J. Jang, G.-T. Lin, C.-L. Hsiao, L.W. Tu, M.-E. Lee, Auger recombination in InN thin films. Appl. Phys. Lett. **92**, 042101 (2008)
12. J.R. Hauser, M.A. Littlejohn, Sol. Sta. Elect. **11**, 667 (1968)
13. P. Pereyra, Semicond. Sci. Technol. **39**, 095006 (2024)
14. H.C. Pao, C.T. Sah, Sol. Sta. Elect. **9**, 927 (1966)
15. S. Khandelwal et al., BSIM-IMG: a compact model for ultrathin-body SOI MOSFETs with back-gate control. IEEE Trans. Elect. Dev. **59**, 2019 (2012)
16. P.A. Mello, P. Pereyra, N. Kumar, Ann. Phys. **181**, 290 (1988)
17. L.K. Hua, *Harmonic Analysis of Functions of Several Complex Variables in the Classical Domains* ed. by l. Ebner, A. Koranyi. (Transactions of the American Mathematical Society, Providence, RI, 1963)

P. Pereyra, *Contemporary Theory of Semiconductors and Devices*,
Graduate Texts in Physics, https://doi.org/10.1007/978-3-032-00296-9

Index

P. Pereyra, *Contemporary Theory of Semiconductors and Devices*,
Graduate Texts in Physics, https://doi.org/10.1007/978-3-032-00296-9

MIX
Papier aus verantwortungsvollen Quellen
Paper from responsible sources
FSC® C105338
FSC
www.fsc.org

If you have any concerns about our products,
you can contact us on
ProductSafety@springernature.com

In case Publisher is established outside the EU,
the EU authorized representative is:
Springer Nature Customer Service Center GmbH
Europaplatz 3, 69115 Heidelberg, Germany

Printed by Libri Plureos GmbH
in Hamburg, Germany